Nematicons

WILEY SERIES IN PURE AND APPLIED OPTICS

Founded by Stanley S. Ballard, University of Florida

EDITOR: Glenn Boreman, University of Central Florida, CREOL & FPCE

A complete list of the titles in this series appears at the end of this volume.

Nematicons

Spatial Optical Solitons in Nematic Liquid Crystals

Edited by

GAETANO ASSANTO

WILEY

A JOHN WILEY & SONS, INC., PUBLICATION

Published by John Wiley & Sons, Inc., Hoboken, New Jersey.
Published simultaneously in Canada.

Library of Congress Cataloging-in-Publication Data:

Nematicons : spatial optical solitons in nematic liquid crystals / [edited by]
Gaetano Assanto.
pages cm. – (Wiley series in pure and applied optics ; 74)
Includes bibliographical references.
ISBN 978-0-470-90724-5
1. Solitons. 2. Nematic liquid crystals. 3. Liquid crystals–Spectra. I.
Assanto, Gaetano, 1958-
QC174.26.W28N46 2012
530.12′4–dc23

2012010716

Printed in the United States of America

10 9 8 7 6 5 4 3 2 1

To my parents

Contents

Preface

Solitons in physics and solitons in optics are well-established contemporary topics, addressed in a large number of scientific papers and several books. Spatial optical solitons form a specific class, as optics in space is characterized by diffraction rather than dispersion, beam size rather than pulse duration, one or two transverse dimensions rather than one in the temporal domain. For a long time, the available experimental observations of optical solitons in space were limited by the magnitude of the material nonlinearities, until molecular and photorefractive media allowed investigating them at low power and with continuous-wave sources, including incoherent ones. Among the well-known molecular dielectrics exhibiting a large optically nonlinear response were liquid crystals, typically employed in thin samples. It was realized in the early days of both nonlinear optics and liquid crystals that the reorientational response of nematic liquid crystals could lead to quite impressive effects, both in the electro-optic and all-optical domains. Later on, beam propagation over extended distances in nematic liquid crystals was exploited to demonstrate self-focusing and related phenomena, until it became clear that optical spatial solitons could be supported by such a response at the molecular level. I came across light self-localization in nematic liquid crystals during international meetings, where I attended the inspiring presentations by Prof. M. Karpierz (Poland) and Prof. M. Warenghem (France) on light self-confinement in nematic liquid crystals, and decided to get involved in research on nematicons. The discussions with Prof. I. C. Khoo were enlightening and the collaboration with Prof. C. Umeton allowed the program to get started on the right foot. The term "nematicon" was actually coined during a car trip in Poland as I was having a conversation on the topic with M. Karpierz and G. I. Stegeman. The Greek root $\nu\epsilon\mu\alpha\tau\iota\kappa o\sigma$ means "filament-like" or "spaghetti-like," appropriate to both the topic and the culinary culture of someone like me, of Italian birth and upbringing.

This is the first book specifically dealing with spatial optical solitons in nematic liquid crystals. It is a multiauthor contribution to the field and contains review as well as original (previously unpublished) material, from theoretical models to advanced numerical simulations and from experimental observations to applications. The various contributors and chapters have been selected and invited in order to cover most of the relevant activities in this field over the past 12 years.

G. ASSANTO

Italy
February 2012

Acknowledgments

Prof. Glenn Boreman and his wife, Maggie, friends since my PhD studies at the University of Arizona in Tucson, Arizona, encouraged me to consider preparing a Wiley book on nematicons. George Telecki soon joined them in keeping up the necessary pressure. Thanks a lot. I hope you were right and that readers will enjoy this book.

I thank all the authors who kindly accepted my invitation to contribute one or more chapters, and to subject themselves to a number of requests concerning contents, style, mode of presentation, and deadlines. I express my gratitute to all the students and colleagues who do not appear as book contributors but are coauthors of papers and precious actors inspiring various portions of the scientific activities. They include R. Asquini, R. Barboza, I. Burgess, O. Buchnev, G. Coschignano, D. Christodoulides, A. d'Alessandro, A. de Luca, R. Dabrowski, A. Dyadyusha, A. Fratalocchi, M. Kaczmarek, I. C. Khoo, M. Kwasny, L. Lucchetti, R. Morandotti, E. Nowinowski-Kruszelnicki, A. Pasquazi, K. A. Rutkowska, S. V. Serak, F. Simoni, G. I. Stegeman, N. Tabiryan, M. Trotta, and C. Umeton.

Finally, I pay a special tribute to Alessandro Alberucci and Armando Piccardi for greatly supporting me in the no less important task of arranging, organizing, managing, and editing the manuscript.

GA

Contributors

Alessandro Alberucci, Nonlinear Optics and OptoElectronics Lab, University ROMA TRE, Rome, Italy

Gaetano Assanto, Nonlinear Optics and OptoElectronics Lab, University ROMA TRE, Rome, Italy

Jeroen Beeckman, Department of Electronics and Information Systems, Ghent University, Ghent, Belgium

Milivoj R. Belić, Science Program, Texas A&M University at Qatar, Doha, Qatar

Jean-Francois Blach, Unité de Catalyse et de Chimie du Solide, Faculté des Sciences, Université d'Artois, Lens, France

Umberto Bortolozzo, INLN, Université de Nice-Sophia Antipolis, CNRS, Valbonne, France

Dongmei Deng, Laboratory of Nanophotonic Functional Materials and Devices, School of Information and Photoelectronic Science and Engineering, South China Normal University, Guangzhou, China

Anton S. Desyatnikov, Nonlinear Physics Centre, Research School of Physics and Engineering, The Australian National University, Canberra, ACT, Australia

Catherine García-Reimbert, Department of Mathematics and Mechanics, IIMAS, Fenomenos Nonlineales y Mecánica, Universidad Nacional Autónoma de México, Mexico D.F., Mexico

Qi Guo, Laboratory of Nanophotonic Functional Materials and Devices, School of Information and Photoelectronic Science and Engineering, South China Normal University, Guangzhou, China

Jean-Francois Henninot, Unité de Catalyse et de Chimie du Solide, Faculté des Sciences, Université d'Artois, Lens, France

Wei Hu, Laboratory of Nanophotonic Functional Materials and Devices, School of Information and Photoelectronic Science and Engineering, South China Normal University, Guangzhou, China

Yana V. Izdebskaya, Nonlinear Physics Centre, Research School of Physics and Engineering, The Australian National University, Canberra, ACT, Australia

Miroslaw A. Karpierz, Warsaw University of Technology, Warsaw, Poland

Yuri S. Kivshar, Nonlinear Physics Centre, Research School of Physics and Engineering, The Australian National University, Canberra, ACT, Australia

Urszula A. Laudyn, Warsaw University of Technology, Warsaw, Poland

Daquan Lu, Laboratory of Nanophotonic Functional Materials and Devices, School of Information and Photoelectronic Science and Engineering, South China Normal University, Guangzhou, China

Tim R. Marchant, School of Mathematics and Applied Statistics, University of Wollongong, Wollongong, New South Wales, Australia

Antonmaria A. Minzoni, Department of Mathematics and Mechanics, IIMAS, Fenomenos Nonlineales y Mecánica, Universidad Nacional Autónoma de México, Mexico D.F., Mexico

Kristiaan Neyts, Department of Electronics and Information Systems, Ghent University, Ghent, Belgium

Shigen Ouyang, Laboratory of Nanophotonic Functional Materials and Devices, School of Information and Photoelectronic Science and Engineering, South China Normal University, Guangzhou, China

Marco Peccianti, Institute for Complex Systems, ISC-CNR, Rome, Italy

Armando Piccardi, Nonlinear Optics and OptoElectronics Lab, University ROMA TRE and CNISM, Rome, Italy

Stefania Residori, INLN, Université de Nice-Sophia Antipolis, CNRS, Valbonne, France

Katarzyna A. Rutkowska, Warsaw University of Technology, Warsaw, Poland

Luke W. Sciberras, School of Mathematics and Applied Statistics, University of Wollongong, Wollongong, New South Wales, Australia

Noel F. Smyth, School of Mathematics and Maxwell Institute for Mathematical Sciences, University of Edinburgh, Edinburgh, Scotland, United Kingdom

Marc Warenghem, Unité de Catalyse et de Chimie du Solide, Faculté des Sciences, Université d'Artois, Lens, France

Annette L. Worthy, School of Mathematics and Applied Statistics, University of Wollongong, Wollongong, New South Wales, Australia

Wei-Ping Zhong, Department of Electronic and Information Engineering, Shunde Polytechnic, Guangdong Province, Shunde, China

1

Nematicons

Gaetano Assanto, Alessandro Alberucci, and Armando Piccardi

Nonlinear Optics and OptoElectronics Lab, University ROMA TRE, Rome, Italy

1.1 INTRODUCTION

The term *nematicon* was coined to denote the material, nematic liquid crystals (NLC), supporting the existence of optical spatial solitons via a molecular response to light, a *reorientational* nonlinearity. *Nematicons* was first used in the title of Reference 1, after three years since the first publication on reorientational spatial optical solitons in NLC [2]. Since then, a large number of results, including experimental, theoretical, and numerical, have been presented in papers and conferences and formed a body of literature on the subject. In this chapter we attempt to summarize the most important among them, leaving the details to the specific articles but trying to provide a feeling of the amount of work carried out in slightly more than a decade.

1.1.1 Nematic Liquid Crystals

Liquid crystals are organic mesophases featuring various degrees of spatial order while retaining the basic properties of a fluid. In the absence of absorbing dopants, they are excellent dielectrics, transparent from the ultraviolet to the mid-infrared, with highly damaged thresholds, relatively low electronic susceptibilities, and significant birefringence at the molecular level and in the nematic phase. In the latter phase, their elongated molecules have the same average angular orientation, although their individual location is randomly distributed as they are free to move (Fig. 1.1a). NLC exhibit a molecular nonlinearity; when an electric field is present, the electrons in the molecular orbitals tend to oscillate with it and give rise to dipoles which, in turn, react to and tend to align with the field in order to minimize the resulting Coulombian torque [3–5] (Fig. 1.1b–c). This torque is

Nematicons: Spatial Optical Solitons in Nematic Liquid Crystals, First Edition.
Edited by Gaetano Assanto.

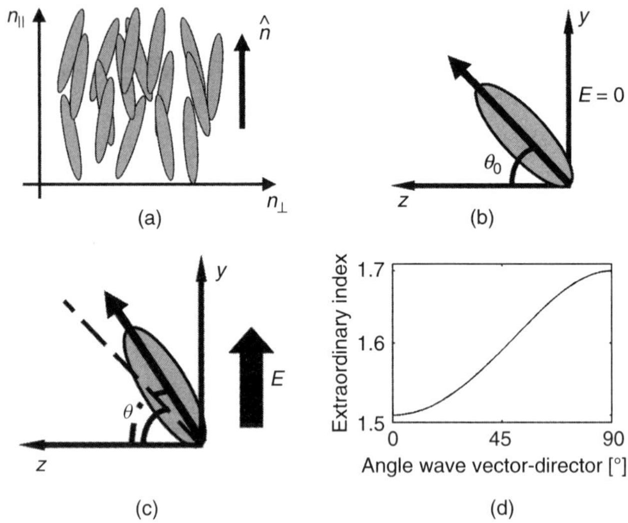

FIGURE 1.1 (a) Sketch of molecular distribution in the nematic phase and definition of director $\hat{n}$; the ellipses represent NLC molecules. (b) Director orientation in the absence of electric field: the angle θ_0 is determined by anchoring at the boundaries. (c) In a positive uniaxial NLC, a linearly polarized electric field can induce dipoles and rotate the molecular director towards its vector; the resulting stationary angle θ is determined by the equilibrium between the electric torque and the elastic intemolecular links. (d) Extraordinary refractive index versus angle between wave vector and director for a positive uniaxial NLC with $n_{\|} = 1.7$ and $n_{\perp} = 1.5$.

counteracted by the elastic forces stemming from intermolecular links: equilibrium is established when the free energy of the system is minimized, as modeled by a set of Euler–Lagrange equations. Because the polarizability of the molecules is higher along their major axes, their reorientation toward the field will increase the optical density, both at the microscopic and macroscopic levels. It is noteworthy that an initial orthogonality between the field and the induced molecular dipoles corresponds to a threshold effect known as *Freedericksz transition* [3]. For static or low frequency fields, reorientation leads to a large electro-optic response with a positive refractive index variation for light polarized in the same plane of the field lines and the long molecular axes [3]. For fields at optical frequencies, the average angular orientation or molecular director in the nematic phase corresponds to the optic axis of the equivalent uniaxial crystal; hence, the refractive index for extraordinarily polarized electric fields (i.e., with field vector coplanar with both optic axis and wave-vector) will increase with the orientation angle θ (Fig. 1.1c–d for wave-vectors along z).

The reorientational mechanism described above is neither instantaneous nor fast (see Chapter 13), but can be very large, with effective Kerr coefficients n_2 of about 10^{-4} cm/W^2 [6], that is, eight to twelve orders of magnitude larger than that in CS_2 and in electronic media, respectively [7]. Therefore, nonlinear effects can be observed in NLC even with continuous wave lasers, at variance with many other nonlinear dielectrics often requiring pulsed excitations.

Nevertheless, the reorientational response is not the only available response in NLC. Owing to their fluidic nature, a high electric field can change the portion of molecules aligned to the director, that is, can affect the order parameter [8], particularly in the presence of dye dopants [9]. Doped NLC also features an enhanced reorientational nonlinearity because of the Janossy effect [10]. As a result of thermo-optic effect, a nonlinear response also stems from temperature changes, modifying the refractive indices mainly via the order parameter in phase transitions [6] (see Chapter 9). Moreover, NLC can show the photorefractive effect [4] and fast electronic nonlinearities (see Chapter 14).

1.1.2 Nonlinear Optics and Solitons

In nonlinear optics, the basic example of an intensity-dependent refractive index is the Kerr response $n(I) = n_0 + n_2 I$. When n_2 is positive, the index increases with the light intensity and, in the case of a finite beam, it gives rise to a lens-like refractive distribution, which is capable of self-focusing the excitation. Such a mechanism can actually compensate for the natural diffraction of the beam, resulting (in the simplest case) in a size/profile-invariant spatial soliton. Otherwise stated, the excitation beam deforms the refractive index distribution of the nonlinear (initially uniform) dielectric, generating a transverse graded-index profile that acts as a waveguide, that is, confines the field into a guided mode. The fundamental soliton in space is the lowest order mode guided by the self-induced dielectric waveguide. Spatial solitons of a Kerr nonlinearity, the so-called Townes solitons [11], tend to be unstable in two transverse dimensions because the exact balance of diffraction and self-focusing is achieved at a critical power [12, 13]. They are stable in one dimension (e.g., in planar waveguides [14]) or in the presence of higher order effects as compared to the Kerr law, such as saturation of the nonlinear change in index [15, 16], multiphoton absorption [17], discreteness [18, 19], and nonlocality [20]. In most cases they are observable in actual media although, being no longer exact solutions of an integrable differential system, they should be rigorously referred to as *spatial solitary waves* [21]. The terms soliton and solitary wave are interchangeably used throughout this chapter.

1.1.3 Initial Results on Light Self-Focusing in Liquid Crystals

As discussed in Section 1.1.1, several terms can contribute to the nonlinear response of NLC. Experiments conducted in the early 1980s demonstrated that, in undoped NLC, the dominant contribution is the reorientational nonlinearity [6, 22, 23]. An equivalent Kerr response was measured with light beams passing through the thickness of a planar cell, the latter behaving as a lens, the focus of which is dependent on the input power. For Rayleigh distances much smaller than the NLC layer thickness, rings could be observed in the diffraction pattern [24].

An experiment on self-focusing in the bulk of a dye-doped NLC layer was carried out in 1993 by Braun et al. [25], who imaged the scattered light from a beam propagating in a cylindrical geometry with NLC subject to Freedericksz

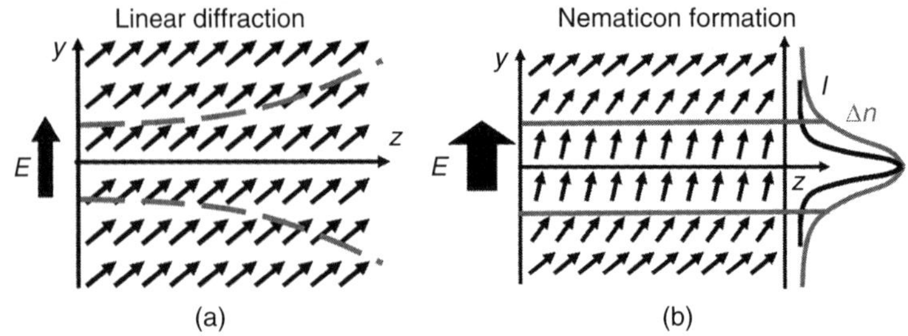

FIGURE 1.2 Basic physics of nematicons. An extraordinarily polarized bell-shaped beam with wave-vector along $\hat{z}$ is launched in an NLC layer with director lying in the plane yz. The major axes of the molecules are at an angle $\neq 90^\circ$ with the wave-vector, thanks to a pretilt (the arrows indicate the molecular director). (a) In the linear regime light does not affect the angular distribution of the director: the beam diffracts as in homogeneous media. (b) Conversely, at high powers the director is perturbed and reorientates toward $\hat{y}$, increasing θ and thus the refractive index (Fig. 1.1d). The perturbation is stronger where the intensity I is higher; hence, an index well is created by the light beam itself, leading to the formation of a waveguide and a self-trapped nematicon. Noticeably, the perturbation extends far beyond the beam profile owing to the elastic links between molecules. For the sake of simplicity, in this illustration the role of walk-off is ignored (Section 1.2.1).

threshold. Various phenomena were observed, including undulation, filamentation, and nonstationary evolution along the capillary; they were interpreted and modeled with joint reorientational and nonlinear Schrödinger equations [26–27]. After such a pioneering work, self-localization of light as a consequence of thermo-optic effects in capillaries was reported by Derrien et al. [28]; the interplay between thermal and reorientational responses was addressed by Warenghem et al. [29] (see Chapter 9). The use of suitably built planar cells with the director tilted by an external bias to avoid the Freedericksz threshold allowed Peccianti et al. to observe the profile-invariant spatial solitons at a few milliWatts [2]. Unbiased planar cells with pretilt determined by rubbing permitted the detailed study of walk-off [30] (see Chapter 6). Figure 1.2 sketches the basic mechanism of nematicon formation via a purely reorientational response.

Finally, nematicons were also reported in slab waveguides with homeotropically aligned NLC [31], in one-dimensional arrays of coupled waveguides [18, 32] (see Chapter 10) and in twisted/chiral NLC [33, 34] (see Chapter 12).

1.2 MODELS

In this section, we review the main theoretical results concerning nonlinear light propagation in NLC cells, with specific reference to a reorientational response supporting optical spatial solitons as well as modulational instability. We first discuss scalar geometries (voltage-biased cells), that is, those in which the role of birefringent walk-off can be left aside. Afterward, we consider the most general case of cells where the walk-off has a substantial effect.

The director distribution can be described by the two polar angles ξ (tilt from the plane yz) and ζ (in the plane yz) (Fig. 1.3a). In addition, θ is the angle

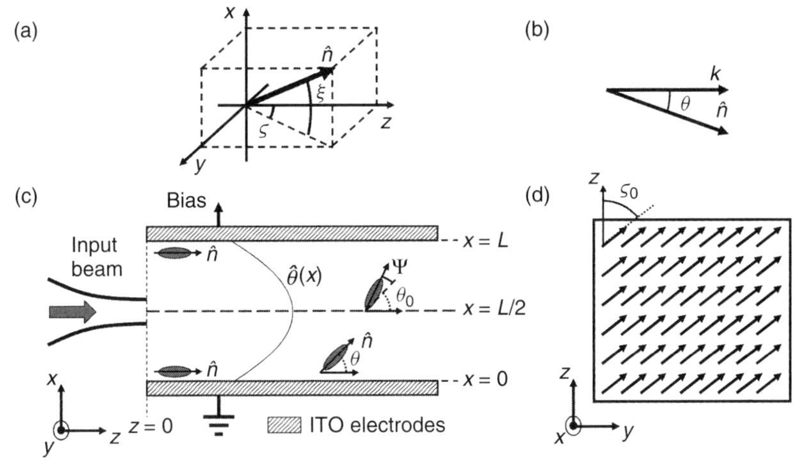

FIGURE 1.3 (a) Definition of polar angles describing the director in the xyz space. (b) Definition of the angle θ between the wave-vector $\boldsymbol{k}$ and the director $\hat{n}$. (c) Side view sketch of a biased planar NLC cell with anchoring condition at the interfaces such that $\hat{n}\|\hat{z}$ (i.e., $\theta = \xi$) and a focused light beam launched along z. The structure is assumed to be infinitely extended along y. (d) Top view of a planar cell showing the rubbing angle ζ_0 in the plane yz; the arrows represent the director distribution in the absence of external excitations (neither bias nor illumination).

between the beam wave-vector $\boldsymbol{k}$ and the molecular director $\hat{n}$ (Fig. 1.3b). In scalar geometries $\zeta = 0$, the latter implying $\theta = \xi$ (Fig. 1.3c). In general, $\zeta \neq 0$ owing to the anchoring at the (glass/NLC) interfaces parallel to the plane yz; at rest the director $\hat{n}$ lies in the plane yz at an angle ζ_0 with z (Fig. 1.4d).

We stress that the equations and the results shown hereby hold valid in the limit of small optical perturbations; the highly nonlinear case is dealt with in Chapter 11.

1.2.1 Scalar Perturbative Model

We consider the configuration of Fig. 1.3c: a finite light beam is launched in the planar NLC cell with wave-vector along the z axis and the field linearly polarized along the x axis. Two parallel glass plates contain the NLC, with molecular director $\hat{n}$ lying in the plane xz (i.e., $\hat{n} \cdot \hat{y} = 0$) at an angle θ with $\hat{z}$ (i.e., $\hat{n} \cdot \hat{z} = \cos\theta$). A low frequency electric field E_{LF} is applied (via transparent electrodes on the plates) across $\hat{x}$ to overcome the Fréedericksz threshold and pretilt the molecules in the plane xz via the electro-optic response, creating a potential $\hat{\theta}(x)$ in the absence of illumination; $\hat{\theta}$ depends only on x due to the symmetry of the problem.

In this configuration the beam excites only the extraordinary component, generally at the walk-off δ with respect to $\hat{x}$, owing to birefringence. Hereby, we scalarize the problem and assume the electric field E_{opt} of the beam to be linearly polarized along $\hat{x}$, leaving the vectorial case to Section 1.2.2. The use of a scalar model also implies neglecting the tilt between Poynting and wave-vectors. Let us define A, the slowly varying envelope of E_{opt}, that is,

$$E_{\mathrm{opt}} = A\left(x, y, z\right) \exp\left[ik_0 n_{\mathrm{e}}\left(\theta_0\right) z\right]$$

with θ_0 the orientation without light and $n_e(\theta) = \left(\frac{\cos^2\theta}{\epsilon_\perp} + \frac{\sin^2\theta}{\epsilon_\parallel}\right)^{-1/2}$ the extraordinary wave refractive index, where $\epsilon_\perp$ ($\epsilon_\parallel$) is the electronic susceptibility perpendicular (along) to $\hat{n}$. In the paraxial approximation, light propagation is ruled by [2]

$$2ik\frac{\partial A}{\partial z} + \nabla_{xy}^2 A + k_0^2\epsilon_a \sin(2\theta_0)\Psi A = 0, \tag{1.1}$$

where $k = k_0 n_e(\theta_0)$ and we set $\theta = \theta_0 + \Psi$, with Ψ being the light-induced perturbation on θ.

As we are interested in the reorientational nonlinearity, we need a further equation describing how the angle θ varies under the application of both E_{opt} and E_{LF}. To this extent, minimization of the NLC free energy that assumes a single constant to describe the elastic (intermolecular) forces, yields the Euler–Lagrange equation [3, 35]

$$\begin{aligned} &K\frac{\hat{\theta}}{\theta_0}\left[\nabla^2\Psi + \frac{\cos(2\theta)}{K}\left(\Delta\epsilon_{\mathrm{LF}}E_{\mathrm{LF}}^2 + \frac{\epsilon_0\epsilon_a|A|^2}{2}\right)\Psi\right] \\ &\quad + K\frac{\Psi}{\theta_0}\frac{\mathrm{d}^2\hat{\theta}}{\mathrm{d}x^2} + \frac{2K}{\theta_0}\frac{\mathrm{d}\Psi}{\mathrm{d}x}\frac{\mathrm{d}\hat{\theta}}{\mathrm{d}x} + \frac{\epsilon_0\epsilon_a}{4}|A|^2\sin(2\hat{\theta}) = 0, \end{aligned} \tag{1.2}$$

with $\Delta\epsilon_{\mathrm{LF}}$ being the anisotropy and K the Frank elastic constant, and $\theta(x, y, z) = \hat{\theta}(x) + \frac{\hat{\theta}(x)}{\theta_0}\Psi(x, y, z)$ being $\mathrm{d}^2\hat{\theta}/\mathrm{d}x^2 + (\Delta\epsilon_{\mathrm{LF}}E_{\mathrm{LF}}^2/2)\sin(2\hat{\theta}) = 0$ [35]. Equation 1.2 is obtained when $(\hat{\theta}/\theta_0)\Psi \ll \hat{\theta}$, that is, in the perturbative limit.

For straight beam trajectories (i.e., homogeneous medium, uniform director distribution, no walk-off) we can set $\hat{\theta} \approx \theta_0$. When $\epsilon_0\epsilon_a|A|^2 \ll \Delta\epsilon_{\mathrm{LF}}E_{\mathrm{LF}}^2$ and the beam axis is in the cell mid-plane $x = 0$ with $\mathrm{d}\hat{\theta}/\mathrm{d}x = 0$, the light-induced reorientation is governed by [35]

$$K\nabla^2\Psi - \frac{\sin(2\theta_0)}{2\theta_0}\left[1 - 2\theta_0\frac{\cos(2\theta_0)}{\sin(2\theta_0)}\right]\Delta\epsilon_{\mathrm{LF}}E^2\Psi + \frac{\epsilon_0\epsilon_a}{4}\sin(2\theta_0)|A|^2 = 0, \tag{1.3}$$

that is, by a Yukawa (or screened Poisson) equation, with forcing term given by the light intensity and screening length l equal to

$$l = \sqrt{\frac{2\theta_0}{\sin(2\theta_0) - 2\theta_0\cos(2\theta_0)}\frac{K}{\Delta\epsilon_{\mathrm{LF}}E^2}} = c_l(\theta_0)l_{\mathrm{LC}}, \tag{1.4}$$

where we set $c_l(\theta_0) = \sqrt{\frac{2\theta_0}{\sin(2\theta_0)-2\theta_0\cos(2\theta_0)}}$ and $l_{\mathrm{LC}} = \sqrt{\frac{K}{\Delta\epsilon_{\mathrm{LF}}E^2}}$. It is straightforward to obtain $\lim_{\theta_0\to 0} c_l = \infty$ and $c_l(\theta_0 = \pi/2) = 1$; between these two extrema c_l decreases monotonically. We note that θ_0 depends only on the applied voltage

$V \approx E_{\mathrm{LF}}L$; hence, Equation 1.4 provides $l = c_l[\theta_0(V)]\sqrt{K/\Delta\epsilon_{\mathrm{LF}}}L/V$ and for a given bias V, the spatial width of the nonlinear response is proportional to the cell thickness L.

The system formed by Equations 1.1 and 1.3 governs nonlinear light propagation in biased NLC cells; for any NLC and cell size (i.e., thickness L), the parameters depend on the bias via the applied electric field and θ_0, that is, on low frequency reorientation bias, including pretilt. Such a feature allows for electrically tuning both nonlinearity and nonlocality of the medium [36].

To quantify the nonlinearity, let us define the material-dependent parameter $\gamma = \epsilon_0\epsilon_a/(4K)$; using Equation 1.3, Ψ can be expressed via the Green formalism as

$$\Psi = \gamma \sin(2\theta_0) \int G(\boldsymbol{r} - \boldsymbol{r}')|A(\boldsymbol{r}')|^2 \mathrm{d}^3\boldsymbol{r}', \tag{1.5}$$

where $G(\boldsymbol{r} - \boldsymbol{r}')$ is the Green function of the Yukawa equation (Eq. 1.3).

Using Equation 1.5 the *photonic potential* [defined as $V_{\mathrm{ph}} = k_0^2\epsilon_a \sin(2\theta_0)\Psi$ and corresponding to the potential with a change in sign] reads $V_{\mathrm{ph}} = k_0^2\epsilon_a\gamma \sin^2(2\theta_0) \int G(\boldsymbol{r} - \boldsymbol{r}')|A(\boldsymbol{r}')|^2\mathrm{d}^3\boldsymbol{r}'$. We can thus write the effective (nonlocal) Kerr coefficient as

$$n_2^{\mathrm{eff}} = \frac{\epsilon_0\epsilon_a^2}{4K} l^2(\theta_0) \sin^2(2\theta_0) = \frac{\epsilon_0\epsilon_a^2}{4K} l_{\mathrm{LC}}^2 c_l^2(\theta_0) \sin^2(2\theta_0). \tag{1.6}$$

The square dependence on the screening length l^2 stems from the integral in Equation 1.5: for intensity distributions maintaining their transverse size with respect to l (i.e., $|A(\boldsymbol{r}/l)|^2$ invariant), the perturbation Ψ scales with l^2. Conversely, the magnitude of the nonlocality, that is, the ratio between the widths of the photonic potential and the intensity profile, is determined by l. In fact, in the limit $l \to \infty$, Equation 1.3 becomes a Poisson equation, with degree of nonlocality fixed by the boundaries (see Chapter 11 and references therein). After setting $|A|^2 = |B|^2/l^2$, for $l \to 0$ we get $\Psi \propto |B|^2$: in this regime NLC resemble local Kerr media.

1.2.1.1 Solitary Waves Let us define the normalized coordinates $X = \sqrt{2k}x$, $Y = \sqrt{2k}y$, and $Z = z$; we also introduce the normalized quantities $\psi = k_0^2 \sin(2\theta_0)/(2K)\Psi$ and $a = \sqrt{1/(k_0^2 n_2^{\mathrm{eff}})}A$, with $\alpha = 1/(2kl^2)$. The parameter α is inversely proportional to nonlocality, that is, α is equal to zero if the nonlocality range is infinite, whereas it tends to ∞ in the local (Kerr) case. Equations 1.1 and 1.3 now read [35]

$$\frac{1}{2k}\frac{\partial^2\psi}{\partial Z^2} + \nabla_{XY}^2\psi - \alpha\psi + |a|^2 = 0, \tag{1.7}$$

$$i\frac{\partial a}{\partial Z} + \nabla_{XY}^2 a + \psi a = 0. \tag{1.8}$$

Suppose $\partial^2\psi/\partial Z^2 = 0$. From Equation 1.7 we can write $\psi = -\left(\nabla_{XY}^2 - \alpha\right)^{-1}|a|^2$. For large α we can write $\left(\nabla_{XY}^2/\alpha - 1\right)^{-1} \approx -\left(1 + \nabla_{XY}^2/\alpha\right)$; hence, light propagation is governed by the single equation [35]

$$i\frac{\partial a}{\partial Z} + \nabla_{XY}^2 a + \frac{a}{\alpha}\left(\frac{\nabla^2|a|^2}{\alpha} + |a|^2\right) = 0. \tag{1.9}$$

Equation 1.9 describes nonlinear light propagation in a weakly nonlocal medium; it was shown in References 20, 35, and 37 how nonlocality inhibits catastrophic beam collapse. Conversely, if terms proportional to α^{-2} are neglected, Equation 1.9 transforms into a local NLSE (NonLinear Schrödinger Equation): solitary waves in (2+1)D are Townes-like and undergo catastrophic collapse [13]. Therefore, spanning the free parameter from zero to infinity, solitary waves evolve from accessible solitons ($\alpha = 0$) [38] to Townes solitons [11] ($\alpha \to \infty$).

To confirm this qualitative assessment, let us look for soliton-like solutions of Equations 1.7 and 1.8 after setting $\partial\psi/\partial Z = 0$ and $a = v(X, Y)\exp(i\beta Z)$. We obtain

$$\nabla_{XY}^2\psi_v - \alpha\psi_v + v^2 = 0, \tag{1.10}$$

$$\nabla_{XY}^2 v - \beta v + \psi_v v = 0, \tag{1.11}$$

with ψ_v the optical perturbation in presence of solitary waves. Interestingly, the system of Equations 1.10 and 1.11 determines the profile of solitary waves in parametric crystals as well [35, 39]. If the boundaries are circularly symmetric or if their influence can be neglected (see Chapter 11), the lowest order (i.e., single hump) solitary wave solutions of system (1.10) and (1.11) are radially symmetric. Thus, without loss of generality, we can expand ψ_v in a MacLaurin series around $R = \sqrt{(X - L/2)^2 + Y^2} = 0$, that is, we write $\psi_v = \psi_0 + \psi_2 R^2 + \psi_4 R^4 + \cdots$, where the odd terms in R are zero due to the symmetry in the problem. In the highly nonlocal case [38, 40], the soliton waist is small compared to the size of the self-induced index well; hence, it is possible to set $\psi_v \approx \psi_0 + \psi_2 R^2$. After substitution into Equation 1.11, the latter becomes the well-known model of the quantum harmonic oscillator, with oscillator strength depending on the beam power via Equation 1.10 [38, 40].

Let us set $f = |v|^2$ and, in analogy with what had been done above for ψ_v, expand f into $f = f_0 + f_2 R^2 + f_4 R^4 + \cdots$; Equation 1.10 then gives

$$\psi_2 = \frac{\alpha\psi_0 - f_0}{4}, \tag{1.12}$$

$$\psi_4 = \frac{\alpha^2\psi_0 - \alpha f_0 - 4f_2}{64}, \tag{1.13}$$

having retained terms up to R^2. In the highly nonlocal limit the soliton profile is Gaussian, that is, $f = 2\mathcal{P}/(\pi w^2)\exp(-2R^2/w^2)$, where $\mathcal{P} = \iint |v|^2 \mathrm{d}x\mathrm{d}y$ is the normalized power; hence, it is straightforward to get $f_0 = 2\mathcal{P}/(\pi w^2)$. At the same time, from quantum mechanics $w^2 = \sqrt{-2/\psi_2}$. In the simplest case $\alpha = 0$, Equation 1.12 provides $\psi_2 = -\mathcal{P}/(2\pi w^2)$, yielding the condition for nematicon existence [40]:

$$\mathcal{P}w^2 = 4\pi. \tag{1.14}$$

1.2.1.2 Modulational Instability In Section 1.2.1.1, we focused on bell-shaped wavepackets undergoing self-confinement through a reorientational response, underlining the stabilizing effect of nonlocality. In self-focusing media, spatial solitons are states of minimum energy; hence, lightwaves will evolve to these configurations whenever possible. An example of this is modulational instability: a plane wave in a self-focusing material is unstable and evolves first into a periodically modulated wavefront before it eventually forms multiple solitons. The spatial frequency components generated (i.e., amplified from noise at the expense of the zero frequency component) during propagation are dependent on the input power, with a spectral gain $G(k_y)$.

To model such processes, we can refer to Equations 1.7 and 1.8 in the one-dimensional limit (i.e., we set $\partial/\partial x = 0$); we consider the plane wave solution $a_{\mathrm{PW}} = a_0\sqrt{\alpha}\exp(\mathrm{i}\beta z)$ to be consistent with the discussion in Section 1.2.1 (the power density is $\mathcal{P}_{1D} = \alpha|a_0|^2$) with $\psi_{\mathrm{PW}} = |a_0|^2$ and $\beta_{\mathrm{PW}} = |a_0|^2$, and introduce a small perturbation by $a = (a_{\mathrm{PW}} + \Delta a)\exp(\mathrm{i}\beta_{\mathrm{PW}} z)$ and $\psi = \psi_{\mathrm{PW}} + \Delta\psi$ for amplitude and reorientation, respectively. For $\angle a_0 = 0$, at the first order the gain coefficient $G(k_y)$ (i.e., the perturbation evolves as $\propto \exp[G(k_y)z]$) reads [41]

$$G\left(k_y\right) = |k_y|\sqrt{\frac{2a_0^2\alpha}{\alpha + k_y^2} - k_y^2}. \tag{1.15}$$

As anticipated, the modulational instability gain is dependent on the transverse frequency k_y, with amplification changing with the input power density $\mathcal{P}_{1D}$. In the limit $\alpha \to \infty$ Equation 1.15 yields $G = |k_y|\sqrt{2a_0^2 - k_y^2}$, retrieving the correct expression for local Kerr media.

1.2.2 Anisotropic Perturbative Model

In a generic (nonmagnetic) optically uniaxial medium, the electric field obeys [42]

$$\nabla \times \left[\nabla \times \mathbf{E}_{\mathrm{opt}}\right] = \nabla(\nabla\cdot\boldsymbol{E}_{\mathrm{opt}}) - \nabla^2\boldsymbol{E}_{\mathrm{opt}} = k_0^2\boldsymbol{\epsilon}(\boldsymbol{r})\cdot\boldsymbol{E}_{\mathrm{opt}}, \tag{1.16}$$

where $\epsilon_{ij} = \epsilon_\perp\delta_{ij} + \epsilon_a n_i n_j$ [4, 5], are the dielectric tensor components, δ_{ij} is the Kronecker function, n_j is the jth component of the director $\hat{\boldsymbol{n}}$, $\epsilon_a = \epsilon_{||} - \epsilon_\perp$ is the anisotropy.

Let us consider the extraordinary solution with wave-vector along z and write $E_{\mathrm{opt}} = A\mathrm{e}^{ik_0 n_\mathrm{e} z}$; if A is a constant, the solution is a plane wave satisfying the tensorial equation [42]

$$\mathbf{L}(n_\mathrm{e}) \cdot \mathbf{A} \equiv \left[n_\mathrm{e}^2(\hat{z}\hat{z} - \mathbf{I}) + \boldsymbol{\epsilon}\right] \cdot \mathbf{A} = 0, \quad (1.17)$$

where the last equivalence stems from Equation 1.16 and $\mathbf{I}$ is the identity matrix. The effect of optical reorientation is to perturb the dielectric tensor so that $\boldsymbol{\epsilon} = \boldsymbol{\epsilon_0} + \eta^2\delta\boldsymbol{\epsilon}$, where $\boldsymbol{\epsilon_0}$ is the unperturbed tensor in the absence of light, and $\delta\boldsymbol{\epsilon}$ is the light-induced perturbation; η plays the role of a *small* parameter, to be set to unity at the end of the derivation. We assume that the optic axis is at an angle θ with respect to the z axis (Fig. 1.3c). We use for convenience a new reference system xts obtained by rotating the original xyz around x by the walk-off angle δ; therefore, we get $\hat{\boldsymbol{s}} = \hat{z}\cos\delta - \hat{\boldsymbol{y}}\sin\delta$ and $\hat{\boldsymbol{t}} = \hat{\boldsymbol{y}}\cos\delta + \hat{\boldsymbol{s}}\sin\delta$.

After defining the slow scales $\mathbf{r} = \mathbf{r}_0 + \eta\mathbf{r_1} + \cdots + \eta^n\mathbf{r}_n$ with $\mathbf{r} = x\hat{\boldsymbol{x}} + t\hat{\boldsymbol{t}} + s\hat{\boldsymbol{s}}$, the electric field in xts can be expanded as $\mathbf{E}_{\mathrm{opt}} = [\hat{\boldsymbol{t}}E_\mathrm{e} + \eta\mathbf{F}_\mathrm{e} + \eta^2\mathbf{G}_\mathrm{e} + o(\eta^3)]\mathrm{e}^{ik_0 n_\mathrm{e} z_0}$; finally, the differential operator can be cast as $\nabla = \nabla_0 + \eta\nabla_1 + \cdots + \eta^n\nabla_n$. Imposing the solvability conditions (up to order η^2) for E_{opt} and letting $\eta \to 1$, we get [42]

$$2ik_0 n_\mathrm{e}\cos\delta\frac{\partial A}{\partial s} + D_t\frac{\partial^2 A}{\partial t^2} + D_x\frac{\partial^2 A}{\partial x^2} + \frac{in_\mathrm{e}\sin(2\delta)}{2k_0\lambda_s}\frac{\partial^2 A}{\partial s\partial t} + k_0^2\delta\epsilon_{tt}A = 0, \quad (1.18)$$

where $\delta\epsilon_{tt} = \hat{\boldsymbol{t}}\cdot\delta\boldsymbol{\epsilon}\cdot\hat{\boldsymbol{t}}$, $D_x = 1 + \frac{n_\mathrm{e}^2\sin^2\delta}{-n_\mathrm{e}^2+\epsilon_\perp}$, and $D_t = \frac{n_\mathrm{e}^2\cos^2\delta}{-n_\mathrm{e}^2\sin^2\delta+\epsilon_\perp+\epsilon_a\cos^2(\theta-\delta)}$ are the diffraction coefficients, differing from unity because of anisotropy.

If the mixed derivative can be neglected (small walk-off δ), Equation 1.18 is an NLSE modeling light propagation in the walk-off reference system. Equation 1.18 is valid in the perturbative regime, that is, when nonlinear variations on the dielectric tensor are small compared with its linear value; in this limit, photons propagate along the Poynting vector of the carrier plane wave, that is, walk-off does not depend on power. Furthermore, under these approximations the beam remains linearly polarized (other components appear at the next order, i.e., when $\mathbf{F}_\mathrm{e}$ is accounted for) and paraxial with respect to s [43].

Equation 1.18 has to be solved with the reorientational equation for the nonlinear changes in director profile, allowing in turn to compute the photonic potential $V_{\mathrm{ph}} = k_0^2\delta\epsilon_{tt}$. For the configuration in Figure 1.3, the simplest case is when the cell is unbiased, with $\xi = 0$, $\theta = \zeta$, and consequently $\theta_0 = \zeta_0$; in fact, molecular rotation takes place only in yz, with θ governed by [44]

$$\nabla^2\Psi + \gamma\sin\left[2\left(\theta_0 + \Psi - \delta\right)\right]|A|^2 = 0. \quad (1.19)$$

The photonic potential is

$$V_{\mathrm{ph}} = k_0^2\epsilon_a\left[\sin^2(\theta_0 + \Psi - \delta) - \sin^2(\theta_0 - \delta)\right] \approx k_0^2\epsilon_a\sin\left[2(\theta_0 - \delta)\right]\Psi, \quad (1.20)$$

taking into account the smallness of Ψ. V_{ph} has the same expression calculated in Section 1.2.1 in the limit $\delta \to 0$. By using elementary trigonometry, we can eliminate δ from Equation 1.20 and get [42]

$$V_{\text{ph}} = k_0^2 \epsilon_a T(\theta_0)\Psi = k_0^2 \epsilon_a \frac{2\epsilon_\perp \left(\epsilon_\perp + \epsilon_a\right) \sin\left(2\theta_0\right)}{\left(\epsilon_\perp + \epsilon_a\right)^2 + \epsilon_\perp^2 + \left[\left(\epsilon_\perp + \epsilon_a\right)^2 - \epsilon_\perp^2\right] \cos\left(2\theta_0\right)} \Psi. \tag{1.21}$$

For $\zeta_0 = 0$ and applied bias, the configuration corresponds to the one in Section 1.2.1 with director moving in the plane xz; thus, Equation 1.3 is valid if the walk-off δ is included in the term modeling the torque; hence, all results derived in that section remain valid, specifically Equation 1.3 (with the torque correction just pointed out), which, jointly with Equations 1.18 and 1.20, is a complete model for vectorial nematicons propagating in the perturbative regime when the director reorients in the single plane xz.

When a bias is applied to the cell in Figure 1.3 with $\zeta_0 \neq 0$, the reorientation dynamics becomes more complicated, as light and voltage induced torques tend to move the molecules in two different planes and two angles are needed to describe the director distribution (Fig. 1.3a) [3]. Owing to the symmetry, bias acts only on angle ξ, inducing in the absence of light a director profile equal to the case $\zeta_0 = 0$, that is, $\xi = \hat{\theta}$. To first approximation, the result is to modulate the magnitude of the screening term depending on the low frequency field, with an expected dependence on $\sin\hat{\theta}$ [42]: in fact, at zero bias ($\hat{\theta} = 0$) the screening term is absent (see Section 1.2.2.1), whereas for $\hat{\theta} \to \pi/2$ we retrieve the case of Section 1.2.1.

1.2.2.1 Nematicon Breathers In an unbiased cell with $\zeta_0 \neq 0$, for small Ψ Equation 1.19 becomes

$$K\nabla_{xt}^2 \Psi + \frac{\epsilon_0 \epsilon_a}{4} |A|^2 \sin\left[2(\theta_0 - \delta_0)\right] + \frac{\epsilon_0 \epsilon_a}{2} |A|^2 \cos\left[2(\theta_0 - \delta_0)\right] \Psi = 0, \tag{1.22}$$

where $\delta_0 = \delta(\theta_0)$. Assuming a parabolic index well of the form $\psi \approx \Psi_0 + \Psi_2 \left[(x - L/2)^2 + t^2\right]$, from Equation 1.22 the coefficient Ψ_2 is [40]

$$\Psi_2 = -\frac{\epsilon_0 \epsilon_a}{8K} f_0 \left\{ \frac{\sin\left[2(\theta_0 - \delta_0)\right]}{2} + \Psi_0 \cos\left[2(\theta_0 - \delta_0)\right] \right\}, \tag{1.23}$$

where $f_0 = |A(x = L/2, t = 0)|^2$. We remark that the Equation 1.23 is obtained from Eq. 1.22 with no approximation, and thus is valid whenever the intensity profile and the nonlinear perturbation are radially symmetric.

The photonic potential becomes

$$V_{\text{ph}} \approx k_0^2 \epsilon_a \sin\left[2(\theta_0 - \delta_0)\right] \left\{\Psi_0 + \Psi_2 \left[(x - L/2)^2 + t^2\right]\right\}. \tag{1.24}$$

In analogy with quantum mechanics, the term proportional to Ψ_0 is a rest energy, depending on the beam shape and boundary conditions and generally varying with s, that is, an equivalent time; in optics, it is responsible for nonlinear changes in the propagation constant. Conversely, the term Ψ_2 determines the resonance frequency of the equivalent harmonic oscillator. Taking—for the sake of simplicity—$D_x = D_t = D$, from quantum harmonic oscillator theory we infer that the lowest order solitary wave of power P_s has the shape [38]

$$A(x,t,s) = \left(\frac{4Z_0 P_s}{\pi n_e \cos\delta_0 w_s^2}\right)^{1/2} e^{-\frac{(x-L/2)^2+t^2}{w_s^2}} e^{i(\beta_0-\beta_2)s}, \tag{1.25}$$

where $\beta_0 = k_0\epsilon_a T(\theta_0)\Psi_0/\left(2n_e\cos\delta_0\right)$ and $\beta_2 = \sqrt{-\epsilon_a DT(\theta_0)\Psi_2}/\left(2n_e\cos\delta_0\right)$ are the nonlinear correction of propagation constant β due to the equivalent rest energy (increase in β) and the parabolic potential (decrease in β), respectively. The soliton width w_s and power P_s are not independent variables owing to the dependence of Ψ_2 on power; we find [38, 40, 42]

$$P_s = \frac{8\pi K D n_e \cos\delta_0}{k_0^2 Z_0 \epsilon_0 \epsilon_a^2 \sin[2(\theta_0-\delta)]\left\{\frac{\sin[2(\theta_0-\delta_0)]}{2} + \Psi_0\cos[2(\theta_0-\delta_0)]\right\}}\frac{1}{w_s^2}. \tag{1.26}$$

For small Ψ_0, Equation 1.26 reduces to $P_s = \left(\frac{16\pi K D n_e}{\epsilon_0\epsilon_a^2 Z_0 k_0^2 \sin^2[2(\theta_0-\delta)]}\right)\frac{1}{w_s^2}$; thus, the nematicon properties do not depend on the boundary conditions. Equation 1.26 is the existence curve for single-hump nematicons in unbiased cells and in the highly nonlocal approximation.

The next step is to investigate light propagation for a Gaussian beam launched in the NLC cell, but with waist w and power P not satisfying Equation 1.26. Assuming a parabolic approximation for the index distribution, we can still refer to the harmonic oscillator to model self-confinement. Using a generalization of the Ehrenfest theorem, the waist w is governed by [45]

$$\frac{d^4 w^2}{ds^4} + \frac{\sigma P}{w^2}\frac{d^2 w^2}{ds^2} - \frac{\sigma P}{w^4}\left(\frac{dw^2}{ds}\right)^2 = 0, \tag{1.27}$$

where $\sigma = \frac{\epsilon_0\epsilon_a^2 D Z_0 T(\theta_0) F(\theta_0)}{4\pi K n_e^3 \cos^3\delta_0}$ with $F(\theta_0) = \sin[2(\theta_0-\delta_0)] + \Psi_0\cos[2(\theta_0-\delta_0)]$. If w changes only slightly in propagation ($w - w_s \ll w_s$) and w_s is large enough to neglect the terms depending on w_s^{-4} with respect to those proportional to w_s^{-2}, Equation 1.27 can be recast as [40]

$$\frac{d^2}{ds^2}(w^2 - w_s^2) + \frac{\sigma P}{w_s^2}\left(w^2 - w_s^2\right) = 0. \tag{1.28}$$

Equation 1.28 can be analytically solved, yielding [40]

$$\frac{w^2}{w_s^2} = 1 + \left(\frac{w_0^2}{w_s^2} - 1\right)\cos\left(\frac{\sqrt{\sigma P}}{w_s}s\right), \tag{1.29}$$

with w_0 being the initial waist and $w_s = w_s(P)$ the soliton waist corresponding to an input power P through Equation 1.26. We assumed a flat phasefront at the input. Equation 1.29 predicts a sinusoidal oscillation of the waist along propagation, that is, breathing self-confined wavepackets [38, 40, 42]. In particular, if $w_0 > w_s$, the nonlinearity dominates on diffraction at the input, the beam initially focuses; if $w_0 < w_s$, the roles are inverted and the beam initially expands. The oscillation amplitude is proportional to the distance from the solitary wave condition (i.e., $|w_0 - w_s|$), and the oscillation period Λ is $2\pi w_s/(\sqrt{\sigma P})$. From Equations 1.29 and 1.26, the breathing periodicity is inversely related to the input power P ($\Lambda \propto P^{-1}$), that is, larger powers correspond to faster changes in the nematicon waist.

We stress that all results derived in this section for unbiased cells apply as well to biased cells with $\zeta_0 = 0$, with straightforward corresponding expressions, once the strength of the nonlinear harmonic oscillator is calculated from Equation 1.23.

1.3 NUMERICAL SIMULATIONS

1.3.1 Nematicon Profile

The profile of paraxial nematicons in unbiased cells can be evaluated numerically from Equations 1.18 and 1.19 by setting $\partial_s \Psi = 0$, $\theta = \theta_u(x, t)$, and $A = \sqrt{2Z_0 P/n_e}u(x, t)\exp(i\beta s)$, where Z_0 is the vacuum impedance, P the soliton power, and β the nonlinear correction to the propagation constant. We obtain the nonlinear eigenvalue problem

$$\left\{k_0^2\epsilon_a\left[\sin^2(\theta_u - \delta) - \sin^2(\theta_0 - \delta)\right] - 2k_0 n_e\beta\cos\delta\right\}u + D_t\frac{\partial^2 u}{\partial t^2} + D_x\frac{\partial^2 u}{\partial x^2} = 0, \tag{1.30}$$

$$\nabla_{xt}^2\theta_u + \frac{2\gamma Z_0 P}{n_e}\sin\left[2(\theta_u - \delta)\right]|u|^2 = 0. \tag{1.31}$$

The system of Equations 1.30 and 1.31 can be solved using a standard relaxation scheme. We consider the NLC mixture E7, often employed in experiments. Its refractive indices versus wavelength are plotted in Figure 1.4, and the Frank elastic constant is 12×10^{-12}N.

Figure 1.5a–d shows the numerically computed results for a single-hump soliton at $P = 1$ mW. Even if the diffraction coefficients D_x and D_t are unequal, the intensity profile is radially symmetric to a very good approximation. Moreover, if the nematicon waist is small with respect to the cell thickness L, the shape of θ_u

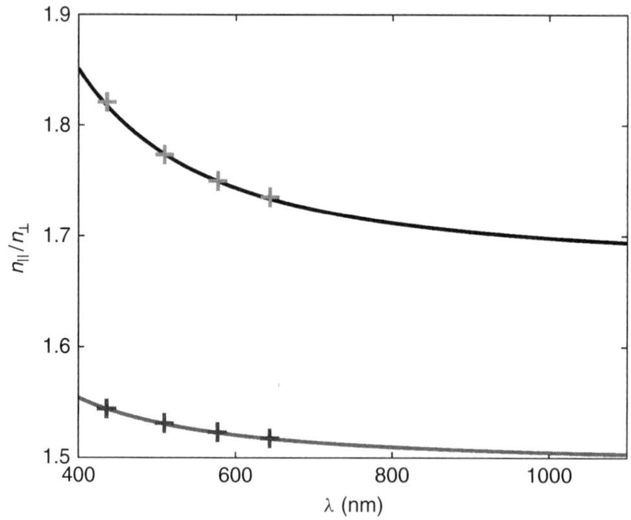

FIGURE 1.4 Refractive index versus wavelength in the NLC mixture E7, for electric field parallel (top line) and perpendicular (bottom line) to the director (optic axis). Symbols are data and solid lines represent the interpolations using a dispersion model.

is nearly symmetric around the beam, despite the boundary conditions with strong anchoring at the glass interfaces. Figure 1.5e and f plots light and director along s computed with a BPM (Beam Propagation Method) code for $P = 1$ mW nematicon profile at the input: both θ and u are invariant in propagation.

Figure 1.6 displays nematicon profiles for various input powers. The transverse intensity distribution is quite close to Gaussian, with small departures in the tails (Figure 1.6a–d). The nematicon existence curve versus power and waist is approximately given by $P \propto w^{-2}$, consistently with the model of accessible solitons (Fig. 1.6e). For a given power, the nematicon is narrower at shorter wavelengths due to reduced diffraction, as predictable; such a stronger confinement is also demonstrated by higher intensity peaks and larger maxima in director orientation, as in Figure 1.6f and g, respectively.

1.3.2 Gaussian Input

For a direct comparison with experimental data, it is important to investigate light evolution in the case of a Gaussian beam injected into the NLC layer.

We take an input excitation

$$A(x, t, s = 0) = \sqrt{\frac{4Z_0 P}{\pi n_e w_0^2}} e^{-\frac{(x-L/2)^2+t^2}{w_0^2}}. \tag{1.32}$$

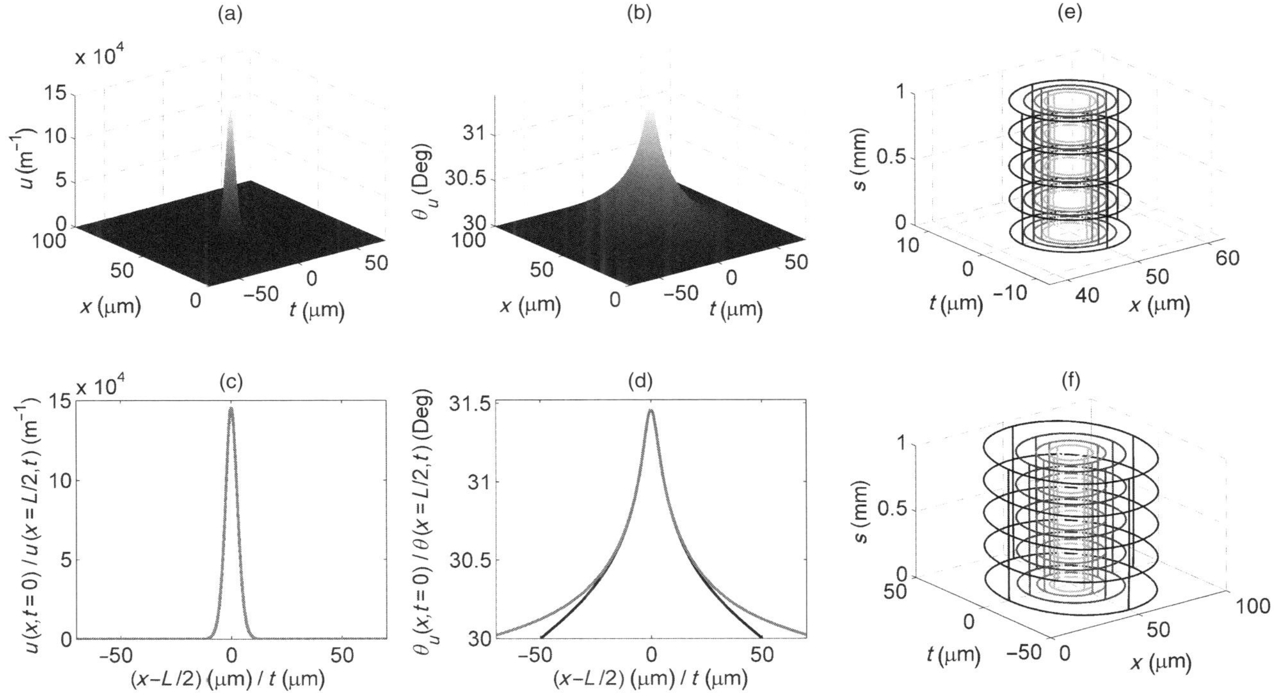

FIGURE 1.5 (a) Electric field profile u and (b) director distribution θ_u in the plane xt for a nematicon of power 1 mW. Graphs of (c) u and (d) θ_u in plane $x = L/2$ (gray line) versus t and in plane $t = 0$ versus x (black line). Computed 3D (e) intensity profile and (f) director distribution for an input beam as in (a). The wavelength is 632.8 nm, $\theta_0 = \pi/6$, and the cell thickness L is 100 μm.

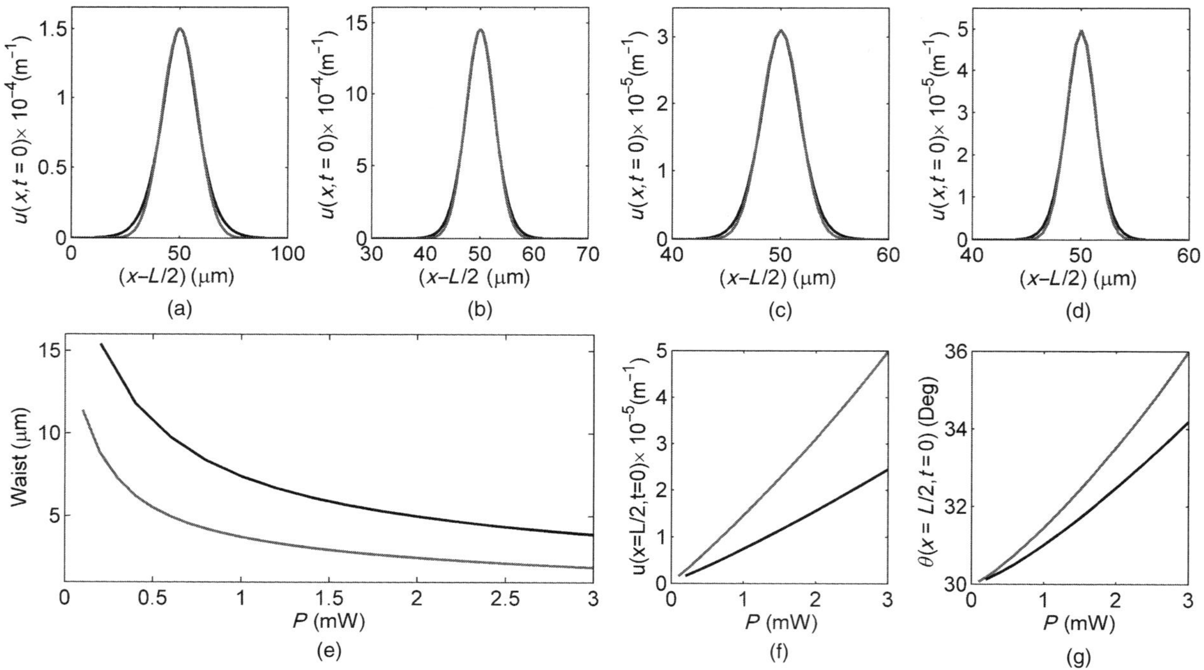

FIGURE 1.6 Nematicon profile for (a) $P = 0.1$, (b) 1, (c) 2, and (d) 3 mW versus $x - L/2$ at $\lambda = 632.8$ nm; dark and gray lines are exact numerical results and Gaussian best-fit, respectively. (e) Waist, (f) peak of the electric field u, and (g) maximum of θ_u for solitary waves versus power P at $\lambda = 632.8$ nm (gray line) and $\lambda = 1064$ nm (black line). Here, $\theta_0 = \pi/6$ and $L = 100$ μm.

of power P and waist w_0. We numerically solve the system [46]

$$2ik_0n_e\cos\delta\frac{\partial A}{\partial s}+D_t\frac{\partial^2 A}{\partial t^2}+D_x\frac{\partial^2 A}{\partial x^2}+k_0^2\epsilon_a\left[\sin^2(\theta-\delta)-\sin^2(\theta_0-\delta)\right]A=0,$$
$$K\nabla_{xt}^2\theta+\frac{\epsilon_0\epsilon_a}{4}\sin\left[2(\theta-\delta)\right]|A|^2=0, \tag{1.33}$$

where we neglected longitudinal derivatives in the reorientational equation. Figure 1.7 plots the evolution of a red beam ($\lambda = 632.8$ nm) for $\theta_0 = \pi/6$, $P = 1$ mW, and $w_0 = 3.5$ μm; we define $I_x(x,s) = \int_{\infty}^{\infty}|A|^2 \mathrm{d}t$ and $I_t(t,s) = \int_0^L |A|^2 \mathrm{d}x$, the transverse integrals of the intensity. In experiments, I_x and I_t are proportional to the scattered light acquired by a camera to monitor the beam evolution in the sample.

Because the initial conditions do not correspond to a point in the existence curve, a breather gets excited, with a quasi-sinusoidal oscillation in the waist along s; by the entrance (i.e., near $s = 0$) the beam narrows as in the plane waist-power the input profile lies above the existence curve: self-focusing prevails on diffractive spreading. We also note that the self-confined wave retains its radial symmetry, consistently with the results of Section 1.3.1. Finally, even if the waist oscillates with s, the distribution of θ remains nearly constant (Fig. 1.7f).

Figure 1.8a–c illustrates nematicon evolution for a given waist w_0 and various powers P at the input: at low power ($P = 0.1$ mW, panel a) the beam diffracts. As power increases ($P = 1$ mW, panel b) a nematicon forms, with a long breathing period and large waist oscillations. Further increases in power ($P = 2$ mW, panel c) lead to shorter periods and smaller oscillations in the waist.

For initial conditions corresponding to a point below the existence curve, the solitary wave initially increases the waist and undergoes slow and large oscillations. As the initial conditions approach the existence curve, the nematicon remains almost undistorted against propagation; it is worth noting that small changes persist as the exact soliton is not Gaussian (Fig. 1.8). As the initial conditions cross the existence curve, the nematicon breathes again, with both period and amplitude of oscillations related to the departure from the exact soliton. An aperiodic breathing appears for large oscillations, as the self-induced parabolic index well varies in propagation (see Eq. 1.27). Finally, Figure 1.8 also shows that self-confinement is eased at lower wavelengths.

1.4 EXPERIMENTAL OBSERVATIONS

The typical setup for the observation of nematicons is sketched in Figure 1.9, which shows an optical microscope imaging, the light outscattered by the NLC as the beam propagates in a planar cell. In a planar cell, two glass plates are held parallel to one another by spacers at a distance of $75 \div 120$ μm (across x), with the inner surfaces treated by mechanical rubbing of a coating to induce the desired

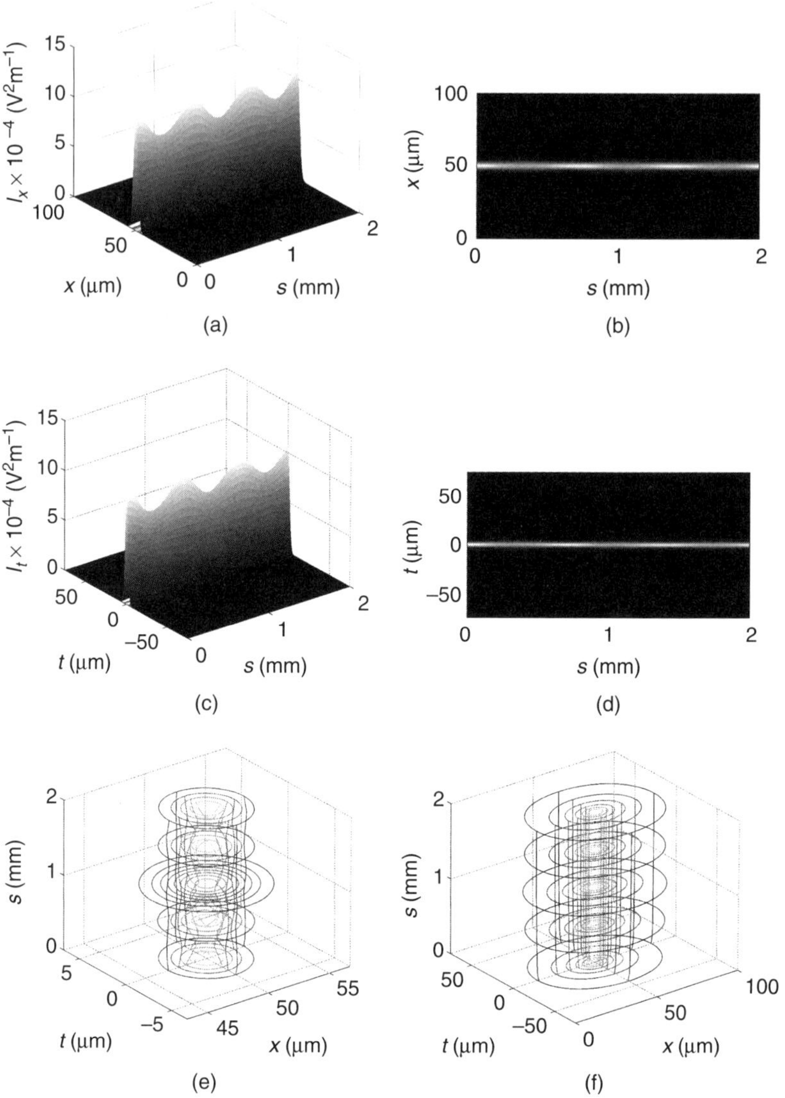

FIGURE 1.7 Beam evolution for $P = 1$ mW, $w_0 = 3.5$ μm, and $\lambda = 632.8$ nm. (a and b) I_x in the plane xs. (c and d) I_t in the plane ts. 3D profiles of (e) intensity $|A|^2$ and (f) director perturbation. Here, $L = 100$ μm and $\theta_0 = \pi/6$.

alignment of the molecules. A thin film of indium tin oxide (ITO) can also be predeposited on the plates for the application of an electric potential across the NLC thickness. Planar cells are normally filled by capillarity and glued together using the spacers. One (or two) other glass slide(s)—treated to ensure molecular anchoring, as well—is (are) arranged perpendicular to the cell (i.e., normal to z) in

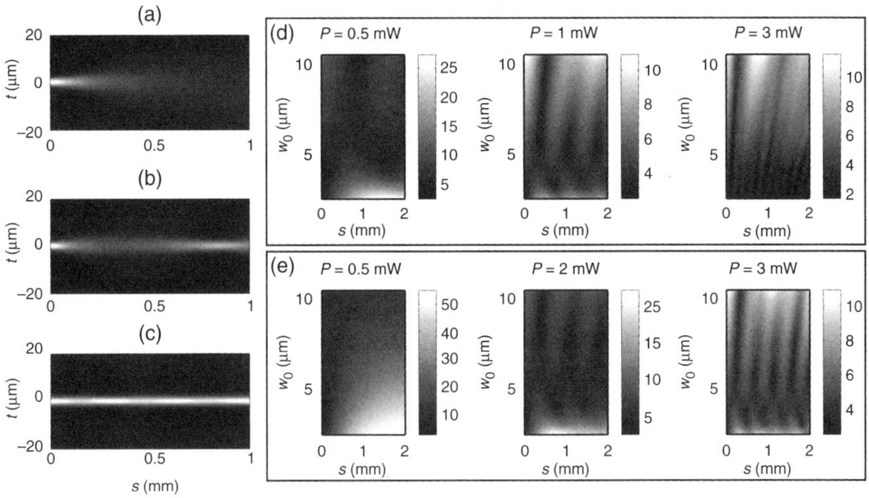

FIGURE 1.8 Beam evolution in the plane ts for $w_0 = 2.5$ μm, $\lambda = 632.8$ nm, and (a) $P = 0.1$ mW; (b) $P = 1$ mW; (c) $P = 2$ mW. Gray maps of beam waist at (d) $\lambda = 632.8$ nm and (e) $\lambda = 1064$ nm against (Poynting) propagation coordinate s and input waist w_0 for $P = 0.5$ mW, $P = 1$ mW and $P =$ 3 mW, respectively. The units of colorbar legends are micrometers; here $L = 100$μm and $\theta_0 = \pi/6$.

order to seal the entrance (and the exit) of the cell and prevent the formation of a meniscus with the consequent unpredictable depolarization of the input excitation. An optical microscope and a camera can image the light scattered by a beam out of one of the glass plates, thus allowing the real-time monitoring of the evolution of the excitation in the observation plane yz [1, 2, 47–49]. In cells with reduced propagation length (as compared to the attenuation distance) along z, the output profile can be acquired at the exit with a microscope objective and another camera [50–53]. At the input, a light beam linearly polarized in the plane containing the molecular director and the wave-vector, that is, an extraordinary wave, is focused with a microscope objective at the cell entrance, ensuring a transverse size and a direction of propagation suitable for the excitation of a self-trapped nematicon in the NLC bulk [54]. A weak copolarized collinear probe at a different wavelength can also be colaunched to monitor the formation of a waveguide and its signal transmission [48, 53].

Figure 1.10 shows a set of photographs displaying a 2 mW Gaussian beam of waist $w_0 = 4$ μm at $\lambda = 514.5$ nm propagating along a 100-μm -thick cell filled with the NLC mixture E7 and with director planarly oriented along z at the glass interfaces but pretilted in the xz plane by an external voltage across x (Fig. 1.3): for (ordinary) input polarization along y the beam diffracts, as its intensity cannot overcome the Freedericksz threshold (Fig. 1.10a); for (extraordinary) input polarization along x the beam undergoes self-focusing and becomes a spatial soliton, with an invariant profile over several Rayleigh lengths [1–2] (Fig. 1.10b). Figure 1.10c and d shows the corresponding output beam profiles after propagation over 1.5 mm in linear and solitary cases, respectively; the nematicon remains bell

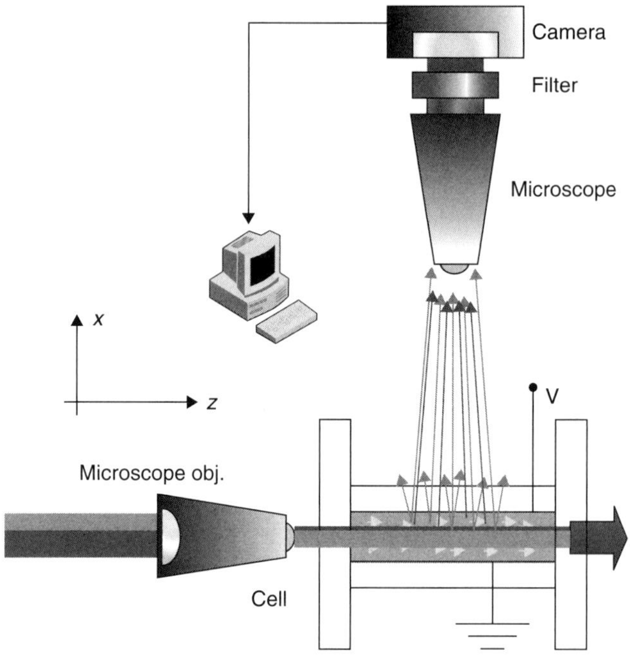

FIGURE 1.9 Typical experimental setup and planar cell, with copolarized collinear pump and probe launched in the NLC and propagating along z. The outscattered light at each wavelength is acquired with an optical microscope, a filter, and a camera, before digital processing.

shaped and circularly symmetric despite the NLC anisotropy and the presence of boundaries across x. Similarly, a colaunched probe at 632.8 nm (its power is negligible with respect to the green beam) either diffracts when the nematicon is not excited, or gets confined in the nematicon waveguide despite its wavelength (Fig. 1.10e and f). The nonlocal character of the nematicon, in fact, allows light at longer wavelengths being trapped in the nematicon refractive potential [2, 48].

Taking advantage of the nonlocal response of NLC [35, 40, 49], nematicons can also be excited by spatially incoherent beams that are bell shaped with a speckle structure as obtained when a coherent beam is launched through a rotating diffuser. The nonlocality, in fact, acts as a low pass filter and allows the reorientational nonlinearity to respond according to the excitation envelope. Because the wave-vector spectrum is wider, however, a higher power is required to compensate the increased diffraction. Figure 1.11a shows the input profile of a speckled 514.5 nm beam generated with the aid of a diffuser and Figure 1.11b is the corresponding spatial spectrum; for an incoherent beam as in (a) and an input power of 2.7 mW in the same cell as in Figure 1.10, Figure 1.11c shows the diffraction in the y polarization (ordinary) and Figure 1.11d the nematicon in the x polarization (extraordinary) [47, 55, 56].

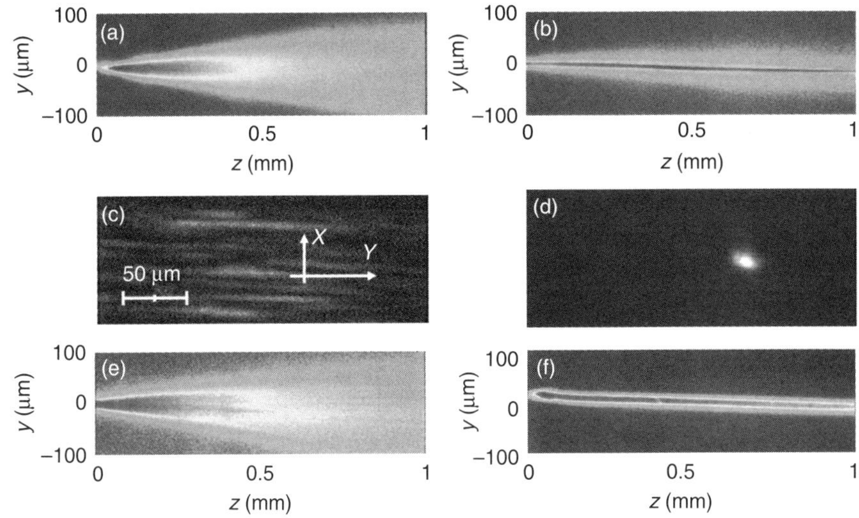

FIGURE 1.10 Behavior of light inside the NLC sample acquired through the setup shown in Figure 1.9. Ordinary (a) and extraordinary (b) propagation when the input power is 2 mW. The corresponding output beam profiles are shown in panels (c) and (d), for the ordinary and extraordinary wave, respectively. Propagation of an extraordinarily polarized red probe corresponding to the absence (e) and presence (f) of the soliton.

Nonlocal character of nematicons is demonstrated also by their breathing in propagation, as theoretically discussed in Section 1.2.2.1. Figure 1.12 shows experimental results obtained exciting the NLC E7 with an NIR beam ($\lambda = 1064$ nm). As predicted, beam width oscillates in a periodic manner in propagation, with both amplitude and period of oscillation decreasing with input power: in the plotted cases, diffraction prevails at the early stage, that is, we have $w_0 < w_s$ for all range of used powers (Section 1.2.2.1) [40].

As the NLC behave as birefringent uniaxials, their optic axis coincides with the molecular director $\hat{n}$, and because all-optical reorientation and self-focusing are driven (below the Freedericksz threshold) by extraordinary waves, nematicons are subject to walk-off; that is, their Poynting vector is not collinear with the wave-vector but walks off at an angle. The latter depends on the angle θ and the dispersion, and can reach several degrees in NLC owing to the large birefringence [57]. A nematicon excited as discussed in the above examples, that is, with electric field linearly polarized in xz, experiences walk-off in the same plane and, if launched with wave-vector along z in the mid-plane $x = 0$, it travels out of it towards one of the glass plates, until it interacts with the (repulsive) boundaries. Therefore, a typical nematicon trajectory in a planar cell with a voltage bias is oscillating in xz, unless the walk-off is compensated for by an input wave-vector tilt. The role of walk-off is more apparent in planar cells with director in yz but at an angle with respect to z (Fig. 1.3 for $\zeta_0 \neq 0$), in order to have noncollinear wave-vector and z axis. Let us consider a planar cell with director lying in yz but at

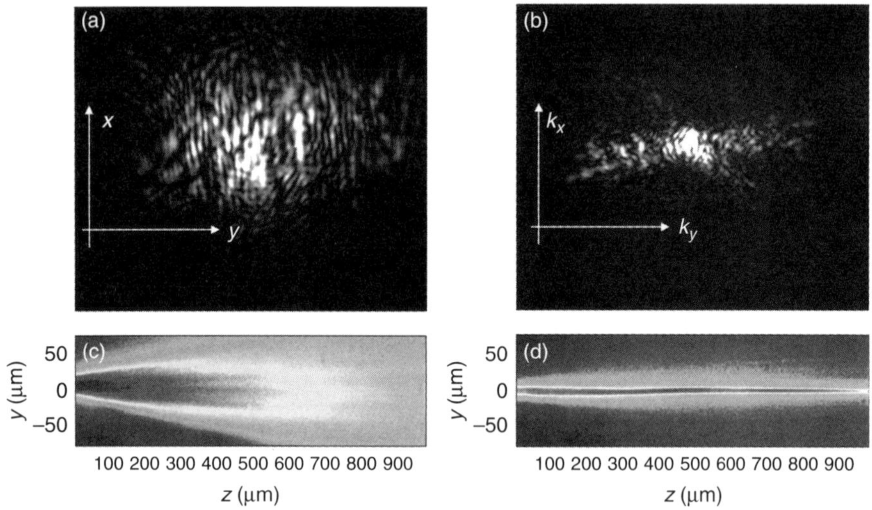

FIGURE 1.11 Speckled beam at the input of the NLC cell (a) and its spatial spectrum (b). Intensity distribution inside the sample for input polarizations corresponding to the ordinary (c) and extraordinary (d) component; input power is 2.7 mW.

$\pi/4$ with respect to $\hat{z}$ for zero bias, that is, $\zeta_0 = \pi/4$: a nematicon launched by a y-polarized input beam with $\mathbf{k}||\hat{z}$ experiences a walk-off of about 8° at $\lambda = 632.8$ nm; hence, its Poynting vector is slanted with respect to $\hat{z}$, as visible in the observation plane yz and shown in the photograph Figure 1.13a and b [30].

The cell sketched in Figure 1.3 for $\zeta_0 \neq 0$ can also be employed to alter the walk-off by applying a voltage across its thickness. The electro-optic NLC response can make the molecules reorientate out of the plane yz, correspondingly changing the principal plane $(\mathbf{k},\hat{n})$ where the Poynting vector lies; in the limit of a molecular director $\hat{n}||\hat{x}$, the extraordinary wave will take an ordinary configuration, that is, the walk-off angle goes to zero as the Poynting vector becomes collinear with the wave-vector. Therefore, the application of a bias across the thickness of this cell can progressively reduce the walk-off observable in yz and modify the nematicon direction of propagation in the NLC volume (Fig. 1.13c and Chapter 5) [30, 58]. Such an approach to nematicon steering, that is, one based on electro-optic changes in walk-off, is just an example of a variety of strategies that can be implemented to modify the trajectory of a spatial soliton (and its waveguide) and therefore route/readdress the guided signal(s) [59–62]. Several cases are discussed in this book, see, for example, Chapters 5, 6, 11 [30, 63–67]. Given their robustness, the path of spatial solitons in NLC can also be affected by mutual interactions between two (or more) of them, as briefly addressed in the following section.

1.4.1 Nematicon–Nematicon Interactions

Because NLC exhibits a large nonlocality, nematicons tend to behave as incoherent entities and, as pointed out above, they can also be excited by spatially

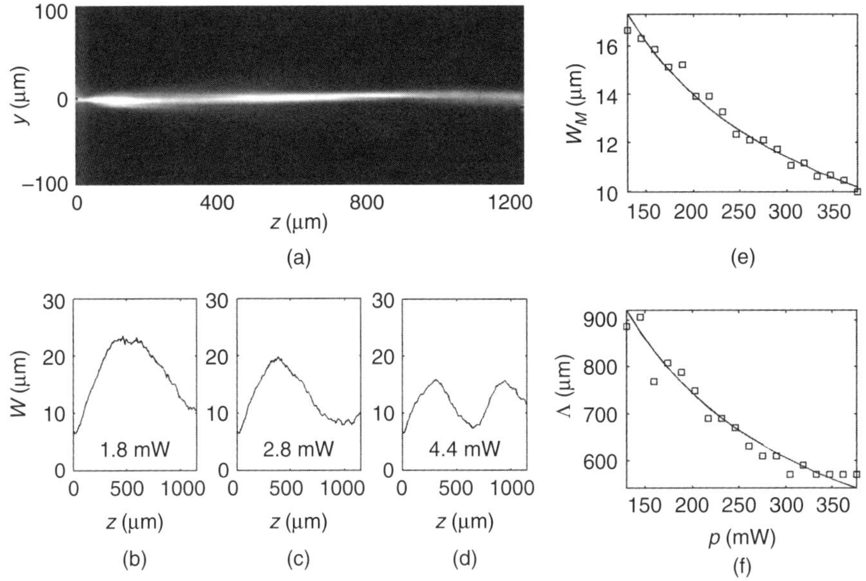

FIGURE 1.12 (a) Example of acquired intensity profile in plane yz. (b–d) Waist versus z. (e) Maximum waist and (f) breathing period versus input power (symbols are experimental data and solid lines are fits from theory).

incoherent excitations. Similarly, the interaction between two or more nematicons is essentially incoherent, that is, it largely depends on the interaction between the associated refractive index deformations or waveguides. In all cases, where the nonlocality range extends beyond the details of the field distribution, this interaction is attractive and two (or more) nematicons tend to get closer to one another as they propagate (see also Chapter 2) [68–71]. This concept is illustrated in Figure 1.14a–d showing the simulated attraction of two nematicons launched in plane yz at a relative angle of 2°: the refractive index perturbation (shown in Fig. 1.14b and c at the input and by the maximum separation, respectively) provides a transverse link between the two solitons, which attracts and bends their paths toward one another (Fig. 1.14d), eventually crossing and interlacing. The results of an actual experiment carried out in a biased cell with $\zeta_0 = 0^\circ$ at $\lambda - 514.5$ nm are displayed in Figure 1.14e–g for increasing input power, from 1.3 mW (e) to 1.7 mW (f) and 4.3 mW (g), respectively.

As not only the nematicons but also the transmitted signal(s) change trajectories when the interaction takes place, attraction between nematicons can be exploited to implement various analog or digital processors/routers, from power-driven switches to logic gates [72, 73]. Figure 1.15 is an example of excitation-dependent X-junction performing a power-controlled interchange of the output channels.

In the experiments shown in Figures 1.14 and 1.15 the actual motion of nematicons inside the samples are 3D, given the not null velocity along the x axis owing to walk-off (see the discussion for a single nematicon). To study pure planar

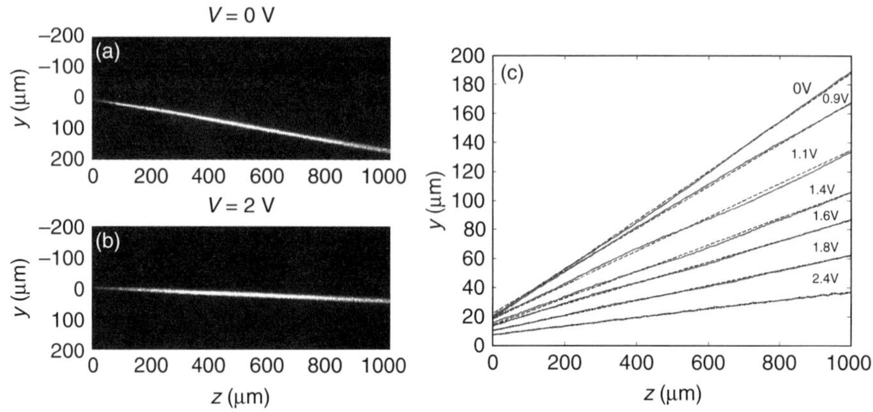

FIGURE 1.13 Direct observation of nematicon walk-off. Extraordinary wave beam with 3 mW initial power propagating in a cell (a) with zero bias and (b) with 2V applied across the NLC layer. (c) Nematicon trajectories in the observation plane yz for several applied voltages; solid and dashed lines are measured paths and corresponding linear best-fits, respectively. Here, the wavelength is 632.8 nm.

interaction between nematicons, an unbiased cell with $\zeta_0 \neq 0$ can be employed. To investigate the interaction of solitons we take the input beam in the form

$$A(x, y) = \sqrt{\frac{4Z_0}{\pi n_e w_0^2}}\sqrt{P}\left[e^{-\frac{(x-L/2)^2+(y-d/2)^2}{w_0^2}} + e^{i\phi}e^{-\frac{(x-L/2)^2+(y+d/2)^2}{w_0^2}}\right], \quad (1.34)$$

that is, two fundamental Gaussian beams launched in the mid-plane with wavevectors parallel to $\hat{z}$ and separated by d along y. Each beam carries the same power P and ϕ is the phase difference between them. We are interested in long-range interactions, that is, w_0 much smaller than the initial separation d: in this limit, mutual interaction forces are independent from ϕ due to nonlocality, at variance with (local) Kerr media [13] (see Chapter 2). Figure 1.16a–d shows numerical simulations of $\lambda = 1064$ nm nematicons interacting: as the power increases, the attraction becomes stronger, giving rise to multiple crossings along propagation. For a better comparison with experiments the scattering losses were accounted for, yielding a decreasing attraction versus z. Figure 1.16e and f shows the corresponding data: as the input power increases, the attraction pulls more and more the beams toward each other, up to three crossings when $P = 10$ mW. Finally, Figure 1.16g displays the trend of the first crossing point versus initial separation d for several excitations: the attraction is larger for both closer beams and higher powers [71].

In a thick NLC cell, in general, the attractive interaction between nematicons is obviously not limited to the plane yz of observation. If two nematicons are launched skewed to one another with initial momenta out of yz, the mutual attraction can combine the two solitons in a dipolar structure with angular momentum, in such a way that the two-nematicon cluster rotates as it propagates down the cell, maintaining a constant profile as the centrifugal force is balanced by the nonlinear

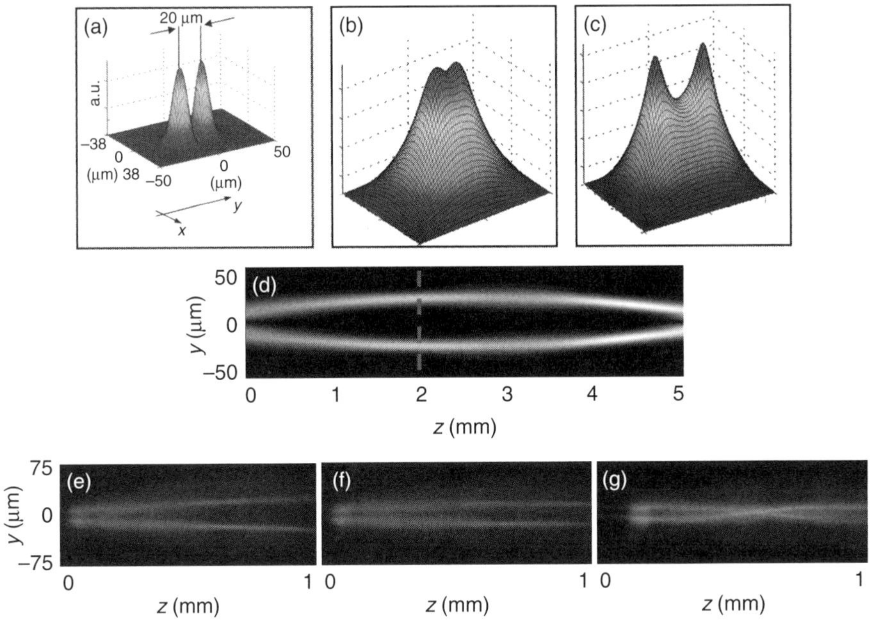

FIGURE 1.14 Mutual interaction between two solitons. Intensity profile at the input section (a) and the corresponding computed self-induced index landscape (b). (c) Refractive index distribution when the distance between solitons is maximum. (d) Interaction of numerically computed nematicons in the plane yz. Experimental interaction between nematicons for input power of 1.3 (e), 1.7 (f), and 4.3 mW (g). Wavelength is 514.5 nm.

attraction [50, 52, 74]. The two nematicons will therefore spiral in a rigid manner around the straight trajectory of the center of mass as sketched in Figure 1.17a and their angular velocity will be dependent on the initial angular momentum, that is, components of the wave-vectors launched out of yz (i.e., $\mathbf{k} \cdot \hat{x}$) and the effective masses associated to the input powers. Therefore, the position of the two spots at the output of the cell is dependent on the input powers, as visible in the experimental results shown in Figure 1.17b–g, obtained in a planar cell (unbiased) with molecular director at $\pi/6$ with respect to $\hat{z}$ in the plane yz (i.e., $\zeta_0 = 30^\circ$). Each nematicon carries the same power and input wave-vectors are tilted in the plane yz, so that the Poynting vector of each beam is parallel to $\hat{z}$. Moreover, to avoid velocity drifts along x the wave-vector components along $\hat{x}$ are equal in modulus and opposite in sign. The rate of angular rotation of the nematicon cluster is proportional to the total excitation of the two identical solitons, as theoretically expected [52].

A special case of nonlinear incoherent interaction between the beams in NLC is the formation of a vector soliton, that is, a self-confined wave where diffraction is balanced by the combined intensities of two color components in the extraordinary polarization. If the power in each input beam is too low to excite a nematicon by itself but their coaction is sufficient to induce self-trapping, the two components are

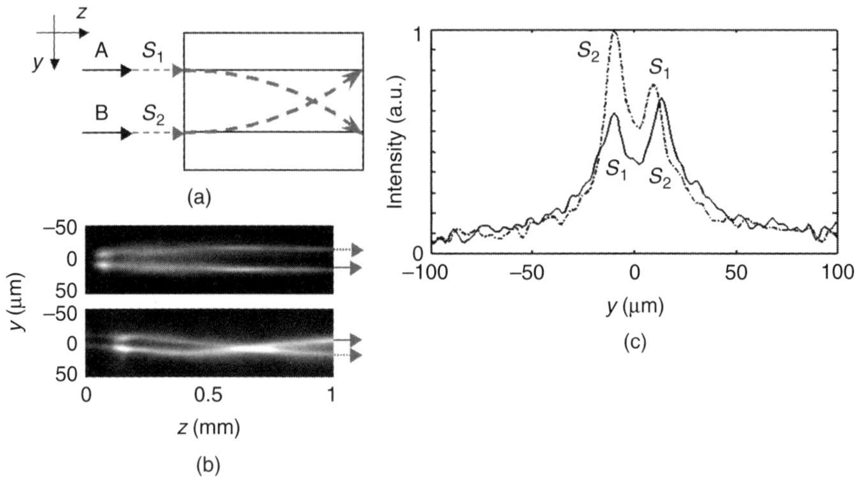

FIGURE 1.15 All-optical X-junction. Principle of the device: two nematicons (labeled as A and B) are launched inside the NLC cell with equal initial powers: if input power is small, the interaction is weak and no appreciable motion takes place in the light path, whereas for high power nonlinear interaction gives rise to mutual attraction, up to crossing and position exchange with respect to the initial configuration, thereby forming an X-junction. (a) Signals launched on light-induced guides A and B are indicated as S_1 and S_2, respectively. (b) Measured light distribution on the plane yz for an initial power of 1.73 mW (top panel) and 4.3 mW (bottom panel). (c) Scattered light distribution on the section $z = 1$ mm.

capable of spatially confining each other forming a vectorial nematicon, undergoing the weighted walk-off between the two colors, as dispersion can be significant and affect birefringence as well as diffraction. Figure 1.18 shows experimental results obtained in a planar (unbiased) NLC cell with molecular director anchored at $\pi/4$ with respect to $\hat{z}$ in the plane yz (in other words $\zeta_0 = \pi/4$), with input wave-vector tilted such that the two Poynting vectors at the two different colors overlap; the two components were in the red at 632.8 nm and in the near-infrared at 1064 nm [44]. More in general, the simultaneous launch of two beams at different wavelength implies a reciprocal influence on the breathing in propagation, that is, the waist behavior versus z. Noticeably, the effect of each wavelength is different for three reasons: a diverse amount of diffraction, a different optical perturbation for a fixed intensity profile, and finally a different refractive index profile for a given reorientation angle, the last two are caused by dispersion in the birefringence.

1.4.2 Modulational Instability

As anticipated in Section 1.2.1.2, transverse patterns with a dominant harmonic component in space can emerge in the propagation of a wide beam in a self-focusing reorientational medium and in the presence of nonlocality [75, 76]. Nonlocal modulational instability can indeed be observed in NLC, both in the spontaneous case originating from noise [41] and in the "seeded" case in the presence of an input (transverse) modulation [77, 78]. Modulational instability can

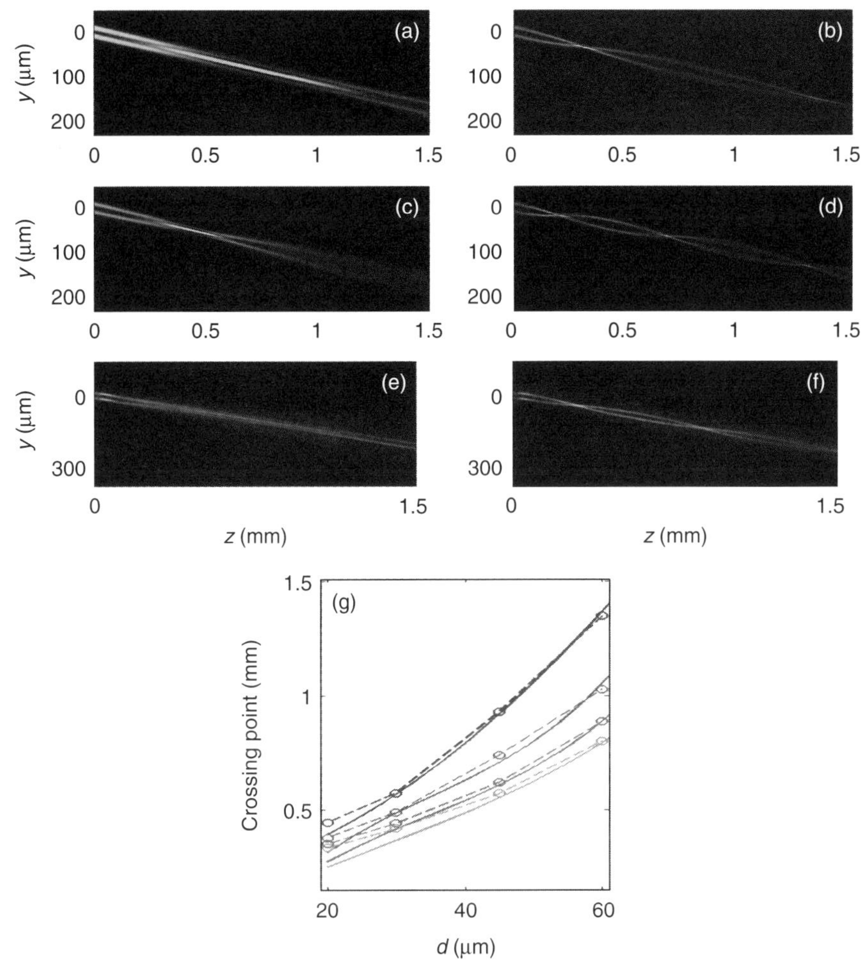

FIGURE 1.16 Planar interactions between nematicons. Numerically calculated nematicon path for $d = 25$ μm and (a) $P = 0.2$, (b) 0.5, (c) 1, and (d) 3 mW. Actual nematicon trajectories for the same d and (e) $P = 2$ mW, (f) $P = 10$ mW. (g) Position of the first crossing versus d; solid ($P =$ 0.7, 1, 1.3, 1.7 mW from top to bottom, respectively) and dashed lines with symbols (4, 6, 8, 10 mW from top) are calculated and experimental, respectively. Small discrepancies between theoretical and experimental power values can be ascribed to the 2D nature of the code (see Chapter 11). Here, $w_0 = 7$ μm and $\theta_0 = 45°$.

be investigated versus propagation distance and versus input power, eventually resulting in a number of filaments or solitary waves emerging from the wide input beam. Figure 1.19 shows typical results in a planar NLC cell with a bias $V = 1.6$V, using a highly elliptical input beam at 514.5 nm: the transverse modulation is clearly visible in the plane yz for an input power of 17 mW (Fig. 1.19a), but gets more and more pronounced at 88 mW with the emergence of filaments (Fig. 1.19b); at 193 mW the nematicons attract one another and the whole cluster deforms undergoing "global" self-focus (Fig. 1.19c) [41, 42, 79].

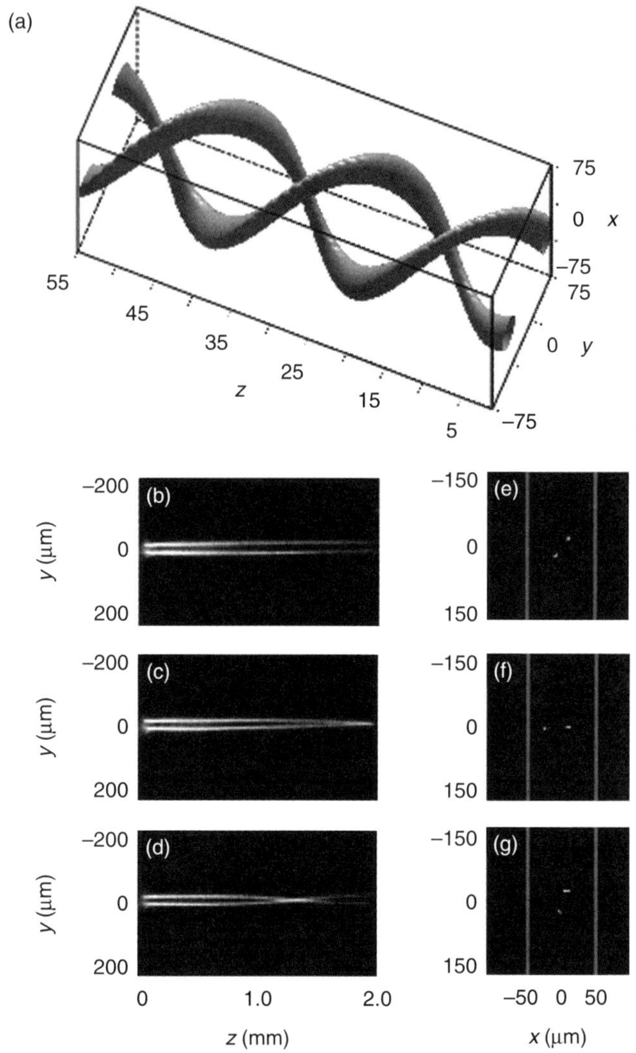

FIGURE 1.17 (a) Artist's 3D sketch of spiraling nematicons. Measured (middle column) evolution in the plane yz and (right panels) corresponding output distribution in plane xy for initial powers of (b–e) 2.1, (c–f) 3.3, and (d–g) 3.9 mW. Here, the wavelength is $\lambda = 1064$ nm, the cell thickness is 100 µm, and the propagation length is 2 mm.

In analogy with nematicons of which is considered a precursor, modulational instability can also be generated with a spatially incoherent beam at the expense of the gain, with a resulting path contrast which at a given propagation distance is dependent on the spatial spectrum and on the excitation, as apparent in Figure 1.20 [80]. Owing to the larger diffraction counteracting nonlinear self-focusing, amplification of high frequencies is weakened as beam incoherence increases. Modulational instability patterns and/or clusters of nematicons can even be steered in the

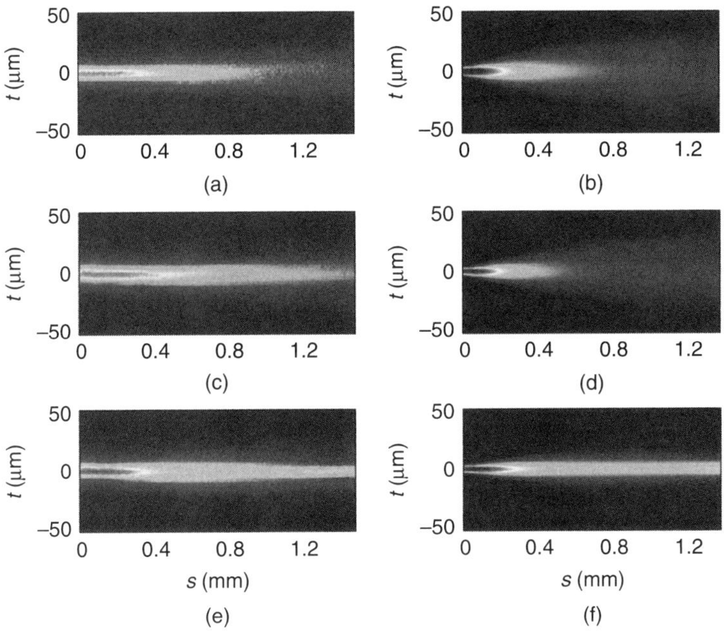

FIGURE 1.18 (Left) Acquired and (right) calculated evolution of the red beam component in the plane *st*. (a and b) A 0.1 mW red beam at 632.8 nm is colaunched with a 1.2 mW near-infrared beam; (c and d) a 0.4 mW red beam is injected by itself; (e and f) 0.4 mW red and 1.2 mW infrared beams are colaunched and a vector nematicon is generated. Both beams are extraordinary waves with comparable Rayleigh lengths. Calculations were carried out for in-coupling efficiencies of 40% and 62% for red and near-infrared, respectively.

direction of propagation by modifying the walk-off with a voltage bias, that is, by changing the principal plane of propagation [30].

While the model discussed in Section 1.2.1.2 is valid in the early stages of propagation when the perturbations are actually small, as solitons begin to emerge (Fig. 1.19b) their mutual interactions become relevant and need to be accounted for (Fig. 1.19c). In this regime, a large number of solitons can be represented as a *gas* of interacting particles that are capable of undergoing clustering and dynamic phase transitions [81].

The generation of multiple solitons from a wide beam can be enhanced by imposing a phasefront curvature on the input beam. In fact, by tailoring the size of a converging (focused) input beam, modulational instability can trigger the generation of multiple nematicons from the focal point, as, for example, shown in Figure 1.21 [82]. The number of nematicons increases with an increase in power, whereas their exact location depends on noise-induced fluctuations.

Modulational instability in biased NLC cells with $\zeta_0 = 0$ can also be employed to investigate the interplay between nonlocality and nonlinearity, theoretically discussed in Section 1.2.1.1. As the voltage applied to the cell increases, both nonlinearity n_2^{eff} (Eq. 1.6) and nonlocality l (Eq. 1.4) decrease. This is included

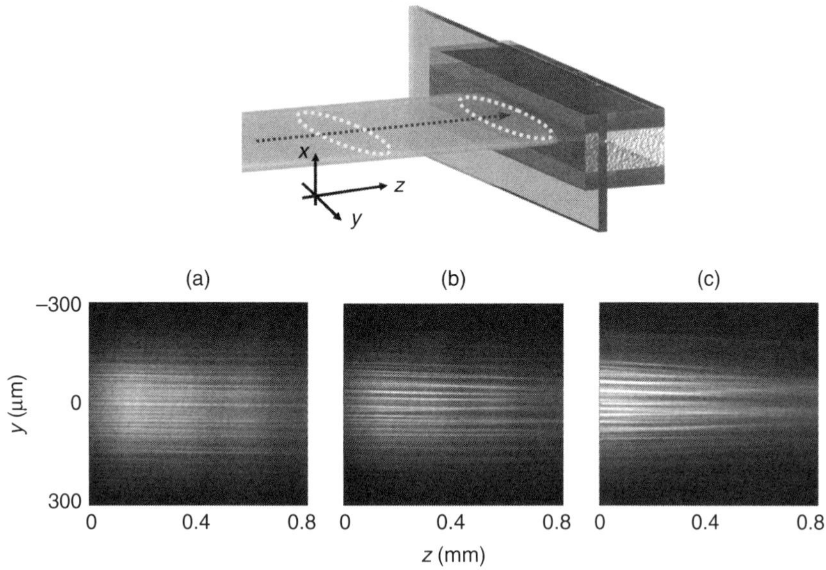

FIGURE 1.19 Sketch of the experiment on modulational instability, and pictures of the emerging pattern in yz for an input power of (a) 17 mW, (b) 88 mW, and (c) 193 mW.

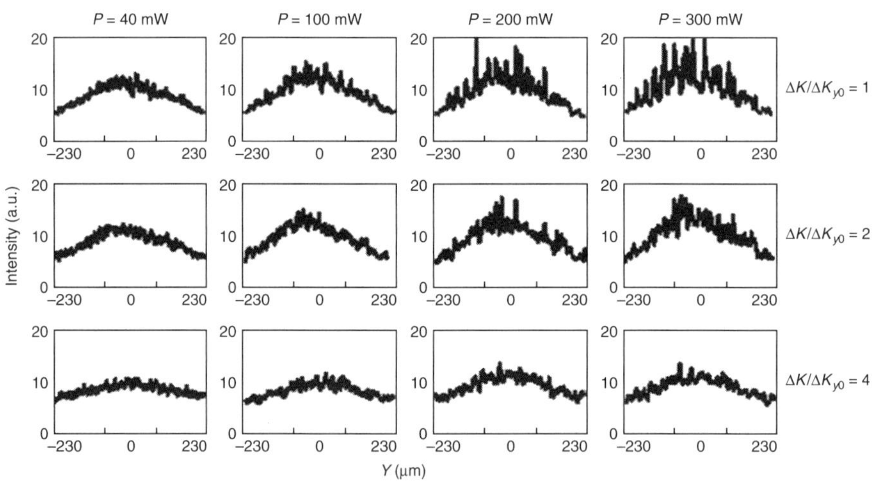

FIGURE 1.20 Synopsis of experimental results on modulational instability for various degrees of incoherence (ratio $\Delta K/\Delta K_{y0}$ between the incoherent and a coherent spatial spectrum) and input power. The graphs show the transverse pattern after propagation along z for 200 μm.

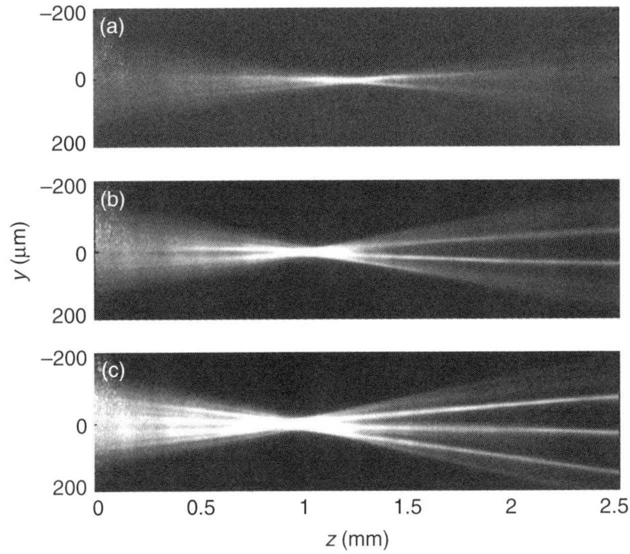

FIGURE 1.21 Generation of multiple nematicons from a focused input beam at 1064 nm in a planar NLC cell for various input powers: (a) 5 mW, (b) 30 mW, and (c) 47.5 mW.

in Equation 1.15, which predicts that the nonlinear gain reduces in both peak and bandwidth as α (inversely proportional to nonlocality) goes up (Fig. 1.22). On physical grounds, nonlocality effectively acts as an integrator and therefore, filters out the response at high spatial frequencies. Figure 1.22 shows some experimental results: the modulational instability gain spectrum lowers in peak and bandwidth (in wave-vector space) at higher voltages [36].

1.5 CONCLUSIONS

In this chapter, we tried to provide a brief (and incomplete) overview of the basic features of (bright) optical spatial solitons in NLC, touching upon pertinent models, numerics, and experiments in an attempt to stimulate the interest of the readers and encourage them to look up the growing literature and the following chapters of this book.

Acknowledgments

We wish to express our gratitude to all those who have contributed to the work presented in this chapter, in particular, to Marco Peccianti, Claudio Conti, and Cesare Umeton. A.A. thanks Regione Lazio for the financial support provided.

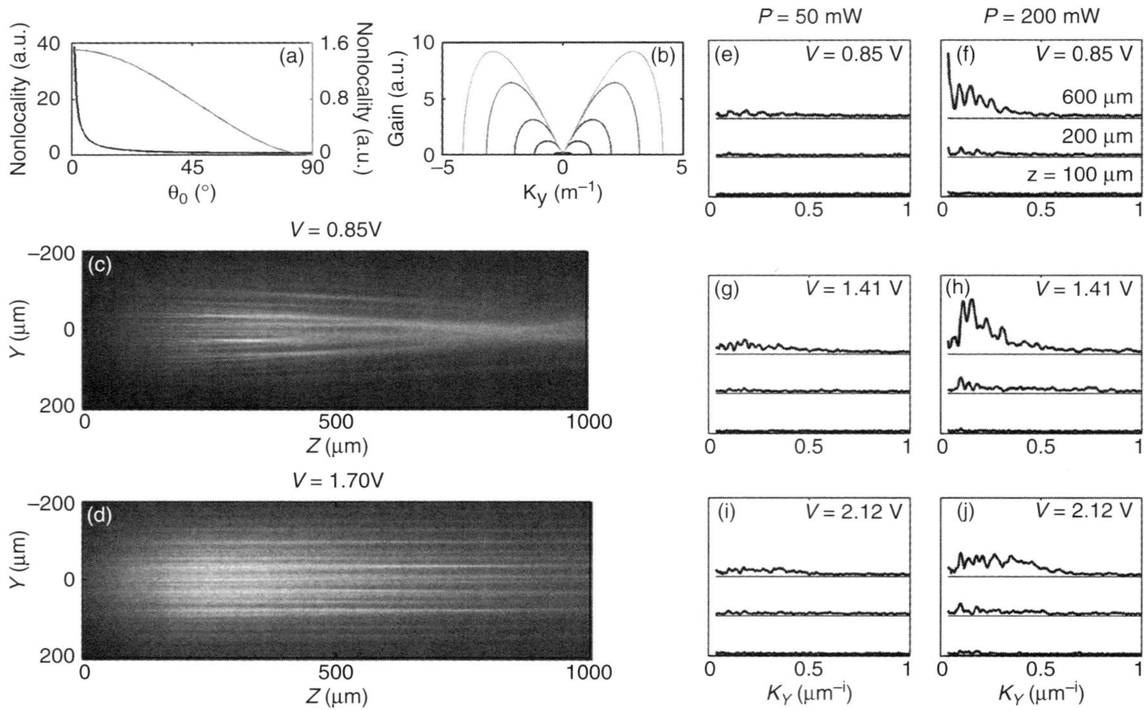

FIGURE 1.22 Calculated (a) trend of nonlocality (black line) and nonlinearity (gray line) versus θ_0 and (b) gain versus transverse wave-vector k_y for $\alpha = 0.001, 0.1, 1, 10, 100$ from top to bottom, respectively. The gain is computed for $\alpha a_0^2 = 1\text{W/m}$. Light intensity in plane yz when a 150 mW elliptic Gaussian beam is launched in the cell with $\zeta_0 = 0$ and bias (c) 0.85 V and (d) 1.70 V. Measured spectral gain (in three longitudinal sections as marked) versus wave-vector k_y for input power (e,g,i) 50 mW or (f,h,j) 200 mW and voltages as indicated. Here, the wavelength is 1064 nm and the NLC thickness is 100 μm.

REFERENCES

1. G. Assanto, M. Peccianti, and C. Conti. Nematicons: Optical spatial solitons in nematic liquid crystals. *Opt. Photon. News*, 14(2):44–48, 2003.
2. M. Peccianti, G. Assanto, A. De Luca, C. Umeton, and I. C. Khoo. Electrically assisted self-confinement and waveguiding in planar nematic liquid crystal cells. *Appl. Phys. Lett.*, 77(1):7–9, 2000.
3. P. G. DeGennes and J. Prost. *The Physics of Liquid Crystals*. Oxford Science, New York, 1993.
4. I. C. Khoo. *Liquid Crystals: Physical Properties and Nonlinear Optical Phenomena*. Wiley, New York, 1995.
5. F. Simoni. *Nonlinear Optical Properties of Liquid Crystals*. World Scientific, Singapore, 1997.
6. I. C. Khoo. Nonlinear optics of liquid crystalline materials. *Phys. Rep.*, 471:221–267, 2009.
7. I. C. Khoo. Extreme nonlinear optics of nematic liquid crystals. *J. Opt. Soc. Am. B*, 471:221–267, 2009.
8. G. K. L. Wong and Y. R. Shen. Optical-field-induced ordering in the isotropic phase of a nematic liquid crystal. *Phys. Rev. Lett.*, 30:895, 1973.
9. D. Paparo, L. Marrucci, G. Abbate, E. Santamato, M. Kreuzer, P. Lehnert, and T. Vogeler. Molecular-field-enhanced optical kerr effect in absorbing liquids. *Phys. Rev. Lett.*, 78:38–41, 1997.
10. I. Janossy, A. D. Lloyd, and B. S. Wherrett. Anomalous optical Freedericksz transition in an absorbing liquid crystal. *Mol. Cryst. Liq. Cryst.*, 179:1–12, 1990.
11. R. Y. Chiao, E. Garmire, and C. H. Townes. Self-trapping of optical beams. *Phys. Rev. Lett.*, 13(15):479–482, 1964.
12. P. L. Kelley. Self-focusing of optical beams. *Phys. Rev. Lett.*, 15(26):1005–1008, 1965.
13. G. I. Stegeman and M. Segev. Optical spatial solitons and their interactions: Universality and Diversity. *Science*, 286(5444):1518–1523, 1999.
14. G. Baruch, S. Maneuf, and C. Froehly. Propagation soliton et auto-confinement de faisceaux laser par non linearite optique de Kerr. *Opt. Commun.*, 55:201, 1985.
15. J. E. Bjorkholm and A. A. Ashkin. cw self-focusing and self-trapping of light in sodium vapor. *Phys. Rev. Lett.*, 32(4):129–132, 1974.
16. G. C. Duree, J. L. Shultz, G. J. Salamo, M. Segev, A. Yariv, B. Crosignani, P. Di Porto, E. J. Sharp, and R. R. Neurgaonkar. Observation of self-trapping of an optical beam due to the photorefractive effect. *Phys. Rev. Lett.*, 71(4):533–536, 1993.
17. S. Tzortzakis, L. Sudrie, M. Franco, B. Prade, A. Mysyrowicz, A. Couairon, and L. Bergé. Self-guided propagation of ultrashort IR laser pulses in fused silica. *Phys. Rev. Lett.*, 87(21):213902, 2001.
18. A. Fratalocchi, G. Assanto, K. A. Brzdakiewicz, and M. A. Karpierz. Discrete propagation and spatial solitons in nematic liquid crystals. *Opt. Lett.*, 29(13):1530–1532, 2004.
19. F. Lederer, G. I. Stegeman, D. N. Christodoulides, G. Assanto, M. Segev, and Y. Silberberg. Discrete solitons in optics. *Phys. Rep.*, 463(1–3):1–126, 2008.
20. O. Bang, W. Krolikowski, J. Wyller, and J. J. Rasmussen. Collapse arrest and soliton stabilization in nonlocal nonlinear media. *Phys. Rev. E*, 66:046619, 2002.

21. Y. S. Kivshar and G. P. Agrawal. *Optical Solitons*. Academic, San Diego, CA, 2003.
22. B. Y. Zeldovich, N. F. Pilipetskii, A. V. Sukhov, and N. V. Tabiryan. Giant optical nonlinearity in the mesophase of a nematic liquid crystals (nlc). *JETP Letters*, 31:263–269, 1980.
23. S. D. Durbin, S. M. Arakelian, and Y. R. Shen. Optical-field-induced birefringence and Freedericksz transition in a nematic liquid crystal. *Phys. Rev. Lett.*, 47:1411–1414, 1981.
24. S. D. Durbin, S. M. Arakelian, and Y. R. Shen. Laser-induced diffraction rings from a nematic-liquid-crystal film. *Opt. Lett.*, 6(9):411–413, 1981.
25. E. Braun, L. P. Faucheux, and A. Libchaber. Strong self-focusing in nematic liquid crystals. *Phys. Rev. A*, 48(1):611–622, Jul 1993.
26. D. W. McLaughlin, D. J. Muraki, M. J. Shelley, and X. Wang. A paraxial model for optical self-focussing in a nematic liquid crystal. *Physica D*, 88(1):55–81, 1995.
27. D. W. McLaughlin, D. J. Muraki, and M. J. Shelley. Self-focussed optical structures in a nematic liquid crystal. *Physica D*, 97(4):471–497, 1996.
28. F. Derrien, J. F. Henninot, M. Warenghem, and G. Abbate. A thermal (2D+1) spatial optical soliton in a dye doped liquid crystal. *J. Opt. A: Pure Appl. Opt.*, 2:332, 2000.
29. M. Warenghem, J. F. Blach, and J. F. Henninot. Thermo-nematicon: an unnatural coexistence of solitons in liquid crystals? *J. Opt. Soc. Am. B*, 25:1882–1887, 2008.
30. M. Peccianti, C. Conti, G. Assanto, A. De Luca, and C. Umeton. Routing of anisotropic spatial solitons and modulational instability in nematic liquid crystals. *Nature*, 432:733, 2004.
31. M. A. Karpierz. Solitary waves in liquid crystalline waveguides. *Phys. Rev. E*, 66(3):036603, 2002.
32. A. Fratalocchi, G. Assanto, K. Brzdakiewicz, and M. A. Karpierz. Discrete light propagation and self-trapping in liquid crystals. *Opt. Express*, 13(6):1808–1815, 2005.
33. U. A. Laudyn, M. Kwasny, and M. A. Karpierz. Nematicons in chiral nematic liquid crystals. *Appl. Phys. Lett.*, 94(9):091110, 2009.
34. U. A. Laudyn, M. Kwasny, K. Jaworowicz, K. A. Rutkowska, M. A. Karpierz, and G. Assanto. Nematicons in twisted liquid crystals. *Photon. Lett. Pol.*, 1(1):7, 2009.
35. C. Conti, M. Peccianti, and G. Assanto. Route to nonlocality and observation of accessible solitons. *Phys. Rev. Lett.*, 91:073901, 2003.
36. M. Peccianti, C. Conti, and G. Assanto. The interplay between non locality and nonlinearity in nematic liquid crystals. *Opt. Lett.*, 30:415, 2005.
37. I. A. Kol'chugina, V. A. Mironov, and A. M. Sergeev. On the structure of stationary solitons in systems with nonlocal nonlinearity. *JETP Letters*, 31(6):333–7, 1980.
38. A. W. Snyder and D. J. Mitchell. Accessible solitons. *Science*, 276:1538, 1997.
39. G. Assanto and G. Stegeman. Simple physics of quadratic spatial solitons. *Opt. Express*, 10(9):388–396, 2002.
40. C. Conti, M. Peccianti, and G. Assanto. Observation of optical spatial solitons in a highly nonlocal medium. *Phys. Rev. Lett.*, 92:113902, 2004.
41. M. Peccianti, C. Conti, and G. Assanto. Observation of optical modulational instability in a non-local medium. *Phys. Rev. E*, 68:025602(R), 2003.
42. C. Conti, M. Peccianti, and G. Assanto. Spatial solitons and modulational instability in the presence of large birefringence: the case of highly non-local liquid crystals. *Phys. Rev. E*, 72:066614, 2005.

43. A. Alberucci and G. Assanto. Nonparaxial (1+1)D spatial solitons in uniaxial media. *Opt. Lett.*, 36(2):193–195, Jan 2011.

44. A. Alberucci, M. Peccianti, G. Assanto, A. Dyadyusha, and M. Kaczmarek. Two-color vector solitons in nonlocal media. *Phys. Rev. Lett.*, 97:153903, 2006.

45. A. Messiah. *Quantum Mechanics*. Dover Publications, New York, 1999.

46. A. Alberucci and G. Assanto. Dissipative self-confined optical beams in doped nematic liquid crystals. *J. Nonlin. Opt. Phys. Mater.*, 16(3):295–305, 2007.

47. M. Peccianti and G. Assanto. Incoherent spatial solitary waves in nematic liquid crystals. *Opt. Lett.*, 26:1791, 2001.

48. M. Peccianti and G. Assanto. Signal readdressing by steering of spatial solitons in bulk nematic liquid crystals. *Opt. Lett.*, 26(21):1690–1692, 2001.

49. M. Peccianti, C. Conti, G. Assanto, A. De Luca, and C. Umeton. Nonlocal optical propagation in nonlinear nematic liquid crystals. *J. Nonlin. Opt. Phys. Mater.*, 12:525, 2003.

50. A. Fratalocchi, A. Piccardi, M. Peccianti, and G. Assanto. Nonlinearly controlled angular momentum of soliton clusters. *Opt. Lett.*, 32:1447, 2007.

51. A. Alberucci, M. Peccianti, and G. Assanto. Nonlinear bouncing of nonlocal spatial solitons at the boundaries. *Opt. Lett.*, 32(19):2795–2797, 2007.

52. A. Fratalocchi, A. Piccardi, M. Peccianti, and G. Assanto. Nonlinear management of the angular momentum of soliton clusters: Theory and experiment. *Phys. Rev. A*, 75(6):063835, 2007.

53. Y. V. Izdebskaya, A. S. Desyatnikov, G. Assanto, and Y. S. Kivshar. Multimode nematicon waveguides. *Opt. Lett.*, 36(2):184–186, 2011.

54. G. Assanto and M. Peccianti. Spatial solitons in nematic liquid crystals. *IEEE J. Quantum Electron.*, 39:13–21, 2003.

55. M. Peccianti and G. Assanto. Nematic liquid crystals: a suitable medium for self-confinement of coherent and incoherent light. *Phys. Rev. E*, 65:035603(R), 2002.

56. K. G. Makris, H. Sarkissian, D. N. Christodoulides, and G. Assanto. Non local incoherent spatial solitons in liquid crystals. *J. Opt. Soc. Am. B.*, 22 (7):1371–1377, 2005.

57. M. Peccianti, A. Fratalocchi, and G. Assanto. Transverse dynamics of nematicons. *Opt. Express*, 12:6524, 2004.

58. A. Alberucci, M. Peccianti, G. Assanto, G. Coschignano, A. De Luca, and C. Umeton. Self-healing generation of spatial solitons in liquid crystals. *Opt. Lett.*, 30:1381–1383, 2005.

59. G. Assanto and M. Peccianti. Routing light at will. *J. Nonlin. Opt. Phys. Mater.*, 16 (1):37–48, 2007.

60. G. Assanto, C. Umeton, M. Peccianti, and A. Alberucci. Nematicons and their angular steering. *J. Nonlin. Opt. Phys. Mater.*, 15(1):33–42, 2006.

61. G. Assanto and M. Peccianti. Guiding and routing light with nematicons. *Mol. Cryst. Liq. Cryst.*, 488:163–178, 2008.

62. G. Assanto, A. Alberucci, and A. Piccardi. Transverse self-acceleration of nematicons: can a self-confined beam change its own path? *J. Nonlin. Opt. Phys. Mater.*, 20(3):237–242, 2011.

63. A. Piccardi, G. Assanto, L. Lucchetti, and F. Simoni. All-optical steering of soliton waveguides in dye-doped liquid crystals. *Appl. Phys. Lett.*, 93:171104, 2008.

64. A. Piccardi, M. Peccianti, G. Assanto, A. Dyadyusha, and M. Kaczmarek. Voltage-driven in-plane steering of nematicons. *Appl. Phys. Lett.*, 94(9):091106, 2009.

65. A. Alberucci, A. Piccardi, U. Bortolozzo, S. Residori, and G. Assanto. Nematicon all-optical control in liquid crystal light valves. *Opt. Lett.*, 35(3):390–392, 2010.

66. A. Piccardi, A. Alberucci, and G. Assanto. Power-dependent nematicon steering via walk-off. *J. Opt. Soc. Am. B*, 27:2398–2404, 2010.

67. R. Barboza, A. Alberucci, and G. Assanto. Large electro-optic beam steering with nematicons. *Opt. Lett.*, 36(14):2725–2727, 2011.

68. M. Peccianti, K. Brzadkiewicz, and G. Assanto. Nonlocal spatial soliton interactions in nematic liquid crystals. *Opt. Lett.*, 27:1460, 2002.

69. W. Hu, T. Zhang, Q. Guo, L. Xuan, and S. Lan. Nonlocality-controlled interaction of spatial solitons in nematic liquid crystals. *Appl. Phys. Lett.*, 89(7):071111, 2006.

70. Y. V. Izdebskaya, V. Shvedov, A. S. Desyatnikov, Y. S. Kivshar, W. Krolikowski, and G. Assanto. Incoherent interaction of nematicons in bias-free liquid crystal cells. *J. Eur. Opt. Soc.*, 5:10008, 2008.

71. M. Kwasny, A. Piccardi, A. Alberucci, M. Peccianti, M. Kaczmarek, M. Karpierz, and G. Assanto. Nematicon-nematicon interactions in a medium with tunable nonlinearity and fixed nonlocality. *Opt. Lett.*, 36 (13)(1):2566–2568, 2011.

72. M. Peccianti, C. Conti, G. Assanto, A. De Luca, and C. Umeton. All optical switching and logic gating with spatial solitons in liquid crystals. *Appl. Phys. Lett.*, 81:3335, 2002.

73. A. Piccardi, A. Alberucci, U. Bortolozzo, S. Residori, and G. Assanto. Readdressable interconnects with spatial soliton waveguides in liquid crystal light valves. *IEEE Photon. Technol. Lett.*, 22:694–696, 2010.

74. G. Assanto, N. F. Smyth, and A. L. Worthy. Two colour, nonlocal vector solitary waves with angular momentum in nematic liquid crystals. *Phys. Rev. A*, 78(1):013832, 2008.

75. W. Krolikowski, O. Bang, J. J. Rasmussen, and J. Wyller. Modulational instability in nonlocal nonlinear Kerr media. *Phys. Rev. E*, 64:016612, 2001.

76. J. Wyller, W. Krolikowski, O. Bang, and J. J. Rasmussen. Generic features of modulational instability in nonlocal Kerr media. *Phys. Rev. E*, 66:066615, 2002.

77. J. Beeckman, X. Hutsebaut, M. Haelterman, and K. Neyts. Induced modulation instability and recurrence in nematic liquid crystals. *Opt. Express*, 15:11185–11195, 2007.

78. J. Beeckman, K. Neyts, and M. Haelterman. Induced modulation instability and recurrence in nonlocal nonlinear media. *J. Phys. B: At. Mol. Opt. Phys.*, 41:065402, 2008.

79. C. Conti, M. Peccianti, and G. Assanto. Complex dynamics and configurational entropy of spatial optical solitons in nonlocal media. *Opt. Lett.*, 31(13):2030–2032, 2006.

80. M. Peccianti, C. Conti, E. Alberici, and G. Assanto. Spatially incoherent modulational instability in a nonlocal medium. *Laser Phys. Lett.*, 2:25, 2005.

81. C. Conti. Complex light: Dynamic phase transitions of a light beam in a nonlinear nonlocal disordered medium. *Phys. Rev. E*, 72:066620, 2005.

82. M. Peccianti, C. Conti, and G. Assanto. Optical multisoliton generation in nematic liquid crystals. *Opt. Lett.*, 28(22):2231–2233, 2003.

2

Features of Strongly Nonlocal Spatial Solitons

QI GUO, WEI HU, DONGMEI DENG, DAQUAN LU, AND SHIGEN OUYANG

Laboratory of Nanophotonic Functional Materials and Devices, School of Information and Photoelectronic Science and Engineering, South China Normal University, Guangzhou, China

2.1 INTRODUCTION

Spatial optical solitons are self-trapped optical beams that exist by virtue of the balance between diffraction and nonlinearity. There are various members in the spatial optical soliton family [1–4]. Among them, nonlocal spatial solitons—that is, spatial optical solitons existing in nonlocal nonlinear media—have greatly held one's interest during the recent years [5–8]. Nonlocal spatial solitons can be modeled by the nonlocal nonlinear Schrödinger equation (NNLSE) [5, 9, 10], where the nonlinear term assumes a nonlocal form (convolution integral) with a real-valued response kernel, whereas the NNLSE also describes several other physical situations [11, 12]. Snyder and Mitchell [5] simplified the NNLSE to a linear model in the strongly nonlocal limit, and found an exact Gaussian-shaped stationary solution known as *accessible soliton*. Their work was highly appreciated by Shen [13], who pointed out that "Such theoretical advances will undoubtedly encourage more experimental research on solitons. Thus Snyder and Mitchell's work could be the stimulant for a new surge of soliton activities in the near future." So far, it has been confirmed that nematic liquid crystals (NLC) [14, 15], lead glasses with a self-focusing thermal nonlinearity [16], and liquids with a self-defocusing thermal response [17] are nonlinear materials whose optical nonlinear responses can exhibit extended characteristic lengths typical of strong nonlocality.

This chapter deals with the work related to Snyder and Mitchell's paper in *Science* [5], discussing the phenomenological theory in the framework of the Snyder–Mitchell model (Section 2.2) and some results of studies on nematicons

Nematicons: Spatial Optical Solitons in Nematic Liquid Crystals, First Edition.
Edited by Gaetano Assanto.

and nonlocal spatial solitons in NLC (Section 2.3; see also Chapter 1). For other materials, different models need to be taken into consideration, as analyzed in References 16–19.

2.2 PHENOMENOLOGICAL THEORY OF STRONGLY NONLOCAL SPATIAL SOLITONS

Spatial light localization is one of the self-action effects in nonlinear optics. Self-action effects are those in which optical beams modify their own propagation behavior by means of the medium nonlinear response [20]. Therefore, in this section, we will first address the nonlocal nonlinear response, then talk about the NNLSE, the model describing optical beams in these media and its simplification into the Snyder–Mitchell model. In the rest of the section, some results from the use of the Snyder–Mitchell model are provided.

2.2.1 The Nonlinearly Induced Refractive Index Change of Materials

A sufficiently intense laser beam can induce a significant change in the refractive index of the medium. The refractive index change in turn affects the propagation of light and leads to a new class of nonlinear optical effects characteristically different from wave mixing (sum-frequency generation, harmonic generation, parametric conversion, and so on) [21]. The former is a self-action at the same frequency, whereas the latter results in new frequency generation.

The refractive index of the medium can be generally expressed as

$$n(\mathbf{r}, z) = n_0(\mathbf{r}, z) + \Delta n(\mathbf{r}, z), \tag{2.1}$$

where n_0 is its linear part and Δn is the nonlinearly induced change (the optical-field-induced perturbation in the profile of the refractive index). A number of physical mechanisms, such as molecular reorientation, thermal nonlinearity, photorefrative effect, electronic response, and electrosuction can contribute to the nonlinearly induced refractive index [20, 21]. No matter what the physical mechanism is, however, Δn can be phenomenologically expressed as for a linearly polarized electric field in an infinite material

$$\Delta n(\mathbf{r}, z) = n_2 \int_{-\infty}^{\infty} R(\mathbf{r} - \mathbf{r}')|E(\mathbf{r}', z)|^2 \mathrm{d}^D\mathbf{r}', \tag{2.2}$$

where $R(\mathbf{r})$ is the real nonlinear response function of the medium, n_2 is the nonlinear-index coefficient, $\mathbf{r}$ ($\mathbf{r}'$) is the D-dimension ($D = 1$ or 2) transverse coordinate vector (when $D = 1$, $\mathbf{r} = x\hat{x}$, when $D = 2$, $\mathbf{r} = x\hat{x} + y\hat{y}$), and $\mathrm{d}^D\mathbf{r}'$ is a D-dimensional volume element at $\mathbf{r}'$. The normalization condition, $\int_{-\infty}^{\infty} R(\mathbf{r})\mathrm{d}^D\mathbf{r} = 1$, is chosen to render the nonlinear-index coefficient n_2 of the same dimensionality of the standard optical (spatially local)

Kerr effect [20, 22]. If $R(\mathbf{r})$ is a delta function, that is, $R(\mathbf{r}) = \delta(\mathbf{r})$, we can write

$$n(\mathbf{r}, z) = n_0(\mathbf{r}, z) + n_2|E(\mathbf{r}, z)|^2. \tag{2.3}$$

A material with refractive index described by Equations 2.1 and 2.2 is a (spatially) nonlocal Kerr medium, whereas a material with refractive index described by Equation 2.3 is a local Kerr medium. According to the sign of the nonlinear-index coefficient n_2, Kerr media can be divided into two categories: self-focusing with $n_2 > 0$ and self-defocusing with $n_2 < 0$.

Nonlinear nonlocality means that the medium nonlinear polarization (nonlinear response) at a certain point within a small volume of radius r_0 (r_0 is far smaller than any wavelength involved) depends not only on the value of the electric field inside this volume but also on the electric field outside it. The stronger the nonlocality, the more extended the field distribution contributing to the polarization. In local nonlinear media, conversely, the nonlinear polarization at a certain point is determined only by the electric field at that point. In other words, in nonlocal nonlinear media, the nonlinear response forced at a certain point diffuses to the surrounding regions. This way, the electric field at a certain point can affect the behavior of other electric fields in the surroundings by inducing a spatially broad response. The stronger the nonlocality, the larger volume the source field can impact on.

2.2.2 From the Nonlocal Nonlinear Schrödinger Equation to the Snyder–Mitchell Model

2.2.2.1 The Nonlocal Nonlinear Schrödinger Equation Let us assume that the time-harmonic electric field $\mathbf{E}(\mathbf{r}, z, t)$ with finite transverse cross section in space is linearly polarized in the transverse direction perpendicular to a propagation (longitudinal) (z axis),[1] and has the form $\mathbf{E}(\mathbf{r}, z, t) = \hat{e}\Psi(\mathbf{r}, z)\exp[\mathrm{i}(kz - \omega t)]$, where $\hat{e}$ is a unit vector along some direction in the transverse plane and k is the wave number ($k = \omega n_0/c$). Then, the dynamics of a paraxial light beam with slowly varying electric field envelope $\Psi(\mathbf{r}, z)$ in space is modeled in nonlocal Kerr media by the NNLSE [5, 9, 10]

$$\mathrm{i}\frac{\partial \Psi}{\partial z} + \frac{1}{2k}\nabla_\perp^2 \Psi + \frac{kn_2}{n_0}\Psi \int_{-\infty}^{\infty} R(\mathbf{r} - \mathbf{r}')|\Psi(\mathbf{r}', z)|^2 \mathrm{d}^D\mathbf{r}' = 0, \tag{2.4}$$

where $\nabla_\perp$ is the D-dimension transverse *nabla* operator ($\nabla_\perp = \hat{x}\partial/\partial x + \hat{y}\partial/\partial y$) [26]. According to the relative scale of w_m and w, where w_m is the characteristic length of the response function R and w is the width of the light beam, the nonlocality can belong to four classes [10, 12]: local, weakly nonlocal, generally

[1]Rigorously, the hypothesis of a time-harmonic electric field that has finite transverse cross section in space and is linearly polarized contradicts the law that an electric field must be divergence free in free space. The electric field, however, can be considered to be linearly polarized to the first-order approximation. For detail, see References 23–25.

nonlocal, and strongly nonlocal. When R is a delta function, the response function is local. When the characteristic length is much smaller than the beam width, that is, $w/w_m \gg 1$, the nonlocality is weak; when the characteristic length is much larger than the beam width, that is, $w/w_m \ll 1$, the nonlocality is strong. The remaining case is that of a general nonlocality.

In the local case, Equation 2.4 becomes the well-known nonlinear Schrödinger equation:

$$\mathrm{i}\frac{\partial \Psi}{\partial z} + \frac{1}{2k}\nabla_{\perp}^{2}\Psi + \frac{kn_2}{n_0}|\Psi|^2\Psi = 0. \tag{2.5}$$

The wave equation (Eq. 2.5) in bulk (three-dimensional medium) predicts [20] the occurrence of beam self-focusing for $n_2 > 0$ and beam self-defocusing for $n_2 < 0$.

The NNLSE (Eq. 2.4) has several well-known invariant integrals [27]; two of them are the beam power integral[2]

$$P_0 = \int_{-\infty}^{\infty} |\Psi(\mathbf{r}, z)|^2 \mathrm{d}^D\mathbf{r}, \tag{2.6}$$

which results from energy conservation of the optical beam propagating in a lossless medium, and the linear momentum

$$\mathbf{M} = \frac{\mathrm{i}}{2k}\int_{-\infty}^{\infty} (\Psi\nabla_{\perp}\Psi^* - \Psi^*\nabla_{\perp}\Psi)\mathrm{d}^D\mathbf{r}, \tag{2.7}$$

where the superscript $*$ denotes the complex conjugate. By using the Ehrenfest theorem of quantum mechanics, from Equation 2.4 we can obtain an equation for the trajectory of "the center of mass" of the light beam

$$\frac{\mathrm{d}\mathbf{r}_c(z)}{\mathrm{d}z} = \frac{\mathbf{M}}{P_0}, \tag{2.8}$$

where

$$\mathbf{r}_c(z) = \frac{1}{P_0}\int_{-\infty}^{\infty} \mathbf{r}|\Psi(\mathbf{r}, z)|^2 \mathrm{d}^D\mathbf{r} \tag{2.9}$$

is the beam center of mass. As $\mathbf{M}$ and P_0 are conserved constants, Equation 2.8 yields

$$\mathbf{r}_c(z) = \frac{\mathbf{M}}{P_0}z + \mathbf{r}_{c0}, \tag{2.10}$$

[2]Whether the "power" defined by Equation 2.6 is the actual power carried by the electromagnetic field depends on the dimensions of the function Ψ. If Ψ is scaled by a factor $(\varepsilon_0 n_0 c/2)^{1/2}$ such that $|\Psi|^2$ represents the optical intensity, Equation 2.6 gives the actual power; if Ψ represents the electric field with units V/m, the actual optical power is $P_0 = (\varepsilon_0 n_0 c/2)\int_{-\infty}^{\infty} |\Psi(\mathbf{r}, z)|^2 \mathrm{d}^D\mathbf{r}$, as seen in Section 2.3. For the (1 + 1)-dimensional case ($D = 1$), the power $P_0 = \int_{-\infty}^{\infty} |\Psi(x, z)|^2 \mathrm{d}x$ is the power per unit length in the y direction.

where $\mathbf{r}_{c0} = \mathbf{r}_c(0)$ is the position of the center of mass at $z = 0$. Equation 2.10 implies that the trajectory of the center of mass is a straight line[3] with slope with respect to the z axis determined by $\mathbf{M}/P_0$.

2.2.2.2 The Snyder–Mitchell Model The Snyder–Mitchell model is a simplified version of the NNLSE (Eq. 2.4) for the limit of a strong nonlocality and a response function $R(\mathbf{r})$ symmetric and regular (or at least twice differentiable) at $\mathbf{r} = 0$. In this section, the procedure to get the Snyder–Mitchell model from the NNLSE is developed, as first reported in Reference 28 for the (1 + 1)-dimensional case and then in Reference 29 for the (1 + 2)-dimensional case, respectively, but with something unclear in the former and something wrong in the latter.

For the strongly nonlocal case, $w/w_m \ll 1$, if the response function $R(\mathbf{r})$ is symmetric and twice differentiable at $\mathbf{r} = 0$, then one can expand $R(\mathbf{r})$ in a Taylor's series keeping only the first two nonzero terms [30]. As a result, Equation 2.4 can be reduced to the strongly nonlocal model [8, 30]

$$\mathrm{i}\frac{\partial \Psi}{\partial z} + \frac{1}{2k}\nabla_\perp^2 \Psi + \frac{kn_2}{n_0}\Psi \int_{-\infty}^{\infty} \left[R_0 + \frac{R_0''}{2}(\mathbf{r} - \mathbf{r}')^2 \right] |\Psi(\mathbf{r}', z)|^2 d^D\mathbf{r}' = 0, \quad (2.11)$$

where $R_0 = R(0)$ and $R_0'' = \partial^2 R(\mathbf{r})/\partial^2 x|_{\mathbf{r}=0}$ [$R_0'' < 0$, because R_0 is a maximum of $R(\mathbf{r})$]. By adding $-\mathbf{r}_c$ and $\mathbf{r}_c$ as defined in Equation 2.9, Equation 2.11 takes the form

$$\begin{aligned} \mathrm{i}\frac{\partial \Psi}{\partial z} &+ \frac{1}{2k}\nabla_\perp^2 \Psi + \frac{kn_2}{n_0} R_0 P_0 \Psi + \frac{kn_2}{n_0}\frac{R_0'' P_0}{2}(\mathbf{r} - \mathbf{r}_c)^2 \Psi \\ &+ \frac{kn_2}{n_0}\frac{R_0''}{2}\Psi \int_{-\infty}^{\infty} (\mathbf{r}' - \mathbf{r}_c)^2 |\Psi(\mathbf{r}', z)|^2 d^D\mathbf{r}' = 0. \end{aligned} \quad (2.12)$$

By introducing coordinate and function transformations [28, 29],

$$\begin{cases} \mathbf{s} = \mathbf{r} - \mathbf{r}_c(z), \\ \zeta = z, \end{cases} \quad (2.13)$$

$$\Psi(\mathbf{r}, z) = \psi(\mathbf{s}, \zeta) \exp[\mathrm{i}\phi(s, \zeta)], \quad (2.14)$$

respectively, where the phase $\phi(\mathbf{s}, \zeta)$ is expressed as

$$\begin{aligned} \phi = k\frac{\mathbf{M}}{P_0} \cdot \left[\mathbf{s} + \mathbf{r}_c(\zeta)\right] - k\frac{M^2}{2P_0^2}\zeta + \frac{kn_2}{n_0}\{R_0 P_0 \zeta \\ + \frac{R_0''}{2}\int_0^\zeta \mathrm{d}\zeta' \int_{-\infty}^{\infty} \left[\mathbf{r}' - \mathbf{r}_c(\zeta')\right]^2 |\psi(\mathbf{r}', \zeta')|^2 \mathrm{d}^D\mathbf{r}'\}, \end{aligned} \quad (2.15)$$

[3]The hypothesis here is an infinitely extended medium, necessary to guarantee linear momentum conservation. Boundary effects in a finite medium, which result in an oscillating soliton trajectory, are addressed in Chapter 11.

and using Equation 2.10, we can show that $\psi(\mathbf{s}, \zeta)$ satisfies the following (details are given in Appendix 2.A)

$$2ik\frac{\partial \psi}{\partial \zeta} + \nabla_\perp^2 \psi - \eta^2 s^2 \psi = 0, \tag{2.16}$$

where $\eta = k\sqrt{n_2(-R_0'')P_0/n_0}$.

When $\mathbf{M} = 0$ and $\mathbf{r}_{c0} = 0$, Equation 2.16 simplifies into the Snyder–Mitchell model [5]

$$2ik\frac{\partial \psi}{\partial z} + \nabla_\perp^2 \psi - \eta^2 r^2 \psi = 0. \tag{2.17}$$

Equation 2.16 has a form similar to Equation 2.17, but in a different coordinate system. The reference frame for the Snyder–Mitchell model (Eq. 2.17) is at rest (a laboratory frame), whereas the reference frame for Equation 2.16 moves with the center of mass. The center-of-mass trajectory from the solution of Equation 2.17 is a straight line parallel to the z axis, whereas that from Equation 2.16 is a straight line with a slope given by Equation 2.8.

In this sense, Equation 2.16 can be considered a modified Snyder–Mitchell model. When both $\mathbf{M}$ and $\mathbf{r}_{c0} = 0$, the modified Snyder–Mitchell model reduces to the Snyder–Mitchell model.

By looking at the function transform (Eq. 2.14), we can conclude that there is a phase difference ϕ defined by Equation 2.15 between the solution of Equation 2.11 and that of the modified Snyder–Mitchell model (Eq. 2.16). Even in the case $\mathbf{M} = 0$ and $\mathbf{r}_{c0} = 0$, the phase difference $\phi(z)$ between Ψ in Equation 2.11 and ψ in the Snyder–Mitchell model (Eq. 2.17) is nonzero; moreover,

$$\phi(z) = \frac{kn_2}{n_0}\left[R_0 P_0 z + \frac{R_0''}{2}\int_0^z d\zeta' \int_{-\infty}^{\infty} \mathbf{r}'^2 |\psi(\mathbf{r}', \zeta')|^2 d^D\mathbf{r}'\right]. \tag{2.18}$$

In summary, the Taylor expansion of $R(\mathbf{r})$ and the function transform (Eq. 2.14) render the NNLSE (Eq. 2.4) a linear and a readily solvable equation, the Snyder–Mitchell model. Physically, the Snyder–Mitchell model transforms a complex nonlinear problem into a simple case of linear propagation of light in a waveguide [13]. After solving the Snyder–Mitchell model (Eq. 2.16 or 2.17), one can get an approximate solution of the NNLSE (Eq. 2.4) for the case of strong nonlocality via the function transform (Eq. 2.14).

2.2.3 An Accessible Soliton of the Snyder–Mitchell Model

Let us assume that a solution of Equation 2.17 has the Gaussian form

$$\psi(\mathbf{r}, z) = \frac{\sqrt{P_0}\exp[i\theta(z)]}{[\sqrt{\pi}w(z)]^{D/2}}\exp\left[-\frac{r^2}{2w^2(z)} + ic(z)r^2\right], \tag{2.19}$$

where w is the beam width, c is the phase-front curvature of the beam, and θ is the phase of the complex amplitude, and they all can vary with propagation distance z. The real amplitude of the solution has the form $\sqrt{P_0}/(\sqrt{\pi}w)^{D/2}$ due to power conservation. In self-focusing materials ($n_2 > 0$), we get (for details see References 5, 8, and 30)

$$w(z) = w_0\left[\cos^2(\beta_0 z) + \frac{P_c}{P_0}\sin^2(\beta_0 z)\right]^{1/2}, \tag{2.20}$$

$$c(z) = \frac{\beta_0 k(P_c/P_0 - 1)\sin(2\beta_0 z)}{4[\cos^2(\beta_0 z) + (P_c/P_0)\sin^2(\beta_0 z)]}, \tag{2.21}$$

$$\theta(z) = \theta^{(D)}(z) = -\frac{D}{2}\arctan\left[\sqrt{\frac{P_c}{P_0}}\tan(\beta_0 z)\right] \ (D = 1 \text{ or } 2), \tag{2.22}$$

where P_c is the critical (input) soliton power

$$P_c = -\frac{n_0}{k^2 w_0^4 R_0'' n_2}, \tag{2.23}$$

$\beta_0 = (P_0/P_c)^{1/2}/kw_0^2$ and $w_0 = w(0)$ is the initial beam width at $z = 0$.

Expression 2.19 with Equations 2.20–2.22 is an exact solution of the Snyder–Mitchell model (Eq. 2.17); via the function transform (Eq. 2.14), one can obtain an approximate z axial symmetric solution of the NNLSE (Eq. 2.4):

$$\Psi(r, z) = \frac{\sqrt{P_0}}{(\sqrt{\pi}w)^{D/2}}\exp\left(-\frac{r^2}{2w^2}\right)\exp[\mathrm{i}(cr^2 + \overline{\theta})], \tag{2.24}$$

where the expression of $\overline{\theta}$ is [30]

$$\overline{\theta} = -\frac{D}{2}\arctan\left[\sqrt{\frac{P_c}{P_0}}\tan(\beta_0 z)\right] + \sigma\frac{P_0}{P_c}\frac{z}{kw_0^2}$$
$$+\frac{D}{8kw_0^2}\left[\frac{1}{2\beta_0}\left(1 - \frac{P_0}{P_c}\right)\sin(2\beta_0 z) - \left(\frac{P_0}{P_c} + 1\right)z\right], \tag{2.25}$$

and $\sigma = R_0/(-R_0'' w_0^2)$. Equation 2.20 shows that [5] when $P_0 < P_c$ the beam diffraction initially overcomes the beam-induced index well: the beam initially expands, with w/w_0 oscillating between a maximum $(P_c/P_0)^{1/2}$ and a minimum equal to one; when $P_0 > P_c$, the reverse happens and the beam initially contracts, with w/w_0 breathing between a maximum (unity) and a minimum $(P_c/P_0)^{1/2}$.

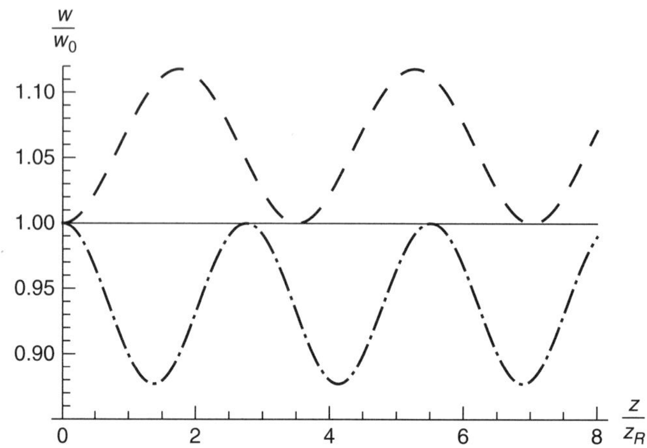

FIGURE 2.1 Evolution of Gaussian beams in a strongly nonlocal Kerr medium. The initial beam width is the same in all cases, but the input powers are different: dashed line $P_0/P_c = 0.8$, solid line $P_0/P_c = 1.0$, dashed-dot line $P_0/P_c = 1.3$.

These two cases correspond to optical breathers[4]. When $P_0 = P_c$, diffraction is exactly balanced by nonlinearity, and the Gaussian-shaped beam preserves its width as it travels in a straight path along z. This is an optical soliton. The evolution of an individual Gaussian beam for various powers P_0 is shown in Figure 2.1.

For $P_0 = P_c$, $w = w_0$, $c = 0$, and $\beta_0 = 1/z_R$, where $z_R = kw_0^2$ is the Rayleigh distance; Equation 2.24 simplifies to the expression of an *accessible soliton* [5][5]

$$\Psi_s(x, y, z) = \frac{\sqrt{P_c}}{(\sqrt{\pi}w_0)^{D/2}} \exp\left(-\frac{r^2}{2w_0^2}\right) \exp(i\phi z), \tag{2.26}$$

where $\phi = (\sigma - 3D/4)/kw_0^2$. ϕz is the phase shift of the slowly varying envelope Ψ after propagating over a distance z. An accessible (spatial) soliton with arbitrary width can propagate as long as its power P_0 exactly equals the critical value P_c defined in Equation 2.23.

The phase shift of an *accessible soliton* can be very large [30], as it stems from the phase difference ϕ given by Equation 2.18, as discussed in Section 2.2.2.2. The

[4]Here, we use the two terms *optical breather* and *optical soliton* to distinguish two states of nonlinear beam propagation. The breather and the soliton resemble a pair of twins born in several nonlinear physical systems. The breather [31, 32] is a localized solution that periodically oscillates versus propagation (in space or time), whereas the soliton is a localized solution that travels in propagation without changes in either shape or size. Although they are often referred to as optical solitons in most nonlinear optics literature, they are physically distinct: in the soliton, diffraction (or dispersion) is exactly balanced by nonlinearity, whereas in the breather [33, 34], diffraction (or dispersion) is only partly balanced by nonlinearity. In this sense, high order solitons of the nonlinear Schrödinger equation [22] should be referred to as *optical breathers* rather than optical solitons.

[5]Snyder and Mitchell in Reference 5 pinpointed the evolution of the beam width, but not the phase evolution.

large phase shift of nonlocal spatial optical solitons in lead glasses, media with an infinite range of nonlocality [16], was recently confirmed experimentally [35].

The comparison between the analytical solutions and the numerical simulations of Equation 2.4 for various w_0/w_m and P_0 shows that [30] the analytical predictions are close (the absolute values of the relative errors are within 10%) to the simulations for w_0/w_m of about 0.5. For the same w_0/w_m, the higher the input power, the better the approximation. It was also proved by both the variational approach [36] and a perturbation method [37] that the single soliton solution of the NNLSE (Eq. 2.4) with the regular response function is exactly the Gaussian soliton (Eq. 2.26) when $w_0/w_m \rightarrow 0$. However, this is not the case for an irregular response function, as discussed in Section 2.3.

2.2.4 Breather and Soliton Clusters of the Snyder–Mitchell Model

We search for solutions of the Snyder–Mitchell model (Eq. 2.17) by writing it as a product of $\psi_F(\mathbf{r}, z)$ and the Gaussian function $\psi_G(\mathbf{r}, z)$ given by Equation 2.19

$$\psi(\mathbf{r}, z) = \psi_F(\mathbf{r}, z)\psi_G(\mathbf{r}, z). \tag{2.27}$$

Substitution of Equation 2.27 into Equation 2.17 yields

$$2ik\frac{\partial \psi_F}{\partial z} + \nabla_\perp^2 \psi_F + 2\left(2ic - \frac{1}{w^2}\right)\mathbf{r} \cdot \nabla_\perp \psi_F = 0. \tag{2.28}$$

The solutions ψ_F of Equation 2.28 have different forms in different coordinate systems and constitute breather and soliton clusters of the Snyder–Mitchell model: the Hermite–Gaussian (HG) cluster in the Cartesian coordinate system [38, 39], the Laguerre–Gaussian (LG) cluster in the cylindrical coordinate system [40–42], and the Ince–Gaussian (IG) cluster [43, 44] in the elliptic coordinate system, as well as the Hermite–Laguerre–Gaussian (HLG) cluster [45].

Soliton clusters have been obtained not only exactly analytically in the framework of the Snyder–Mitchell model (Eq. 2.17) [strong nonlocality with a regular response function $R(\mathbf{r})$], as mentioned in the paragraph earlier, but also approximately analytically and numerically in the other cases with arbitrary degrees of nonlocality modeled by the NNLSE (Eq. 2.4), regardless of a regular or irregular response function $R(\mathbf{r})$. Among them, we mention LG_{0m}-type ring vortex solitons [46], LG_{nm}-type solitons [27], HG_{nm}-type and LG_{nm}-type solitons as well as the transformations between them [47], and rotating HLG-type solitons [48, 49]. All of them are soliton clusters in nonlocal nonlinear media with a phenomenological regular Gaussian response function. In physically real media with an irregular response function, there also exist the (1 + 1)-D HG-type multipole soliton cluster [50–52], the (1 + 2)-D vortex soliton cluster[6] [16, 53], and HG_{nm}-type soliton cluster [54]. The stability of these structures critically depends on the spatial profile of the response function $R(\mathbf{r})$ [12, 55].

[6]The model in Reference [53] (Equation 4 in the reference) is equivalent to an NNLSE (Eq. 2.4) with the response function R being the zeroth order modified Bessel function of the second kind; see Section 2.3.

2.2.5 Complex-Variable-Function Gaussian Breathers and Solitons

A complex-variable-function (CVF) Gaussian beam is the solution of Equation 2.17 in the form $\psi(\mathbf{r}, z) = \psi_C(\mathbf{r}, z)\psi_G(\mathbf{r}, z)$, where ψ_G is the Gaussian function given by Equation 2.19 and ψ_C satisfies Equation 2.28. To obtain this solution, we first introduce a rotating coordinate system (x', y', z) in the transverse plane perpendicular to the z axis: $x' = w_0[x\cos(\vartheta(z)) + y\sin(\vartheta(z))]/w(z)$ and $y' = w_0[y\cos(\vartheta(z)) - x\sin(\vartheta(z))]/w(z)$, where $\mathrm{d}\vartheta(z)/\mathrm{d}z$ denotes the angular velocity (the angular rotation per unit propagation distance) and $w(z)$ is given by Equation 2.20. In the rotating coordinate system, Equation 2.28 can be written as [56, 57]

$$\frac{1}{w^2}\nabla_\perp'^2\psi_c + 2k\frac{\mathrm{d}\vartheta}{\mathrm{d}z}\hat{L}_z'\psi_c - \frac{2}{w^2}\left(x'\frac{\partial\psi_c}{\partial x'} + y'\frac{\partial\psi_c}{\partial y'}\right) = 0, \tag{2.29}$$

where $\nabla_\perp'^2 = \partial^2/\partial x'^2 + \partial^2/\partial y'^2$ and $\hat{L}_z' = -\mathrm{i}(x'\partial/\partial y' - y'\partial/\partial x')$. The solution of Equation 2.29 is found to be [57]

$$\psi_c^{(\pm)}(\mathbf{r}', z) = f\left(\frac{x' \pm \mathrm{i}y'}{b\,w_0}\right),\ \vartheta(z) = \mp\theta^{(2)}(z), \tag{2.30}$$

where $f(\zeta_+')$ denotes an arbitrary analytical function $[\zeta_\pm' = (x' \pm \mathrm{i}y')/(b\,w_0)]$, $b \neq 0$ is an arbitrary real parameter, $\theta^{(2)}(z)$ is given by Equation 2.22, and "$+ (-)$" represents the rotation direction consistent with the right-hand (left-hand) rule relative to the direction of propagation.

From the solution (Eq. 2.30) of Equation 2.29, one can obtain the solution of Equation 2.17 in the laboratory frame [57]:

$$\psi_\pm(\mathbf{r}, z) = \frac{C_0 f(\zeta_\pm)}{w(z)}\exp\left[\frac{-r^2}{2w^2(z)} + \mathrm{i}c(z)r^2 + \mathrm{i}\theta^{(2)}(z)\right], \tag{2.31}$$

where C_0 is a constant to be determined from $\int_{-\infty}^{\infty}|\psi_\pm|^2\mathrm{d}^2\mathbf{r} = P_0$ and $\zeta_\pm = (x \pm \mathrm{i}y)\exp(\mp\mathrm{i}\theta^{(2)})/(bw)$. The solution is referred to as a *CVF Gaussian beam* owing to the fact that $f(\zeta_+)$ is an arbitrary analytical CVF. The structure of the beam is determined by the product of the function $f(\zeta)$ and a Gaussian function. If observed in the laboratory frame (x, y, z), the trajectories of the corresponding points of the beam generally rotate during propagation.

The parameter b, a distribution factor, describes the transverse distribution of the CVF Gaussian beam. The distribution is farther from the beam center for smaller b and is clustered more closely around the beam center for larger b [57].

The CVF Gaussian beam can form either breathers or solitons for different input powers P_0; it evolves similarly to the Gaussian beam [5, 30]. When $P_0 = P_c$, Equation 2.31 can reduce to [56]

$$\psi_s(\mathbf{r}, z) = \frac{C_0 f(\zeta_\pm^{(s)})}{w_0}\exp\left(-\frac{r^2}{2w_0^2} - \mathrm{i}\frac{z}{z_R}\right), \tag{2.32}$$

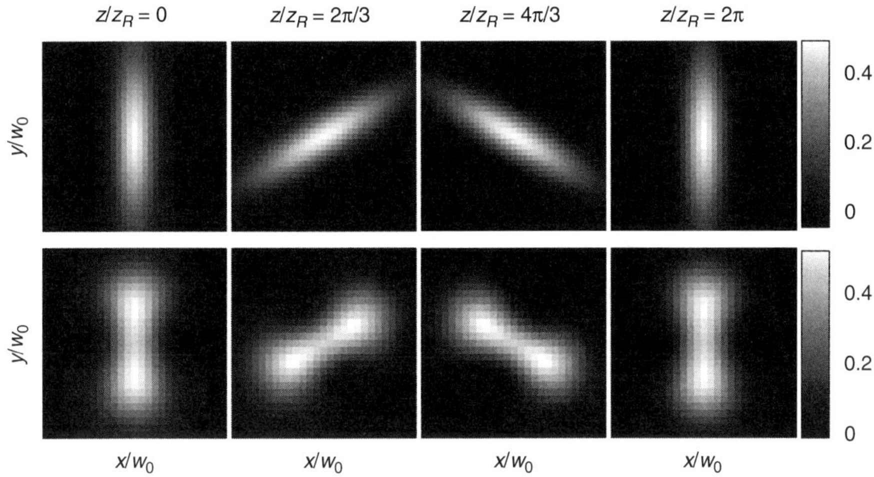

FIGURE 2.2 Dynamics of the Gaussian CVF Gaussian soliton (the upper panels) and the fourth-order Hermite CVF Gaussian soliton (the lower panels) for $b = 1.5$.

where $\zeta_{\pm}^{(s)} = (x \pm \mathrm{i}y)\exp(\pm \mathrm{i}z/z_R)/(bw_0)$. This is a CVF Gaussian soliton, where diffraction is exactly balanced by nonlinearity. Its transverse distribution is preserved but rotates as the CVF Gaussian soliton travels along the propagation axis.

Figure 2.2 shows the evolution of the Gaussian CVF Gaussian soliton $[f(x) = \exp(-x^2/2)]$ and the fourth-order Hermite CVF Gaussian soliton $[f(x) = H_4(x)$, where $H_n(x)$ is the nth-order Hermite polynomial]. It can be found that the pattern of the Gaussian CVF Gaussian soliton consists of a rotating elliptic Gaussian soliton and that of the fourth-order Hermite CVF Gaussian soliton is something similar to a rotating figure 8 when $b = 1.5$.

There also exists another kind of stable rotating mode in nonlocal nonlinear media, the azimuthons [58]. These are not discussed in detail here. An example of azimuthons is briefly discussed in Chapter 6.

2.2.6 Self-Induced Fractional Fourier Transform

The fractional Fourier transform (FRFT), which can show the characteristics of the signal continuously changing from the spatial domain to the spectral domain, was introduced in optics in 1993 [59]. Traditionally, the FRFT is optically performed by linear devices such as lenses and quadratic graded-index media. Here, we show that the FRFT naturally exists in strongly nonlocal nonlinear media described by the Snyder–Mitchell model; the propagation of optical beams in these media can be simply regarded as a self-induced FRFT [60].

For the (1 + 2)-D case, in cartesian coordinates the eigen-solution of Equation 2.17 is the HG soliton cluster [38, 39], that is, $\psi_{mp}^{(q)}(\mathbf{r}, z) = \psi_1^{(q)}(\mathbf{r})\exp(-\mathrm{i}\beta^{(q)}z)$, and

$$\psi_1^{(q)} = C_{mp} H_m\left(\frac{x}{w_0}\right) H_p\left(\frac{y}{w_0}\right) \exp\left(-\frac{r^2}{2w_0^2}\right), \qquad (m+p=q). \tag{2.33}$$

In Equation 2.33, $q(= 0, 1, 2, \ldots)$ is the order of the solution corresponding to the soliton eigenvalue [60]:

$$\beta^{(q)} = (q+1)/z_R. \tag{2.34}$$

Mathematically, $\psi_1^{(q)}$ is the eigen function of the FRFT (represented as $\hat{F}_\gamma\{\cdot\}$) [59, 61]: $\hat{F}_\gamma\{\psi_1^{(q)}(\mathbf{r}_1)\} = \psi_1^{(q)}(\mathbf{r}_2)\mathrm{e}^{-\mathrm{i}q\gamma}$, where the order of the FRFT is $\gamma = z/z_R$. It is well known that an arbitrary square-integrable input field can be expressed as a linear superposition of the eigen-soliton solutions, that is, $\psi(\mathbf{r}_1, 0) = \sum_{q=0}^{\infty} C_q \psi^{(q)}(\mathbf{r}_1, 0)$. According to the linearity of Equation 2.17 and the FRFT, the propagation of an arbitrary beam in strongly nonlocal nonlinear media described by the Snyder–Mitchell model can be regarded as the self-induced FRFT of the input field

$$\psi(\mathbf{r}_2, z) = \hat{F}_\gamma\{\psi(\mathbf{r}_1, 0)\} \exp(-\mathrm{i}\gamma). \tag{2.35}$$

The physical origin of this effect is as follows: when a beam is launched in the medium, it would induce a quadratic graded-index channel owing to strong nonlocality. The propagation in the channel then performs the FRFT [59], as shown in Figure 2.3. As the gradient of the refractive index distribution can be controlled by the power, the order of the self-induced FRFT can be controlled by the input power in addition to the propagation distance z, quite differently from the traditional linear FRFT devices.

According to the properties of the FRFT, the behavior of beams in strongly nonlocal nonlinear media can be conveniently predicted. Here, we discuss three special cases:

1. *Solitons*. As shown in Equation 2.34, during propagation every degenerate eigen-soliton of order q has the same phase $\beta^{(q)}z$ [or, in other words, the

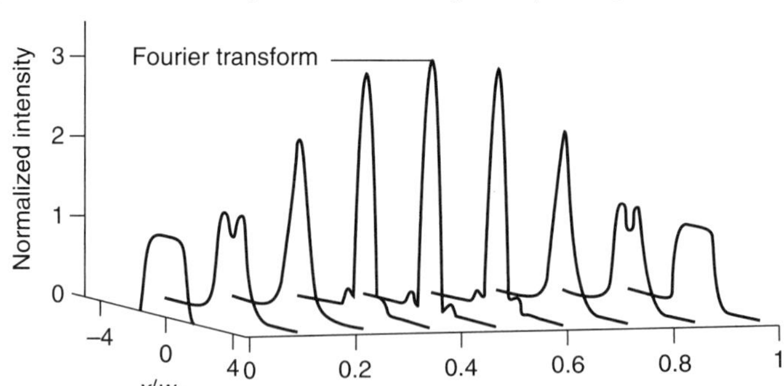

FIGURE 2.3 Propagation dynamics of the super–Gaussian field $\exp[-(x/\sqrt{2}w_0)^8]$ in strongly nonlocal nonlinear media with a Gaussian response function.

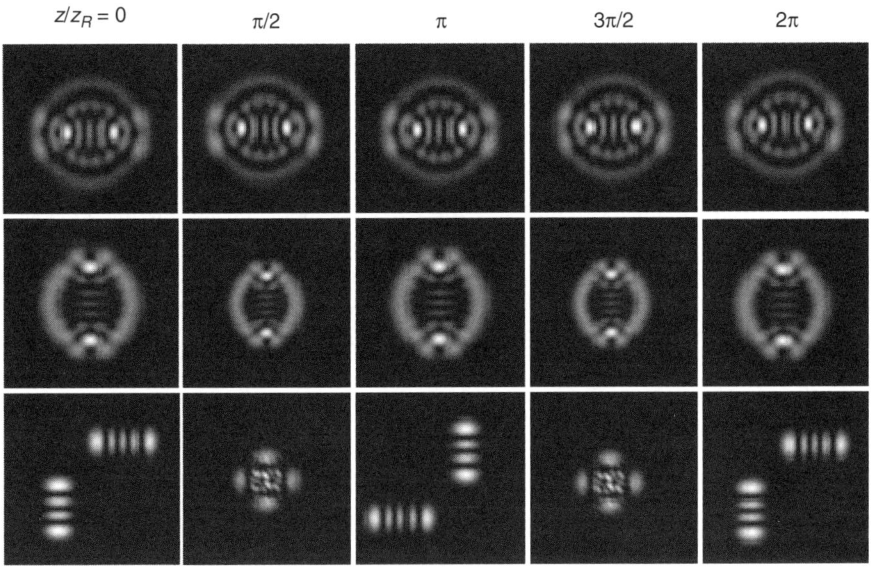

FIGURE 2.4 Dynamics of building-block-like solitons (the first row), building-block-like breathers (the second row), and multi-soliton interactions (the third row) in strongly nonlocal nonlinear media with Gaussian response function based on numerical simulation of the NNLSE (Eq. 2.4).

same FRFT eigenvalue $\exp(-iq\gamma)$]. Therefore, when (i) the input field is the linear superposition of the eigen-soliton solutions of the same order q, beam center, and width w_0 and (ii) the power is the critical value supporting the eigen-solitons with the width w_0, then the field would propagate stationarily, that is, a soliton occurs (first row in Figure 2.4).

2. *Breathers*. When all parameters are the same as in the soliton except that the input power P_0 deviates from the critical value P_c, the FRFT makes the shape of the optical beam unchanged and its width vary periodically with period $\Delta\gamma = \pi$. Subsequently, the evolution gives rise to a breather with period $\Delta z = \pi z_R$ (second row in Figure 2.4).
3. *Multi-Soliton Interactions*. By introducing the FRFT, the evolution of the orbits corresponds simply to a shift in the FRFT, that is, $\psi(\mathbf{r}_2, z) = \hat{F}_\gamma\{\psi(\mathbf{r}_1 - \mathbf{r}_0, 0)\}\exp(-i\gamma)$, where $\mathbf{r}_0$ is the initial deviation of the beam center of the interacting soliton from the center of mass. Under vertical incidence (third row in Figure 2.4), the solitons intersect one another at $z = (n + 1/2)\pi z_R$, evolve to the inversion $\psi(\mathbf{r}_0 - \mathbf{r}, 0)$ at $z = (2n + 1)\pi z_R$, and recur to the input field $\psi(\mathbf{r} - \mathbf{r}_0, 0)$ at $z = 2n\pi z_R$ $(n = 0, 1, 2, \ldots)$.

2.3 NONLOCAL SPATIAL SOLITONS IN NEMATIC LIQUID CRYSTALS

NLC is the first optical nonlinear material with a large enough characteristic length to mimic the strongly nonlocal regime [14, 15]. Spatial solitons in NLC, so-called nematicons [6], are the first *accessible soliton* observed in experiments [15]. The nonlocal nonlinearity in NLC comes from optically induced molecular reorientation. Although some features of nematicons can be explained by the Snyder–Mitchell model [5], others are unique, for instance the voltage-controllable degree of nonlocality [62] and mutual interactions [63].

In this section we discuss the properties of optical beams propagating in NLC-filled planar cells with an external bias voltage but without considering the boundary effects; we address the model and the approximate analytic solution for a single nematicon, as well as nonlocality-controlled interactions [63], short-range interactions [29], and long-range interactions [64] between two nematicons. The boundary effects of the cell on nematicon propagation are addressed in Chapters 11 and 15.

2.3.1 Voltage-Controllable Characteristic Length of NLC

The physical mechanism of the nonlinearity in NLC is optically induced molecular reorientation; the nonlocality comes from interactions between NLC molecules. The dynamics of reorientation in NLC-filled cells is somewhat complicated; with some approximations, however, the NLC nonlinear response function can be found [63, 65].

Let us consider the planar geometry of NLC-filled cells as described in previous works [14, 66]. NLC, anchored at the boundaries along the x coordinate (thickness), is a positive uniaxial with extraordinary index $n_{\|}$ and ordinary index $n_{\perp}$ ($n_{\perp} < n_{\|}$). In the presence of an externally applied (low-frequency) electric field $E_{\rm rf}$, the propagation of the slowly varying envelope Ψ of a light beam linearly polarized along x (an extraordinary wave) and propagating along z can be described by the system [14, 66]

$$2ik\frac{\partial\Psi}{\partial z} + \nabla_{\perp}^{2}\Psi + k_0^2\epsilon_a^{\rm op}(\sin^2\theta - \sin^2\theta_0)\Psi = 0, \tag{2.36}$$

$$2K\left(\frac{\partial^2\theta}{\partial z^2} + \nabla_{\perp}^2\theta\right) + \varepsilon_0\left(\epsilon_a^{\rm rf}E_{\rm rf}^2 + \epsilon_a^{\rm op}\frac{|\Psi|^2}{2}\right)\sin(2\theta) = 0, \tag{2.37}$$

where θ is the tilt angle of the NLC molecules and θ_0 is the peak-tilt in the absence of light, K is the NLC average elastic constant, $k = k_0 n_0$ with k_0 the vacuum wavenumber and $n_0 = n_{\perp}n_{\|}/(n_{\|}^2\cos^2\theta_0 + n_{\perp}^2\sin^2\theta_0)^{1/2} \approx (n_{\perp}^2 + \epsilon_a^{\rm op}\sin^2\theta_0)^{1/2}$ is the refractive index of the extraordinary light, $\epsilon_a^{\rm op} = n_{\|}^2 - n_{\perp}^2$ and $\epsilon_a^{\rm rf} = (\epsilon_{\|} - \epsilon_{\perp})$ are optical and low frequency dielectric anisotropies, respectively. The term $\partial_z^2\theta$ in Equation 2.37 was proven to be negligible compared to $\nabla_{\perp}^2\theta$ [65, 66]; therefore it can be removed. The planar boundaries and anchoring at the interfaces define $\theta|_{x=-L/2} = \theta|_{x=L/2} = 0$, where L is the cell thickness. In the absence of light, the pretilt angle $\hat{\theta}$ is symmetric along x about $x = 0$ (the cell center) and depends only on x [66]

$$2K\frac{\partial^2\hat{\theta}}{\partial x^2} + \varepsilon_0\epsilon_a^{\rm rf}E_{\rm rf}^2\sin(2\hat{\theta}) = 0. \tag{2.38}$$

Furthermore, we can set $\theta = \hat{\theta} + (\hat{\theta}/\theta_0)\Phi$, with Φ being the optically induced perturbation. Noting that $\hat{\theta} \approx \theta_0$ and $\partial_x \hat{\theta} \approx 0$ in the middle of the cell when the beam width is far smaller than the cell thickness, we can simplify Equations 2.36 and 2.37 into the following system, which describes the coupling between Ψ and Φ [62, 66]

$$2ik\frac{\partial \Psi}{\partial z} + \nabla_\perp^2 \Psi + k_0^2 \epsilon_a^{\mathrm{op}} \sin(2\theta_0)\Phi\Psi = 0, \quad (2.39)$$

$$\nabla_\perp^2 \Phi - \frac{1}{w_m^2}\Phi + \frac{\varepsilon_0 \epsilon_a^{\mathrm{op}}}{4K} \sin(2\theta_0)|\Psi|^2 = 0, \quad (2.40)$$

where the parameter w_m ($w_m > 0$ for $|\theta_0| \le \pi/2$), that is, the characteristic length of the nonlinear response function [63], reads

$$w_m = \frac{1}{E_{\mathrm{rf}}} \left\{ \frac{2\theta_0 K}{\varepsilon_0 \epsilon_a^{\mathrm{rf}} \sin(2\theta_0)\left[1 - 2\theta_0 \cot(2\theta_0)\right]} \right\}^{1/2}. \quad (2.41)$$

For a symmetric geometry and ignoring the boundary effects, Equation 2.40 has a particular solution in the form of a convolution integral of $|\Psi|^2$ with the function R

$$\Phi(\mathbf{r}, z) = \frac{\varepsilon_0 \epsilon_a^{\mathrm{op}} \sin(2\theta_0) w_m^2}{4K} \int_{-\infty}^{\infty} R(\mathbf{r} - \mathbf{r}')|\Psi(\mathbf{r}', z)|^2 \mathrm{d}^D \mathbf{r}', \quad (2.42)$$

with R in the (1 + 2)-D case given by Hu et al. [63]

$$R(\mathbf{r}) = \frac{1}{2\pi w_m^2} K_0 \left(\frac{\sqrt{x^2 + y^2}}{w_m} \right), \quad (2.43)$$

where K_0 is the zeroth order modified Bessel function of the second kind; R in the (1 + 1)-D case is [65]

$$R(x) = \frac{1}{2w_m} \exp\left(-\frac{|x|}{w_m} \right). \quad (2.44)$$

Equations 2.39 and 2.40 can be integrated into the NNLSE (Eq. 2.4) with the nonlinear-index coefficient n_2 given by Peccianti et al. [62, 64][7]

$$n_2 = \frac{\varepsilon_0 [\epsilon_a^{\mathrm{op}} w_m \sin(2\theta_0)]^2}{8 n_0 K} = \frac{(\epsilon_a^{\mathrm{op}})^2 \theta_0 \sin(2\theta_0)}{4 n_0 \epsilon_a^{\mathrm{rf}} E_{\mathrm{rf}}^2 \left[1 - 2\theta_0 \cot(2\theta_0)\right]}, \quad (2.45)$$

and the nonlinear response function R expressed by Equation 2.43 or 2.44.

[7]Both results reported {the first of Equation 4 in Reference [62] and Equation 2 in Reference [64]} missed a factor $1/(2n_0)$.

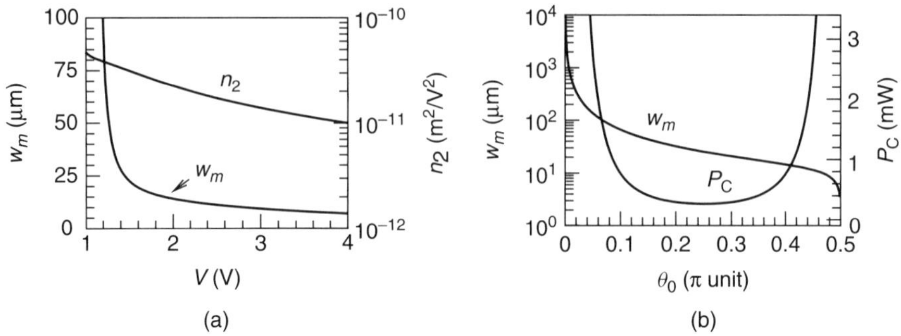

FIGURE 2.5 (a) NLC characteristic length w_m and nonlinear-index coefficient n_2 versus bias voltage V. (b) Characteristic length w_m and critical power of a single soliton versus pretilt angle θ_0. The parameters are for an 80-μm-thick cell filled with TEB30A, as in Reference [63].

A monotonous function of θ_0 on E_{rf} is described by Equation 2.38, and it can be approximated as $\theta_0 \approx (\pi/2)[1 - (E_{\mathrm{FR}}/E_{\mathrm{rf}})^3]$ when E_{rf} is larger than the Fréedericksz threshold E_{FR} [62]. Therefore, we can clearly see from Equations 2.41 and 2.45 that w_m and n_2 are determined by E_{rf} (or the bias V), or equivalently by the peak-pretilt θ_0 for a given NLC cell configuration, as shown in Figure 2.5. Increasing the bias above threshold, θ_0 grows monotonously from 0 to $\pi/2$, w_m decreases monotonously from infinite to 0, and n_2 decreases monotonously, as well. This corresponds to a voltage-controlled change of the NLC characteristic length. As a result, a voltage-controlled degree of nonlocality through the pretilt θ_0 can be conveniently achieved for a fixed w_0. It is important that the characteristic length of the NLC nonlinear optical response can be varied by changing the pretilt angle via a bias voltage [62]. The typical w_m for an 80-μm-thick NLC cell is $w_m = 25.3$ μm for $\theta_0 = \pi/4$ [63], with a nematicon width about $2 \sim 4$μm in experiments.

In summary, if the beam width is far smaller than the thickness of the NLC-filled cell with a bias-induced pretilt, the middle region of the cell can be considered as an infinite NLC medium with a uniform but electrically adjustable pretilt θ_0; the behavior of a paraxial light beam in this region can be described by the NNLSE (Eq. 2.4) with a nonlinear coefficient n_2 given by Equation 2.45 and a nonlinear response function R represented by Equation 2.43 or 2.44.

2.3.2 Nematicons as Strongly Nonlocal Spatial Solitons

As discussed in Section 2.3.1, the propagation of nematicons can be described by the NNLSE (Eq. 2.4) and the characteristic length of the NLC nonlinear optical response can be of the order of 10 μm. Therefore, a strong nonlocality can be achieved for beams of micron-scale width. The crucial difference, however, between the phenomenological response function and the NLC response function is that the NLC response function $R(\mathbf{r})$ for both the (1 + 2)-D (Eq. 2.43) and the (1 + 1)-D cases (Eq. 2.44) has a singularity at the origin $\mathbf{r} = 0$. As a matter of fact, the phenomenological regular response function, for example, Gaussian, is nonphysical although extremely instructive, whereas the response functions of physically real materials are irregular. A regular response function for a real material has not been found so far, to say the least. Here, we rigorously show that the nonlinear induced refractive index Δn determined by Equation 2.2 cannot be simplified as a quadratic self-induced index well if the response function R is irregular, no matter how strong the nonlocality is; in other words, the NNLSE (Eq. 2.4) with an irregular response function cannot be generally reduced to the Snyder–Mitchell model (Eq. 2.17 or 2.16).

Let us take the (1 + 1)-D case as an example. For a symmetric response function and a strong nonlocality, one can expand Equation 2.2 about $x = 0$ and obtain

$$\frac{\Delta n(x,z)}{n_2} = \Delta n(0) + \Delta n'(0)x + \frac{1}{2}\Delta n''(0)x^2 + \cdots, \tag{2.46}$$

where $\Delta n(0) = \int |\Psi(\xi,z)|^2 R(\xi)\mathrm{d}\xi \approx R_0 P_0$, $\Delta n'(0) = -\int |\Psi(\xi,z)|^2 R'(\xi)\mathrm{d}\xi = 0$ [$\Psi(\xi,z)$ is symmetric or antisymmetric about ξ], and $\Delta n''(0) = \int |\Psi(\xi,z)|^2 R''(\xi)\mathrm{d}\xi$. If $R(x)$ is regular, it can be found that $\Delta n''(0) \approx R''(0)P_0$, then we get a quadratic self-induced index well. If $R(x)$ has a singularity in $x = 0$, for example, in NLC, $R''(x)$ satisfies $R''(x) - R(x)/w_m^2 = -\delta(x)/w_m^2$, then we can have $\Delta n''(0) \approx (R_0 P_0 - |\Psi(0,z)|^2)/w_m^2$. It can also be found that $(R_0 P_0)/|\Psi(0,z)|^2 \sim (R_0|\Psi(0,0)|^2 w_0)/|\Psi(0,z)|^2 \sim w_0/w_m \ll 1$ (strong nonlocality),[8] then

$$\frac{\Delta n(x,z)}{n_2} \approx R_0 P_0 - \frac{1}{2w_m^2}|\Psi(0,z)|^2 x^2.$$

Therefore, only for symmetric soliton solutions ($|\Psi(0,z)|^2$ is nonzero and not a function of z), we can have a quadratic self-induced index well. Apart from this, either $\Delta n(x,z)$ will be a function of z for breather solutions or a higher order nonzero term should be taken into account in the expansion (Eq. 2.46) for antisymmetric solutions, because $|\Psi(0,z)|^2 = 0$. The discussion earlier above for the (1 + 1)-D case is instructive, although the (1 + 2)-D case is somewhat more complicated.

In order to investigate this issue further, we take the higher order term of the expansion into consideration and obtain, for both the (1 + 1)-D and the (1 + 2)-D cases [37,67–69],

$$\frac{\Delta n(\mathbf{r},z)}{n_2} = V(0) + \chi_2 r^2 + \chi_4 r^4 + \chi_6 r^6 + \cdots, \tag{2.47}$$

where $V(r) = \int R(|\mathbf{x}-\mathbf{x}'|)|\Psi(r',z)|^2 \mathrm{d}^D\mathbf{x}'$ and $\chi_n = V^{(n)}(r)/n!$ It was shown that, if the response function is regular, and if χ_4 and χ_6 in Equation 2.47 approach zero

[8] For the (1 + 1)-D case, one can have that $P_0 \sim |\Psi(x,0)|^2 w_0$ from the definition of "power," $P_0 = \int |\Psi(x,0)|^2 \mathrm{d}x$ and that $R_0 \sim 1/w_m$ because $R(x)$ should be satisfied $\int R(x)\mathrm{d}x = 1$.

when $w_m/w_0 \to \infty$, Equation 2.4 rigorously converges to Equation 2.17 [37, 67]. However, for the response function with a singularity, that is, Equation 2.43 or 2.44 in NLC, χ_4 and χ_6 are free from the characteristic length w_m [37, 69]. Even when the characteristic length w_m approaches infinity, χ_4 and χ_6 still remain finite and do not tend to zero. It can be found that the ratio of the third term to the second term in Equation 2.47, $\chi_4 r^2/\chi_2$, is about $-r^2/(6w_0^2)$ for (1 + 1)-D geometries [$\chi_2 = 1/(2w_0^2)$ and $\chi_4 = -1/(12w_0^6)$, as given in Reference 37] and $-r^2/(4w_0^2)$ for (1 + 2)-D geometries [$\chi_2 = 1/(2w_0^2)$ and $\chi_4 = -1/(8w_0^6)$, as in Reference 69] in the strong nonlocal case, which means that the third term in Equation 2.47 cannot be ignored when $r \sim w_0$. Therefore, the influence of χ_4 should be accounted for and the profile of the fundamental nematicon does not approach a Gaussian function even if the nonlocality is strong.

A perturbation method can be used to find an approximate analytical solution irrespective of whether the response function is regular or not [37, 67]. To the first-order perturbation, the (1 + 2)-D perturbed solution of a single soliton of Equation 2.4 in NLC is (details can be found in Appendix 2.B)

$$|\Psi(x, y, z)| = \frac{1}{\sqrt{n_0 n_2} k_0} \frac{\sigma w_m}{w_0^2} \exp\left(-\frac{r^2}{2w_0^2}\right)\left(1 + a\frac{r^2}{w_0^2} + b\frac{r^4}{w_0^4}\right), \tag{2.48}$$

where $\sigma \approx 1.44$, $a \approx 0.076$, and $b \approx 0.022$. The profile of the solution (Eq. 2.48) is not Gaussian except for both $a = 0$ and $b = 0$, and its FWHM [full width at half (intensity) maximum] is $w_{\rm FWHM} \approx 1.841 w_0$, whereas the FWHM of the Gaussian function is $w_{\rm FWHM}^{\rm G} = 2(\ln 2)^{1/2} w_0 \approx 1.665 w_0$. The relative error between the corresponding powers carried by the soliton of Equation 2.48 and by the Gaussian soliton ($a = b = 0$ in Eq. 2.48) with the same FWHM is as high as 50%, and the profile given by Equation 2.48 is more accurate than Gaussian, as discussed in the Appendix 2.B.

2.3.3 Nematicon–Nematicon Interactions

Nematicon–nematicon interactions are drastically influenced by the nonlocality of the nonlinear response. For a local Kerr-type nonlinearity, two coherent bright solitons attract (or repel) each other when they are in phase (or out of phase), and the interaction only occurs when the two solitons overlap [1]. In the strongly nonlocal case, it was shown theoretically and experimentally that attraction can take place between bright solitons with any phase difference [5, 63, 65, 70]. Additionally, two solitons can be mutually trapped via the strong nonlocality when their fields do not overlap, which is called *long-range interaction*. Short-range interactions can describe interacting solitons with nonzero overlap.

Similar to strongly nonlocal solitons, interacting nematicons can exhibit some peculiar features, such as a voltage-controllable attraction/repulsion [63]. In this section, we summarize a few experimental results on nematicon–nematicon interactions, including nonlocality-controlled interactions [63] and short- [29] and long-range interactions [64].

2.3.3.1 Voltage-Controllable Interaction The interaction of two spatial solitons depends on the phase difference between them, their coherence and the nonlinear nonlocality [1, 5]. In the local case, two in-phase solitons attract each other and two out-of-phase solitons repel. On the other hand, if the nonlocality of the nonlinear material is strong enough, the soliton interaction is always attractive, independent of their phase difference [5, 70, 71]. Thus, two out-of-phase solitons can repel or attract one another, depending on whether the nonlocality degree is below or above a threshold. The interaction dependence on nonlocality was theoretically described by Rasmussen et al. [65].

As shown in Equation 2.41 and Figure 2.5, the characteristic length of the NLC nonlinear response can vary by acting on the pretilt angle via a bias voltage. For a given beam width, the degree of nonlocality can be changed continuously by w_m through the bias. As a result, a voltage-controlled nonlocality can be conveniently achieved through the NLC pretilt θ_0. Then, a voltage-controlled interaction between nematicons can be performed, with potential applications in all-optical signal processing devices.

The voltage-controllable interaction was first observed experimentally in Reference 63. After Hu et al. [63], Figure 2.6 shows the influence of the pretilt θ_0 (or equivalently the degree of nonlocality for a fixed w_0) on the interaction between two nematicons. One can see that for $\theta_0 \leq \pi/4$ the nonlocality is strong enough to ensure attraction of both in-phase and out-of-phase solitons. However, when $\theta_0 = 0.45\pi$ and the degree of nonlocality reduces, out-of-phase solitons begin to repel each other whereas in-phase solitons keep attracting.

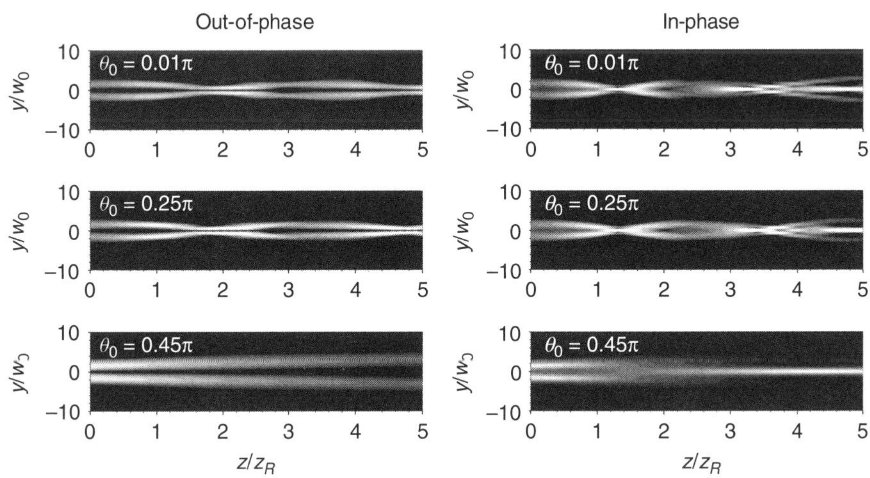

FIGURE 2.6 Numerical simulations of the interaction between in-phase and out-of-phase solitons based on Equations 2.36 and 2.37. The width of each soliton is 4 μm and the input power is 1.1mW. The separation and the relative angle between the two solitons are 12 μm and 0.57°(tan 0.57° = 0.01), respectively. *Source*: Reprinted with permission from W. Hu, et al. *Appl. Phys. Lett.*, 89:071111, 2006. Copyright 2006, American Institute of Physics.

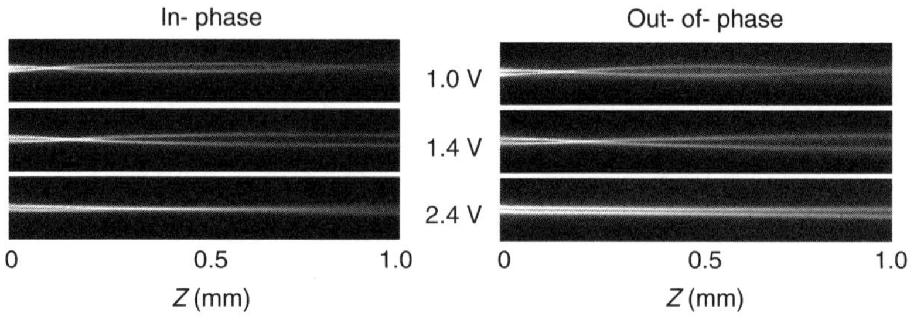

FIGURE 2.7 Photos of soliton pair propagating in the NLC cell. The applied biases are 1.0 V, 1.4 V, and 2.4 V, corresponding to pretilt angles of 0.01π, 0.25π, and 0.45π, respectively. *Source*: Reprinted with permission from Hu et al. *Appl. Phys. Lett.*, 89:071111, 2006. Copyright 2006, American Institute of Physics.

The experimental results are shown in Figure 2.7. When the bias $V = 1.4$ V ($\theta_0 \approx \pi/4$), the photos of in-phase and out-of-phase solitons are almost the same. It means for $\theta_0 = \pi/4$, the degree of nonlocality is strong enough to eliminate the interaction dependence on the phase difference between solitons. In this case, $w_m \approx 25.3$ μm, which is larger than the separation of the two beams.

For a bias $V = 1.0$ V slightly lower than the Fréedericksz threshold, $V_t = 1.14$ V (a small tilt eases reorientation even at voltages below the threshold [62]), the pretilt θ_0 is nearly zero, and the nonlocality is much stronger than when $V = 1.4$ V. For this reason, a second crossing point is observed for both in-phase and out-of-phase solitons.

When the bias V (pretilt angle θ_0) is increased, the degree of nonlocality w_m/w_0 and the characteristic length w_m decrease. For $V = 2.4$ V ($\theta_0 \approx 0.45\pi$), we have $w_m \approx 11$ μm, which approximately equals the separation between the two solitons. In this case, we observe attraction of in-phase solitons and repulsion of out-of-phase solitons. We also see the two in-phase solitons merge into one soliton, qualitatively similar to that observed in the numerical simulations in Figure 2.6.

2.3.3.2 Short-Range Interaction For the strongly nonlocal case, two solitons always attract each other regardless of the phase difference between them, at variance with the local solitons. However, the phase difference can still influence the interaction when two solitons have a nonzero overlap, that is, for short-range interactions. From conservation of linear momentum, the trajectory of two nematicons can be controlled by the phase difference between them. This steering phenomenon controlled by the phase difference could be used in all-optical information processing.

Phase-dependent short-range interactions were first predicted and experimentally confirmed by Hu et al. [29]. For short-range interactions, the two solitons can attract each other and propagate together. Equation 2.10 implies that the trajectory of the center of mass is a straight line with slope determined by $\mathbf{M}/P$ with respect to the z axis. For the NNLSE (Eq. 2.4), let us take two simultaneously incident Gaussian solitons of width w_0 that are coplanar in the x–z plane, with a phase difference γ

and a separation $d(=2h)$, that is,

$$\Psi(x, y, 0) = \Psi_0 \exp\left[-\frac{(x+h)^2+y^2}{2w_0^2} + ik(x+h)\tan\alpha\right] + \Psi_0 e^{i\gamma} \exp\left[-\frac{(x-h)^2+y^2}{2w_0^2} - ik(x-h)\tan\alpha\right], \tag{2.49}$$

where α is the incident angle with respect to the z axis and the amplitude Ψ_0 is sufficiently large for the two beams to propagate as solitons. For the initial condition (Eq. 2.49), the total beam power is

$$P_0 = 2\pi\Psi_0^2 w_0^2 \left[1 + \cos\gamma \exp\left(-h^2/w_0^2 - k^2 w_0^2 \tan^2\alpha\right)\right], \tag{2.50}$$

the momentum

$$\mathbf{M} = \mathbf{e}_x 2\pi h A_0^2 \exp\left(-h^2/w_0^2 - k^2 w_0^2 \tan^2\alpha\right) \sin\gamma / k, \tag{2.51}$$

and the initial position of the center of mass $\mathbf{r}_{c0} = 0$. Let β_x be the angle of the trajectory of the center of mass with respect to the z axis; then the slope $\tan\beta_x = M_x/P$ is

$$\frac{\tan\beta_x}{\Theta} = \frac{(h/w_0)\exp\left[-\left(h/w_0\right)^2 - (\tan\alpha/\Theta)^2\right]\sin\gamma}{1 + \exp\left[-\left(h/w_0\right)^2 - (\tan\alpha/\Theta)^2\right]\cos\gamma}, \tag{2.52}$$

and $\tan\beta_y = 0$, where $\Theta = 1/kw_0$ is the far-field divergence of a Gaussian beam.

As visible in Figure 2.8a and b, the slope of trajectory of the center of mass is highly dependent on the separation $2h$ and the phase difference γ. It shows that $\tan\beta_x = 0$ only when $\gamma = 0$ or π for $h/w_0 \leq 2$, and $\tan\beta_x$ goes to zero when $h/w_0 \geq 2$. $\tan\beta_x$ has a significant value when h is comparable or smaller than the beam width w_0. It can be seen in Figure 2.8c and 2.8d that the maximum tilting angle occurs when γ approximates π for a small distance h. Therefore, the steering angle of the whole beam is significant only for thin beams. It is important to emphasize that the analytical result for the motion of the center of mass (Eq. 2.52) is universal and independent of the form of the nonlinear response function R. This means that, no matter what the material, the degree of nonlocality, and the input power of the beams are, the motion of the center of mass is the same as that in the initial condition (Eq. 2.49).

The experimental results are shown in Figure 2.9; see Reference 29. In Figure 2.9a and b, each of the two solitons is launched alone in the NLC cell and their trajectories are found to be straight and horizontal. When two solitons are injected simultaneously, they propagate as a whole, and tilt (c) up or (d) down in the x direction. As the separation is so small that the two solitons cannot be

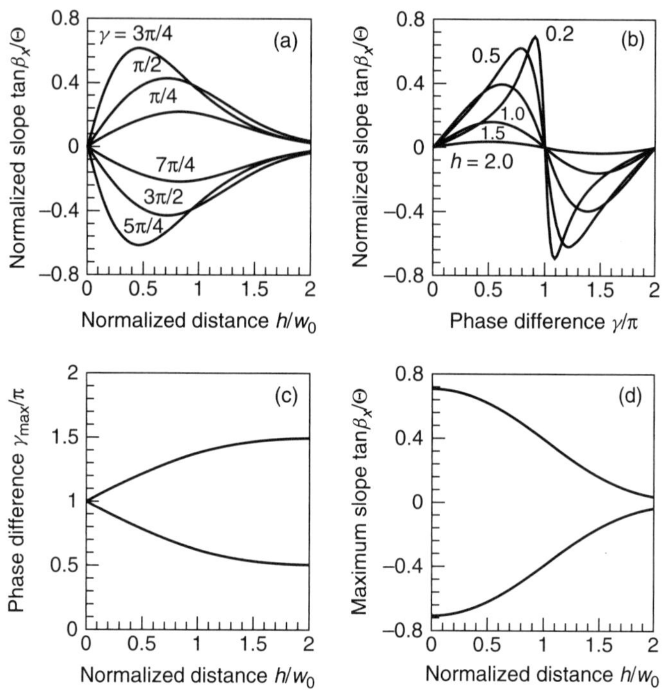

FIGURE 2.8 Dependence of the slope on distance h (a) and phase difference γ (b) for two parallel-injected solitons. (c) Phase difference $\gamma_{\max}$ for the maximum slope angle versus distance h and (d) maximum tilt angle $\tan(\beta_x)$ versus distance h. $\alpha = 0$ for all of the figures.

distinguished with the microscope, one can see a bound-beam state steered by the phase difference γ.

The quantitative comparison between the experimental measurements and the theoretical predictions are shown in Figure 2.9e. One can see that the experimental points are located around the theoretical prediction with a relatively small random error. The latter is mainly due to slight variations in phase difference γ. Except for these random errors, the experimental results agree well with the theoretical prediction.

2.3.3.3 Long-Range Interaction The long-range interaction is a feature of strongly nonlocal solitons, which was first observed in NLC [70] and in lead glass [72]. In strongly nonlocal media, two solitons separated far away can attract via the nonlocal nonlinear response, no matter what their phase difference is. The range of interaction is limited by the range of the nonlocal nonlinearity. The interaction between nematicons separated by 43 times the beam width (the full width at half-maximum) was demonstrated experimentally and reported by Cao et al. [64].

The distance Γ from the input plane to the first collision point of two solitons is used to indicate the strength of the interaction between them. The shorter Γ

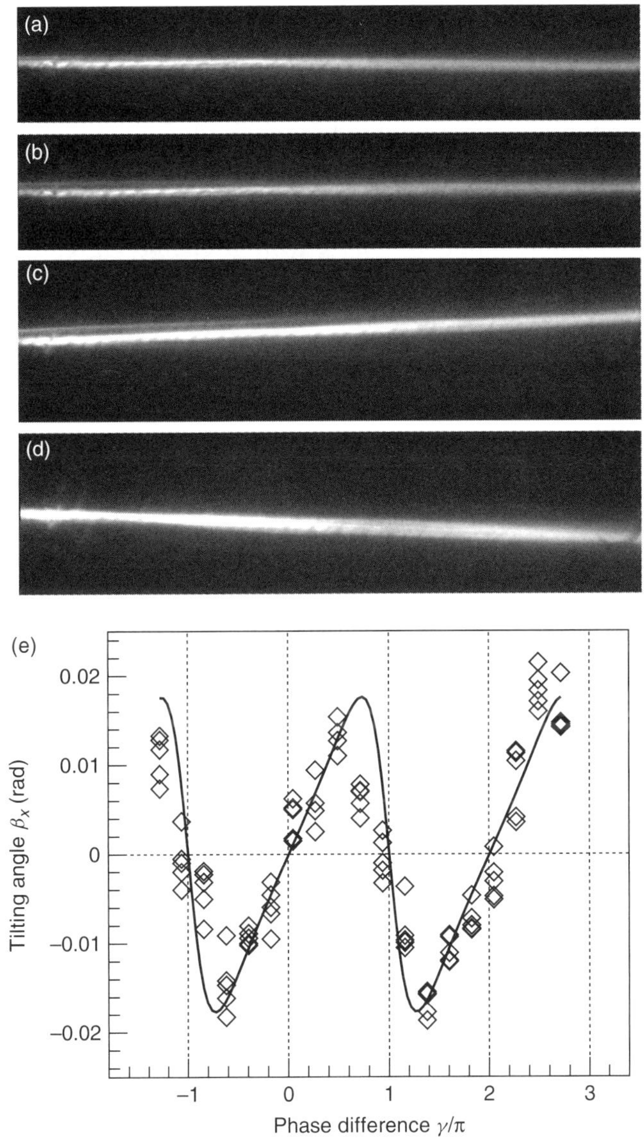

FIGURE 2.9 Photos of beam trajectories for single solitons [(a) and (b)] and two solitons injected together [(c) and (d)] propagating in the NLC cell. The phase differences between the two solitons for (c) and (d) are about $\pi/2$ and $3\pi/2$, respectively. (e) Tilting angle of two beams versus phase difference between them. Square points: experiment results; solid curve: theoretical fitting from Equation 2.52.

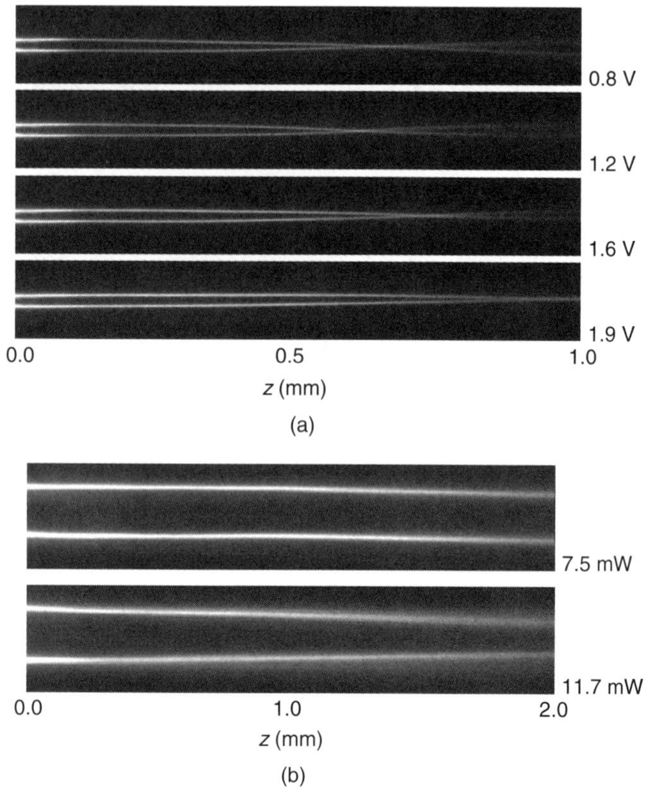

FIGURE 2.10 Photos of beam traces for solitons colliding in NLC: the crossing points vary with the bias voltage V (a) or the incident power P_0 (b). The initial separation between beams is 27.3 μm in (a) and 115 μm in (b).

is, the stronger is the interaction. Figure 2.10a shows experimental results for an initial separation $d_0 = 27.3$ μm, that is, about 10 times the initial beam width w_0 ($w_0 = 3.0$ μm). The launch power of a single beam is maintained at 9.4 mW. The trajectories of two solitons are acquired by a CCD camera as the bias voltage is increased from 0.8 to 1.9 V. When the voltage is about 1.2 V, that is, close to $V_{\pi/4} \approx 1.4$ V to make the pretilt $\theta_0 = \pi/4$, Γ reaches its minimum. It indicates that the interaction between nematicons is the strongest when the pretilt is nearly $\pi/4$.

Then the separation between nematicons is increased to implement long-range interactions at the optimum bias, which is set to 1.2 V. As shown in Figure 2.10b, when d_0 is up to 115 μm, that is, $d_0 \approx 72w_0$ ($w_0 = 1.6$ μm, and the full width at half-maximum is about 2.67 μm), at low input power ($P_0 = 7.5$ mW) the attraction is almost absent. When the input power is high enough, that is, $P_0 = 11.7$ mW, attraction is observed. Please note that $d_0 = 115$ μm is larger than both the thickness of the NLC cell (80 μm) and the characteristic length $w_m = 25.3$ μm when $\theta_0 = \pi/4$ [63]. As the characteristic length w_m is limited by the thickness of the cell, the long-range interaction between nematicons is also limited by it.

2.4 CONCLUSION

In this chapter, we discussed the propagation of spatial optical solitons in nonlocal nonlinear media. The dynamics of paraxial optical beams is modeled in nonlocal Kerr media by the NNLSE (Eq. 2.4).

The Snyder–Mitchell model is the simplified version of the NNLSE for the limit of the strong nonlocality and with the response function symmetric and regular at its origin. The Snyder–Mitchell model can support HG breathers and solitons, one specific case of which is the *accessible soliton* suggested by Snyder and Mitchell, LG breathers and solitons, and IG breathers and solitons in various coordinate systems, as well as various stable rotating modes. The propagation of the optical beams in strongly nonlocal nonlinear media described by the Snyder–Mitchell model can also be simply regarded as the self-induced FRFT.

NLC is the first-found optical nonlinear material with a larger enough characteristic length to mimic the strong nonlocality. The spatial optical solitons in NLC, the so-called nematicons, are the first *accessible solitons* observed in experiments. If the beam width is far smaller than the thickness of the NLC-filled cell with a bias-induced pretilt, the middle region of the cell can be considered as if it were an infinite NLC medium with a uniform but electrically adjustable pretilt; the behavior of the paraxial optical beams in this region can be described by the NNLSE with a nonlinear coefficient n_2 given by Equation 2.45 and a nonlinear response function being a zeroth order modified Bessel function of the second kind (Eq. 2.43) for (1 + 2)-D case or exponential-decay function (Eq. 2.44) for (1 + 1)-D case, both of which are irregular at their origins. The NNLSE with an irregular response function can generally not be reduced to the Snyder–Mitchell model. The nematicons can exhibit various interactions, including the nonlocality-controlled interactions, the short-range interactions, and the long-range interaction, all of which were experimentally confirmed.

APPENDIX 2.A: PROOF OF THE EQUIVALENCE OF THE SNYDER–MITCHELL MODEL (EQ. 2.16) AND THE STRONGLY NONLOCAL MODEL (EQ. 2.11)

From Equations 2.13 and 2.14, one can have the following operations:

$$\nabla_{\perp}^{(\mathrm{r})2}\Psi = \left[\nabla_{\perp}^{(\mathrm{s})2}\psi + \mathrm{i}2\nabla_{\perp}^{(\mathrm{s})}\psi\cdot\nabla_{\perp}^{(\mathrm{s})}\phi + \mathrm{i}\psi\nabla_{\perp}^{(\mathrm{s})2}\phi - \psi\left(\nabla_{\perp}^{(\mathrm{s})}\phi\right)^2\right]\exp(\mathrm{i}\phi), \tag{2.A.1}$$

$$\frac{\partial\Psi}{\partial z} = \left[\nabla_{\perp}^{(\mathrm{s})}\psi\cdot\frac{\partial \mathbf{s}}{\partial\zeta} + \frac{\partial\psi}{\partial\zeta} + \mathrm{i}\psi\left(\nabla_{\perp}^{(\mathrm{s})}\phi\cdot\frac{\partial \mathbf{s}}{\partial\zeta} + \frac{\partial\phi}{\partial\zeta}\right)\right]\exp(\mathrm{i}\phi), \tag{2.A.2}$$

where $\nabla_{\perp}^{(\mathbf{r})}$ and $\nabla_{\perp}^{(\mathbf{s})}$ are the nabla operators in the **r**-coordinate and the **s**-coordinate, respectively. Noting that

$$\frac{\partial \mathbf{s}}{\partial \zeta} = -\frac{\mathbf{M}}{P_0}, \nabla_{\perp}^{(\mathbf{s})}\phi = k\frac{\mathbf{M}}{P_0}, \nabla_{\perp}^{(\mathbf{s})2}\phi = 0$$

and

$$\frac{\partial}{\partial \zeta}\phi(\mathbf{s}, \zeta) = k\frac{\mathbf{M}^2}{2P_0^2} + \frac{kn_2}{n_0}\left[R_0 P_0 + \frac{R_0''}{2}\int_{-\infty}^{\infty}(\mathbf{r}' - \mathbf{r}_c)^2 |\psi(\mathbf{r}', \zeta)|^2 \mathrm{d}^D\mathbf{r}'\right],$$

one can get

$$\nabla_{\perp}^{(\mathbf{r})2}\Psi = \left(\nabla_{\perp}^{(\mathbf{s})2}\psi + 2\mathrm{i}k\frac{\mathbf{M}}{P_0}\cdot\nabla_{\perp}^{(\mathbf{s})}\psi - k^2\frac{\mathbf{M}^2}{P_0^2}\psi\right)\exp(\mathrm{i}\phi), \tag{2.A.3}$$

and

$$\frac{\partial \Psi}{\partial z} = \exp(\mathrm{i}\phi)\left\{-\frac{\mathbf{M}}{P_0}\cdot\nabla_{\perp}^{(\mathbf{s})}\psi + \frac{\partial \psi}{\partial \zeta} - \mathrm{i}k\frac{\mathbf{M}^2}{2P_0^2}\psi\right.$$
$$\left. + \mathrm{i}\frac{kn_2}{n_0}\psi\left[R_0 P_0 + \frac{R_0''}{2}\int_{-\infty}^{\infty}(\mathbf{r}' - \mathbf{r}_c)^2 |\psi(\mathbf{r}', \zeta)|^2 \mathrm{d}^D\mathbf{r}'\right]\right\}. \tag{2.A.4}$$

Substitution of Equations 2.A.3 and 2.A.4 into Equation 2.12 provides Equation 2.16.

APPENDIX 2.B: PERTURBATIVE SOLUTION FOR A SINGLE SOLITON OF THE NNLSE (EQ. 2.4) IN NLC

First, we give the dimensionless (normalization) transform of Equations 2.39 and 2.40. Introducing the dimensionless functions u and q, as well as the dimensionless variables X, Y, and Z,

$$u = \frac{\Psi}{\Psi_0}, q = \frac{\Phi}{\Phi_0}, X = \frac{x}{w_0}, Y = \frac{y}{w_0}, Z = \frac{z}{kw_0^2}, \tag{2.B.1}$$

where

$$\Psi_0 = \frac{1}{\sqrt{n_0 n_2}k_0 w_0} = \frac{2\sqrt{2K/\epsilon_0}}{k_0 w_0 w_m \epsilon_a^{\mathrm{op}}\sin(2\theta_0)}, \Phi_0 = \frac{2}{k_0^2 w_0^2 \epsilon_a^{\mathrm{op}}\sin(2\theta_0)},$$

we get the dimensionless system

$$\mathrm{i}\partial_Z u + \frac{1}{2}\nabla^2_{XY}u + qu = 0, \tag{2.B.2}$$

$$\nabla^2_{XY}q - \alpha_d^2 q + \alpha_d^2|u|^2 = 0, \tag{2.B.3}$$

where $\nabla^2_{XY} = \partial^2_X + \partial^2_Y$, and $\alpha_d = w_0/w_m$. The substitution of the solution of Equation 2.B.3 into Equation 2.B.2 yields the dimensionless NNLSE:

$$\mathrm{i}\frac{\partial u}{\partial Z} + \frac{1}{2}\nabla^2_{XY}u + u\int_{-\infty}^{\infty}\overline{R}(\mathbf{r}_d - \mathbf{r}'_d)|u(\mathbf{r}'_d, Z)|^2 \mathrm{d}X'\mathrm{d}Y' = 0, \tag{2.B.4}$$

where $r_d^2 = X^2 + Y^2$ and $\overline{R}(\mathbf{r}_d) = \alpha_d^2 K_0(\alpha_d r_d)/2\pi$ is the response function in the dimensionless system. Note that we can also obtain Equation 2.B.4 directly from Equation 2.4 with Equation 2.43 via the transform (Eq. 2.B.1) and $\overline{R} = w_0^2 R$.

In order to find an approximate solution for a single soliton of Equation 2.B.3, we use the perturbation method widely employed in quantum mechanics [37, 67, 68]; see also bibliography in Reference 37, for example, which can check whether the response function is regular. The perturbed solution of Equation 2.B.4 is (see Eq. 36 in Reference 68)

$$|u(X, Y, Z)| = \frac{\sigma}{\mu^2\alpha_d}\exp\left(-\frac{r_d^2}{2\mu^2}\right)\left(1 + a\frac{r_d^2}{\mu^2} + b\frac{r_d^4}{\mu^4} + c\frac{r_d^6}{\mu^6} + d\frac{r_d^8}{\mu^8}\right), \tag{2.B.5}$$

where the perturbed parameters $\sigma \approx 1.44$, $a \approx 0.076$, $b \approx 0.022$, $c \approx 0.00022$, and $d \approx 0.00037$, and all of them are free from the degree of nonlocality α_d. Figures 2.B.1 and 2.B.2 compare the numerical simulations of Equation B.4 for the various inputs $u(X, Y, 0)$ given by Equation 2.B.5 with different order perturbation and α_d. It can be observed that the Gaussian function (zero-order perturbation) has a much larger error, and the relative error $(|u(0, 0, 0)| - |u(0, 0, z)|_{\min})/|u(0, 0, 0)|$ is about 46% even if $\alpha_d = 0.01$. The first-order perturbation, however, is accurate enough with a relative error of about 12% when $\alpha_d = 0.1$.

The critical power for the first-order perturbation soliton is

$$P_\mathrm{c} = \frac{1}{2}\varepsilon_0 n_0 c\int_{-\infty}^{\infty}|\Psi|^2\mathrm{d}x\mathrm{d}y = \frac{P_{\mathrm{c,min}}}{\sin^2(2\theta_0)}, \tag{2.B.6}$$

where $P_{\mathrm{c,min}} = (1 + 2a + 4b)\, 2cn_0K\lambda^2/\pi w_0^2(\epsilon_a^{\mathrm{op}})^2 = 1.24 \times 2cn_0K\lambda^2/\pi w_0^2 (\epsilon_a^{\mathrm{op}})^2$, we take $\sigma^2 \approx 2$ and keep only the linear terms in a and b to obtain the expression of $P_{\mathrm{c,min}}$. When $\theta_0 = \pi/4$, P_c reaches its minimum $P_{\mathrm{c,min}}$, as shown in Figure 2.5b, which is the case of the first nematicon observed in experiments [73]. By making $a = 0$ and $b = 0$, we obtain the minimum critical power for the soliton with Gaussian profile $P^{\mathrm{G}}_{\mathrm{c,min}} = 2cn_0K\lambda^2/\pi w_0^2(\epsilon_a^{\mathrm{op}})^2$, which equals exactly

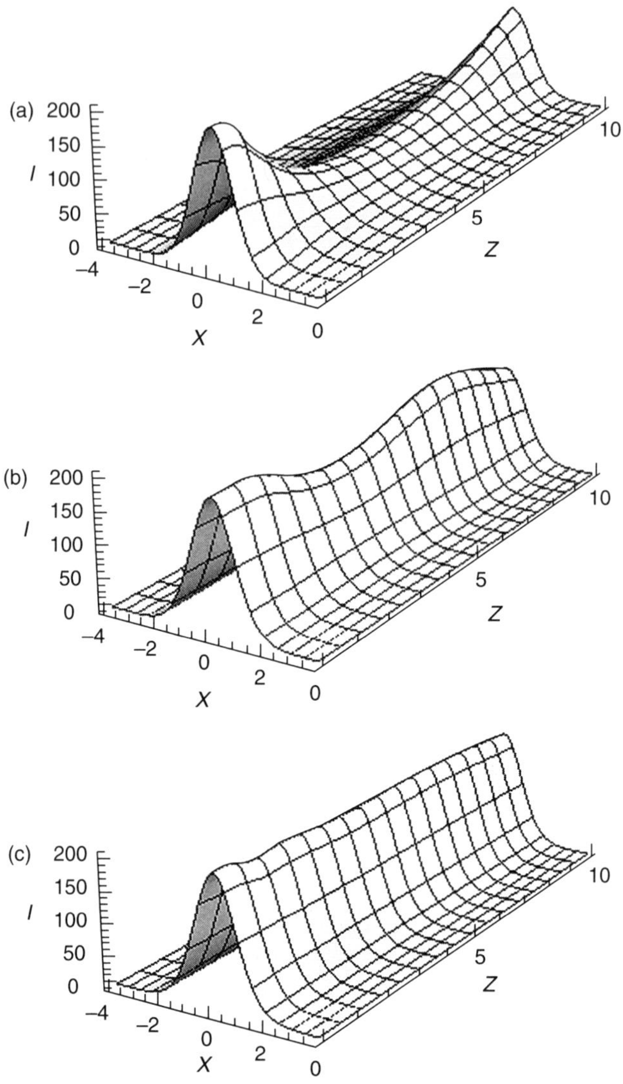

FIGURE 2.B.1 Comparison of propagation for various inputs with different order perturbation, $\alpha_d = 0.1$ and $\mu = 1$. (a) Zero-order perturbation (Gaussian function with a, b, c, and d set to zero); (b) first-order perturbation ($a \neq 0, b \neq 0$, but $c = d = 0$); (c) second-order perturbation (a, b, c, and d nonzero). $I = |u(X, 0, Z)|^2$ in all figures.

the soliton power P_S obtained in Reference 15 and the reference power $\mathcal{P}_c$ given in Reference 14.[9] To check the difference between $P_{c,\min}$ and $P^G_{c,\min}$, we reexpress

[9]In Reference 14, $\mathcal{P}_c = 16cn_0\Delta\epsilon_{RF}E^2/k_0^2(\epsilon_a^{op})^2$ and $R_c^2 = \pi K/2\Delta\epsilon_{RF}E^2$. Then we can have $\mathcal{P}_c = 8\pi cn_0 K/k_0^2 R_c^2(\epsilon_a^{op})^2$. Noting the equivalence of w_0 to R_c, we obtain that $P^G_{c,\min} = \mathcal{P}_c$.

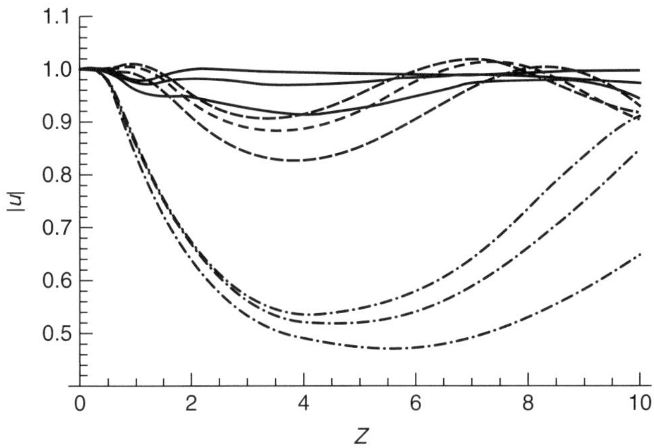

FIGURE 2.B.2 On-axis amplitude $|u(0, 0, Z)|$ (normalized by $|u(0, 0, 0)|$) as a function of Z for the various inputs with different order perturbation. Dash-dot lines are for the zero-order perturbation, dashed lines for the first-order perturbation, solid lines for the second-order perturbation. From up to down, three perturbation cases with $\alpha_d = 0.01$, $\alpha_d = 0.1$, and $\alpha_d = 0.2$, respectively, and $\mu = 1$.

$P_{\mathrm{c,min}}$ and $P^{\mathrm{G}}_{\mathrm{c,min}}$ with w_{FWHM} and $w^{\mathrm{G}}_{\mathrm{FWHM}}$, respectively, and obtain

$$P_{\mathrm{c,min}} = 1.24 \times 1.841^2 \times \frac{\overline{P}}{w^2_{\mathrm{FWHM}}}, \; P^{\mathrm{G}}_{\mathrm{c,min}} = 4(\ln 2) \times \frac{\overline{P}}{(w^{\mathrm{G}}_{\mathrm{FWHM}})^2}, \tag{2.B.7}$$

where $\overline{P} = 2cn_0K\lambda^2/\pi(\epsilon_a^{\mathrm{op}})^2$. Having $w_{\mathrm{FWHM}} = w^{\mathrm{G}}_{\mathrm{FWHM}}$, we obtain

$$P_{\mathrm{c,min}} = \Lambda P^{\mathrm{G}}_{\mathrm{c,min}}, \tag{2.B.8}$$

where $\Lambda = 1.24 \times 1.841^2/4\ln 2 \approx 1.52$. Therefore, we conclude that the relative error between $P_{\mathrm{c,min}}$ and $P^{\mathrm{G}}_{\mathrm{c,min}}$ is as high as 50%, with $P_{\mathrm{c,min}}$ more precise than $P^{\mathrm{G}}_{\mathrm{c,min}}$.

Acknowledgments

This work was supported by the National Natural Science Foundation of China, Grant Nos.10474023, 10674050, 10804033, 10904041, 60908003, 61008007, 11074080.

The authors appreciate contributions from their coworker, Dr. Q. Shou, and all of their students involved in the recent work. One of the authors, Q. Guo, would like to thank Prof. Sien Chi of National Chiao Tung University (Taiwan, China) for valuable discussions at the beginning of this research.

REFERENCES

1. G. I. Stegeman and M. Segev. Optical spatial solitons and their interactions: University and diversity. *Science*, 286:1518–1523, 1999.
2. G. I. Stegeman, D. N. Christodoulides, and M. Segev. Optical spatial solitons: Historical perspectives. *IEEE J. Sel. Top. Quantum Electron.*, 6:1419–1427, 2000.
3. S. Trillo and W. Torruellas. *Spatial Solitons*. Springer-Verlag, Berlin, 2001.
4. Y. S. Kivshar and G. P. Agrawal. *Optical Solitons: From Fibers to Photonic Crystals*. Elsevier, New York, 2003.
5. A. W. Snyder and D. J. Mitchell. Accessible solitons. *Science*, 276:1538–1541, 1997.
6. G. Assanto, et al. Nematicons. *Opt. Photon. News*, 14:45–48, 2003; Spatial solitons in nematic liquid crystals. *IEEE J. Quantum Electron.*, 39:13–21, 2003; Routing light at will, *J. Nonlin. Opt. Phys. Mater.*, 6:37–47, 2007; Nematicons: Self-localized beams in nematic liquid crystals, *Liq. Cryst.*, 36:1161–1172, 2009.
7. W. Królikowski, O. Bang, N. I. Nikolov, D. Neshev, J. Wyller, J. J. Rasmussen and D. Edmundson. Modulational instability, solitons and beam propagation in spatially nonlocal nonlinear media. *J. Opt. B: Quantum Semiclass. Opt.*, 6:288–294, 2004.
8. Q. Guo. Nonlocal spatial solitons and their interactions, *Proceedings on Optical Transmission, Switching, and Subsystems*, eds. C. F. Lam, C. Fan, N. Hanik, and K. Oguchi. *Proceedings of SPIE (Asia-Pacific Optical and Wireless Communications Conference, November 2–6, Wuhan, P.R.China 2003)*. 5281: 581–594, 2004.
9. D. J. Mitchell and A. W. Snyder. Soliton dynamics in a nonlocal medium. *J. Opt. Soc. Am. B*, 16:236–239, 1999.
10. W. Królikowski, O. Bang, J. J. Rasmussen, and J. Wyller. Modulational instability in nonlocal nonlinear Kerr media. *Phys. Rev. E*, 64:016612, 2001.
11. W. Królikowski and O. Bang. Solitons in nonlocal nonlinear media: Exact solutions. *Phys. Rev. E*, 63:016610, 2001.
12. O. Bang, W. Królikowski, J. Wyller, and J. J. Rasmussen. Collapse arrest and soliton stabilization in nonlocal nonlinear media. *Phys. Rev. E*, 66:046619, 2002.
13. Y. R. Shen. Solitons made simple. *Science*, 276:1520–1520, 1997.
14. C. Conti, M. Peccianti, and G. Assanto. Route to nonlocality and observation of accessible solitons. *Phys. Rev. Lett.*, 91:073901, 2003.
15. C. Conti, M. Peccianti, and G. Assanto. Observation of optical spatial solitons in highly nonlocal medium. *Phys. Rev. Lett.*, 94:113902, 2004.
16. C. Rotschild, O. Cohen, O. Manela, M. Segev, and T. Carmon. Solitons in nonlinear media with an infinite range of nonlocality: First observation of coherent elliptic solitons and of vortex-ring solitons. *Phys. Rev. Lett.*, 95:213904, 2005.
17. A. Dreischuh, D. Neshev, D. E. Peterson, O. Bang, and W. Królikowski, Observation of attraction between dark soliton. *Phys. Rev. Lett.*, 96:043901, 2006.
18. S. Skupin, M. Saffman, and W. Królikowski. Nonlocal stabilization of nonlinear beams in a self-focusing atomic vapor. *Phys. Rev. Lett.*, 98:263902, 2007.
19. C. Conti, A. Fratalocchi, M. Peccianti, G. Ruocco, and S. Trillo. Observation of a gradient catastrophe generating solitons. *Phys. Rev. Lett.*, 102:083902, 2009.
20. R. W. Boyd. *Nonlinear Optics*, 3rd edn, Chapters 4–5 and 7. Academic Press, Amsterdam, 2008.

21. Y. R. Shen. *The Principles Of Nonlinear Optics*, Chapter 16. John Wiley & Sons, New York, 1984.
22. G. P. Agrawal. *Nonlinear Fiber Optics*, 3rd edn. Academic Press, San Diego, CA, 2001.
23. M. Lax, W. H. Louisell, and W. B. McKnight. From Maxwell to paraxial wave optics. *Phys. Rev. A*, 11:1365–1370, 1975.
24. S. Chi and Q. Guo. Vector theory of self-focusing of an optical beam in Kerr media. *Opt. Lett.*, 20:1598–1600, 1995.
25. H. A. Haus. *Waves and Fields in Optoelectronics*, Chapter 4. Prentice-Hall, New Jersey, 1984.
26. I. N. Bronshtein and K. A. Semendyayev. Handbook of Mathematics, 4.2.2.9, Leipzig, 539, 1985.
27. A. I. Yakimenko, V. M. Lashkin, and O. O. Prikhodko. Dynamics of two-dimensional coherent structures in nonlocal nonlinear media. *Phys. Rev. E*, 73:066605, 2006.
28. S. Ouyang, W. Hu, and Q. Guo. Light steering in strongly nonlocal nonlinear medium. *Phys. Rev. A*, 76:053832, 2007
29. W. Hu, S. Ouyang, P. Yang, Q. Guo, and S. Lan. Short-range interactions between strongly nonlocal spatial solitons. *Phys. Rev. A*, 77:033842, 2008.
30. Q. Guo, B. Luo, F. Yi, S. Chi, and Y. Xie. Large phase shift of nonlocal optical spatial solitons. *Phys. Rev. E*, 69:016602, 2004.
31. G. L. Lamb Jr. *Elements of Soliton Theory*, Chapter 5. John Wiley & Sons, New York, 133–168, 1980.
32. S. Flach and C. R. Willis. Discrete breathers. *Phys. Rep.*, 295:181–264, 1998.
33. R. Michalska-Trautman. Formation of an optical breather. *J. Opt. Soc. Am. B*, 6:36–44, 1989.
34. J. N. Kutz, P. Holmes, S. G. Evangelides Jr., and J. P. Gordon. Hamiltonian dynamics of dispersion-managed breathers. *J. Opt. Soc. Am. B*, 15:87–96, 1998.
35. Q. Shou, X. Zhang, W. Hu, and Q. Guo. Large phase shift of spatial solitons in lead glass. *Opt. Lett.*, 36:4194–4196, 2011.
36. Q. Guo, B. Luo, and S. Chi. Optical beams in sub-strongly non-local nonlinear media: A variational solution. *Opt. Commun.*, 259:336–341, 2006.
37. S. Ouyang, Q. Guo, and W. Hu. Perturbative analysis of generally nonlocal spatial optical solitons. *Phys. Rev. E*, 74:036622, 2006.
38. X. Zhang and Q. Guo. Analytical solution in the Hermite-Gaussian form of the beam propagating in the strong nonlocal media. *Acta. Phys. Sin.*, 54:3178–3182, 2005 (in Chinese).
39. D. Deng, X. Zhao, Q. Guo, and S. Lan. Hermite-Gaussian breathers and solitons in strongly nonlocal nonlinear media. *J. Opt. Soc. Am. B*, 24:2537–2544, 2007.
40. X. Zhang, Q. Guo, and W. Hu. Analytical solution to the spatial optical solitons propagating in the strong nonlocal media. *Acta. Phys. Sin.*, 54:5189–5193, 2005 (in Chinese).
41. D. Deng and Q. Guo. Propagation of Laguerre-Gaussian beams in nonlocal nonlinear media. *J. Opt. A.: Pure Appl. Opt.*, 10:035101, 2008.
42. W. Zhong and Y. Lin. Two-dimensional Laguerre-Gaussian soliton family in strongly nonlocal nonlinear media. *Phys. Rev. A*, 75:061801, 2007.
43. D. Deng and Q. Guo. Ince-Gaussian solitons in strongly nonlocal nonlinear media. *Opt. Lett.*, 32:3206–3208, 2007.

44. D. Deng and Q. Guo. Ince-Gaussian beams in strongly nonlocal nonlinear media. *J. Phys. B: At. Mol. Opt.*, 41:145401, 2008.
45. D. Deng, Q. Guo, and W. Hu. Hermite-Laguerre-Gaussian beams in strongly nonlocal nonlinear media. *J. Phys. B: At. Mol. Opt.*, 41:225402, 2008.
46. D. Briedis, D. E. Petersen, D. Edmundson, W. Królikowski, and O. Bang. Ring vortex solitons in nonlocal nonlinear media. *Opt. Express*, 13:435–443, 2005.
47. D. Buccoliero, A. S. Desyatnikov, W. Królikowski, and Y. S. Kivshar. Laguerre and Hermite soliton clusters in nonlocal nonlinear media. *Phys. Rev. Lett.*, 98:053901, 2007.
48. D. Buccoliero, A. S. Desyatnikov, W. Królikowski, and Y. S. Kivshar. Spiraling multivortex solitons in nonlocal nonlinear media. *Opt. Lett.*, 33:198–200, 2008.
49. D. Buccoliero and A. S. Desyatnikov. Quasi-periodic transformations of nonlocal spatial solitons. *Opt. Express*, 17:9608–9613, 2009.
50. D. W. McLaughlin, D. J. Muraki, and M. J. Shelley. Self-focused optical structures in a nematic liquid crystal. *Physica D*, 97:471–497, 1996.
51. Z. Xu, Y. V. Kartashov, and L. Torner. Upper threshold for stability of multipole-mode solitons in nonlocal nonlinear media. *Opt. Lett.*, 30:3171–3173, 2005.
52. L. Dong and F. Ye. Stability of multipole-mode solitons in thermal nonlinear media. *Phys. Rev. A*, 81:013815, 2010.
53. A. I. Yakimenko, Y. A. Zaliznyak, and Y. S. Kivshar. Stable vortex solitons in nonlocal self-focusing nonlinear media. *Phys. Rev. E*, 71:065603, 2005.
54. C. Rotschild, M. Segev, Z. Xu, Y. V. Kartashov, L. Torner, and O. Cohen. Two-dimensional multipole solitons in nonlocal nonlinear media. *Opt. Lett.*, 31:3312–3314, 2006.
55. S. Skupin, O. Bang, D. Edmundson, and W. Królikowski. Stability of two-dimensional spatial solitons in nonlocal nonlinear media. *Phys. Rev. E*, 73:066603, 2006.
56. D. Deng, Q. Guo, and W. Hu. Complex-variable-function Gaussian solitons. *Opt. Lett.*, 34:43–45, 2009.
57. D. Deng, Q. Guo, and W. Hu. Complex-variable-function Gaussian beam in strongly nonlocal nonlinear media. *Phys. Rev. A*, 79:023803, 2009.
58. S. Lopez-Aguayo, A. S. Desyatnikov, and Y. S. Kivshar. Azimuthons in nonlocal nonlinear media. *Opt. Express*, 14:7903–7908, 2006.
59. D. Mendlovic and H. M. Ozaktas. Fractional Fourier transforms and their optical implementation. *J. Opt. Soc. Am. A*, 10:1875–1881, 1993.
60. D. Lu, W. Hu, Y. Zheng, Y. Liang, L. Cao, S. Lan, and Q. Guo. Self-induced fractional Fourier transform and revivable higher-order spatial solitons in strongly nonlocal nonlinear media. *Phys. Rev. A*, 78:043815, 2008.
61. V. Namias. The fractional order Fourier transform and its application to quantum mechanics. *J. Inst. Math. Appl.*, 25:241–265, 1980.
62. M. Peccianti, C. Conti, and G. Assanto. Interplay between nonlocality and nonlinearity in nematic liquid crystals. *Opt. Lett.*, 30:415–417, 2005.
63. W. Hu, T. Zhang, Q. Guo, L. Xuan, and S. Lan. Nonlocality-controlled interaction of spatial solitons in nematic liquid crystals. *Appl. Phys. Lett.*, 89:071111, 2006.
64. L. Cao, Y. Zheng, W. Hu, and Q. Guo. Long-range interactions between nematicons. *Chin. Phys. Lett.*, 26:064209, 2009.
65. P. D. Rasmussen, O. Bang, and W. Królikowski. Theory of nonlocal soliton interaction in nematic liquid crystals. *Phys. Rev. E*, 72:066611, 2005.

66. M. Peccianti, C. Conti, G. Assanto, A. De Luca, and C. Umeton. Nonlocal optical propagation in nonlinear nematic liquid crystals. *J. Nonlin. Opt. Phys. Mater.*, 12:525–538, 2003.
67. H. Ren, S. Ouyang, Q. Guo, and L. Wu. (1+2)-Dimensional sub-strongly nonlocal spatial optical solitons: perturbation method. *Opt. Commun.*, 275:245–251, 2007.
68. S. Ouyang and Q. Guo. (1+2)-dimensional strongly nonlocal solitons. *Phys. Rev. A*, 76:053833, 2007.
69. H. Ren, S. Ouyang, Q. Guo, W. Hu, and L. Cao. A perturbed (1+2)-Dimensional soliton solution in nematic liquid crystals. *J. Opt. A: Pure Appl. Opt.*, 10:025102, 2008.
70. M. Peccianti, K. A. Brzdakiewicz, and G. Assanto. Nonlocal spatial soliton interactions in bulk nematic liquid crystals. *Opt. Lett.*, 27:1460–1462, 2002.
71. Y. Xie and Q. Guo. Phase modulations due to collisions of beam pairs in nonlocal nonlinear media. *Opt. Quantum Electron.*, 36:1335–1351, 2004.
72. C. Rotschild, B. Alfassi, O. Cohen, and M. Segev. Long-range interactions between optical solitons. *Nat. Phys.*, 2:769–774, 2006.
73. M. Peccianti, A. De Rossi, G. Assanto, A. De Luca, C. Umeton, and I. C. Khoo. Electrically assisted self-confinement and waveguiding in planar nematic liquid crystal cells. *Appl. Phys. Lett.*, 77:7–9, 2000.

3

Theoretical Approaches to Nonlinear Wave Evolution in Higher Dimensions

ANTONMARIA A. MINZONI

Fenomenos Nonlineales y Mecánica, Department of Mathematics and Mechanics, Universidad Nacional Autónoma de México, Mexico D.F., Mexico

NOEL F. SMYTH

School of Mathematics and Maxwell Institute for Mathematical Sciences, University of Edinburgh, Edinburgh, Scotland, United Kingdom

3.1 SIMPLE EXAMPLE OF MULTIPLE SCALES ANALYSIS

One of the main ideas used to understand the evolution of coherent structures is that of modulations, provided by Whitham [1, 2], which is related to the perturbation theory technique of multiple scales [3]. Modulation theory was developed in the context of slowly varying waves, both linear and nonlinear, which are governed by partial differential equations. It is illustrative to understand the basic principles of modulation theory in terms of a linear oscillator as it is governed by an ordinary differential equation. The basic ideas and concepts of modulation theory will then become clear, without the complexity of the solutions of partial differential equations.

Let us begin by recalling the ordinary differential equation for a linear, simple harmonic oscillator

$$\frac{d^2 y}{dt^2} + \omega^2 y = 0, \tag{3.1}$$

where ω is the constant frequency of the oscillator and y is some characteristic displacement of the oscillatory motion. The solution of this equation is

$$y = A\cos(\omega t + \delta). \tag{3.2}$$

Nematicons: Spatial Optical Solitons in Nematic Liquid Crystals, First Edition.
Edited by Gaetano Assanto.

Here, A and δ are constants related to the initial condition for Equation 3.1. This solution can be reinterpreted in the form

$$y(t) = A\cos(\psi(t)), \tag{3.3}$$

where the phase ψ is

$$\psi = \int_0^t \omega(s)\,\mathrm{d}s + \delta. \tag{3.4}$$

The time derivative of the phase gives the local frequency of the oscillator $\psi' = \omega$. The key advance of Green and Liouville was to use the same formula for a variable frequency $\omega(t)$, for example, for a pendulum with varying length. In this case of variable frequency, the solution of the oscillator equation (Eq. 3.1) will have the form

$$y(t) = A(t)\cos\psi(t). \tag{3.5}$$

As the frequency is now varying, the amplitude of the oscillation must vary as well. This amplitude variation can be interpreted as follows. In polar coordinates, the harmonic oscillator orbits in a circle of constant radius A when the frequency ω is constant. However, when the frequency ω varies, the radius of this circle must also vary.

Substituting the assumed solution (Eq. 3.3) into the simple harmonic oscillator Equation 3.1 gives

$$A\left(-\dot{\psi}^2 + \omega^2\right)\cos\psi - \left(2\dot{A}\dot{\psi} + A\ddot{\psi}\right)\sin\psi + \ddot{A}\cos\psi = 0. \tag{3.6}$$

In this equation, the dots refer to differentiation with respect to t. It can then be seen that Equation 3.5 cannot be a solution of this equation, so that it is not a solution of the simple harmonic oscillator equation (Eq. 3.1) with varying frequency. However, if we neglect the term $\ddot{A}\cos\psi$, we have that Equation 3.6 is satisfied, provided

$$\dot{\psi}^2 = \omega^2, \quad 2\dot{A}\dot{\psi} + A\ddot{\psi} = 0. \tag{3.7}$$

The first of these equations can be integrated to give

$$\psi(t) = \pm\int_0^t \omega(s)\,\mathrm{d}s + \delta, \tag{3.8}$$

so that the second becomes

$$A^2\dot{\psi} = \text{constant}. \tag{3.9}$$

Hence, both the amplitude and phase of the oscillation are determined within the approximation that $\ddot{A}$ can be neglected.

Let us now determine when the approximation that $\ddot{A}$ can be neglected could be valid. From the solution (Eq. 3.9), we have that

$$2\dot{A}^2\dot{\psi} + 2A\ddot{A}\dot{\psi} + 4A\dot{A}\ddot{\psi} + A^2\dddot{\psi} = 0. \tag{3.10}$$

Thus,

$$\ddot{A} = -\frac{\dot{A}^2}{A} - 2\dot{A}\frac{\ddot{\psi}}{\dot{\psi}} - \frac{A}{2}\frac{\dddot{\psi}}{\dot{\psi}}. \tag{3.11}$$

Then, $|\ddot{A}| \ll 1$ provided $|\ddot{\psi}| = |\dot{\omega}| \ll 1$ and $|\dddot{\psi}| = |\ddot{\omega}| \ll 1$. It is then apparent that neglecting $\ddot{A}$ and the Equations 3.8 and 3.9 are valid provided that $\omega(t)$ is a slowly varying function of t.

This slowly varying solution could be derived on a more formal basis by constructing a solution of the form

$$y(t) = A(t)\cos(\psi(t)) + r(t), \tag{3.12}$$

where the remainder term r has to be estimated. Substituting this solution form into the simple harmonic oscillator equation (Eq. 3.1), we have

$$\ddot{r} + \omega^2(t)r = A\left(\dot{\psi}^2 - \omega^2\right)\cos\psi + \left(2\dot{A}\dot{\psi} + A\ddot{\psi}\right)\sin\psi - \ddot{A}\cos\psi. \tag{3.13}$$

As ω is slowly varying, we see that the oscillator equation for r is forced at resonance by the variations in A and ψ. Therefore, to avoid resonant growth in the solution for r, we need to eliminate these resonant terms, also called secular terms. This is achieved by requiring A to satisfy (3.6) and neglecting $\ddot{A}$. This idea of eliminating secular terms can be extended by taking $r = p(\theta(t))$, where the functions p and θ are to be determined. Substituting this form into Equation 3.13, we obtain

$$\begin{aligned}\dot{\theta}^2\frac{\mathrm{d}^2p}{\mathrm{d}\theta^2} + \omega^2(t)p = -\ddot{\theta}\frac{\mathrm{d}p}{\mathrm{d}\theta} &+ A\left(\dot{\psi}^2 - \omega^2\right)\cos\psi + \left(2\dot{A}\dot{\psi} + A\ddot{\psi}\right)\sin\psi \\ &- \ddot{A}\cos\psi. \end{aligned}\tag{3.14}$$

We now take $\dot{\theta}^2 = \omega^2(t)$. The equation for p is then a simple harmonic oscillator equation in its argument θ with periodic solutions of period 2π. The right-hand side of Equation 3.14 is also a periodic function of θ multiplied by slowly varying coefficients.

To solve Equation 3.14, we assume that $\dot{\theta} = \omega \neq 0$. We can then use the variable θ instead of t as an independent variable. The equation to be solved for p is therefore

$$\omega^2\left[\frac{\mathrm{d}^2p}{\mathrm{d}\theta^2} + p(\theta)\right] = F_1(\theta)\cos\theta + F_2(\theta)\sin\theta - \ddot{A}(t(\theta))\cos\theta + p'(\theta)\ddot{\theta}(t(\theta)), \tag{3.15}$$

where

$$F_1(\theta) = A(t(\theta))\left[\dot{\psi}^2(t(\theta)) - \omega^2(t(\theta))\right],$$
$$F_2(\theta) = 2\dot{A}(t(\theta))\dot{\psi}(t(\theta)) + A(t(\theta))\ddot{\psi}(t(\theta)). \tag{3.16}$$

The dominant terms in the forcing for p are the forcing terms F_1 and F_2 as $\ddot{A}$ is small as ω is assumed to be slowly varying. Equation 3.15 with the forcing terms F_1 and F_2 can be solved using variation of parameters as

$$p = \sin\theta \int_0^\theta \frac{F_1(s)}{\omega^2(s)} \cos^2 s \, \mathrm{d}s - \cos\theta \int_0^\theta \frac{F_1(s)}{\omega^2(s)} \sin s \cos s \, \mathrm{d}s$$
$$+ \sin\theta \int_0^\theta \frac{F_2(s)}{\omega^2(s)} \sin s \cos s \, \mathrm{d}s - \cos\theta \int_0^\theta \frac{F_2(s)}{\omega^2(s)} \sin^2 s \, \mathrm{d}s. \tag{3.17}$$

For this to be a valid approximate solution, we need $p(\theta)$ to remain bounded as θ increases. To study the behavior of p, let us study the first term in this solution, written as

$$\sin\theta \int_0^\theta \left(\frac{1}{2} + \frac{1}{2}\cos 2s\right) \frac{F_1(s)}{\omega^2(s)} \, \mathrm{d}s$$
$$= \frac{1}{2}\sin\theta \int_0^\theta \frac{F_1(s)}{\omega^2(s)} \, \mathrm{d}s + \frac{1}{2}\sin\theta \int_0^\theta \cos 2s \frac{F_1(s)}{\omega^2(s)} \, \mathrm{d}s$$
$$= \frac{1}{2}\sin\theta \int_0^\theta \frac{F_1(s)}{\omega^2(s)} \, \mathrm{d}s + \frac{1}{2}\sin\theta \left[\frac{1}{2}\sin 2s \frac{F_1(s)}{\omega(s)}\right]_0^\theta$$
$$- \frac{1}{4}\sin\theta \int_0^\theta \sin 2s \frac{\mathrm{d}}{\mathrm{d}s}\left(\frac{F_1(s)}{\omega(s)}\right) \mathrm{d}s \tag{3.18}$$

after an integration by parts. The second and third terms on the right-hand side are bounded functions of θ, leading to an acceptable solution for $p(\theta)$. However, the first term grows proportional to θ, unless $F_1 = 0$. This requirement gives Equation 3.6 for the amplitude A. A similar argument shows that we require $F_2 = 0$ for a bounded solution. The term proportional to $\ddot{A}$ in Equation 3.15 is also resonant, but smaller, giving a contribution of $O(\ddot{A}\theta)$. The correction $p'\ddot{\theta}$ in this equation is of $O(\ddot{A}\dot{\theta}\theta)$ and is also growing but is again of higher order. This analysis gives an estimate of the validity of the approximation, which shows that it is valid to a range of θ inversely proportional to $\ddot{A}$, which is large since $\ddot{A}$ is assumed small. This procedure can be carried to all orders, and it was shown to be convergent by Hale [4]. The modulated solutions can therefore be constructed, taking advantage of the difference between the timescale of the period of the oscillator and the scale of the variation of the parameters in the simple harmonic oscillator equation (Eq. 3.1), which is slow compared to the oscillation frequency.

The results obtained here can be phrased in a "two timing" or " multiple scales" form, which is usual in the applied mathematics literature [3], in the following manner. Let us again consider the dominant terms in Equation 3.15 for p

$$\frac{d^2 p}{d\theta^2} + p(\theta) = \frac{F_1(\theta)}{\omega^2(\theta)} \cos\theta + \frac{F_2(\theta)}{\omega^2(\theta)} \sin\theta. \tag{3.19}$$

Now the function $F_1(\theta)$ is slowly varying in θ and so can be considered to be independent of θ in a cycle $0 \le \theta \le 2\pi$, which is what was described earlier. Equation 3.19 then becomes

$$\frac{d^2 p}{d\theta^2} + p(\theta) = \frac{F_1(t)}{\omega^2(t)} \cos\theta + \frac{F_2(t)}{\omega^2(t)} \sin\theta. \tag{3.20}$$

This assumption is known as the two timing assumption, which is to treat the slow time t and the fast variable θ as independent variables. We require a periodic solution of Equation 3.20. The only way to obtain a periodic solution of this boundary value problem with periodic boundary conditions is to require the forcing term to be orthogonal to the corresponding null space. This condition gives the same result as the one obtained earlier through variation of parameters, as we obtain

$$\frac{F_1(t)}{\omega^2(t)} \int_0^{2\pi} \cos^2\theta \, d\theta = 0 \quad \text{and} \quad \frac{F_2(t)}{\omega^2(t)} \int_0^{2\pi} \sin^2\theta \, d\theta = 0. \tag{3.21}$$

Hence, $F_1 = F_2 = 0$, as before. The separation of scales then provides a way of constructing approximate solutions.

A fundamental formulation by Whitham [1, 2] is based on the variational form of the equation to be solved. Let us revisit this alternative formulation. The simple harmonic oscillator equation (Eq. 3.1) is derived from the Lagrangian

$$L = \int_0^T \left(\frac{1}{2}\dot{y}^2 - \frac{1}{2}\omega^2(t) y^2 \right) dt. \tag{3.22}$$

Whitham's idea was to use a Galerkin approximation, but with a trial function in the form of a modulated oscillation. In this case, this trial function is

$$y(t) = A(t)\cos\theta(t), \tag{3.23}$$

so that

$$\dot{y} = \dot{A}(t)\cos\theta(t) - A(t)\dot{\theta}(t)\sin\theta(t), \tag{3.24}$$

$$\dot{y}^2 = A^2\dot{\theta}^2 \sin^2\theta(t) - 2A\dot{A}\dot{\theta}\sin\theta(t)\cos\theta(t) + \dot{A}^2(t)\cos^2\theta(t)$$

and neglect, as before, terms involving $\dot{A}$. Substituting these expressions into the Lagrangian, Equation 3.22 gives

$$L = \int_0^T \left[\frac{1}{2}A^2\dot{\theta}^2 \sin^2\theta(t) - \frac{1}{2}\omega^2 A^2 \cos^2\theta(t)\right] \mathrm{d}t \tag{3.25}$$
$$= \int_0^T \frac{1}{4}A^2\left(\dot{\theta}^2 - \omega^2\right) \mathrm{d}t - \frac{1}{4}\int_0^T \left(A^2\dot{\theta}^2 + \omega^2 A^2\right)\cos 2\theta(t)\, \mathrm{d}t.$$

Let us change the variable of integration to θ in the second integral, to give

$$\int_0^T \left[\frac{1}{2}A^2\dot{\theta}^2 \sin^2\theta(t) - \frac{1}{2}\omega^2 A^2 \cos^2\theta(t)\right] \mathrm{d}t = \int_0^T \frac{1}{4}A^2\left(\dot{\theta}^2 - \omega^2\right) \mathrm{d}t$$
$$- \frac{1}{4}\int_0^{2\pi} \frac{A^2(t(\theta))\dot{\theta}^2(t(\theta)) + \omega^2(t(\theta))A^2(t(\theta))}{\dot{\omega}(t(\theta))} \cos 2\theta\, \mathrm{d}\theta. \tag{3.26}$$

In the second integral on the right-hand side, the function $t(\theta)$ is slowly varying and is assumed to be constant in a cycle, which is equivalent to saying that t is independent of θ. The $\cos 2\theta$ term then integrates to 0, and the Lagrangian (Eq. 3.25) becomes

$$L = \frac{1}{4}\int_0^T A^2\left(\dot{\theta}^2 - \omega^2\right) \mathrm{d}t. \tag{3.27}$$

The modulation equations obtained by taking variations of the Lagrangian (Eq. 3.27) are again (Eq. 3.7) in the form $\dot{\theta}^2 = \omega^2$, which is obtained by varying with respect to A, and $\mathrm{d}/\mathrm{d}t(A^2\dot{\theta}) = 0$, which is obtained by varying with respect to θ. The modulation equations are then the variational equations of the averaged Lagrangian. The procedure outlined here can be phrased as saying that the modulation equations are obtained from the averaged Lagrangian

$$L = \int_0^T \int_0^{2\pi} \frac{1}{2}\left[A^2\dot{\theta}^2 \sin^2\theta(t) - A^2\omega^2\cos^2\theta(t)\right] \mathrm{d}\theta \mathrm{d}t, \tag{3.28}$$

where the variable θ is assumed to be independent of t, as before.

This method can be further developed to show how the orthogonality idea leads to the averaged Lagrangian formulation in general [1, 2]. Indeed, the method can be extended to dispersive wave equations, leading to clearly interpretable modulation equations for the slowly varying wave parameters [1, 2]. In the next section, we show how suitable trial functions with prescribed spatial profiles modulated in the evolution variable with slowly varying parameters can be used effectively to describe the evolution of coherent structures of interest. Moreover, we show how the modulation theory for coherent structures can be coupled with the radiation emitted as they evolve.

3.2 SURVEY OF PERTURBATION METHODS FOR SOLITARY WAVES

Many nonlinear dispersive wave equations describing physical processes have solitary wave solutions, standard examples with applicability to a wide range of physical systems being the Korteweg-de Vries (KdV) equation, the nonlinear Schrödinger (NLS) equation, and the Sine–Gordon equation [1, 5]. In standard form, these equations are derived for nonlinear wave propagation in uniform media. In most applications, however, the propagation medium is rarely uniform. In the case in which the medium is slowly varying, so that the length scale for the variation of the medium is much greater than a characteristic length of the nonlinear wave—for instance, the width of a solitary wave or the wavelength of a periodic wave—perturbation methods can be employed to determine the changes in the properties of the wave as it propagates through the changing medium, much as for the simple harmonic oscillator example of the previous section [3]. As the medium is slowly varying, the method of multiple scales [3] is usually employed. The actual method of multiple scales can be cast in many forms and is also called the method of averaging [3] and modulation theory [1]. It should be noted that the standard equations mentioned earlier, the KdV, NLS, and Sine–Gordon equations, are also standard examples of completely integrable equations that can be solved using the method of inverse scattering [6, 7]. In the special case in which an inverse scattering solution is available, soliton evolution in slowly varying media can be studied using perturbed inverse scattering theory [8]. However, as the emphasis of this chapter is nonlinear beam evolution in nematic liquid crystals, for which the governing equations do not have an inverse scattering solution, the method of perturbed inverse scattering is not discussed here.

To motivate the extension of modulation theory, which forms the core of the methods used to analyze the evolution of nonlinear beams in nematic liquid crystals, let us start with a simple problem of soliton evolution in a medium with slowly varying linear dispersion, governed by an NLS equation with slowly varying coefficients

$$i\frac{\partial u}{\partial z} + \frac{\beta(Z)}{2}\frac{\partial^2 u}{\partial x^2} + |u|^2 u = 0. \tag{3.29}$$

Here, β is the linear diffraction and is a function of $Z = \epsilon z$, $\epsilon \ll 1$, $\epsilon > 0$. The evolution of a solitary wave will be analyzed using three variations of the method of multiple scales, the standard form of multiple scales, the method of averaging and modulation theory. For constant β, the NLS Equation 3.29 has the soliton solution

$$u = a \operatorname{sech}\,(ax/\sqrt{\beta})\, e^{ia^2 z/2}. \tag{3.30}$$

Let us first use the standard method of multiple scales to analyze the evolution of the soliton (Eq. 3.30) for slowly varying β [3]. We therefore seek a perturbation series solution of the form

$$u = u_0 + \epsilon u_1 + \cdots = a(Z)\ \mathrm{sech}\ (a(Z)x/\sqrt{\beta(Z)})\ \mathrm{e}^{i\sigma(Z)/\epsilon} + \epsilon u_1 + \cdots. \tag{3.31}$$

We now substitute the perturbation series (Eq. 3.31) into the NLS equation (Eq. 3.29) and separate terms in powers of ϵ. At $O(1)$, we obtain the soliton solution (Eq. 3.30) with $\sigma' = a^2/2$. At $O(\epsilon)$, we obtain

$$P(u_1) = i\frac{\partial u_1}{\partial z} + \frac{\beta(Z)}{2}\frac{\partial^2 u_1}{\partial x^2} + u_0^2 u_1^* + 2|u_0|^2 u_1 = -i\left[a'\ \mathrm{sech}\ \varphi + \left(\frac{a^2\beta'}{2\beta^{3/2}} - \frac{aa'}{\sqrt{\beta}}\right) x\ \mathrm{sech}\,\varphi\tanh\varphi\right]\mathrm{e}^{i\sigma(Z)/\epsilon}, \tag{3.32}$$

where the asterisk superscript denotes the complex conjugate and

$$\varphi = a(Z)x/\sqrt{\beta(Z)}. \tag{3.33}$$

To eliminate secular terms, that is, terms that grow in x, from the solution of Equation 3.32 for u_1, the right-hand side of this equation must be orthogonal to the bounded solutions of the adjoint of $P(u_1)$ [3]. There are two bounded solutions of the adjoint, u_{0x} and u_{0z}. The right-hand side of Equation 3.32 must then be orthogonal to $\mathrm{sech}\,\varphi\tanh\ \varphi$ and $\mathrm{sech}\,\varphi$. As the right-hand side is even in x, it is automatically orthogonal to the first of these. Enforcing orthogonality to the second solution, that is, the integral from $x = -\infty$ to $x = \infty$ of the right-hand side times $\mathrm{sech}\,\varphi$ is zero, gives the differential equation for the amplitude a of the soliton

$$\frac{a'}{a} + \frac{\beta'}{2\beta} = 0, \tag{3.34}$$

which has the solution

$$a = a(0)\left[\frac{\beta(0)}{\beta}\right]^{1/2}, \tag{3.35}$$

with $a(0)$ and $\beta(0)$ being a and β at $z = 0$. This equation gives the evolution of the amplitude of the soliton as it propagates through the medium with slowly varying linear diffraction.

Let us now analyze the evolution of the soliton using the method of averaging [3]. Multiplying the NLS equation (Eq. 3.29) by u^* and subtracting the complex conjugate of the resulting equation gives, after some manipulation,

$$i\frac{\partial}{\partial z}|u|^2 + \frac{\beta}{2}\frac{\partial}{\partial x}\left(u^* u_x - u u_x^*\right) = 0. \tag{3.36}$$

This equation is formally termed conservation of mass, as it corresponds to invariance in phase of the Lagrangian for the NLS equation (Eq. 3.29) [8]. Whereas it physically corresponds to conservation of mass in the application of the NLS equation to water waves, in the optical context, it corresponds to conservation of power [5]. Integrating this mass conservation in x from $-\infty$ to ∞, which is termed as averaging in x, gives

$$\frac{d}{dZ}\int_{-\infty}^{\infty}|u|^2\,dx = 0. \tag{3.37}$$

Substituting the soliton solution (Eq. 3.30) into this integral gives the differential equation for the amplitude

$$\frac{d}{dZ}\left(a\sqrt{\beta}\right) = 0. \tag{3.38}$$

This is the same differential equation (Eq. 3.34) as that obtained using a multiple scales analysis, and so has the same solution (Eq. 3.35). It is readily apparent that the use of averaging has resulted in a much more rapid and simple derivation of the equation governing the evolution of the soliton than the multiple scales analysis. Its simplicity is its key attraction. In fact, the orthogonality condition resulting from the multiple scales analysis is nothing more than the conservation of mass relation in integral form for the basic solution u_0. A word of caution needs to be noted, though. As the NLS equation has an inverse scattering solution, it has an infinite number of conservation laws [1, 6, 7]. The question then arises as to why the conservation of mass equation, the lowest order conservation equation, was used to derive the equation for the evolution of the amplitude of the soliton. The use of this equation, of course, agrees with the formal multiple scales analysis. In general, the orthogonality relation from a multiple scales analysis corresponds to the lowest order conservation law(s). The use of the correct conservation laws for an averaging analysis of a slowly varying equation can be quite subtle. In the case of the KdV equation, the use of the lowest order conservation law, that of mass, does not lead to the correct equation for the evolution of a slowly varying KdV soliton [9]. This is because as the soliton evolves, it sheds an $O(1)$ amount of mass, which must be included in the mass conservation law [9]. The correct conservation law to use in this case is that of energy, also called momentum. This is because the shed radiation has an $O(\epsilon)$ amount of energy. A full multiple scales analysis will lead to the correct orthogonality relation being the energy conservation (momentum conservation) integral. A slowly varying NLS soliton does not shed an $O(1)$ amount of mass, so mass conservation gives the correct orthogonality relation. The topic of the radiation shed by slowly varying solitons is taken up later in this chapter.

Modulation theory was originally developed by Whitham [1, 2] to analyze slowly varying nonlinear, dispersive wave trains. This method is based on averaging a Lagrangian for the governing equation, in the same sense as the method of averaging averages conservation laws. The two averaging techniques are related, as by Nöther's Theorem conservation equations arise from invariances of the Lagrangian [10]. Although the modulation theory was originally developed for periodic wave trains, it can still be used for solitary waves as a solitary wave is a limiting case of a periodic wave for which the period goes to infinity, within which limit an isolated wave is obtained. The Lagrangian for the NLS equation (Eq. 3.29) is

$$L = i\left(u^* u_z - u u_z^*\right) - \beta |u_x|^2 + |u|^4. \tag{3.39}$$

The key to modulation theory is to take the exact soliton, or periodic wave, solution and assume that its parameters are slowly varying functions. If the soliton u_0 in Equation 3.31 is now substituted into this Lagrangian and the Lagrangian averaged by integrating in x from $-\infty$ to ∞, the averaged Lagrangian

$$\mathcal{L} = -4\sqrt{\beta} a \sigma' + \frac{2\sqrt{\beta}}{3} a^3 \tag{3.40}$$

results, where the derivative is with respect to $Z = \epsilon z$. Taking variations of this averaged Lagrangian with respect to a and σ gives

$$\delta\sigma : \frac{\mathrm{d}}{\mathrm{d}Z}\left(\sqrt{\beta} a\right) = 0 \tag{3.41}$$

$$\delta a : \sigma' = a^2/2. \tag{3.42}$$

These two equations lead to the same results as the multiple scales and averaging analyses.

The averaged Lagrangian method can be extended by letting the slowly varying soliton have an arbitrary width and then letting the modulation equations determine the amplitude/width relation. We then set

$$u = a \ \mathrm{sech}\frac{x}{w}\ \mathrm{e}^{i\sigma}, \tag{3.43}$$

where, as before, a, w, and σ are functions of Z. Substituting this solution form into the Lagrangian (Eq. 3.39) and averaging in x results in the averaged Lagrangian

$$\mathcal{L} = -4a^2 w \sigma' - \frac{2\beta}{3}\frac{a^2}{w} + \frac{4}{3} a^4 w. \tag{3.44}$$

Taking variations with respect to the soliton parameters then results in the modulation equations

$$\begin{aligned} \frac{\mathrm{d}}{\mathrm{d}Z}\left(a^2 w\right) &= 0, \\ a^2 w^2 &= \beta, \\ \frac{\mathrm{d}\sigma}{\mathrm{d}Z} &= \frac{a^2}{2}. \end{aligned} \tag{3.45}$$

These modulation equations lead to the same results as before. The appropriate use of the variational principle and the averaged Lagrangian will then lead to the correct soliton relations [1].

3.3 LINEARIZED PERTURBATION THEORY FOR NONLINEAR SCHRÖDINGER EQUATION

Before proceeding to develop modulation theory for the evolution of a soliton from an initial condition for the NLS equation, let us consider soliton perturbation theory. The results of this theory will be compared and contrasted with the results of modulation theory in Section 3.4.

A standard approach used to study small perturbations for a soliton is to use perturbation theory based on linearizing about the exact soliton solution [11, 12]. This leads to the study of an initial value problem for a linear equation, which can be solved using an appropriate eigenfunction expansion. This eigenfunction expansion includes both the localized, discrete spectrum and the corresponding projections on the continuous spectrum. It is the spectrum of the linearized equation that determines the behavior of the solution. This equation then needs to be analyzed in detail. Let us now assume that the linear diffraction is constant and set $\beta = 1$ in the NLS equation (Eq. 3.29)

$$i\frac{\partial u}{\partial z} + \frac{1}{2}\frac{\partial^2 u}{\partial x^2} + |u|^2 u = 0. \tag{3.46}$$

This equation has the one parameter family of soliton solutions

$$u_0(x, a, x_0) = a \text{ sech } a(x - x_0)\, e^{ia^2 z/2}. \tag{3.47}$$

To study the linearized stability of the soliton u_0, let us take the perturbation expansion $u = u_0 + v$, where $|v| \ll |u_0|$. The equation for v is then

$$i\frac{\partial v}{\partial z} + \frac{1}{2}\frac{\partial^2 v}{\partial x^2} + 2|u_0|^2 v + u_0^2 v^* = 0, \tag{3.48}$$

with $v(x, 0) = f(x)$ a given initial condition. This equation can be solved using Laplace transforms. The inversion of this Laplace transform solution is calculated in terms of singularities, which are poles corresponding to the eigenfunctions of the operator P defined by

$$Pv = \frac{1}{2}\frac{\partial^2 v}{\partial x^2} + 2|u_0|^2 v + u_0^2 v^*, \tag{3.49}$$

and branch points that give the contribution of the continuous spectrum. It should be noted that the operator P is not self-adjoint, so that the poles can be double poles, giving Jordan blocks [13]. These double poles always lead to linear growth in z of the solution for v, just as for the linear resonance discussed in Section 3.2 when the perturbation solution of Equation 3.29 was developed. These resonant terms were eliminated by modulating the soliton parameters to eliminate this growth. It is clear from the definition (Eq. 3.49) that as $|x| \to \infty$, $Pv \to v_{xx}/2$ and that P has a real continuous spectrum $\lambda = -k^2$, which gives the radiation modes.

The detailed analysis of the point spectrum of P is given in Reference 14. It was shown that the corresponding eigenspace gives a Jordan block of dimension 4. This eigenspace is spanned by the eigenfunctions $\varphi_1(x) = \operatorname{sech} ax$ and $\varphi_2(x) = \operatorname{sech} ax \tanh ax$, which satisfy the eigenvalue equation $P\varphi_1 = 0$ and $P\varphi_2 = 0$, and the functions $\varphi_3(x) = \operatorname{sech} ax(1 - ax \tanh ax)$ and $\varphi_4(x) = x \operatorname{sech} ax$, which satisfy $P\varphi_3 = \varphi_1$ and $P\varphi_4 = \varphi_2$. We again note that Equation 3.34 for the slowly varying parameters of the slowly varying soliton (3.31) was obtained by requiring orthogonality of φ_1 and φ_2 to the right-hand side of the first-order perturbation equation (Eq. 3.32). We note that the solution for v takes the form

$$v = \mathrm{e}^{i\sigma z}\left[A_1\varphi_1 + A_2\varphi_2 + A_3 z\varphi_3 + A_4 z\varphi_4\right] + \mathrm{e}^{i\sigma z}\int_0^\infty A(k)\mathrm{e}^{i(\sigma+k^2)z} f(k, z)\, \mathrm{d}k. \tag{3.50}$$

The coefficients A_1, A_2, A_3, and A_4 and the function $A(k)$ are determined from the initial condition. This solution can be interpreted as follows. The contributions of φ_1 and φ_2 account for a shift in the phase and a position shift, respectively, as can be seen by expanding the soliton solution (Eq. 3.47) in a Taylor series with respect to x_0 and a. The function φ_4 accounts for the soliton's acceleration, whereas φ_3 accounts for the distortions in the soliton's width.

The continuous contribution has the form of a shelf as $k \to 0$, because it takes the form

$$A(0)\mathrm{e}^{2i\sigma z} \operatorname{sech}^2 ax \tag{3.51}$$

for $x \sim 0$. This is a shelf that oscillates with frequency 2σ. We thus see that this shelf can be approximated in the form $A(0)\exp(2i\sigma z)$ for $-1/a \le x \le 1/a$. This suggests a trial function for the varying soliton of the form

$$u = a\mathrm{e}^{i\theta} \operatorname{sech} \frac{x}{w} + \mathrm{i}g\mathrm{e}^{\mathrm{i}\theta}, \tag{3.52}$$

where $g(z) = 0$ for $|x| \ge \ell$ and ℓ is the length of the shelf. To agree with the solution (Eq. 3.50) for v, we can take $\ell = w$. The leading order trial function for the varying soliton then includes a contribution of the continuous spectrum, which is given by the shelf g. If the shelf length ℓ is left as a free parameter in an averaged Lagrangian, we observe that the resulting modulation equations have a fixed point when $a = 1/w$, $g = 0$, and $\dot{\theta} = a^2/2$, which is the exact soliton solution. This trial function and modulation theory are taken up further in Section 3.4.

The solution (Eq. 3.52) must be consistent with the solution (Eq. 3.50) of the linearized equation. As the soliton fixed point is a center, its frequency must be $\dot{\theta}$. This requirement will give the length ℓ of the shelf discussed in the previous paragraph. It should be noted that the shelf arising from the continuous spectrum has the largest overlap with the eigenfunction φ_3, which controls the evolution of the soliton's width. This shows the strong interaction between the width and the continuous spectrum, which in turn shows that w and g are conjugate variables in the sense of classical mechanics, as are a and $\dot{\theta}$.

Finally, using a more conventional perturbation theory approach, it was shown that the nonlinear terms of the NLS equation give the excitation of the radiation due to the appropriate resonance and projection onto the continuous spectrum [14]. The main contribution is from φ_1, which shows that it is a mass imbalance that triggers the shedding of diffractive radiation. In the variational formulation of Section 3.4, this projection is replaced by the boundary condition of the solution of a linear equation that accounts for the imbalance in mass between the exact soliton solution and the modulated one.

3.4 MODULATION THEORY: NONLINEAR SCHRÖDINGER EQUATION

All three techniques used in Section 3.2 to analyze slowly varying waves are based on allowing the exact soliton solution to have slowly varying parameters. Many equations governing nonlinear waves in physical applications do not possess exact periodic or solitary wave solutions, particularly in more than one space dimension. An example of relevance here is nonlinear beam evolution in nematic liquid crystals [15–17]. The question naturally arises as to how to perform any analysis of nonlinear wave evolution in such cases, other than by obtaining full numerical solutions of the governing equations. One approach that is successful in such cases is the "chirp" variational approach developed by Anderson [18]. This approach also solved another problem that arises for soliton evolution. Any initial condition for the NLS equation will evolve into a finite number of solitons, plus diffractive radiation, as long as the mass of the initial condition is above a critical value [6, 7]. The question then arises as to how to analyze this evolution from an initial condition to solitons, in the simplest case for which only one soliton is formed. For the production of a single soliton, the simplest case is a sech initial condition as then the initial condition can be envisaged as "slowly varying" from initial to final steady soliton state.

The standard NLS equation (Eq. 3.46) with constant linear diffraction has the Lagrangian

$$L = i\left(u^* u_z - u u_z^*\right) - |u_x|^2 + |u|^4. \tag{3.53}$$

We shall take an initial condition of the form

$$u = a_0 \operatorname{sech}\frac{x}{w_0}. \tag{3.54}$$

An obvious assumption is to seek a varying soliton-like solution of the form

$$u = a \operatorname{sech}\frac{x}{w}\, \mathrm{e}^{i\sigma}, \tag{3.55}$$

where the amplitude a, width w, and phase σ are functions of z. Substituting this trial function into the Lagrangian (Eq. 3.53) and averaging in x yields the averaged Lagrangian

$$\mathcal{L} = -2a^2 w\sigma' - \frac{1}{3}\frac{a^2}{w} + \frac{2}{3}a^4 w. \tag{3.56}$$

Variations of this averaged Lagrangian, and some manipulation gives the variational, or modulation, equations

$$\frac{\mathrm{d}}{\mathrm{d}z}a^2 w = 0, \tag{3.57}$$

$$aw = 1, \tag{3.58}$$

$$\sigma' = a^2/2. \tag{3.59}$$

It is readily apparent that a and w are then constants and the initial condition does not evolve. The trial function (3.55) is then an inadequate approximation to allow the initial condition to evolve to a soliton. Anderson overcame this problem by adding "chirp" to the trial function (Eq. 3.55)

$$u = a\ \mathrm{sech}\frac{x}{w}\ \mathrm{e}^{i\sigma + icx^2}. \tag{3.60}$$

The addition of this phase contribution was motivated by solutions of the linearized NLS (Schrödinger's) equation (Eq. 3.46). With the addition of this chirp, the resulting modulation equations allowed the initial condition to evolve. The chirp method is not continued further here for one vital reason. The modulation equations arising from the chirp trial function form a nonlinear oscillator and the amplitude and width do not evolve to a steady state but oscillate about a mean [19].

Inverse scattering theory shows that an initial condition evolves in a soliton(s) via the shedding of diffractive radiation [6, 7]. The trial function (Eq. 3.60) does not include this radiation loss and so cannot give modulation equations whose solution evolves to a steady state. To date, there has been no extension of the chirp method that includes this shed radiation. A successful method based on the extension of trial functions of the form (Eq. 3.55) and one that includes radiation loss was developed by Kath and Smyth [19]. The trial function used for the NLS equation was

$$u = a\ \mathrm{sech}\frac{x}{w}\ \mathrm{e}^{i\sigma} + ig\mathrm{e}^{i\sigma}, \tag{3.61}$$

where the parameters a, w, σ, and g are functions of z. The first term is a varying soliton, as discussed earlier. The origin and importance of the second term is less clear. In Section 3.3, we showed how this shelf term arises out of linearized soliton perturbation theory. It can also be shown to arise from perturbed inverse scattering theory, which is taken up later. Before we consider this, it is noted that the physical effect of this shelf term is that it represents the effect of the low wave number radiation that accumulates under the soliton as it evolves. The linearized NLS equation (Eq. 3.46) (in fact Schrödinger's equation) has the dispersion relation $\omega = k^2/2$, so that the group velocity is $c_g = k$. Hence, shed low wave number diffractive waves have low group velocity and so accumulate under the soliton as it evolves, forming a shelf of radiation. This shelf of radiation under a pulse is well known in the context of pulse propagation in optical fibers and is referred to as a pedestal [20, 21]. Figure 3.1 shows a numerical solution of the NLS equation (Eq. 3.46) for a nonsoliton initial condition and the shelf of radiation under the pulse can be clearly seen.

The existence of the shelf of radiation under an evolving soliton can also be demonstrated through a perturbation analysis based on the inverse scattering solution for the NLS equation. Let us linearize about an exact soliton

$$u_0 = a \text{ sech } ax\, \mathrm{e}^{ia^2z/2} \tag{3.62}$$

with $u = u_0 + u_1$, where $|u_1| \ll |u_0|$. Substituting this expansion into the NLS equation (Eq. 3.46) gives

$$u_1 = -\frac{\partial^2 f}{\partial x^2} + 2a \tanh ax \frac{\partial f}{\partial x} - a^2 f \tanh^2 ax + a^2 f^* \text{ sech}^2 ax\, \mathrm{e}^{2i\theta}, \tag{3.63}$$

where the phase $\theta = a^2 z/2$ and f is the solution of

$$\frac{\partial f}{\partial z} = \frac{i}{2}\frac{\partial^2 f}{\partial x^2} \tag{3.64}$$

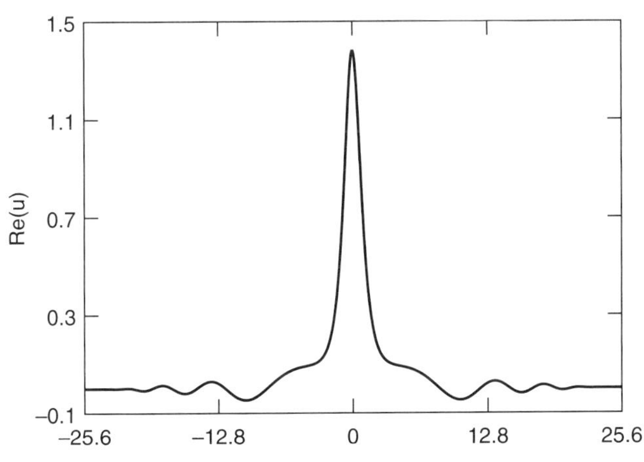

FIGURE 3.1 Numerical solution of NLS equation (Eq. 3.46) for initial condition $u(x, 0) = 1.25$ sech x. *Source:* Reproduced with permission from Fig. 2 in Reference 19.

with the initial condition for f depending on $u_0(x, 0)$ [22]. Owing to diffraction, solutions of Equation 3.64 rapidly become flat as $z \to \infty$. Therefore, for large z

$$u_1 \approx h(1 - \operatorname{sech}^2 ax) - h^* e^{2i\theta} \operatorname{sech}^2 ax, \tag{3.65}$$

with $h = -a^2 f$. Hence, for large z

$$u \approx e^{i\theta} \left[a \operatorname{sech} ax + |h| \left(1 - 2 \operatorname{sech}^2 ax\right) \cos \psi - i|h| \sin \psi \right], \tag{3.66}$$

where $\psi = \theta - \arg h$. This approximation for large z closely resembles the trial function (Eq. 3.61), with the first term the varying soliton and the third term the flat shelf $\pi/2$ out of phase with the soliton. The question then is, what does the second term represent? This can be seen on expanding $a \operatorname{sech} x/w$ around an exact soliton. For an exact soliton, the width–amplitude relation is $w_f = 1/a_f$. Hence, perturbing from an exact soliton with $w = 1/a_f + \delta w$ and $a = a_f + \delta a$ gives

$$a \operatorname{sech} x/w = a_f \operatorname{sech} a_f x + \delta a \operatorname{sech} a_f x + a^2 \delta w \operatorname{sech} ax \tanh ax + \cdots. \tag{3.67}$$

Around $x = 0$ the perturbing term from the exact soliton is δa. Similarly near $x = 0$ the second term in Equation 3.66 is $-|h| \cos \psi$. The second term in (3.66) then represents changes in the amplitude of the soliton as it evolves. Finally, it should be noted that the soliton perturbation theory leading to (3.66) has $aw = 1$, so that $\delta w = -\delta a/a^2$ within this theory. This analysis shows that the second term in Equation 3.66 is incorporated in the trial function (Eq. 3.61) through changes in a and does not need to be separately incorporated, unlike the shelf term, the third term in Equation 3.66. The shelf term adds in and out of phase with the soliton, causing the soliton to oscillate in amplitude as it evolves. Furthermore, the shelf term is $\pi/2$ out of phase with the soliton as the in-phase corrections are accounted for by variations in the amplitude and width of the soliton. The existence of a shelf of low wave number radiation under evolving solitons has also been shown for coupled NLS equations [11, 12].

The shelf of radiation under the evolving soliton, determined by g, must have limited extent and must match the radiation propagating away from the evolving soliton, as seen in Figure 3.1. It is then assumed that the shelf has length ℓ, so that it extends in the region $-\ell/2 \leq x \leq \ell/2$, as previously discussed in Section 3.3. The length ℓ is determined from the modulation equations given later.

Substituting the trial function into the Lagrangian (Eq. 3.53) and averaging in x results in the averaged Lagrangian

$$\mathcal{L} = -\left(2a^2 w + \ell g^2\right) \sigma' + \pi w g a' + \pi a g w' - \pi a w g' - \frac{1}{3}\frac{a^2}{w} + \frac{2}{3} a^4 w. \tag{3.68}$$

Taking variations of this averaged Lagrangian gives the modulation equations

$$\frac{\mathrm{d}}{\mathrm{d}z}\left(2a^2 w + \ell g^2\right) = 0, \tag{3.69}$$

$$\frac{\mathrm{d}}{\mathrm{d}z} a w = \frac{\ell}{\pi} g \sigma', \tag{3.70}$$

$$\frac{\mathrm{d}g}{\mathrm{d}z} = -\frac{2a}{3\pi}\left(a^2 - w^{-2}\right), \tag{3.71}$$

$$\frac{\mathrm{d}\sigma}{\mathrm{d}z} = a^2 - \frac{1}{2w^2}. \tag{3.72}$$

These modulation equations have the fixed point $a_f w_f = 1$ and $g = 0$, which is the soliton solution of the NLS equation, as required.

Using Nöther's Theorem on the NLS Lagrangian (Eq. 3.53) [10], the NLS equation (Eq. 3.46) has the energy conservation equation

$$i\frac{\partial}{\partial z}\left[|u_x|^2 - |u|^4\right] + \frac{1}{2}\frac{\partial}{\partial x}\left[u_x^* u_{xx} - u_x u_{xx}^* - 2|u|^2\left(u^* u_x - u u_x^*\right)\right] = 0. \tag{3.73}$$

Averaging this energy conservation equation gives the modulation equation for conservation of energy

$$\frac{\mathrm{d}H}{\mathrm{d}z} = \frac{\mathrm{d}}{\mathrm{d}z}\left[\frac{a^2}{w} - 2a^4 w\right] = 0. \tag{3.74}$$

As at the soliton steady state $a_f w_f = 1$, this energy conservation equation gives this steady state given the initial values of a and w, $a(0)$ and $w(0)$,

$$a_f^3 = -H(0). \tag{3.75}$$

Inverse scattering gives that for an initial width $w(0) = 1$, the amplitude of the final soliton is [23]

$$a_f = 2a(0) - 1. \tag{3.76}$$

Figure 3.2 shows a comparison between the final soliton amplitudes as given by inverse scattering and modulation theory. It can be seen that modulation theory is in excellent agreement with the inverse scattering result. It should be noted that the inverse scattering result is valid for $a(0) > 1/2$. Below this initial amplitude, no soliton is formed as, unlike for the KdV equation, the NLS equation requires a critical initial mass for a soliton to form [6, 7].

The Lagrangian (Eq. 3.53) also has scale invariance. Applying Nöther's Theorem [10] to this scale invariance leads to the mass conservation equation

$$i\frac{\partial}{\partial z}|u|^2 + \frac{1}{2}\frac{\partial}{\partial x}\left[u^* u_x - u u_x^*\right] = 0. \tag{3.77}$$

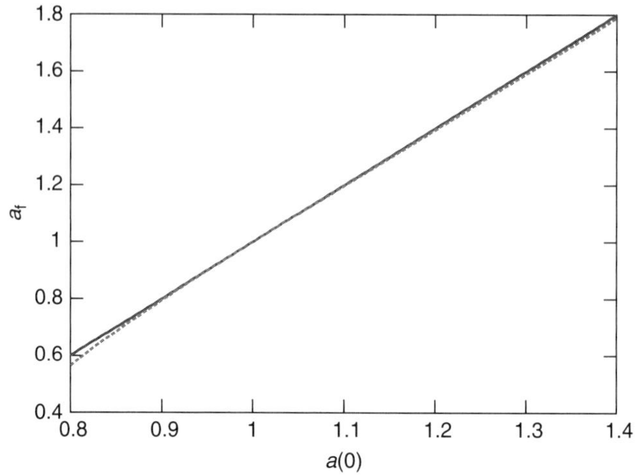

FIGURE 3.2 Comparison of final soliton amplitude a_f for initial condition $a(0)\,\mathrm{sech}\,x$. Inverse scattering (Eq. 3.76): solid line; modulation theory (Eq. 3.75): dashed line.

This is the same as the mass conservation equation (Eq. 3.36) for constant linear diffraction, as expected. Averaging this mass conservation equation using the trial function (3.61) gives the modulation equation (Eq. 3.69).

The final quantity to be determined is the length ℓ of the shelf of radiation under the evolving soliton. Let us linearize the modulation Equations 3.69–3.72 about the steady state with $a = a_f + a_1$, $w = 1/a_f + w_1$, and $g = g_1$, $|a_1| \ll a_f$, $|w_1| \ll w_f$, and $|g_1| \ll 1$. After some algebra, we find $w_1 = -2a_1/a_f$ with g_1 satisfying the simple harmonic oscillator equation

$$g_1'' + \frac{4a_f^3\sigma_f'\ell}{3\pi^2} g_1 = 0. \tag{3.78}$$

As discussed in Section 3.3, near the steady state, the shelf oscillates with the soliton frequency $\sigma_f' = a_f^2/2$ [19]. Hence, we find the shelf length

$$\ell = \frac{3\pi^2}{8a_f}. \tag{3.79}$$

3.5 RADIATION LOSS

The modulation equations (Eqs. 3.69–3.72) of the previous section cannot evolve to a steady state as they form a conservative system. Their solution will oscillate about a mean indefinitely. General initial conditions for solitary wave equations evolve to solitary wave steady states through shedding diffractive radiation [1, 6, 7]. This diffractive radiation is the physical and mathematical effect missing from the modulation equations of Section 3.4. Figure 3.1 shows that the shed radiation has small amplitude relative to the evolving solitary wave. The NLS equation (Eq. 3.46) then shows that it is governed by the linearized NLS (Schrödinger's) equation

$$i\frac{\partial u}{\partial z} + \frac{1}{2}\frac{\partial^2 u}{\partial x^2} = 0 \tag{3.80}$$

subject to a boundary condition that matches the shed radiation to the edge of the shelf, $u = S(z)$ at $x = \pm\ell/2$. As there is no radiation initially, this equation is solved together with the initial condition $u = 0$ at $z = 0$. There are a number of ways to solve Equation 3.80, but the easiest is to use Laplace transforms in z.

The key quantity to determine to account for the effect of the shed diffractive radiation in the modulation equations is the mass flux associated with this shed radiation. This radiation also takes away energy from the soliton, but this loss is second order [19]. Integrating the mass conservation equation (Eq. 3.77) from the edge of the shelf at $x = \ell/2$ to ∞ gives the mass loss flux to $x > \ell/2$ as

$$\frac{\mathrm{d}}{\mathrm{d}z}\int_{\ell/2}^{\infty} |u|^2\,\mathrm{d}x = \mathrm{Im}(u^* u_x)\,\big|_{x=\ell/2}. \tag{3.81}$$

By symmetry, there is an identical mass flux into $x < -\ell/2$. To determine the amount of mass lost by the soliton as it evolves, we then need to determine u and u_x at the edge of the shelf $x = \ell/2$.

Assigning s as the Laplace transform variable and denoting transforms by a ^ superscript, we find Equation 3.80 becomes

$$\hat{u}_{xx} + 2is\hat{u} = 0. \tag{3.82}$$

Taking the solution of this equation that decays at ∞ by taking the correct branch for the square root of $-i$, we have

$$\hat{u} = A\mathrm{e}^{-\sqrt{2s}\mathrm{e}^{-i\pi/4}x}, \tag{3.83}$$

where A is a constant of integration. Hence,

$$\hat{u}_x = -\sqrt{2s}\mathrm{e}^{-i\pi/4}\hat{u}. \tag{3.84}$$

Inverting this Laplace transform using the convolution theorem gives

$$u_x(\ell/2, z) = -\sqrt{2}\mathrm{e}^{-i\pi/4}\frac{\mathrm{d}}{\mathrm{d}z}\int_0^z \frac{u(\ell/2,\zeta)}{\sqrt{\pi(z-\zeta)}}\,\mathrm{d}\zeta. \tag{3.85}$$

We then see that all we need to determine the mass flux to the diffractive radiation in $x > \ell/2$ is $u(\ell/2, z)$. From Equation 3.81, this mass flux to $x > \ell/2$ is

$$\frac{\mathrm{d}}{\mathrm{d}z}\int_{\ell/2}^{\infty} |u|^2 \,\mathrm{d}x = -\sqrt{2}\mathrm{Im}\left[\mathrm{e}^{-i\pi/4} u^*(\ell/2, z)\frac{\mathrm{d}}{\mathrm{d}z}\int_0^z \frac{u(\ell/2, \zeta)}{\sqrt{\pi(z-\zeta)}}\,\mathrm{d}\zeta\right]. \tag{3.86}$$

To complete the radiation analysis, we need to determine the function $S(z)$. In the vicinity of the shelf, the trial function 3.61 can be decomposed into the fixed-point soliton, plus a component that must radiate away, so that

$$u = u_{\text{fixed}} + u_1, \tag{3.87}$$

where u_1 is the radiated component. From Figure 3.1, it can be seen that $|u_1|$ is small relative to $|u_{\text{fixed}}|$. The mass in the combined soliton and shelf is then

$$\int_{-\ell/2}^{\ell/2} |u|^2 \mathrm{d}x = \int_{-\ell/2}^{\ell/2} \left[|u_{\text{fixed}}|^2 + 2\mathrm{Re}(u_{\text{fixed}} u_1) + |u_1|^2\right] \mathrm{d}x. \tag{3.88}$$

If we assume a small overlap between u_{fixed} and u_1, we have from the mass equation (Eq. 3.69) that

$$2a^2 w + \ell g^2 \approx 2a_f^2 w_f + \ell |u_1|^2_{x=\pm\ell/2}, \tag{3.89}$$

near the fixed point, due to symmetry, where the f subscripts denote fixed-point values. Mass conservation then gives

$$|S(z)|^2 = \frac{1}{\ell}\left[2a^2 w - 2a_f + \ell g^2\right], \tag{3.90}$$

as $a_f w_f = 1$ and $S = u_1$ at $x = \pm\ell/2$. Using this value of u at the edges of the shelf in the mass flux equation (Eq. 3.86), we have

$$\frac{\mathrm{d}}{\mathrm{d}z}\int_{\ell/2}^{\infty} |u|^2 \,\mathrm{d}x = -\sqrt{2}\mathrm{Im}\left[\mathrm{e}^{-i\pi/4} S^*(z)\frac{\mathrm{d}}{\mathrm{d}z}\int_0^z \frac{S(\zeta)}{\sqrt{\pi(z-\zeta)}}\,\mathrm{d}\zeta\right]. \tag{3.91}$$

The shelf of radiation under the soliton is flat, so that its phase will be slowly varying. The phase of S will then be slowly varying. If we set S in polar form as $S = R\exp(i\varphi)$, we can approximate the mass flux (Eq. 3.91) by

$$\frac{\mathrm{d}}{\mathrm{d}z}\int_{\ell/2}^{\infty} |u|^2 \,\mathrm{d}x = -\sqrt{2}\mathrm{Im}\left[\mathrm{e}^{-i\pi/4} R(z)\frac{\mathrm{d}}{\mathrm{d}z}\int_0^z \frac{R(\zeta)}{\sqrt{\pi(z-\zeta)}}\,\mathrm{d}\zeta\right], \tag{3.92}$$

with $R = |S|$ determined by Equation 3.90. The mass flux to the left of the soliton will also be given by this expression by symmetry.

Adding the mass flux expression to the mass modulation equation (Eq. 3.69), we obtain

$$\frac{\mathrm{d}}{\mathrm{d}z}\left(2a^2w + \ell g^2\right) = -2R^2\ell\delta, \tag{3.93}$$

where the loss coefficient δ is

$$\delta = \frac{1}{\ell R}\frac{\mathrm{d}}{\mathrm{d}z}\int_0^z \frac{R(z')}{\sqrt{\pi(z-z')}}\,\mathrm{d}z' \tag{3.94}$$

with

$$R^2 = \frac{1}{\ell}\left[2a^2w - 2a_f + \ell g^2\right]. \tag{3.95}$$

To add the mass loss in a consistent manner to the modulation Equations 3.69–3.72, we need to leave the energy Equation 3.74 unchanged as the energy lost in the diffractive radiation is second order [19]. To achieve this, we need to add a loss term to the g equation (Eq. 3.71) [19], so that this equation becomes

$$\frac{\mathrm{d}g}{\mathrm{d}z} = -\frac{2a}{3\pi}\left(a^2 - w^{-2}\right) - 2\delta g. \tag{3.96}$$

The final set of modulation equations governing the evolution of the soliton is then Equations 3.70, 3.72, 3.93, and 3.96. The mass equation (Eq. 3.93) can be replaced by the energy equation (Eq. 3.74), as was done by Kath and Smyth [19], as it is not independent of the modulation Equations 3.69–3.72.

Figure 3.3 shows a comparison between the soliton amplitude as given by the full numerical solution of the NLS equation (Eq. 3.46) and the modulation equations with mass loss to shed diffractive radiation. It can be seen that there is an excellent comparison. The numerical amplitude decays to the steady state at a slightly faster rate than the modulation amplitude. Because the modulation equations form a nonlinear damped oscillator, the difference in decay rate translates into a difference in the period of the oscillation, as seen in the figure. Finally, the mean of the modulation oscillation is slightly lower than that of the numerical solution, which means that the steady state of the numerical solution is slightly higher than that of the modulation solution.

3.6 SOLITARY WAVES IN NEMATIC LIQUID CRYSTALS: NEMATICONS

The ideas and techniques developed in the previous section for the NLS equation can be applied to analyze the motion of solitary wave beams in a nematic liquid crystal, so-called nematicons [15]. These nonlinear beams are two space

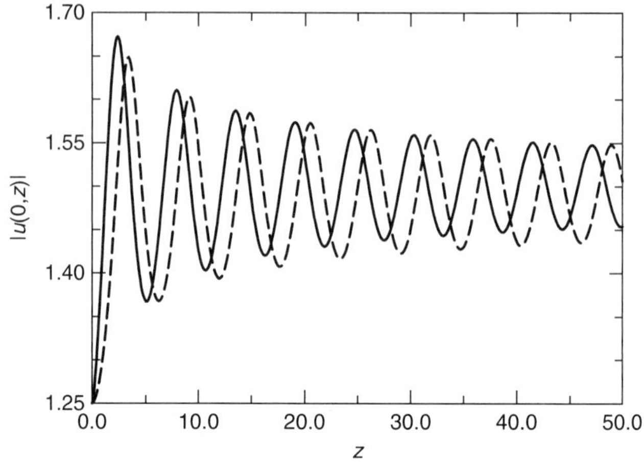

FIGURE 3.3 Comparison of amplitude a of soliton as given by the full numerical solution of the NLS equation (Eq. 3.46) (solid line) and the modulation equations (dashed line) for the initial condition $a = 1.25$ and $w = 1$. *Source:* Reproduced with permission from Fig. 4 in Reference [19].

dimensional, which makes the analysis more complicated than that of the one space dimensional NLS equation. At the simplest level, nematicons can be taken to be circularly symmetric, even though they have slight ellipticity [24].

Let us consider the propagation of a polarized, coherent light beam (laser light) through a cell filled with a nematic liquid crystal. The direction down the cell is the z-direction and the (x, y) plane is orthogonal to this. The beam is polarized in the x-direction and an external static electric field is applied in the x-direction to pretilt the nematic molecules at an angle θ_0 to the z-direction. This external field helps overcome the Freédericksz threshold for the nematic so that a low power light beam can self-focus as the total electric field—external plus the electric field of the electromagnetic radiation—is above the threshold. The perturbation of the director angle from the pretilt is denoted by θ. The envelope of the electric field of the light is denoted by $\tilde{E}$. Typical dimensions are ~100 μm for the cell width, ~1 mm for the cell length, and ~5 μm for the beam width [24]. Owing to the small size of the beam relative to the cell, if the beam is launched near the center of the cell, the effect of the cell boundaries on its propagation can be ignored. In nondimensional form, the equations governing the propagation of the beam are [16, 17, 25]

$$i\frac{\partial \tilde{E}}{\partial z} + i\tan\Delta\frac{\partial \tilde{E}}{\partial x} + \frac{1}{2}\nabla^2\tilde{E} + \sin(2\theta)\tilde{E} = 0, \tag{3.97}$$

$$\nu\nabla^2\theta - q\sin(2\theta) = -2\cos(2\theta)|\tilde{E}|^2. \tag{3.98}$$

Here, the Laplacian ∇^2 is in the (x, y) plane. In mathematical terms, the distance z down the liquid crystal cell is a timelike variable for the NLS-type equation (Eq. 3.97). The parameter ν is related to the elastic properties of the liquid crystal and, in most experimental situations, is large, typically $O(100)$ [15, 16]. The limit of large ν is called the nonlocal limit. The parameter q is related to the square of the external pretilting electric field [16, 17, 26]. Finally, the parameter Δ is the birefringent walk-off of the Poynting vector [1, 25].

If the medium is uniform, the walk-off term in the electric field equation (Eq. 3.97) can be factored out by the phase transformation [27]

$$\tilde{E} = E\mathrm{e}^{(\mathrm{i}/2)z\tan^2\Delta - ix\tan\Delta}. \tag{3.99}$$

The nematicon Equations 3.97 and 3.98 then become

$$i\frac{\partial E}{\partial z} + \frac{1}{2}\nabla^2 E + \sin(2\theta)E = 0, \tag{3.100}$$

$$\nu\nabla^2\theta - q\sin(2\theta) = -2\cos(2\theta)|E|^2. \tag{3.101}$$

The transformation (Eq. 3.99) is, of course, only valid if the liquid crystal is in a uniform state. In Chapters 1, 5, and 11, examples will be considered for which this walk-off term needs to be retained as nonuniformities in the medium due, for example, to a varying static pretilting field or the influence of other light beams, mean that the walk-off varies across the liquid crystal cell, altering the nematicon trajectory. As the nonlocality parameter ν increases, the perturbation θ of the director from the pretilt θ_0 decreases as $1/\sqrt{\nu}$ [28, 29]. In experimental situations, ν is usually in the range $O(10)$–$O(100)$ [16, 24, 26]. So for the milliWatt beam powers employed in experiments, the angle θ is small and the trigonometric functions in the full nematicon equations (Eqs. 3.100, 3.101) can be approximated by the first terms in their Taylor series, giving the simplified nematicon equations

$$i\frac{\partial E}{\partial z} + \frac{1}{2}\nabla^2 E + 2\theta E = 0, \tag{3.102}$$

$$\nu\nabla^2\theta - 2q\theta = -2|E|^2. \tag{3.103}$$

These small deviation nematicon equations are amenable to analysis using the same variational techniques as for the NLS equation in Section 3.4. Numerical solutions have shown that the nematicon equations have a solitary wave solution, termed a nematicon [15–17]. However, to date, no exact nematicon solution has been found for either the full nematicon Equations 3.100 and 3.101 or the simplified Equations 3.102 and 3.103. As discussed previously, this means that standard perturbation techniques cannot be used to analyze nematicon evolution, leaving full numerical solutions and variational techniques the main avenues for such analysis. This section discusses the application of variational techniques to the nematicon Equations 3.102–3.103. For circularly symmetric solutions, the nematicon Equations 3.102 and 3.103 have the Lagrangian

$$L = ir\left(E^*E_z - EE_z^*\right) - r|E_r|^2 + 4r\theta|E|^2 - \nu r\theta_r^2 - 2qr\theta^2. \tag{3.104}$$

In analogy with the trial function (Eq. 3.61) for the NLS equation, we shall develop modulation equations for nematicon evolution based on the trial functions

$$E = a \operatorname{sech}\frac{r}{w}\, \mathrm{e}^{i\sigma} + ig\mathrm{e}^{i\sigma},$$
$$\theta = \alpha \operatorname{sech}^2\frac{r}{\beta} \tag{3.105}$$

for the electric field and the optical axis, where $r = \sqrt{x^2 + y^2}$ [30]. The trial function for the electric field is the same as that for the NLS equation, but in radially symmetric form as the nematicon equations are in two space dimensions. The new trial function is that for the optical axis. In the nonlocal limit with ν large, the optical axis perturbation extends far beyond the waist of the electric field solitary wave [15–17]. Therefore, the optical axis has a different width from that of the electric field. The forcing in the Poisson Equation 3.103 for the director is proportional to $|E|^2$. Hence, the functional form for the optical axis is assumed to be the square of that for the electric field. As for the NLS equation, the shelf of low wave number diffractive radiation under the beam must have limited extent before it matches the radiation propagating away from the beam. It is then assumed that g is nonzero in the disk $0 \le r \le \ell$.

Substituting the trial functions (Eq. 3.105) into the Lagrangian (Eq. 3.104) and averaging by integrating in r from 0 to ∞ gives the averaged Lagrangian [30]

$$\begin{aligned}\mathcal{L} = {} & -2\left(a^2w^2I_2 + \Lambda g^2\right)\sigma' - 2I_1aw^2g' + 2I_1gw^2a' + 4I_1awgw' \\ & - a^2I_{22} - 4\nu I_{42}\alpha^2 - 2qI_4\alpha^2\beta^2 + \frac{2A^2B^2\alpha a^2\beta^2w^2}{A^2\beta^2 + B^2w^2}.\end{aligned} \tag{3.106}$$

Here, $\Lambda = \ell^2/2$. The various integrals I_i and $I_{i,j}$ in this averaged Lagrangian are listed in Appendix 3.A. There is one complication in the calculation of this averaged Lagrangian. Inspection of the Lagrangian (Eq. 3.104) shows that the integral

$$\int_0^\infty r \operatorname{sech}^2\frac{r}{\beta} \operatorname{sech}^2\frac{r}{w}\, \mathrm{d}r \tag{3.107}$$

needs to be evaluated. This integral can only be evaluated exactly for $w = \beta$, which is not the case in the nonlocal limit with $\beta \gg w$. One option is to evaluate the integral (Eq. 3.107) numerically, which also means evaluating its derivatives with respect to w and β numerically too, as both of these will result when variations of the averaged Lagrangian are taken. However, an analytical approximation is possible. This relies on the idea of "equivalent functions" [30]. The sech functions in Equation 3.107 are replaced by Gaussians with adjustable width parameters, and these parameters are determined by requiring that the resulting integral matches with the exact integral (Eq. 3.107) in the nonlocal limit $\beta \gg w$. The idea behind this substitution is that what is required in the averaged Lagrangian is an integral.

If two different integrals have the same value, the averaged Lagrangian will be unchanged. So using the idea of "equivalent functions," we replace $\mathrm{sech}^2 r/\beta$ by $\exp(-r^2/(A\beta)^2)$ and $\mathrm{sech}^2 r/w$ by $\exp(-r^2/(Bw)^2)$. Then,

$$\int_0^\infty r\mathrm{e}^{-r^2/(A\beta)^2}\mathrm{e}^{-r^2/(Bw)^2}\,\mathrm{d}r = \frac{A^2B^2w^2\beta^2}{2\left(A^2\beta^2+B^2w^2\right)}. \tag{3.108}$$

Let us now evaluate the integral (Eq. 3.107) in the limit $\beta \gg w$. Using the change of variable $r = \beta\rho$ and expanding in Taylor series for $w/\beta \ll 1$, this integral becomes

$$\int_0^\infty r\,\mathrm{sech}^2\frac{r}{\beta}\,\mathrm{sech}^2\frac{r}{w}\,\mathrm{d}r = w^2\left(I_2 - I_{x32} + \ldots\right), \tag{3.109}$$

where I_{x32} is given in Appendix 3.A. Then expanding the equivalent integral (Eq. 3.108) in a Taylor series for $w/\beta \ll 1$ and matching with Equation 3.109 gives the values of A and B listed in Appendix 3.A.

Taking variations of the averaged Lagrangian (Eq. 3.106) with respect to the parameters a, w, α, β, σ, and g results in the variational equations

$$\frac{\mathrm{d}}{\mathrm{d}z}\left(I_2a^2w^2 + \Lambda g^2\right) = 0, \tag{3.110}$$

$$\frac{\mathrm{d}}{\mathrm{d}z}\left(I_1aw^2\right) = \Lambda g\frac{\mathrm{d}\sigma}{\mathrm{d}z}, \tag{3.111}$$

$$I_1\frac{\mathrm{d}g}{\mathrm{d}z} = \frac{I_{22}a}{2w^2} - \frac{A^2B^4\alpha aw^2\beta^2}{\left(A^2\beta^2+B^2w^2\right)^2}, \tag{3.112}$$

$$I_2\frac{\mathrm{d}\sigma}{\mathrm{d}z} = -\frac{I_{22}}{w^2} + \frac{A^2B^2\alpha\beta^2\left(A^2\beta^2+2B^2w^2\right)}{\left(A^2\beta^2+B^2w^2\right)^2}, \tag{3.113}$$

plus the algebraic equations

$$\alpha = \frac{A^2B^2\beta^2w^2a^2}{2\left(A^2\beta^2+B^2w^2\right)\left(2\nu I_{42}+qI_4\beta^2\right)},$$

$$\alpha = \frac{A^2B^4w^4a^2}{qI_4\left(A^2\beta^2+B^2w^2\right)^2}. \tag{3.114}$$

These modulation equations are the equations governing the evolution of the nematicon. The variational equation (Eq. 3.110) is the equation for conservation of mass in the sense of the invariance of the Lagrangian (Eq. 3.104) with respect to shifts in the phase [8]. In the present optical context, it is the equation for the conservation of optical power. Eliminating α between the algebraic equations (Eq. 3.114), the director angle width β can be found as

$$\beta^2 = \frac{qI_4B^2w^2 + \sqrt{q^2I_4^2B^4w^4 + 16\nu qI_{42}I_4A^2B^2w^2}}{2qA^2I_4}. \tag{3.115}$$

By Nöther's Theorem [10], the nematicon equations (Eqs. 3.102 and 3.103) have an energy conservation equation that can be obtained from the Lagrangian (Eq. 3.104) based on the invariance of the Lagrangian to shifts in z. This energy conservation equation is

$$i\frac{\partial}{\partial z}\left[r|E_r|^2 - 4r\theta|E|^2 + \nu r\theta_r^2 + 2qr\theta^2\right] + \frac{\partial}{\partial r}\left[\frac{r}{2}\left(E_r^*E_{rr} - E_rE_{rr}^*\right)\right.$$
$$\left. - 2r\theta\left(E^*E_r - EE_r^*\right) - 2i\nu r\theta_r\theta_z\right] = 0. \tag{3.116}$$

Averaging this energy conservation equation gives

$$\frac{\mathrm{d}H}{\mathrm{d}z} = \frac{\mathrm{d}}{\mathrm{d}z}\left(I_{22}a^2 + 4\nu I_{42}\alpha^2 + 2qI_4\alpha^2\beta^2 - \frac{2A^2B^2\alpha a^2w^2\beta^2}{A^2\beta^2 + B^2w^2}\right) = 0. \tag{3.117}$$

As energy is conserved, this energy conservation equation can be used to determine the final steady nematicon state from the input beam.

The final parameter to be determined is the shelf radius ℓ. As in Section 3.4, this is determined by linearizing the modulation equations (Eqs. 3.110–3.114) about the fixed-point values a_f, w_f, σ_f, α_f, β_f, and $g = 0$ to obtain a simple harmonic oscillator equation. The frequency of this oscillator is then equated to the nematicon frequency σ_f'. In principle, the details are as for the NLS equation but are much more involved because of two trial functions being involved. The details are presented in Appendix 3.B.

3.7 RADIATION LOSS FOR THE NEMATICON EQUATIONS

As a nematicon propagates down a cell, it sheds diffractive radiation so as to evolve to a steady state. The contribution of this radiation to the modulation equations of Section 3.6 can be calculated as for the NLS equation in Section 3.5. The major difference is that nematicons are circularly symmetric, so that the shed radiation will be expressed in terms of Bessel functions. Again, the shed radiation has small amplitude relative to the nematicon beam and so is governed by the linearized electric field equation (Eq. 3.102)

$$i\frac{\partial E}{\partial z} + \frac{1}{2}\frac{\partial^2 E}{\partial r^2} + \frac{1}{2r}\frac{\partial E}{\partial r} = 0. \tag{3.118}$$

The solution of this linear equation must be matched with the electric field at the edge of the shelf, so that $E = S(z)$ at $r = \ell$. An identical argument as for the NLS equation in Section 3.5 yields

$$|S(z)|^2 = \frac{1}{\Lambda}\left[I_2 a^2 w^2 - I_2 a_f^2 w_f^2 + \Lambda g^2\right]. \tag{3.119}$$

Again Nöther's Theorem gives the mass conservation equation for the radiation equation (Eq. 3.118) as

$$i\frac{\partial}{\partial z}\left(r|E|^2\right) + \frac{1}{2}\frac{\partial}{\partial r}\left(rE^*E_r - rEE_r^*\right) = 0. \tag{3.120}$$

Integrating this equation from the edge of the shelf to ∞ gives the mass flux to shed diffractive radiation as

$$\frac{\mathrm{d}}{\mathrm{d}z}\int_{\ell}^{\infty} r|E|^2\,\mathrm{d}r = \mathrm{Im}\left(rE^*E_r\right)\big|_{r=\ell} + O(\dot{\ell}(z)). \tag{3.121}$$

To determine this mass flux, we again need to determine E and E_r at the edge of the shelf. To use the mass flux expression, we will assume that the radius of the shelf ℓ is slowly varying relative to the time z scale for propagation of the shed radiation. Then, ℓ can be taken to be constant in the radiation calculation.

The radiation equation (Eq. 3.118) can again be solved using Laplace transforms and the convolution theorem used to relate E_r to E at the edge of the shelf. The details will be omitted here as they are similar to those for the NLS equation in Section 3.5. Using the transforms listed in Reference 31, the final result is

$$E_r\big|_{r=\ell} = -\frac{1}{2\pi i}\int_C \sqrt{2s}\,\mathrm{e}^{-i\pi/4}\frac{K_1\left(\sqrt{2s}\,\mathrm{e}^{-i\pi/4}\ell\right)}{K_0\left(\sqrt{2s}\,\mathrm{e}^{-i\pi/4}\ell\right)}\hat{S}(s)\mathrm{e}^{sz}\,\mathrm{d}s. \tag{3.122}$$

Here, K_0 and K_1 are the modified Bessel functions of orders 0 and 1, respectively, and C is the standard inversion contour for Laplace transforms. The loss flux on the right-hand side of Equation 3.121 is then

$$E^*E_r = S^*(z)\int_0^z G(z-z')S(z')\,\mathrm{d}z', \tag{3.123}$$

where G is the Green's function

$$G(\eta) = -\frac{1}{2\pi i}\int_C \sqrt{2s}\,\mathrm{e}^{-i\pi/4}\frac{K_1\left(\sqrt{2s}\,\mathrm{e}^{-i\pi/4}\ell\right)}{K_0\left(\sqrt{2s}\,\mathrm{e}^{-i\pi/4}\ell\right)}\mathrm{e}^{s\eta}\,\mathrm{d}s. \tag{3.124}$$

For simplicity, let us set S in polar form as $S = R(z)\exp(i\varphi(z))$ and assume that the phase φ is slowly varying, as in Section 3.5. The mass flux can then be approximated as

$$E^* E_r = R(z) \int_0^z G(z - z')R(z')\, dz'. \tag{3.125}$$

The Green's function expression (Eq. 3.124), expressed in terms of an inverse Laplace transform, is too complicated as it stands to be of use in the modulation equations. The most useful approximation to make is to perform an asymptotic calculation of the flux for large z, which corresponds to s small in Laplace transforms [32]. To this end, the integrand in Equation 3.124, excluding $\exp(s\eta)$, can be rewritten as

$$-\frac{2s}{\ell}\frac{\mathrm{d}}{\mathrm{d}s} \log\left(K_0(\sqrt{2s}\, \mathrm{e}^{-i\pi/4}\ell)\right). \tag{3.126}$$

Now as $z \to 0$, $K_0(z) \sim -\ln(z/2)$ [31]. Using this asymptotic value, the Green's function (Eq. 3.124) becomes

$$G(\eta) = -\frac{1}{2\pi i \ell} \int_C \frac{2\mathrm{e}^{s\eta}}{\log s + \log \Lambda - i\pi/2}\, \mathrm{d}s. \tag{3.127}$$

This integral for the Green's function can now be evaluated by completing the contour around the branch point of the logarithmic function, adding the contributions from both sides of the branch cut and changing variables to -Re$(s) = \exp(\xi)$, to give

$$G(\eta) = \frac{1}{4\ell} \int_{-\infty}^{\infty} \frac{\mathrm{e}^{-\mathrm{e}^{\xi}\eta + \xi}\, \mathrm{d}\xi}{\left(\xi/2 + \frac{1}{2}\log \Lambda\right)^2 - i(\pi/2)\left(\xi/2 + \frac{1}{2}\log \Lambda\right) + 3\pi^2/16}. \tag{3.128}$$

This integral can now be evaluated for large η, which corresponds to large z in Equations 3.122 and 3.125, using the method of stationary phase. Adding this asymptotic value of the integral to the flux (Eq. 3.125) gives

$$E^* E_r = -\frac{\sqrt{2\pi} R}{4e\ell} \int_0^z \frac{R(z')}{z - z'} \frac{\mathrm{d}z'}{\left[(1/2)\log((z - z')/\Lambda) - i\pi/4\right]^2 + \pi^2/4}. \tag{3.129}$$

Even though this integral contains the term $1/(z - z')$, a careful examination of the integrand near $z' = z$ will show that the integral is convergent.

The mass flux in Equation 3.121 can now be evaluated by taking the imaginary part of Equation 3.129. As for the NLS modulation equations in Section 3.5, the mass flux can then be added to the mass modulation equation (Eq. 3.110). Again, for consistency, the modulation equation (Eq. 3.112) must be modified so that the mass equations (Eqs. 3.110, 3.112 for g, and 3.111) are consistent with the energy equation (Eq. 3.117). In this manner, it is found that the mass equation (Eq. 3.110) becomes

$$\frac{\mathrm{d}}{\mathrm{d}z}\left(I_2 a^2 w^2 + \Lambda g^2\right) = -\frac{2}{I_1} R^2 \Lambda \delta \tag{3.130}$$

and Equation 3.112 becomes

$$I_1 \frac{\mathrm{d}g}{\mathrm{d}z} = \frac{I_{22} a}{2w^2} - \frac{A^2 B^4 \alpha a w^2 \beta^2}{\left(A^2\beta^2 + B^2 w^2\right)^2} - 2\delta g, \tag{3.131}$$

where the loss coefficient δ is

$$\delta = -\frac{\sqrt{2\pi} I_1}{32 e R \Lambda} \int_0^z \frac{\pi R(z') \log((z-z')/\Lambda)}{\left\{[\frac{1}{4}\log((z-z')/\Lambda)]^2 + \frac{3\pi^2}{16}\right\}^2 + \frac{\pi^2}{16}\left[\log((z-z')/\Lambda)\right]^2} \times \frac{\mathrm{d}z'}{(z-z')} \tag{3.132}$$

and

$$R^2 = \frac{1}{\Lambda}\left[I_2 a^2 w^2 - I_2 a_f^2 w_f^2 + \Lambda g^2\right]. \tag{3.133}$$

There is, however, one subtle difference between the radiative loss from a nematicon and that from an NLS soliton, which can be seen by examining the solutions of the nematicon equations (Eqs. 3.100 and 3.101) shown in Figure 3.4. It can be seen that the shelf under the electric field beam extends well beyond the beam waist, in contrast to the NLS shelf shown in Figure 3.1, and forms a truncated cone. This extension of the shelf is due to the nonlocal nature of the nematic response in that the optical axis response away from electric field beam is forcing a new low wave number radiative mode. This extra shelf of radiation does not affect the frequency of the nematicon and the shelf length ℓ as it is away from the electric field beam and is due to the optical axis. However, it does have the effect of forcing the start of the propagating radiation away from what may be termed an inner shelf at $r = \ell$ to the end of this new outer shelf. This extended shelf of radiation was also found for coupled systems of NLS equations [11]. These two components of the shelf were shown to be due to two distinct types of eigenfunctions of the continuous spectrum for the equations describing the perturbation.

Let us assume that the outer shelf has radius ρ. It was found from numerical solutions that this radius can be estimated as $\rho = 7\beta_{1/2}$, where $\beta_{1/2} = \operatorname{sech}^{-1}(1/\sqrt{2})\beta$ is the half-width of the optical axis [30]. In the loss terms in Equations 3.130–3.133, the inner shelf area $\Lambda = \ell^2/2$ (modulo 2π) is replaced by the outer shelf area

$\tilde{\Lambda} = \rho^2/2$ (modulo 2π), so that the final modulation equations that incorporate mass loss to shed diffractive radiation are

$$\frac{\mathrm{d}}{\mathrm{d}z}\left(I_2 a^2 w^2 + \Lambda g^2\right) = -\frac{2}{I_1} R^2 \tilde{\Lambda}\delta \tag{3.134}$$

and

$$I_1 \frac{\mathrm{d}g}{\mathrm{d}z} = \frac{I_{22} a}{2w^2} - \frac{A^2 B^4 \alpha a w^2 \beta^2}{\left(A^2\beta^2 + B^2 w^2\right)^2} - 2\delta g, \tag{3.135}$$

where the loss coefficient δ is

$$\delta = -\frac{\sqrt{2\pi} I_1}{32 e R \tilde{\Lambda}} \int_0^z \frac{\pi R(z') \log((z-z')/\tilde{\Lambda})}{\left\{[\frac{1}{4}\log((z-z')/\tilde{\Lambda})]^2 + \frac{3\pi^2}{16}\right\}^2 + \frac{\pi^2}{16}\left[\log((z-z')/\tilde{\Lambda})\right]^2} \times \frac{\mathrm{d}z'}{(z-z')} \tag{3.136}$$

and

$$R^2 = \frac{1}{\tilde{\Lambda}}\left[I_2 a^2 w^2 - I_2 a_f^2 w_f^2 + \Lambda g^2\right]. \tag{3.137}$$

The final set of modulation equations governing the evolution of the nematicon, which include loss to shed dispersive radiation, are Equations 3.111, 3.113, 3.114, 3.134, and 3.135.

Figure 3.5 shows a comparison between the full numerical solution of the nematicon Equations 3.100 and 3.101 and the solution of the modulation equations. It can be seen that there is good agreement in the amplitude oscillations, with the mean of the modulation oscillation being the same as that of the numerical oscillation. This means that the modulation steady-state amplitude will be the same as that of the numerical solution. It can also be seen that the decay rate of the modulation solution is the same as that of the numerical solution, so that the radiation calculation of this section gives a good approximation to the flux of radiation from the evolving nematicon. There is a small difference in the periods of the amplitude oscillations. This is because the modulation period is determined by the shelf radius ℓ. The calculation of this radius carried out in the Appendix 3.B is based on linearizing the modulation equations about their fixed point. The evolution shown in Figure 3.5 is far from the steady state, so the difference in the two oscillation periods is not surprising. It can also be seen that there is a phase difference between the numerical and modulation amplitude oscillations. The phase is a higher order quantity in modulation theory and its calculation is, in general, nontrivial. The major difference between the numerical and modulation solutions is that the numerical solution shows a beating. The origin of this beating is that the numerical beam shape is undergoing significant changes during the evolution

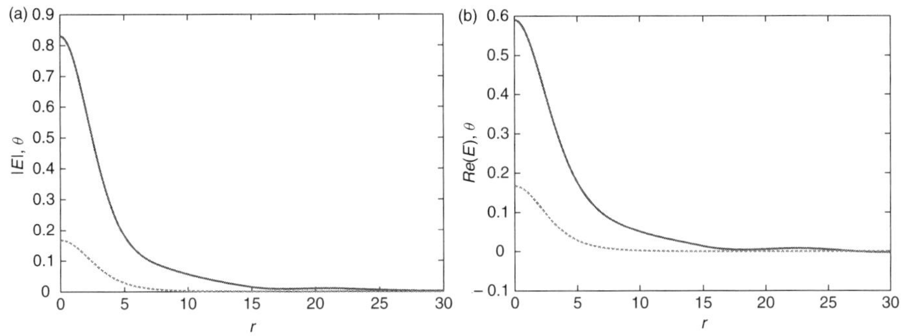

FIGURE 3.4 Numerical solution of nematicon equations (Eqs. 3.100 and 3.101) at $z = 400$ for the initial values $a = 0.5$ and $w = 4$. The parameter values are $q = 2$ and $\nu = 10$. (a) Solution for $|E|$: solid line; solution for θ: dashed line, (b) solution for $Re(E)$: solid line; solution for θ: dashed line.

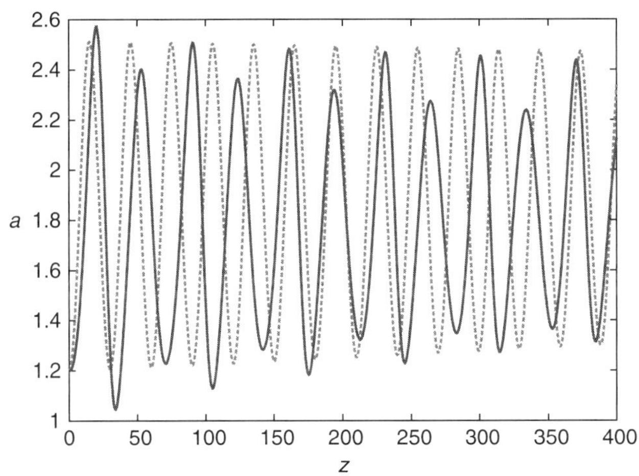

FIGURE 3.5 Comparison of amplitude a of nematicon as given by the full numerical solution of the nematicon equations (Eqs. 3.100 and 3.101) (solid line) and the modulation equations (dashed line). Initial conditions are $a = 1.2$, $w = 3.5$, with α and β determined by the first of Equations 3.114 and (3.115). The parameter values are $q = 2$ and $\nu = 100$.

of the beam, as discussed further in Section 3.8. An important observation from Figure 3.5 is that the finite dimensional modulation theory does a very good job of capturing the dynamics of an infinite dimensional system.

3.8 CHOICE OF TRIAL FUNCTION

As there is no exact solitary wave, or nematicon, solution of the nematicon equations (Eqs. 3.102 and 3.103), the choice of trial function to model the evolution of a nematicon is much less clear than it was for the sech trial function (Eq. 3.61) for the NLS equation. Although a good agreement has been obtained for the evolution of a nematicon in the nonlocal limit with a sech trial function (Eq. 3.105), the exact nematicon profile is not of this form [16]. Near its peak, it has a Gaussian shape, and in its tails, it has the profile of the modified Bessel function K_0 because of circular symmetry [16]. The choice of trial function then warrants further investigation.

If the initial condition for the nematicon equations (Eqs. 3.100 and 3.101 or Eqs. 3.102 and 3.103) is not an exact nematicon solution, the beam evolves to an exact nematicon with its amplitude and width oscillating. Figure 3.6 shows the solution of the nematicon equations (Eqs. 3.100 and 3.101) for a sech initial condition with the same parameter values as the amplitude comparison of Figure 3.5. The solution is shown at a maximum of the amplitude oscillation and its neighboring minimum. To the numerical profile at these values of z, sech and Gaussian profiles have been fitted. It can be seen that at the minimum the beam shape is well presented by a sech, whereas at the maximum, the beam is better approximated by a Gaussian. It is known that in the extreme nonlocal limit, in essence ν infinite, that the steady nematicon profile is Gaussian [33], so the development of the beam shape in the nonlocal limit to a Gaussian during part of its oscillation is not surprising. However, the evolution of the beam is much more complicated than this comparison with the Gaussian of the extreme nonlocal limit, as shown in Figure 3.7, where similar comparisons are shown for a Gaussian initial condition for the same value of the nonlocality ν. Larger values of the initial amplitude and width have been chosen because of the faster decay of a Gaussian over a sech, resulting in lower power for the same amplitude and width. The beam shape at the maximum is again well approximated by a Gaussian. However, the situation at the minimum is more complicated than for the sech initial condition. Neither the fitted Gaussian nor the fitted sech give good approximations to the beam shape. This is mirrored by the fact that the amplitude evolution of the Gaussian initial condition shows the same beating effect as for the sech initial condition of Figure 3.5, again due to the beam

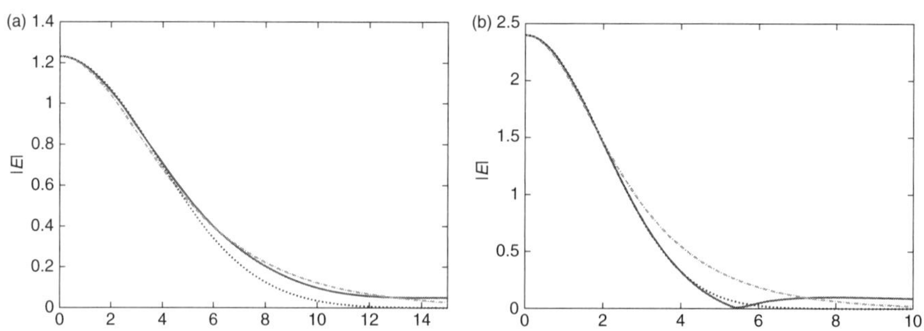

FIGURE 3.6 Numerical solution of nematicon equations (Eqs. 3.100 and 3.101) for $|E|$ for sech initial condition: solid line; sech fitted to numerical solution: dot-dashed line; and Gaussian fitted to numerical solution: dotted line. (a) At minimum of amplitude oscillation and (b) at maximum of amplitude oscillation. The initial conditions are $a = 1.2$, $w = 3.5$ with $q = 2$ and $\nu = 100$.

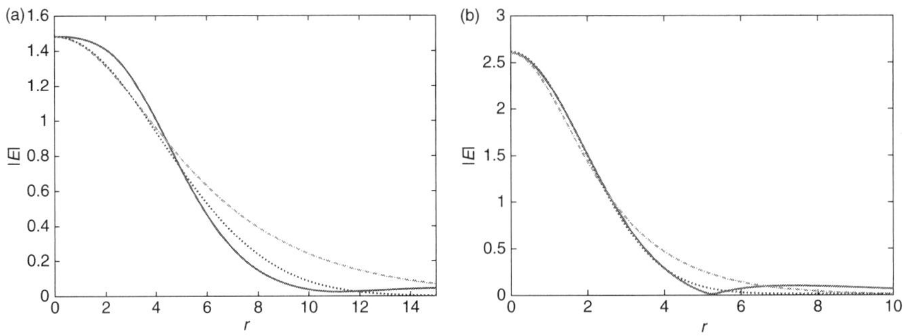

FIGURE 3.7 Numerical solution of nematicon equations (Eqs. 3.100 and 3.101) for $|E|$ for Gaussian initial condition: solid line; sech fitted to numerical solution: dot-dashed line; and Gaussian fitted to numerical solution: dotted line. (a) At minimum of amplitude oscillation and (b) at maximum of amplitude oscillation. The initial conditions are $a = 2.0$, $w = 5.0$ with $q = 2$ and $\nu = 100$.

changing shape as it evolves. The Gaussian is a better approximation at the tail of the profile, but both are equally good approximations around the peak. Indeed, the fitted sech is an equally good approximation as the Gaussian near the peak for the amplitude maximum. As the sole remnants of the trial functions when an averaged Lagrangian is calculated are various integrals of them, the degree of approximation by the trial function around the peak of the profile is more important than that at the tails. In all these comparisons, the numerical solution shows diffractive radiation in the tails, which should be ignored for the beam shape comparisons.

The degree of beam profile evolution as a nematicon evolves depends on the degree of nonlocality ν and the size of the amplitude oscillation. In general, as the amplitude of the oscillation increases, the beam profile changes more, resulting in greater levels of beating in the amplitude oscillation. This result is independent of the initial beam shape chosen. The other main determinant of beam shape is the nonlocality ν. This is illustrated further in Figure 3.8, for which the initial condition is a Gaussian and the nonlocality is $\nu = 10$, which is at the lower end of the experimental range $O(10)$–$O(100)$ [16, 24, 26]. It can be seen that the beam profile oscillates between a sech at an oscillation minimum and a Gaussian at a maximum. The behavior at the maximum is the same as in Figure 3.7, but at the minimum, the profile is a sech. As the nonlocality ν approaches ∞, the nematicon profile approaches a Gaussian [33]. This asymptotic behavior can be seen in the comparisons of Figures 3.6–3.8.

These beam shape comparisons indicate the complex nature of the evolution of a nematicon and show that simple trial functions cannot capture all this complexity. The use of a Gaussian for the beam shape is popular because of the shape of a nematicon in the extreme nonlocal limit [33], but for the values of ν used in experiments, which are in the range $O(10)$–$O(100)$ [16, 24, 26], the comparisons of this section show that a sech provides an equally good approximation. Indeed, the comparisons with modulation theory based on a sech profile shown in Figure 3.5 show that a sech gives results in good agreement with full numerical solutions.

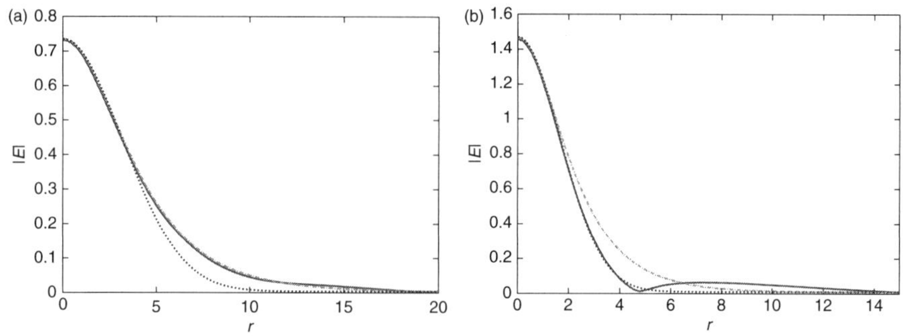

FIGURE 3.8 Numerical solution of nematicon equations (Eqs. 3.100 and 3.101) for $|E|$ for Gaussian initial condition: solid line; sech fitted to numerical solution: dot-dashed line; and Gaussian fitted to numerical solution: dotted line. (a) At minimum of amplitude oscillation and (b) at maximum of amplitude oscillation. $q = 2$ and $\nu = 10$.

Although it may be deduced from the comparisons of Figures 3.6–3.8 that a Gaussian is a better approximation than a sech for ν of $O(100)$, nematicon evolution is much more complicated than this. The best profile approximation depends on the details of the situation at hand. For example, for two-color nematicon evolution in the nonlocal limit, it was found that a Gaussian gave a good approximation to the evolution only for ν of $O(1000)$ and greater, whereas a sech gave a good approximation for ν of $O(100)$ [27]. The NLS soliton in one space dimension has a sech profile. The examples of this section of nematicon evolution in two space dimensions show that the shadow of this exact NLS soliton solution carries through to higher space dimensions and more intricate NLS-type equations. The examples of nematicon evolution discussed so far have the nematicon propagating on a straight-line trajectory. In Chapter 15, examples will be discussed, which have nonstraight nematicon trajectories. It is found that in many of these examples the nematicon trajectory is independent, or nearly independent, of its profile, as also found in the published literature [28, 29, 34]. As in potential technological applications of nematicons, it is their trajectories that are the important quantities and the quantities that need to be controlled [35–37], the present discussion shows that the use of trial functions based on simple profiles is a powerful technique to analyze nematicon evolution.

As previously mentioned, a popular approximation to use for the analysis of nematicon evolution is the so-called accessible soliton limit of Snyder and Mitchell [33]. This limit is essentially that of infinite nonlocality ν and in this limit the nematicon equations reduce to

$$i\frac{\partial E}{\partial z} + \frac{1}{2}\nabla^2 E - 2r^2 PE = 0, \quad P = \int_0^\infty r|E|^2 \, dr. \tag{3.138}$$

P is then the conserved power of the beam and is constant. In this limit, the nematicon equations thus reduce to the linear Schrödinger's equation, so that the techniques and solutions of quantum mechanics become available. However, caution must be applied in making deductions about the evolution of nematicons in experimental situations based on the results of the Snyder–Mitchell limit. The usual range of ν in experiments is $O(10)-O(100)$ [16, 24, 26], which, although large, is not in the "highly nonlocal" limit needed for the Snyder–Mitchell approximation. The finite values of nonlocality valid for experiments have one major consequence. The "potential" in the Schrödinger's equation (Eq. 3.138) is infinite, so that only bound states are possible. This means that any radiation shed by an evolving nematicon is trapped by the potential, so that it cannot evolve to a steady state. The beam solutions in the Snyder–Mitchell limit [33] then oscillate indefinitely, rather than shed radiation, and evolve in an oscillatory manner to a steady nematicon as for finite ν. Cell lengths in experiments are of $O(1mm)$ and so, typically, only two or three oscillations of the beam width are observed [15, 24]. The decay of the nematicon onto the steady state is $1/\sqrt{\nu z}$, which is very slow in the nonlocal limit. Given that it is impossible to experimentally generate an exact nematicon as an input to a cell, it is then impossible to observe an exact steady nematicon. So, although experiments appear to support the idea that the nematicon Equations (Eqs. 3.100 and 3.101) have an oscillatory "breather" solution of Snyder–Mitchell type, mathematically the nematicon equations do not possess such a solution. In general, NLS-type equations do not possess single solitary wave breather solutions [6]. Although for highly nonlocal media single solitary wave breathing solutions for NLS-type equations may appear to be steady solutions for experimentally relevant distances [38], they are not steady solutions in the mathematical sense in that they are not steady indefinitely. They will shed radiation, maybe of very low amplitude on a very long space scale, to evolve to a fixed amplitude solitary wave solution. The distance needed to see this evolution may, of course, be far longer than is experimentally feasible. Another effect of the infinite potential of the Schrödinger's equation (Eq. 3.138) is that it can suppress instabilities that arise for finite values of ν. An example of this is the "crescent vortices" of He et al. [39], which were found to be stable as solutions in the Snyder–Mitchell limit but are unstable for experimental values of ν.

3.9 CONCLUSIONS

There are many nonlinear wave equations and systems of such equations that do not possess exact solitary wave solutions. Of particular relevance for this book is the system of equations governing solitary wave, or nematicon, propagation in nematic liquid crystals. It may then be thought that the only manner in which to gain insight into the dynamics of such solitary waves is through numerical solutions of the governing equations. However, it has been shown that extending Whitham modulation theory [1] results in a powerful approximate technique that allows much insight into the mechanisms governing solitary wave and nematicon evolution, as well as giving solutions in excellent agreement with numerical solutions of the governing equations, as demonstrated for the standard NLS equation and for the equations governing nematicons in a nematic liquid crystal. Indeed, Whitham modulation theory itself is an analytically sophisticated, infinite dimensional extension of multiple scales perturbation theory. It can then include higher order effects, and so can the present extension of this theory, provided that a good trial function is chosen.

As the exact solitary wave profile is unknown, modulation theory relies on assuming an appropriate profile. Although it may be thought that the results of the modulation theory would depend critically on this assumed profile, it has been shown that this is not the case, as long as the assumed profile is reasonable in that it has a solitary wave form. Numerical solutions of the nematicon equations show that the nematicon profile changes form as the nematicon evolves. A popular profile for a nematicon is a Gaussian as in the infinite nonlocality limit it can be shown that the profile is of this form [33]. However, it has been shown that for realistic experimental values of the nonlocality ν an initial Gaussian profile does not stay of this form but oscillates between a Gaussian and a more sech profile.

An added benefit of the present form of modulation theory is that it easily allows for the incorporation of the effect of the diffractive radiation shed by an evolving solitary wave. All solitary waves shed radiation in order to evolve. Without the inclusion of this radiation, the solutions from any perturbation or approximate theory cannot evolve to a steady state as there is no damping term in the equations. For nematicons, as the nonlocality ν increases, the amplitude of the shed radiation decreases, so that the nematicon takes longer to reach a steady state. This distance to reach a steady state is usually longer than the physical length of the cell containing the nematic liquid crystal.

Acknowledgments

This research was supported by the Royal Society of London under Grant No. JP090179.

APPENDIX 3.A: INTEGRALS

The integrals I_i and $I_{i,j}$ in the modulation equations are

$$I_1 = \int_0^\infty \rho f(\rho)\,\mathrm{d}\rho,\ I_2 = \int_0^\infty \rho f^2(\rho)\,\mathrm{d}\rho,$$

$$I_{22} = \int_0^\infty \rho \left[\frac{\mathrm{d}f}{\mathrm{d}\rho}\right]^2 \mathrm{d}\rho,\ I_{x32} = \int_0^\infty \rho^3 f^2(\rho)\,\mathrm{d}\rho,$$

$$I_{42} = \frac{1}{4}\int_0^\infty \rho \left[\frac{\mathrm{d}}{\mathrm{d}\rho} f^2(\rho)\right]^2 \mathrm{d}\rho,\ I_4 = \int_0^\infty \rho f^4(\rho)\mathrm{d}\rho. \qquad (3.A.1)$$

For $f(\rho) = \operatorname{sech} \rho$

$$I_1 = 2C, \quad I_2 = \ln 2, \quad I_{22} = \frac{1}{3}\ln 2 + \frac{1}{6}, \quad I_{x32} = 1.352314016\ldots,$$
$$I_{42} = \frac{2}{15}\ln 2 + \frac{1}{60}, \quad I_4 = \frac{2}{3}\ln 2 - \frac{1}{6}. \tag{3.A.2}$$

Here, C is the Catalan constant $C = 0.915965594\ldots$ [31]. For $f(\rho) = \exp(-\rho^2)$

$$I_1 = \frac{1}{2}, \quad I_2 = \frac{1}{4}, \quad I_{22} = \frac{1}{2}, \quad I_{x32} = \frac{1}{8}, \quad I_{42} = \frac{1}{8}, \quad I_4 = \frac{1}{8}. \tag{3.A.3}$$

The constants A and B arising in the modulation equations are

$$A = \frac{I_2\sqrt{2}}{\sqrt{I_{x32}}} \quad \text{and} \quad B = \sqrt{2I_2}. \tag{3.A.4}$$

APPENDIX 3.B: SHELF RADIUS

As for the NLS equation, the radius ℓ of the shelf of diffractive radiation under an evolving nematicon is determined by linearizing the modulation Equations 3.110–3.114 about the fixed point with

$$a = a_f + a_1, \quad w = w_f + w_1, \quad \alpha = \alpha_f + \alpha_1, \quad \beta = \beta_f + \beta_1,$$
$$g = g_1, \quad \sigma = \sigma_f + \sigma_1, \tag{3.B.1}$$

where $|a_1| \ll a_f$, $|w_1| \ll w_f$, $|\alpha_1| \ll \alpha_f$, $|\beta_1| \ll \beta_f$, $|g_1| \ll 1$, and $|\sigma_1| \ll \sigma_f$. In a straightforward, but tedious, calculation, it is found on eliminating all the derivatives in terms of a_1' and then eliminating a_1' in terms of g_1 that

$$\frac{d^2 g_1}{dz^2} - \frac{Q\Lambda\sigma_f'}{I_1^2\left(w_f^2 + 2a_f w_f \Theta\right)} g_1 = 0, \tag{3.B.2}$$

where

$$\Theta = \frac{2I_{22}D^2 a_f - A^2B^2D\beta_f^2 w_f^2 a_f\left(2\alpha_f + \Gamma_1 a_f\right)}{A^2B^2\beta_f w_f a_f^2\left(2A^2\alpha_f\beta_f^3 + \Gamma_2 D\beta_f w_f + 2B^2\Gamma_3\alpha_f w_f^3\right)},$$

$$Q = Q_1 - \frac{I_{22}}{2A^2B^4\alpha_f\beta_f^2 w_f^4}\left[Q_2 + \Theta Q_3\right],$$

$$Q_1 = \frac{I_{22}}{2w_f^2}\left(1 - \frac{2a_f}{w_f}\Theta\right), \quad Q_2 = \alpha_f^2\beta_f^2 w_f^2 + \Gamma_1\beta_f^2 w_f^2 a_f, \tag{3.B.3}$$

$$Q_3 = 2\alpha_f \beta_f^2 w_f a_f + \Gamma_2 \beta_f^2 w_f^2 a_f + 2\Gamma_3 \alpha_f \beta_f w_f^2 a_f$$
$$- 4D^{-1} \alpha_f \beta_f^2 w_f^2 a_f \left(A^2 \Gamma_3 \beta_f + B^2 w_f\right),$$
$$\Gamma_1 = \frac{2A^2 B^4 w_f^4 a_f}{q I_4 D^2} \quad \Gamma_2 = \frac{4A^4 B^4 \beta_f w_f^3 a_f^2 \left(\beta_f - \Gamma_3 w_f\right)}{q I_4 D^3},$$
$$\Gamma_3 = \frac{B^2 w_f \left(q I_4 \beta_f^2 + 4\nu I_3\right)}{q I_4 \beta_f \left(2A^2 \beta_f^2 - B^2 w_f^2\right)}, \quad D = A^2 \beta_f^2 + B^2 w_f^2. \tag{3.B.4}$$

As for the NLS equation, the frequency of the simple harmonic motion equation (Eq. 3.B.2) is now matched to the steady nematicon oscillation frequency σ'_f, given by Equation 3.113 with a, w, β, and α replaced by their fixed-point values. This results in the radius ℓ of the shelf of low wave number diffractive radiation under the beam being given by

$$\Lambda = -\frac{\sigma'_f I_1^2 \left(w_f^2 + 2a_f w_f \Theta\right)}{Q}. \tag{3.B.5}$$

REFERENCES

1. G. B. Whitham. *Linear and Nonlinear Waves*. John Wiley & Sons, New York, 1974.
2. G. B. Whitham. Non-linear dispersive waves. *Proc. R. Soc. London, Ser. A*, 283: 238–261, 1965.
3. J. D. Cole and J. K. Kevorkian. *Perturbation Methods in Applied Mathematics*. Springer-Verlag, Berlin, 1981.
4. J. Hale. *Ordinary Differential Equations*. Wiley Interscience, New York, 1972.
5. Y. S. Kivshar and G. Agrawal. *Optical Solitons: From Fibers to Photonic Crystals*. Academic Press, San Diego, CA, 2003.
6. G. L. Lamb. *Elements of Soliton Theory*. John Wiley & Sons, New York, 1980.
7. A. C. Newell. *Solitons in Mathematics and Physics*. SIAM, Philadelphia, PA, 1985.
8. D. J. Kaup and A. C. Newell. Solitons as particles, oscillators, and in slowly changing media: a singular perturbation theory. *Proc. R. Soc. A*, 361: 413–446, 1978.
9. C. Knickerbocker and A. C. Newell. Shelves and the Korteweg-de Vries equation. *J. Fluid Mech.*, 98: 803–818, 1980.
10. I. M. Gelfand and S. V. Fomin. *Calculus of Variations*. Prentice-Hall, Englewood Cliffs, NJ, 1963.
11. J. Yang. Vector solitons and their internal oscillations in birefringent nonlinear optical fibers. *Stud. Appl. Math.*, 98: 61–97, 1997.
12. D. E. Pelinovsky and Y. Yang. Internal oscillations and radiation damping of vector solitons. *Stud. Appl. Math.*, 105: 245–276, 2000.
13. E. A. Coddington and N. Levinson. *Theory of Ordinary Differential Equations*. Krieger Pub. Co., Malabar, 1984.

14. R. L. Pego and M. I. Weinstein. Asymptotic stability of solitary waves. *Commun. Math. Phys.*, 164: 305–349, 1994.
15. G. Assanto, M. Peccianti, and C. Conti. Optical spatial solitons in nematic liquid crystals: Nematicons. *Opt. Photon. News*, 14: 45–48, 2003.
16. C. Conti, M. Peccianti, and G. Assanto. Route to nonlocality and observation of accessible solitons. *Phys. Rev. Lett.*, 91:073901, 2003.
17. C. Conti, M. Peccianti, and G. Assanto. Observation of optical spatial solitons in a highly nonlocal medium. *Phys. Rev. Lett.*, 92:113902, 2004.
18. D. Anderson. Variational approach to nonlinear pulse propagation in optical fibers. *Phys. Rev. A*, 27: 3135–3145, 1983.
19. W. L. Kath and N. F. Smyth. Soliton evolution and radiation loss for the nonlinear Schrödinger equation. *Phys. Rev. E*, 51: 1484–1492, 1995.
20. T. Jintao, L. Hongjun, G. Cunxiao, Z. Wei, W. Yishan, and C. Guofu. Spectrum shape compression and pedestal elimination employing a Sagnac loop. *Opt. Commun.*, 282: 74–77, 2009.
21. Q. Li, K. Senthilnathan, K. Nakkeeran, and P. K. A. Wai. Nearly chirp- and pedestal-free pulse compression in nonlinear fiber Bragg gratings. *J. Opt. Soc. Am. B*, 26: 432–443, 2009.
22. J. P. Gordon. Dispersive perturbations of solitons of the nonlinear Schrödinger equation. *J. Opt. Soc. Am. B*, 9: 91–97, 1992.
23. J. Satsuma and N. Yajima. Initial value problems of one-dimensional self-modulation of nonlinear waves in dispersive media. *Prog. Theor. Phys. Suppl.*, 55: 284–306, 1974.
24. M. Peccianti, C. Conti, G. Assanto, A. De Luca, and C. Umeton. Routing of anisotropic spatial solitons and modulational instability in liquid crystals. *Nature*, 432: 733–737, 2004.
25. M. Peccianti, A. Fratalocchi, and G. Assanto. Transverse dynamics of nematicons. *Opt. Express*, 12: 6524–6529, 2004.
26. G. Assanto, A. A. Minzoni, M. Peccianti, and N. F. Smyth. Optical solitary waves escaping a wide trapping potential in nematic liquid crystals: modulation theory. *Phys. Rev. A*, 79:033837, 2009.
27. B. D. Skuse and N. F. Smyth. Interaction of two-color solitary waves in a liquid crystal in the nonlocal regime. *Phys. Rev. A*, 79:063806, 2009.
28. G. Assanto, B. D. Skuse, and N. F. Smyth. Optical path control of spatial optical solitary waves in dye-doped nematic liquid crystals. *Photon. Lett. Poland*, 1: 154–156, 2009.
29. G. Assanto, B. D. Skuse, and N. F. Smyth. Solitary wave propagation and steering through light-induced refractive potentials. *Phys. Rev. A*, 81:063811, 2010.
30. A. A. Minzoni, N. F. Smyth, and A. L. Worthy. Modulation solutions for nematicon propagation in non-local liquid crystals. *J. Opt. Soc. Am. B*, 24: 1549–1556, 2007.
31. M. Abramowitz and I. A. Stegun. *Handbook of Mathematical Functions with Formulas, Graphs and Mathematical Tables*. Dover Publications, New York, 1972.
32. C. García-Reimbert, A. A. Minzoni, and N. F. Smyth. Spatial soliton evolution in nematic liquid crystals in the nonlinear local regime. *J. Opt. Soc. Am. B*, 23: 294–301, 2006.
33. A. W. Snyder and M. J. Mitchell. Accessible solitons. *Science*, 276: 1538–1541, 1997.
34. B. D. Skuse and N. F. Smyth. Two-colour vector soliton interactions in nematic liquid crystals in the local response regime. *Phys. Rev. A*, 77:013817, 2008.

35. A. Fratalocchi, R. Asquini, and G. Assanto. Integrated electro-optic switch in liquid crystals. *Opt. Express*, 13: 32–37, 2005.

36. M. Peccianti, A. Dyadyusha, M. Kaczmarek, and G. Assanto. Tunable refraction and reflection of self-contained light beams. *Nat. Phys.*, 2: 737–742, 2006.

37. A. Piccardi, A. Alberucci, U. Bortolozzo, S. Residori, and G. Assanto. Soliton gating and switching in a liquid crystal light valve. *Appl. Phys. Lett.*, 96:071104, 2010.

38. I. Kaminer, C. Rotschild, O. Manela, and M. Segev. Periodic solitons in nonlocal nonlinear media. *Opt. Lett.*, 32: 3209–3211, 2007.

39. Y. J. He, B. A. Malomed, D. Mihalache, and H. Z Wang. Crescent vortex solitons in strongly nonlocal nonlinear media. *Phys. Rev. A*, 78:023824, 2008.

4

Soliton Families in Strongly Nonlocal Media

WEI-PING ZHONG

Department of Electronic and Information Engineering, Shunde Polytechnic, Guangdong Province, Shunde, China

MILIVOJ R. BELIĆ

Science Program, Texas A&M University at Qatar, Doha, Qatar

4.1 INTRODUCTION

Spatial optical solitons are nondiffracting light beams propagating without change in nonlinear (NL)—and sometimes nonlocal—media. Spatial optical solitons in nematic liquid crystals (NLCs)—nematicons—can propagate unchanged for long distances because the strong nonlocal nonlinearity can saturate the change in the material refractive index; this makes them suitable for all optical applications. In this chapter, we theoretically address and numerically describe various spatial soliton families in such a medium.

Our analysis is based on the widely accepted model for the generation of spatial solitons in NLCs, which consists of scaled partial differential equations (PDEs) describing the scalar field of the director orientation angle θ and the paraxial wave equation for the envelope of the electric field. The equation for θ has the form

$$\tau \frac{\partial \theta}{\partial t} = 4K \nabla_{\perp}^{2} \theta + \varepsilon_0 \Delta\varepsilon \sin(2\theta)|A|^2, \tag{4.1}$$

where τ is the director relaxation time, K is the Frank constant, $\nabla_{\perp}^{2} = \partial_x^2 + \partial_y^2$ is the transverse Laplacian, $\Delta\varepsilon$ is the change in dielectric permittivity, and $0 \leq \theta \leq \pi/2$. From Maxwell's equations, in which the longitudinal field components are neglected, one gets the scalar paraxial propagation equation for the envelope A:

$$2ik \frac{\partial A}{\partial z} + \nabla_{\perp}^{2} A + k_0^2 \Delta\varepsilon \sin(\theta + \theta_0) \sin(\theta - \theta_0) A = 0, \tag{4.2}$$

Nematicons: Spatial Optical Solitons in Nematic Liquid Crystals, First Edition.
Edited by Gaetano Assanto.

where k is the wave number in the medium, k_0 is the vacuum wave number, and θ_0 is the orientation angle in the absence of optical field(s). This system of coupled PDEs is of primary interest here and will be analyzed to various degrees of approximation.

The physical model introduced by Equations 4.1 and 4.2 is the basic scalar model for studying beam propagation in NLCs. Depending on the underlying details, it can take more than one similar forms; being a system of coupled *evolution* PDEs, it is difficult to treat either analytically or numerically. Various simplifying procedures are introduced in the following sections. The basic approximation stems from realizing that the nonlinearity in NLCs is highly nonlocal. As it is known, from Equations 4.1 and 4.2, it follows that the power required to excite nematicons is at a minimum when $\theta_0 = \pi/4$.

We review mathematical models of spatial solitons on the basis of the nonlinear Schrödinger equation (NLSE) derived from Equations 4.1 and 4.2. In strongly nonlocal media, this equation can be simplified to a linear differential equation, leading to the so-called *accessible* solitons. We utilize mathematical tools mainly based on two techniques: the first applies self-similarity to various multidimensional linear models and produces analytical results; the second is a numerical procedure allowing one to find stable or quasi-stable localized solutions.

4.2 MATHEMATICAL MODELS

4.2.1 General

The recent interest in the study of self-trapped optical beams in nonlocal nonlinear (NN) media was fueled by experimental observation of nonlocal spatial solitons in NLCs [1–3] and in lead glasses [4], as well as by a number of interesting theoretical predictions [5–8]. Many of the predicted and demonstrated properties of NN models suggest that in such optical media one should expect stabilization of different types of NL wave structures such as 1D Hermite–Gaussian (HG) solitons [9], necklaces [10–14], and soliton clusters [15–18] in two-dimensional (2D) transverse space. The creation of 3D solitons, built out of matter or optical waves in the form of so-called *light bullets* presents a great challenge to experiments. A possibility is offered by Bose–Einstein condensates (BECs) with attractive interaction between atoms [19]. 3D solitons supported by optical lattices were reported in References 20–23, their form predicted by a variational approximation used as an initial guess to simulate several stable solutions. It was also demonstrated that localized wave packets in cubic materials with a symmetric NN response of arbitrary shape and degree of nonlocality can be described by the general nonlocal nonlinear Schrödinger equation (NNSE). The nonlocality of the nonlinearity prevents beam collapse in optical Kerr media in all physical dimensions, resulting in stable solitary waves under proper conditions [24]. Stable 3D spatiotemporal solitons in cubic NN media were reported in References 25 and 26. Light bullets via the synergy of reorientational and electronic nonlinearities in NLCs were proposed and discussed in References 27 and 28. In the following section we introduce a general NL nonlocal model, described by the general NNSE.

4.2.2 Nonlocality Through Response Function

We consider evolution of a scalar wave envelope field $u(\vec{r}, z)$ governed by the general NNSE in the scaled form, in which the nonlinearity $N(I)$ is assumed of the general NN form [7, 8, 15, 21, 23, 24]

$$\mathrm{i}\frac{\partial u}{\partial z} + \frac{1}{2}\nabla^2 u + V(\vec{r})u + N(I)u = 0, \tag{4.3}$$

$$N(I) = \int R(\vec{r} - \vec{r}')I(\vec{r}')\mathrm{d}^D\vec{r}', \tag{4.4}$$

introduced through the medium response function R. Here, $V(\vec{r})$ is an external potential, z is the longitudinal (propagation) coordinate, $\vec{r}$ and $\vec{r}'$ are the D-dimensional $(D = 1, 2)$ transverse coordinate vectors, $\mathrm{d}^D\vec{r}'$ is a D-dimensional volume element at $\vec{r}'$, and ∇^2 is the D-dimensional transverse Laplacian. This description is standard in NL optics. D might equal 3 when one searches for light bullets (in which case one "transverse" coordinate will be the reduced time) or in BECs, in which case z will be the regular time and the Laplacian will be the full 3D spatial operator.

The integral in Equation 4.4 has taken over the whole space and $I = |u|^2$. The kernel $R(\vec{r})$ is the response function; it is regular, real, and normalized, with $\int R(\vec{r})\mathrm{d}^D\vec{r} = 1$. The NLSE (Eq. 4.3) possesses a number of conserved quantities, among them the power $P = \int I(\vec{r})\mathrm{d}^D\vec{r}$, the angular momentum $L = \vec{r} \times \int(u^*\nabla u - u\nabla u^*)\mathrm{d}^D\vec{r}$, and the Hamiltonian $H = \frac{1}{2}\int(|\nabla u|^2 - 2VI - NI)\mathrm{d}^D\vec{r}$. The linear momentum $M = \int(u^*\nabla u - u\nabla u^*)\mathrm{d}^D\vec{r}$ is also conserved, provided there is no external potential $(V = 0)$.

According to the degree of nonlocality, as determined by the relative width of the response function R with respect to the size of the light beam, there are four categories of nonlocality [6, 7]: local, weakly nonlocal, generally nonlocal, and strongly nonlocal. In the limit that the response function is a delta function, $R(\vec{r}) = \delta(\vec{r})$, the NL response is local (Fig. 4.1a). The weak nonlocality applies when the characteristic length of $R(\vec{r})$ is much smaller than the width of the beam (Fig. 4.1b) and the strong nonlocality corresponds to the case of a characteristic length of the NL response much larger than the beam width (Fig. 4.1d). The case of general nonlocality is in between b and d, as shown in Figure 4.1c.

The general NNSE model naturally arises in NL optics, where it describes the propagation of an electric field envelope in a wave-guiding potential, in the paraxial approximation. The nonlocal index change, induced by the propagating beam, involves some transport-like process, for example, heat transfer in materials with thermal response [29], diffusion of molecules in atomic vapors [30], charge separation in photorefractive crystals [31], or intermolecular links in NLCs [32]. The general NNSE also appears in the description of BECs [33–36], in which

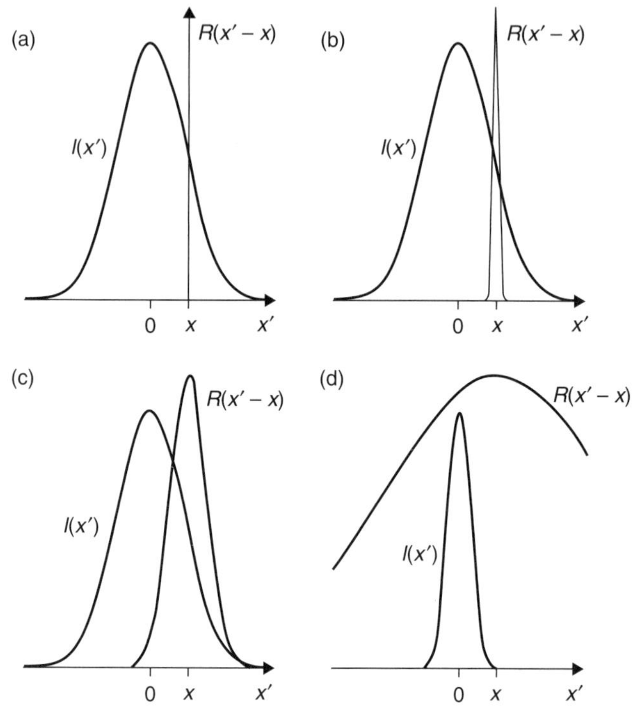

FIGURE 4.1 Different degrees of nonlocality, as given by the width of the response function $R(x)$ and the intensity profile $I(x)$. Shown are the (a) local, (b) weakly nonlocal, (c) general, and (d) strongly nonlocal responses.

case u stands for the collective wave-function and I is the density of atoms in the condensate; V then represents the magnetic trap potential, z is the time, and Equation 4.3 becomes the nonlocal Gross–Pitaevskii (GP) equation [37–39].

In the limit when the response function $R(\vec{r})$ is sharply peaked at point $\vec{r}$ and much narrower than the intensity distribution $I(\vec{r})$, the NL term becomes local, $N(I) \approx I$. In NL optics, the model (Eq. 4.3) then becomes the standard NLSE with an external potential, describing local Kerr media. In BECs, it becomes the standard GP equation. In the opposite limit, when the response function is much broader than the intensity distribution, the NL term becomes proportional to the response function, $N(I) \approx PR$, with P the beam power. Assuming the intensity distribution peaked at the origin, one can expand the response function about the origin to obtain $N(I) \approx P(R_0 + R_2 r^2)$. In this case, the highly nonlocal NLSE becomes the *linear* Schrödinger equation with a harmonic potential. A more general treatment in Reference 24, even without the external potential, leads to an NL optical model in which the change in the NL term is proportional to an NL function of the power, $\Delta N(I) \approx \alpha^2(P)r^2$. Although linear in u, the model still describes highly NL solitons through the NL dependence of the coefficient α on the beam power P [24]. For this reason, the model is referred to as the *strongly nonlocal NLSE*. It was used in References 1–3, for instance, to explain the experimental observation of optical spatial solitons in NLCs. In this chapter, we mostly deal with this limit of the general NNSE.

In the limit of a strongly nonlocal nonlinearity, the evolution of field u in three dimensions is described by the strongly NNSE [10–12, 21, 23, 24]:

$$\mathrm{i}\frac{\partial u}{\partial z} + \frac{1}{2}\nabla^2 u - sr^2 u = 0, \tag{4.5}$$

where $s(>0)$ is a parameter proportional to $\alpha^2(P)$, containing the effect of beam power. Note that P is constant and equal to the total input power P_0. Clearly, the same equation describes the time-dependent linear quantum harmonic oscillator (QHO); hence, in solving Equation 4.5, we also deal with a linear quantum-mechanical problem. Although several solutions to the *z-independent* QHO in various coordinate systems are known, we look for the self-similar z-dependent solutions of Equation 4.5 in the form of localized D-dimensional spatiotemporal solitons. Such solutions will naturally impose certain conditions on the input parameters and those describing the solutions. It should also be underlined that the beam collapse cannot occur in Equation 4.5, as it is a linear PDE. The second term in Equation 4.5 represents diffraction and the third one originates from the optical nonlinearity. We consider here only the case $s>0$.

4.3 SOLITON FAMILIES IN STRONGLY NONLOCAL NONLINEAR MEDIA

We discuss several spatial solitons of different dimensions in NN media, such as 1D HG solitons, 2D Laguerre–Gaussian soliton families, 2D self-similar HG solitons, and 2D Whittaker solitons (WS).

4.3.1 One-Dimensional Hermite–Gaussian Spatial Solitons

The interest in self-similar waves in complex NL optical systems has grown greatly in recent years [37–43]. Although self-similar solutions have been extensively studied in several fields, such as plasma physics and nuclear physics [44, 45], in NL optics this interest is relatively young, with only a few optical self-similar phenomena investigated to date [37, 38, 39]. Specifically, exact one-dimensional self-similar solitary waves were found in optical fibers in which dispersion, nonlinearity, and gain profile are allowed to change with the propagation distance, but with functionally related forms [40–42].

In the 1D case, Equation 4.5 reduces to

$$\mathrm{i}\frac{\partial u}{\partial z} + \frac{1}{2}\frac{\partial^2 u}{\partial x^2} - sx^2 u = 0. \tag{4.6}$$

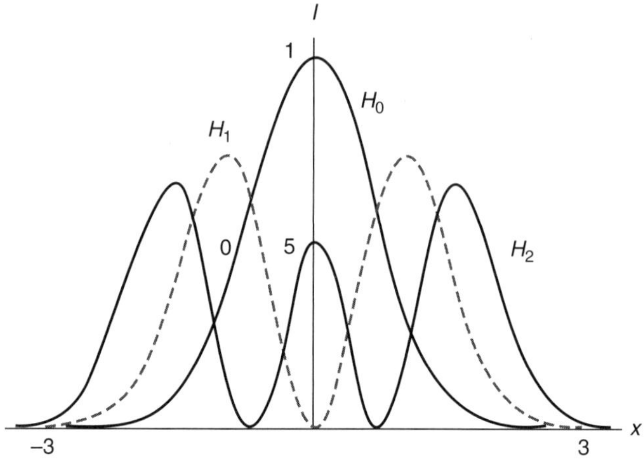

FIGURE 4.2 H_n solitons, for different n. Here, the parameter n has the values 0–2 from top to bottom.

It is easy to obtain the self-similar soliton solution of Equation 4.6:

$$u(z, x) = \sqrt{\frac{1}{2^n n! \sqrt{\pi} w_0}} \mathrm{e}^{-\frac{x^2}{2w_0^2} - \mathrm{i}\frac{(2n+1)z}{w_0^2}} H_n\left(\frac{x}{w_0}\right), \tag{4.7}$$

where $n = (0, 1, 2, \ldots)$ is a nonnegative integer, H_n are Hermite polynomials, and w_0 denotes the initial width of the beam. The power of the light beam $P = \int_{-\infty}^{+\infty} |u(z, x)|^2 \, \mathrm{d}x$ can be calculated using the orthogonality relations between H_n and is a constant of motion, which does not depend on the degree of the Hermite polynomials. We set it equal to 1. Because Equation 4.6 is linear, NL wave collapse cannot occur. A spatial soliton of this kind can exist only in strongly nonlocal media and is the accessible soliton, with the same form as the standard Hermite modes of 1D QHO.

We now illustrate the distributions of amplitude, intensity, and the positions of zeroes and extremal points, where $I = \frac{1}{2^n n! \sqrt{\pi}} H_n^2(x) \exp(-x^2)$. Figure 4.2 shows analytical solutions of several low order solitons along the x axis. Figure 4.3 displays the shapes of a few odd and even soliton distributions. When n is even (odd) the optical intensity is nonzero (zero) in the beam center.

4.3.2 Two-Dimensional Laguerre–Gaussian Soliton Families

As mentioned earlier, NLCs are useful dielectrics exhibiting huge optical nonlinearities [46–50], which makes them suitable for investigating NL phenomena with low power lasers and low cost detection equipment. In this section, we study the propagation of a beam in two-dimensional strongly NN media using the wave equation (Eq. 4.5), showing that there is a class of LG_{nm} solitary waves propagating in a self-similar manner. It is found that the predicted self-similar waves can be regarded as a family of spatial solitons.

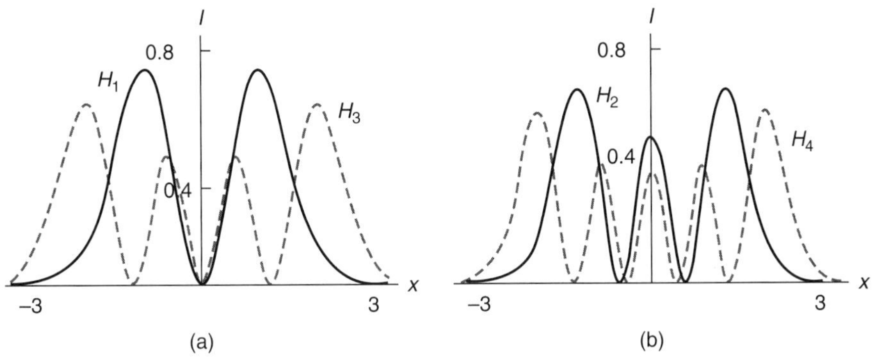

FIGURE 4.3 Optical field distributions, for different n: (a) odd solitons, for n odd; (b) even solitons, for n even.

Exact solutions of Equation 4.5 were obtained in the form of two-dimensional self-similar soliton waves in Reference 17:

$$u^s_{mn}(z,r,\phi)=\frac{kP_0}{w_0}[\cos(m\phi)+\mathrm{i}q\sin(m\phi)]\left(\frac{r}{w_0}\right)^m L_n^{(m)}\left(\frac{r^2}{w_0^2}\right)\mathrm{e}^{-\frac{r^2}{2w_0^2}+\mathrm{i}a(z)}, \tag{4.8}$$

where $a(z)=a_0-(2n+m+1)z/w_0^2$ and $L_n^{(m)}$ are the generalized Laguerre polynomials. Spatial solitons in Equation 4.8 are determined by two parameters, n and m. For a fixed n and various m (or a fixed m and various n), the LG_{nm} solitons share a few characteristics and form a family.

Figure 4.4a shows the radial distribution of low order solitons for $m=0$ and various n. There are n zeroes and $n+1$ extremal points along the radius. Figure 4.4b and c displays the distribution of the optical field and intensity of $\mathrm{LG}_{n0}(n=0,1,2,3)$ solitons. Their peak intensity is on the propagation axis. Physically, $m=0$ indicates that the NL polarization has the symmetry of the electric field due to the strong nonlocality; field and intensity distributions are clearly independent of the azimuthal angle.

Figure 4.5 illustrates some properties of LG_{n2} solitons for $m>0$ and $q=0$. The amplitude η_{n2} has $n+1$ zeroes and $n+1$ extremal points along the radius. Moreover, there are $2m$ zeroes and $2m$ extremal points along the azimuth direction. In contrast to the case $m=0$, the optical intensity is zero at the center. As seen from Figures 4.4a and 4.5a, LG_{nm} solitons decay fast in transverse space. This stems from nonlocality, as the NL polarization in a small volume of radius r_0 ($r_0 \ll$ any wavelength involved) depends not only on the electric field within this volume but also on its value outside and around it within the nonlocal range.

We now discuss multipole solitons ($q=0$) when $n=0,3$ for various m. For the LG_{00} soliton, the amplitude η_{00} has no zeros along the radial direction. Other solitons have one and more zeroes (Figure 4.4a). Figure 4.6 shows intensity distributions for various m when $n=0,3$. The distributions change regularly along the

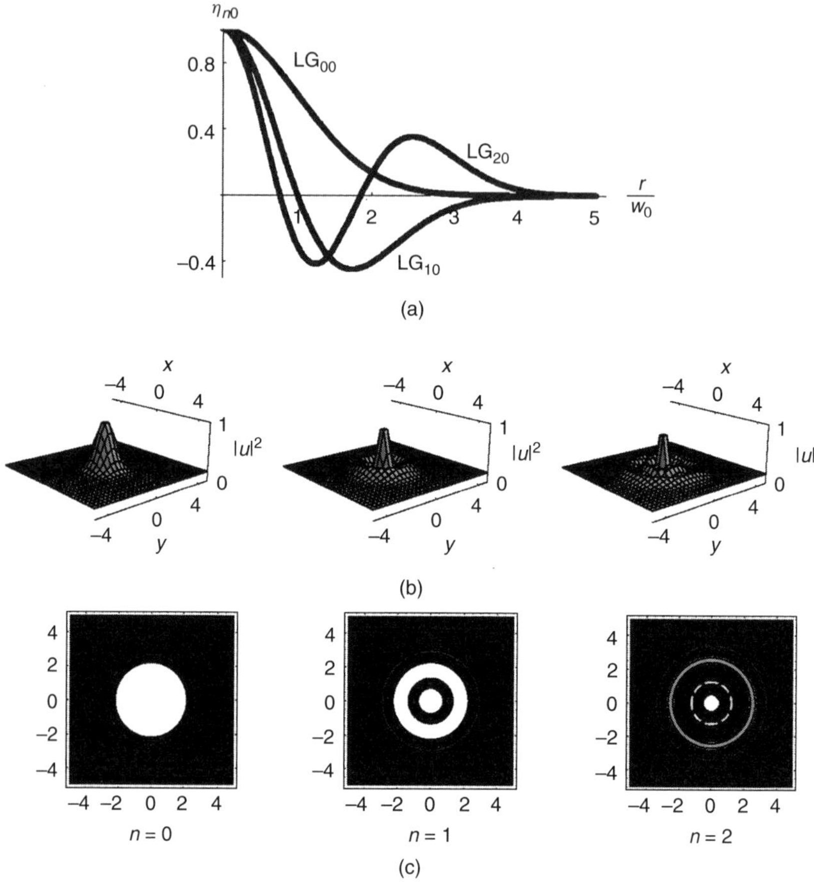

FIGURE 4.4 (a) Amplitudes of radial distributions of LG_{n0} solitons, corresponding to LG_{00}, LG_{10}, and LG_{20}; (b) optical field distribution of LG_{n0} solitons with $w_0 = 1$ and $P_0 = 1$; (c) intensity distributions viewed from above of LG_{n0} solitons with $w_0 = 1$ and $P_0 = 1$.

azimuth; when m is large enough, the beam forms a ring (necklace soliton) [51]. In strongly NN media, the refractive index is determined by the intensity distribution over the entire transverse plane and the nonlocality can lead to a refractive index increase in the region r_0 ($r_0 \ll$ any wavelength involved), supporting multipole solitons. Note that the nonlocal response function is much wider than the beam itself [52–55], thus the width of the index distribution greatly exceeds the size of an individual spot [56].

4.3.3 Accessible Solitons in the General Model of Beam Propagation in NLC

In this section, we present the self-similar method for the general (2 + 1)D model of light propagation in NLC, previously introduced in Equations 4.1 and 4.2 but conveniently simplified with reference to the NLC cell geometry reported in

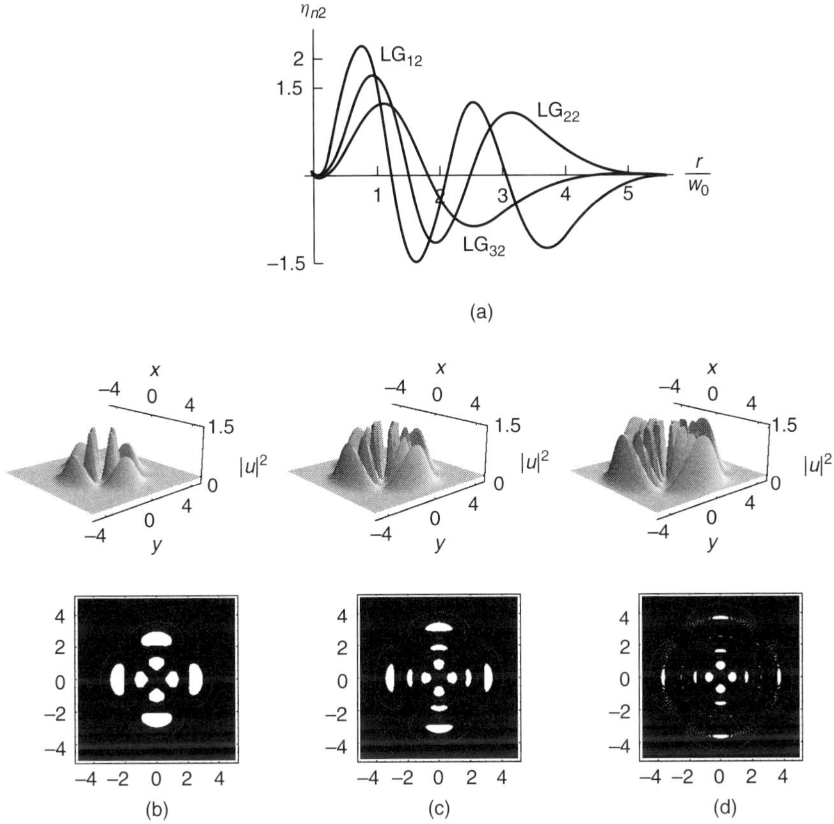

FIGURE 4.5 (a) Amplitude distributions of LG_{n2} solitons. (b–d) Optical field and intensity distributions viewed from above of LG_{12}, LG_{22}, and LG_{32} solitons with $w_0 = 1$ and $P_0 = 1$.

References 48 and 57. The field envelope is polarized along x and propagates along z. The evolution of the paraxial envelope E and the tilt angle θ are described by Peccianti et al. [47, 48, 57]:

$$2ik\frac{\partial E}{\partial z} + \nabla_{\perp}^2 E + k_0^2 \varepsilon_{\alpha} \sin(\theta + \theta_0) \sin(\theta - \theta_0) E = 0, \tag{4.9}$$

$$K\nabla_{\perp}^2 \theta + \frac{\varepsilon_{\alpha}\varepsilon_0}{4} \sin(2\theta)|E|^2 = 0, \tag{4.10}$$

where $\varepsilon_{\alpha}^2 = n_{\|}^2 - n_{\perp}^2$ is the dielectric anisotropy of the medium and $k \approx k_0 [n_{\perp}^2 + \varepsilon_{\alpha} \sin^2(\theta_0)]^{\frac{1}{2}}$. Steady state is assumed. The parameters k and k_0 are the wave numbers in NLC and in vacuum, respectively; ε_0 is the vacuum permittivity; θ_0 is the tilt of NLC molecules in the absence of light (as mentioned, the value $\theta_0 = \pi/4$ corresponds to the minimum Frédericksz threshold); and K is the

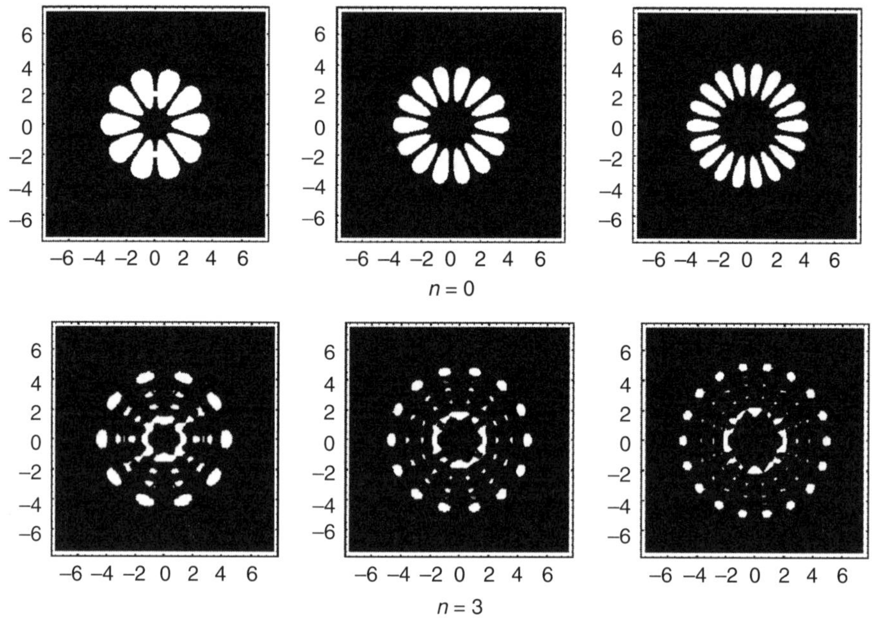

FIGURE 4.6 LG_{nm} solitons when n and m are varied. Upper row $n = 0$, lower row $n = 3$; $m = 1, 2, 3$ from left to right.

average Frank elastic constant. We work in cylindrical coordinates, φ is the azimuthal angle and $r (= \sqrt{x^2 + y^2})$ is the radial distance.

For a finite radial distance from the beam axis ($r = 0$), the dipole-induced perturbation is much wider than the beam itself [48], with $\beta \cong \beta_0 - \frac{\varepsilon_0 \varepsilon_\alpha^2}{16K} |E(r = 0, z)|^2 r^2$. Here, β ($\beta = |\theta - \theta_0| \ll 1$) is a small optically induced correction to the orientation angle of the director; hence, β_0 corresponds to the peak in the director angle distribution and θ_0 to the background value. After the normalization $V = E/E_0$, with $E_0 = E(r = 0, z = 0)$, in the highly nonlocal limit, we can reduce the two coupled Equations 4.9 and 4.10 to one:

$$\mathrm{i}\frac{\partial V}{\partial z} + \mu \nabla_\perp^2 V + \rho V - r^2 \chi |V(r = 0, z)|^2 V = 0, \tag{4.11}$$

where $\mu = \frac{1}{2k}$, $\rho = \mu k_0^2 \varepsilon_\alpha \beta_0$, and $\chi = \frac{k_0^2 \varepsilon_\alpha^2 \varepsilon_0 \mu}{16K}$. The novelty here is that the parabolic term has acquired z dependence, stemming from the on-axis beam intensity. We have found an *approximate analytical* self-similar solution of Equation 4.11:

$$V_{l\frac{m}{2}}(z, r, \varphi) = [\cos(m\varphi) + \mathrm{i}q \sin(m\varphi)] \frac{\alpha P_0}{r} \left(1 + \frac{w^2}{r^2}\right)^{\frac{m}{2}} W_{l\frac{m}{2}} \frac{r^2}{w^2} \mathrm{e}^{\mathrm{i}[a(z) + c(z) r^2]}, \tag{4.12}$$

where α is the normalization constant and P_0 is the beam power. $w(z)$, $a(z)$, and $c(z)$ are determined by Equations 4.20 and 4.21 and other parameters must satisfy the constraints of Equation 4.17. W satisfies

$$\frac{\mathrm{d}^2 W}{\mathrm{d}\tau^2}+\left[-\frac{1}{4}+\frac{l}{\tau}+\left(\frac{1}{4}-\left(\frac{m}{2}\right)^2\right)\frac{1}{\tau^2}\right]W=0. \tag{4.13}$$

Equation 4.13 is the well-known Whittaker differential equation, with solutions known as Whittaker functions (sometimes also called the *parabolic cylinder functions*) [58]:

$$W_{l\frac{m}{2}}(\tau)=\frac{\mathrm{e}^{-\frac{\tau}{2}}\tau^{l}}{\Gamma\left(\frac{1}{2}-l+\frac{m}{2}\right)}\int_0^{\infty}t^{-l-\frac{l}{2}+\frac{m}{2}}\left(1+\frac{t}{\tau}\right)^{-l-\frac{l}{2}+\frac{m}{2}}\mathrm{e}^{-t}\mathrm{d}t, \tag{4.14}$$

where the real part is $Re[l-l/2+m/2]\geq 0$, $l-l/2+m/2$ is not an integer, l is a real number, and Γ is the Gamma function. The parameters have to satisfy the following equations:

$$\frac{w^3}{16\mu^2}\frac{\mathrm{d}^2 w}{\mathrm{d}z^2}+\frac{w^2\alpha^2\chi p_0^2 f_0^2}{4\mu}=\frac{1}{4}, \tag{4.15}$$

$$\frac{w^2\rho}{4\mu}-\frac{w^2}{4\mu}\frac{\mathrm{d}a}{\mathrm{d}z}=l, \tag{4.16}$$

$$\cos^2(m\varphi)+q^2\sin^2(m\varphi)=1. \tag{4.17}$$

Taking $w(z)|_{z=0}=w_0$ and $\frac{\mathrm{d}w(z)}{\mathrm{d}z}|_{z=0}=0$, after integrating Equation 4.15 one gets

$$\frac{1}{2}\left(\frac{\mathrm{d}\eta}{\mathrm{d}z}\right)^2+\frac{2\mu^2(1-\eta^2+\lambda\eta^2\ln\eta^2)}{w_0^4\eta^2}=0, \tag{4.18}$$

where $\eta=w/w_0$ and $\lambda=\alpha^2\chi f_0^2P_0^2/w_0^2$. As $\eta\leq 1$, we have $\ln\eta^2\approx\eta^2-1$. Integrating Equation 4.18, we find

$$w^2=w_0^2\left[1+\frac{1-\lambda}{\lambda}\sin^2\left(\frac{2\mu\sqrt{\lambda}}{w_0^2}z\right)\right]. \tag{4.19}$$

Thus, the phase offset and the wave front curvature of the beam can be obtained:

$$a(z)=a_0+\rho z-2l\arctan\left[\frac{1}{\sqrt{\lambda}}\tan\left(\frac{2\mu\sqrt{\lambda}}{w_0^2}z\right)\right], \tag{4.20}$$

$$c(z)=\frac{(1-\lambda)\sqrt{\lambda}}{2w_0^2}\frac{\sin\left(\frac{4\mu\sqrt{\lambda}}{w_0^2}z\right)}{\lambda+1-(1-\lambda)\cos\left(\frac{4\mu\sqrt{\lambda}}{w_0^2}z\right)}, \tag{4.21}$$

where $a_0 = a(z)|_{z=0}$.

When $\lambda = 1$, the beam size is independent on the propagation distance and becomes a shape-preserving accessible soliton [24]. Its phase offset, width, and the wave front curvature are given by $a(z) = a_0 + \rho z - 4\mu z/w_0^2$, $w(z) = w_0$, and $c(z) = 0$, respectively. Thus, the approximate self-similar soliton solution of Equation 4.11 can be cast as

$$V^{\mathrm{sol}}_{l\frac{m}{2}}(z, r, \varphi) = [\cos(m\varphi) + \mathrm{i}q \sin(m\varphi)]\frac{\alpha P_0}{r}\left(1 + \frac{w_0^2}{r^2}\right)^{\frac{m}{2}}$$
$$\times W_{l\frac{m}{2}}\left(\frac{r^2}{w_0^2}\right) \mathrm{e}^{\mathrm{i}\left(a_0+\rho z-\frac{4\mu z}{w_0^2}\right)}. \tag{4.22}$$

The parameters of this solution must satisfy the important constraint (Eq. 4.17). It is interesting to note that a spatial soliton of any constant width can propagate in NLC as long as $\alpha^2\chi f_0^2 P_0^2 = w_0^2$ (namely, $\lambda = 1$). However, this condition is not usually experimentally met and solitons typically breathe. It should also be noted that χ is a parameter determined by the specific NLC and the initial beam width. We find that the 2D spatial solitons of Equation 4.22 are defined by the parameters (l, m) and q. In what follows, we describe various structures by studying the field distributions, the intensities of radial distributions, the positions of zeroes, and the extremal points of solutions. We choose the initial conditions $w_0 = 1$, $P_0 = 1$.

When m is an integer, introducing the relation $L(\tau) = \tau^{-\frac{m+1}{2}} \mathrm{e}^{\frac{\tau}{2}} W_{l\frac{m}{2}}$, from Equation 4.13, we obtain the generalized Laguerre differential equation:

$$\tau\frac{\mathrm{d}^2 L}{\mathrm{d}\tau^2} + (m + 1 - \tau)\frac{\mathrm{d}L}{\mathrm{d}\tau} + lL = 0. \tag{4.23}$$

Its solutions are the generalized Laguerre polynomials $L_l^m(\tau)$. Therefore, the solution of Equation 4.11 may also be expressed as

$$V_{nm}(z, r, \varphi) = \alpha P_0\big[\cos(m\varphi) + \mathrm{i}q \sin(m\varphi)\big](1 + r)^m L_l^m(r^2)\mathrm{e}^{-\frac{r^2}{2}+\mathrm{i}\left(a_0+\rho z-\frac{4\mu z}{w_0^2}\right)}, \tag{4.24}$$

where $\alpha = \left(\frac{l!}{\Gamma(l+m+1)}\right)^{\frac{1}{2}}$.

4.3.3.1 Gaussian Solitons ($m = 0$) When $m = 0$, the relation (Eq. 4.17) is naturally satisfied, and Equation 4.24 can be simplified to $V_{l0}(z, r) = \alpha P_0 L_l^0(r^2)\exp[-r^2/2 + \mathrm{i}(a_0 + \rho z - 4\mu z/w_0^2)]$. Figure 4.7 shows analytical solutions of Gaussian solitons for various l. There are l zero points (dark rings) and $l + 1$ extreme points (bright rings) along the radial direction for L_{l0} solitons. The peak optical intensity is at the origin.

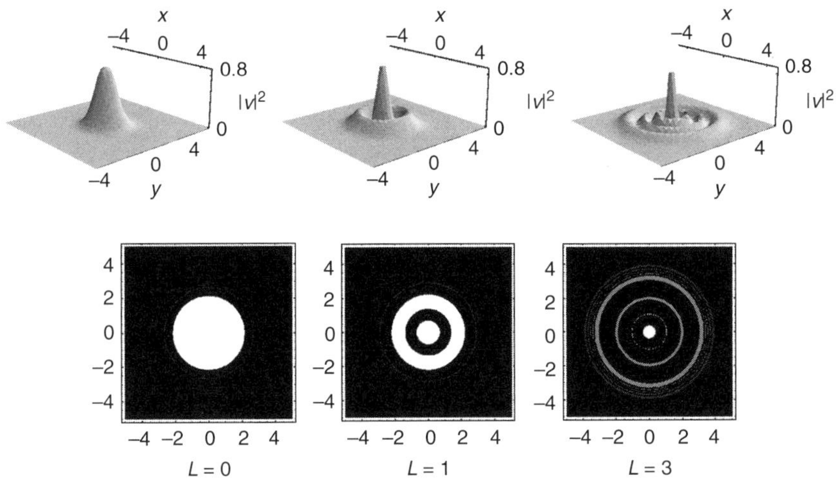

FIGURE 4.7 Gaussian solitons, for different l, when $m = 0$. Top row is the optical field distributions and the bottom row is the view from above. The parameter l is $l = 0, 1, 3$ from left to right.

4.3.3.2 Radially Symmetric Solitons (q = 1, m Positive Integer) In the limit $q = 1$ for $m(>0)$, a nonnegative integer, Equation 4.17 is automatically satisfied. Equation 4.24 generates radially symmetric solitons, which can be expressed as $V_{lm}(z, r, \varphi) = \alpha P_0(1 + r)^m L_l^m(r^2) \exp[-r^2/2 + \mathrm{i}(m\varphi + a_0 + \rho z - 4\mu z/w_0^2)]$. Figure 4.8 illustrates some features of these solitons. For equal l, the larger the m, the faster the field decay. The intensity is zero at $(x, y) = (0, 0)$.

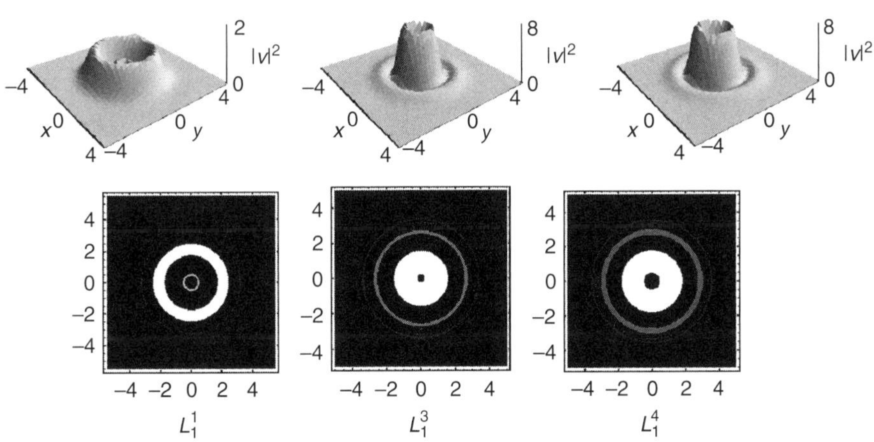

FIGURE 4.8 Radially symmetric solitons for $l = 1$ and $q = 1$. Top row is the optical field distribution and the bottom row is the view from above.

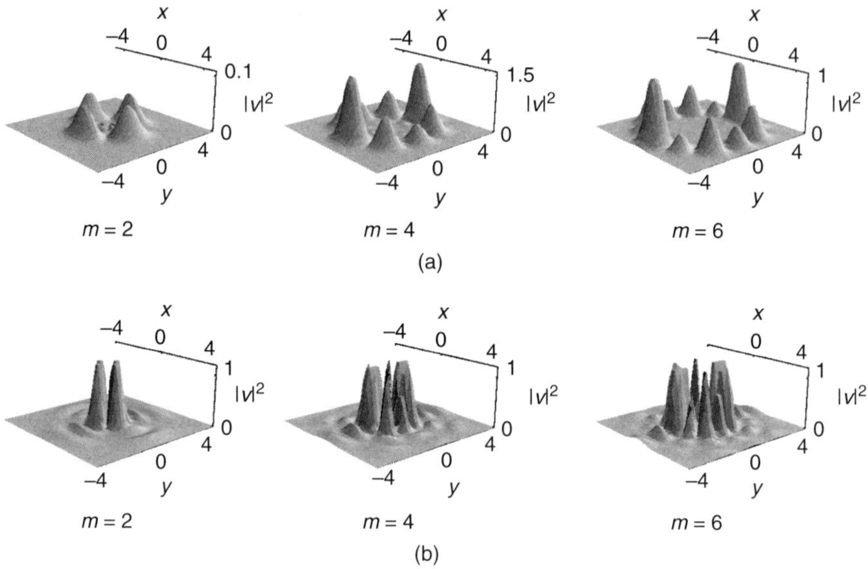

FIGURE 4.9 Multipole solitons. (a) $l = 0$, for different m; (b) $l = 2$, for different m.

4.3.3.3 Multipole Solitons (l Nonnegative Integer, $m \neq 0$, $q = 0$) In the limit $q = 0$, we must take $m\varphi = s\pi(s = 0, \pm 1, \pm 2)$, in order to satisfy Equation 4.17. This case is similar to the previous one. In particular, when $|V(0, z)|^2 = const$. Equation 4.11 reduces to the Snyder–Mitchell model of accessible solitons [24]. From Equation 4.24, we obtain $V_{lm}(z, r, \varphi) = \alpha P_0 \cos(m\varphi)(1 + r)^m L_l^m(r^2) \exp\left[-r^2/2 + \mathrm{i}\left(m\varphi + a_0 + \rho z - 4\mu z/w_0^2\right)\right]$, which forms multipole solitons. These multipole solutions contain single-layer necklace solitons ($l = 0$ but $m \neq 0$) and multilayer necklace solitons (l is a positive integer but $m \neq 0$). Figure 4.9 shows examples of multipole solitons.

Interesting structures can be seen for different multipole solitons. The larger the parameter m, the larger the necklace radius. The distributions change regularly with the azimuth. For a large enough m, the bright spots form a ring. These multipole solitons contain $2m$ spots and $l + 1$ ring layers.

4.3.3.4 Fractional Solitons (m Noninteger) and Soliton Vortices When $q = 1$ and m is a noninteger, interesting structures are formed. Figure 4.10 shows field distributions for $m = 1/2$ and various l. As apparent from Equations 4.20 and 4.21, at the point of phase singularity the phase is undefined and the intensity vanishes, thus yielding a vortex. The soliton wave rotates around the vortex core in a given direction (the sign of m defines the rotation direction), with infinite angular velocity at $(x, y) = (0, 0)$. An interesting feature is the clockwise rotation of the vortex core with the simultaneous relaxation to the radially symmetric profile.

Analytical solutions of Equation 4.11 were obtained exactly and novel solitons found in NLC, including Gaussian solitons, radially symmetric solitons, multipole

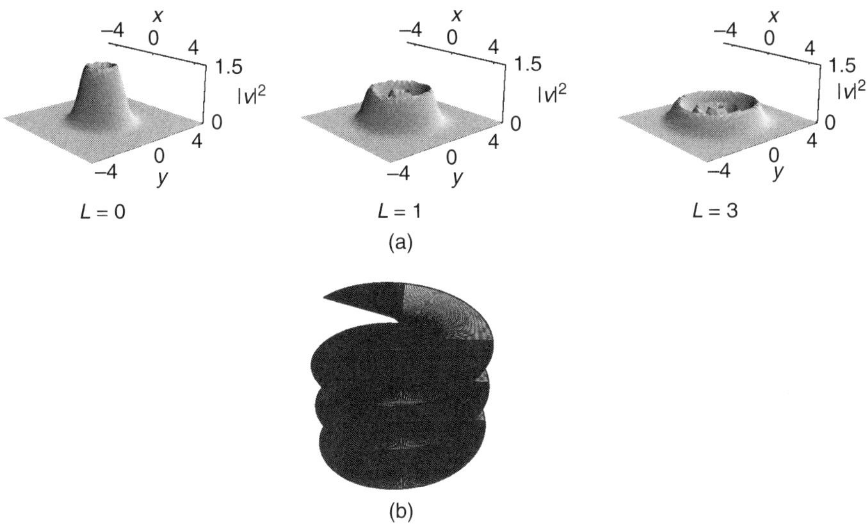

FIGURE 4.10 (a) Field intensity of fractional solitons, showing single, double, and triple summits for $m = 0.5$ and different l; (b) Optical beam helical wave front.

necklace solitons, and vortex solitons. These solitons are solutions to Equation 4.11, the strongly nonlocal approximation to the original model of Equations 4.9 and 4.10. It is plausible, although not yet proven, that these solitons appear in the full model and are stable or quasi-stable for prolonged propagation distances.

4.3.4 Two-Dimensional Self-Similar Hermite–Gaussian Spatial Solitons

In this section, starting from the Snyder–Mitchell model, we construct higher order spatial solitons propagating in a self-similar manner in NN media. We look for these solutions in Cartesian coordinates.

In the limit of a strong nonlocal nonlinearity and in Cartesian coordinates, the wave equation in 2D NN media is the general NNSE [10–12, 24]:

$$\mathrm{i}\frac{\partial u}{\partial z} + \frac{1}{2}\left(\frac{\partial^2 u}{\partial x^2} + \frac{\partial^2 u}{\partial y^2}\right) - s(x^2 + y^2)u = 0. \tag{4.25}$$

This equation is equivalent to the 2D QHO, with several known solutions in the time (or z)-independent case. They range from 1D to 4D (for hydrogen atom) and even higher (in quantum field theories). We have obtained the z-dependent solution to Equation 4.25, using the self-similar method [59]. We provide here the soliton solution:

$$u_{nm}^{\mathrm{sol}}(z, x, y) = \frac{\kappa}{w_0} H_n\left(\frac{x}{w_0}\right) H_m\left(\frac{y}{w_0}\right) \mathrm{e}^{\mathrm{i}[a_0 - \frac{2(n+m+1)z}{w_0^2}]}\, \mathrm{e}^{-\frac{x^2+y^2}{2w_0^2}}, \tag{4.26}$$

where $\kappa = \left(\frac{1}{2^n n! \sqrt{\pi}} \frac{1}{2^m m! \sqrt{\pi}}\right)^{\frac{1}{2}}$. The solution is expressed in terms of 2D HG functions, as one would expect from a QHO problem. The z dependence is explicitly present in the beam width and the wave front curvature of the general solution [59]. Here, for a soliton solution, the phase offset is given by $a = a_0 - 2(n + m + 1)z/w_0^2$, the curvature is zero, and $w = w_0$.

Figure 4.11a illustrates a few properties of solutions for arbitrary n and m. The $(n + 1) \times (m + 1)$ extremal points are arranged in a rectangular matrix. The farther the spot is from the center of the transverse axes, the greater is the intensity. The intensity is zero at the beam center when n (or m) is even; conversely, when n is odd the intensity is the smallest extremum at the center. When $n = m(\neq 0)$, the soliton forms a square matrix of spots. Figure 4.12 displays a few soliton square distributions.

In summary, self-similar waves of HG spatial solitons in strongly NN media were studied analytically. Exact solutions were obtained in the form of HG_{mn} functions with interesting properties, including shape-preserving as well as breathing beams.

4.3.5 Two-Dimensional Whittaker Solitons

In this section, starting from a set of linear Whittaker modes, we construct higher order spatial solitons in NN media in the form of WS. We display different possible WS families: Gaussian solitons, vortex-ring solitons, half-moon solitons, and symmetric and asymmetric single- and multilayer necklace solitons. We find that some classes of WS display well-defined symmetry and give rise to stable solitons, whereas others exhibit an unstable behavior typical of multidimensional soliton clusters [54], although their stability may be improved. Although our conclusions on enhanced stability are based on numerical studies, definitive answers must await a more thorough analysis.

We consider the propagation of beams in an NN medium in the paraxial approximation, described by the generalized NNSE; see Equations 4.3 and 4.4 for the scalar electric field envelope $u(\vec{r}, z)$:

$$\mathrm{i}\frac{\partial u}{\partial z} + \frac{1}{2}\nabla_\perp^2 u + N(I)(\vec{r}, z)u = 0, \tag{4.27}$$

$$N(I)(\vec{r}, z) = \int_{-\infty}^{+\infty} \int_{-\infty}^{+\infty} R(\vec{r} - \vec{r}')|u(\vec{r}', z)|^2 d\vec{r}'. \tag{4.28}$$

As mentioned earlier, in the limit that the response function is a delta function $R(\vec{r} - \vec{r}') = \delta(\vec{r} - \vec{r}')$, the nonlinearity becomes proportional to the intensity, $N(I)(\vec{r}, z) = |u(\vec{r}, z)|^2$ and we recover the local limit of NNSE, that is, the NLSE. In the opposite limit, when the response function is much broader than the intensity distribution, the nonlinearity becomes proportional to the response function, $N(I) = PR(\vec{r})$ with P the beam power. In this *strongly nonlocal* limit, the NNSE becomes the linear SE with a given potential. In this case, the NNSE becomes

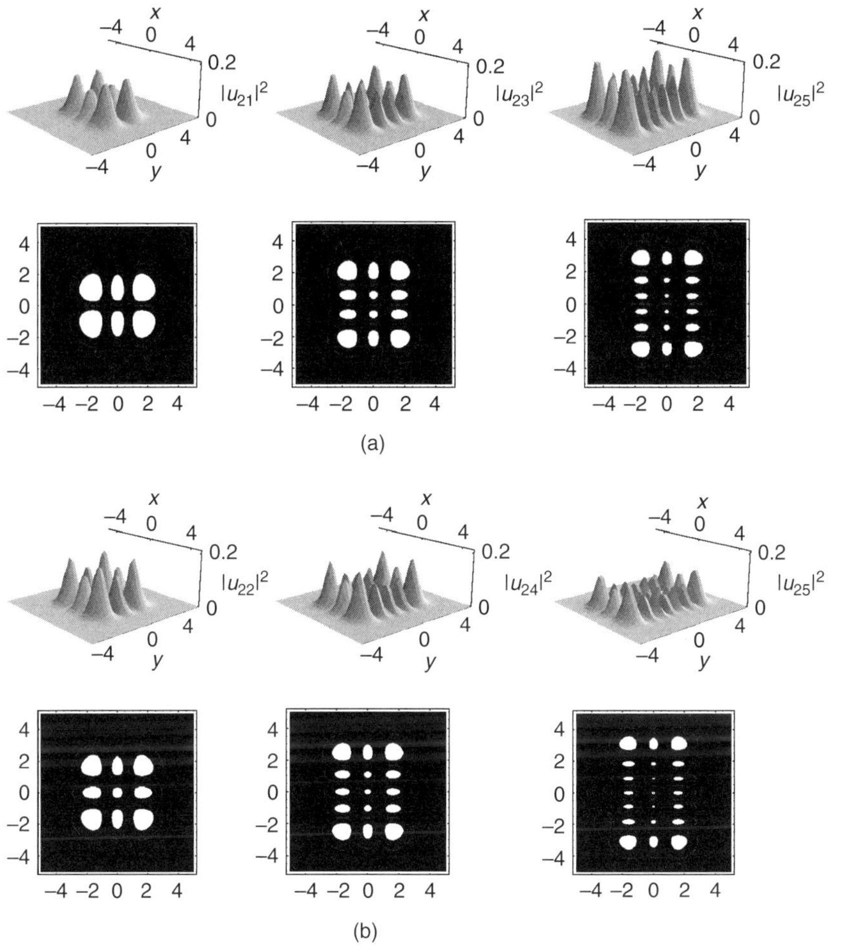

FIGURE 4.11 Optical field distributions of HG solitons, for different parameters. (a) $n = 1$, $m = 1, 3, 5$. (b) $n = 1$, $m = 2, 4, 6$.

the SE for the QHO, supporting accessible solitons [24]. We have shown that the strongly nonlocal NLSE also possesses exact self-similar 2D soliton solutions [21]:

$$u_{nm}(z, r, \varphi) = \frac{\kappa P}{r\sqrt{w_0}}[\cos(m\varphi) + \mathrm{i}q\sin(m\varphi)]W_{nm}(\theta)\mathrm{e}^{\mathrm{i}a(z)} \tag{4.29}$$

where $\kappa = [\frac{n!}{\Gamma(n+m+1)}]^{\frac{1}{2}}$, $a(z) = a_0 - 2nz/w_0^2$, $a_0 = a(z)|_{z=0}$, and $\theta = r^2/w_0^2$. The parameter m is a real constant called the *vorticity* or *topological charge* [12, 60], n is a nonnegative integer, and w_0 is the initial beam width. The parameter $q(0 \leq q \leq 1)$ determines the modulation depth of the beam intensity. $W_{nm}(\theta)$ are the

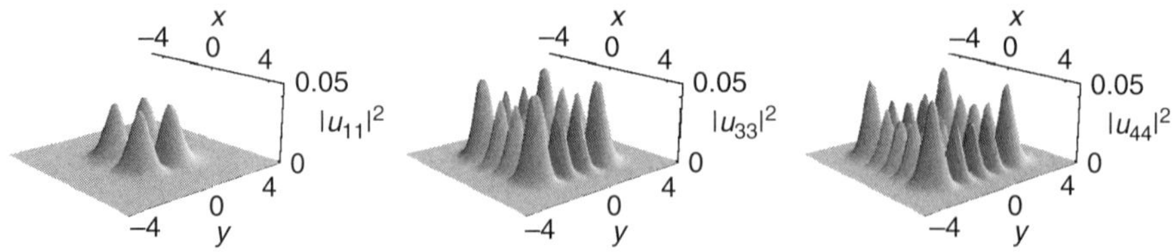

FIGURE 4.12 Typical square matrix HG solitons. The parameters are $n = m = 2$; $n = m = 3$; and $n = m = 4$ from left to right, respectively.

Whittaker functions, defined as [61]

$$W_{nm}(\theta) = \frac{e^{-\frac{\theta}{2}}\theta^n}{\Gamma(\frac{1}{2} - n + 2m)} \int_0^{\infty} \tau^{-n-\frac{1}{2}+2m} \left(1 + \frac{\tau}{\theta}\right)^{n-\frac{1}{2}+2m} e^{-\tau} d\tau, \tag{4.30}$$

with Γ the Gamma function. In the argument of the Gamma function, we assume $\Re[n - 1/2 - 2m] \leq 0$. Note the differences between this solution, presented in terms of Whittaker functions, and the one presented in the previous section, where Whittaker differential equations and functions also appeared. Being solutions to the linear SE, the self-similar WS functions $u_{nm}(z, r, \varphi)$ are stable, without beam collapse. In analogy with these solutions, we look for WS solutions of Equation 4.27 with enhanced stability, in the form $u(z, r, \varphi) = r^m \Phi(\varphi) V(z, r)$, where $\Phi = \cos(m\varphi) + iq \sin(m\varphi)$. Such a choice in the azimuthal function dependence allows for more freedom in the characterization of possible solitons. In this case, Equation 4.27 acquires the form

$$\begin{aligned} & i\frac{\partial V}{\partial z} + \frac{1}{2}\left(\frac{\partial^2 V}{\partial r^2} + \frac{2m+1}{r}\frac{\partial V}{\partial r}\right) + V|\Phi|^2 \int_{-\infty}^{+\infty}\int_{-\infty}^{+\infty} \\ & R(\vec{r} - \vec{r}')|V(\vec{r}', z)|^2 \vec{r}'^{2m} d\vec{r}' = 0. \end{aligned} \tag{4.31}$$

At this point, one has to specify the response function; different possibilities exist and depend on the physical situation [62]. As long as the response function is real, symmetric, positive, and monotonically decaying, the properties of the solutions do not depend much on the shape of the response function [60]. We therefore choose the Gaussian $R(\vec{r} - \vec{r}') = \frac{1}{\pi\sigma^2} \exp[(\vec{r} - \vec{r}')^2/\sigma^2]$ for simplicity [60, 62], and also because it offers stable solutions [63, 64]. The Gaussian width σ controls the nonlocality: when $\sigma \to 0$, we retrieve the local Kerr model. Here, we take $\sigma = 1$ and $P = 1$. This does not imply that we are in the strongly nonlocal regime, because this also depends on the width and the height of the expected solution V; however, by a judicious choice of the parameters and an appropriate numerical procedure one can find solutions that propagate quasi-stably over prolonged distances. For larger values of σ and P, that is, by moving deeper into the nonlocal regime, the stability of the numerical solutions improves.

To find stationary soliton solutions of Equation 4.31, we resort to a variational procedure described in detail in Reference 65. By choosing initial conditions consistent with the linear WS solutions

$$u_{nm}(0, r, \varphi) = \frac{kP}{r\sqrt{w_0}}[\cos(m\varphi) + iq\sin(m\varphi)]W_{nm}\left(\frac{r^2}{w_0^2}\right)e^{ia_0} \tag{4.32}$$

and also for $w_0 = 1$ and $a_1 = 0$, we obtain the propagating WS solutions of Equation 4.31. Such a choice of the initial field allows having the *propagating* fields with *fractional* topological charges m, provided the parameter q is chosen accordingly [66]. Figures 4.13 to 4.16 show contour plots of the real field u distributions in 2D, for some specific values of the above parameters. We also compare the results for $\sigma = 1$ and $P = 1$ with the ones for $\sigma = 100$ and $P = 100$, to ascertain the improved stability.

4.3.5.1 Gaussian and Vortex-Ring Solitons Figure 4.13 presents the stable Gaussian WS family ($m = 0$) and Figure 4.13 the stable vortex-ring soliton family ($q = 1$, $m \neq 0$) for various n. Figure 4.13a shows the radial dependence, including the positions of zeroes and extremal points of Gaussian WS. Figure 4.13a displays the field distribution at $z = 0$. The two rows in Figure 4.13b and c plot field distributions taken at propagation distances $z = 10,\ 100$ (units of diffraction lengths) for $\sigma = 1,\ P = 1$ and for $\sigma = 100,\ P = 100$, respectively. The nearly constant distributions versus z indicate stable propagation. There are n zeroes and $n + 1$ extrema along the radius. The bright spots or rings in Figure 4.13b and c represent the positive extrema, the dark rings represent the negative extrema. The maximum field is located at the origin.

The layout of Figure 4.14 is similar to that shown in Figure 4.13. The field u of the vortex-ring WS is shown *excluding* the azimuthal complex function $\Phi(\phi)$. Similarly, there are $n + 1$ zeroes and $n + 1$ extrema along the radius; the bright ring represents the positive extrema and the dark spots and rings represent the negative extrema. At the origin, where the topological defect is located, the field is zero. The fields of these two classes of solitons are radially symmetric and decaying. We find that 2D Gaussian WS and vortex-ring WS can propagate stably in NN media in the highly nonlocal limit.

4.3.5.2 Half-Moon Solitons When $m \neq 0$ but $q = 0$ $(-1/4 \leq m < 1)$, we obtain the asymmetric *half-moon* WS family. For the parameters considered here, half-moon solitons display modulational instability. Figure 4.15 shows the shapes of typical half-moon solitons for $q = 0$, $m = 1/2$ and for various n. Clearly, the half-moon soliton is stratified in a circular arrangement, because of the cosine azimuthal dependence. Bright (dark) regions represent a positive (negative) field with $u_{n\frac{1}{2}} > 0$ ($u_{n\frac{1}{2}} < 0$). The uniform gray corresponds to zero and the innermost layer is a bright spot. The half-moon soliton alternates bright and dark regions from inside to outside along every radial direction, except along the negative

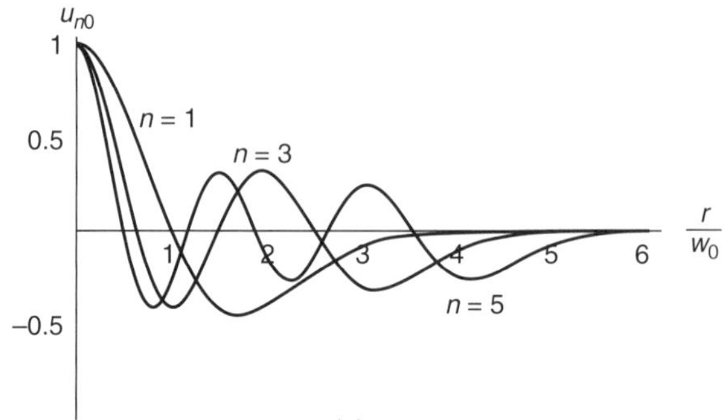

(a)

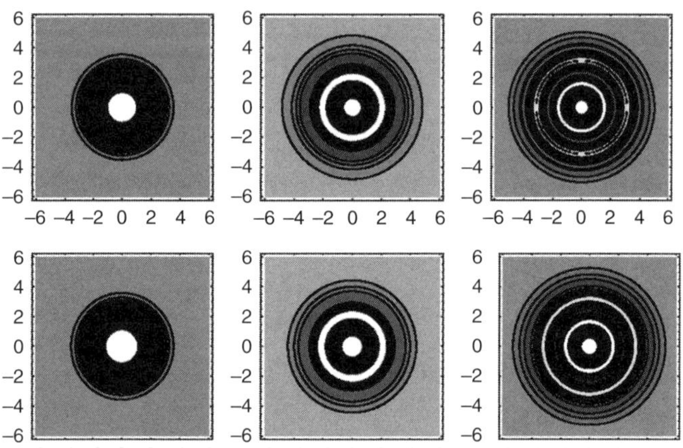

(b) $\sigma = 1$, $p = 1$

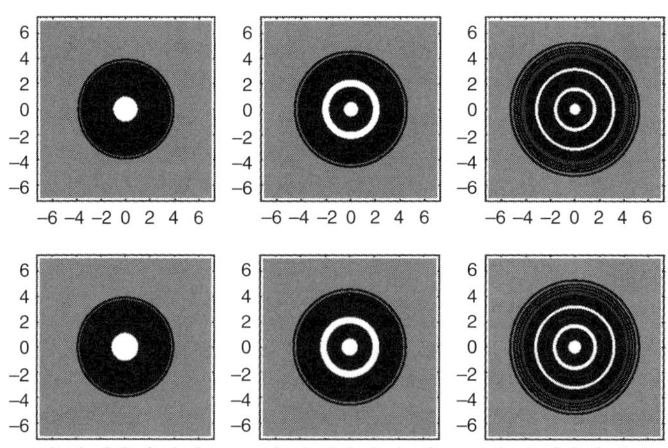

(c) $\sigma = 100$, $p = 100$

FIGURE 4.13 Gaussian WS family ($m = 0$). (a) Dependence along the radial direction, showing the positions of zeroes ($u_{nm} = 0$) and extremal points ($\frac{\mathrm{d}u_{nm}}{\mathrm{d}r} = 0$) for three values of $n = 1, 3, 5$, from left to right. The field distributions in (b) and (c) are at different propagation distances $z = 10, 100$ from top to bottom. In (b) $\sigma = 1$, $P = 1$; in (c) $\sigma = 100$, $P = 100$.

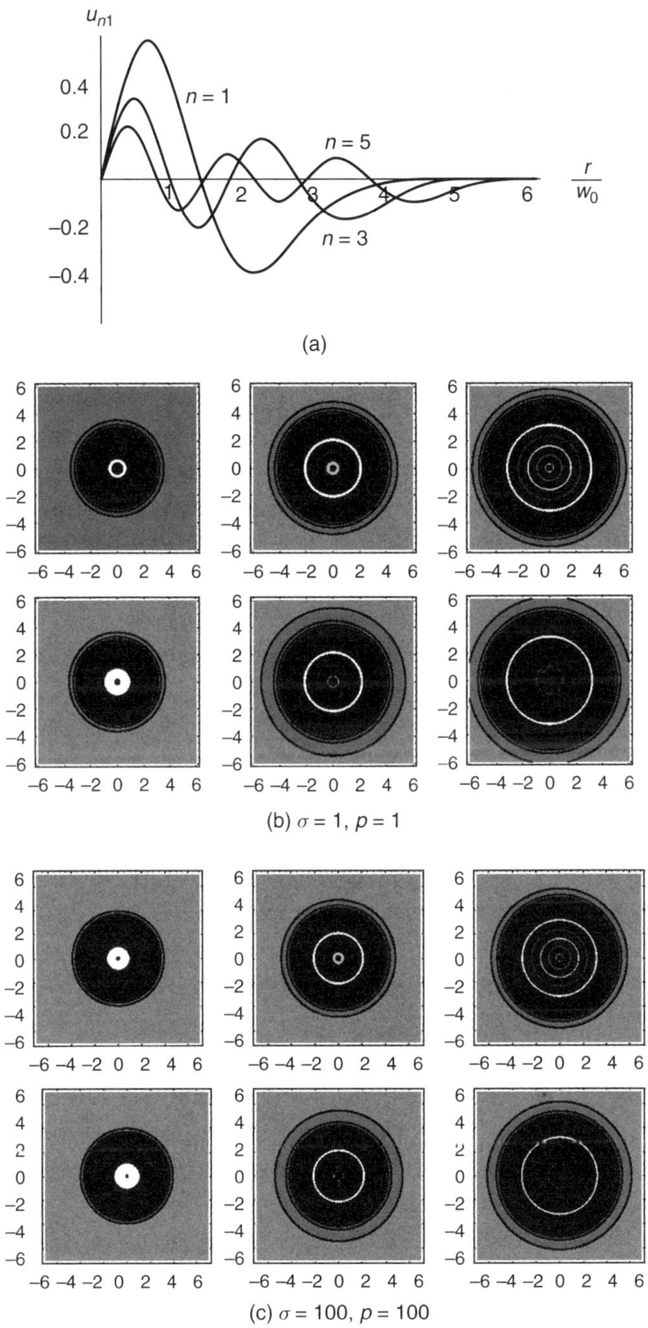

FIGURE 4.14 (a) Vortex-ring WS family without the complex function $\Phi(\varphi)$ (layout as shown in Figure 4.13). Here $m = 1$, $q = 1$, and $n = 1, 3, 5$. The two rows in (b) and (c) present the field distributions at distances $z = 10, 100$, respectively. In (b) $\sigma = 1$, $P = 1$ and in (c) $\sigma = 100$, $P = 100$.

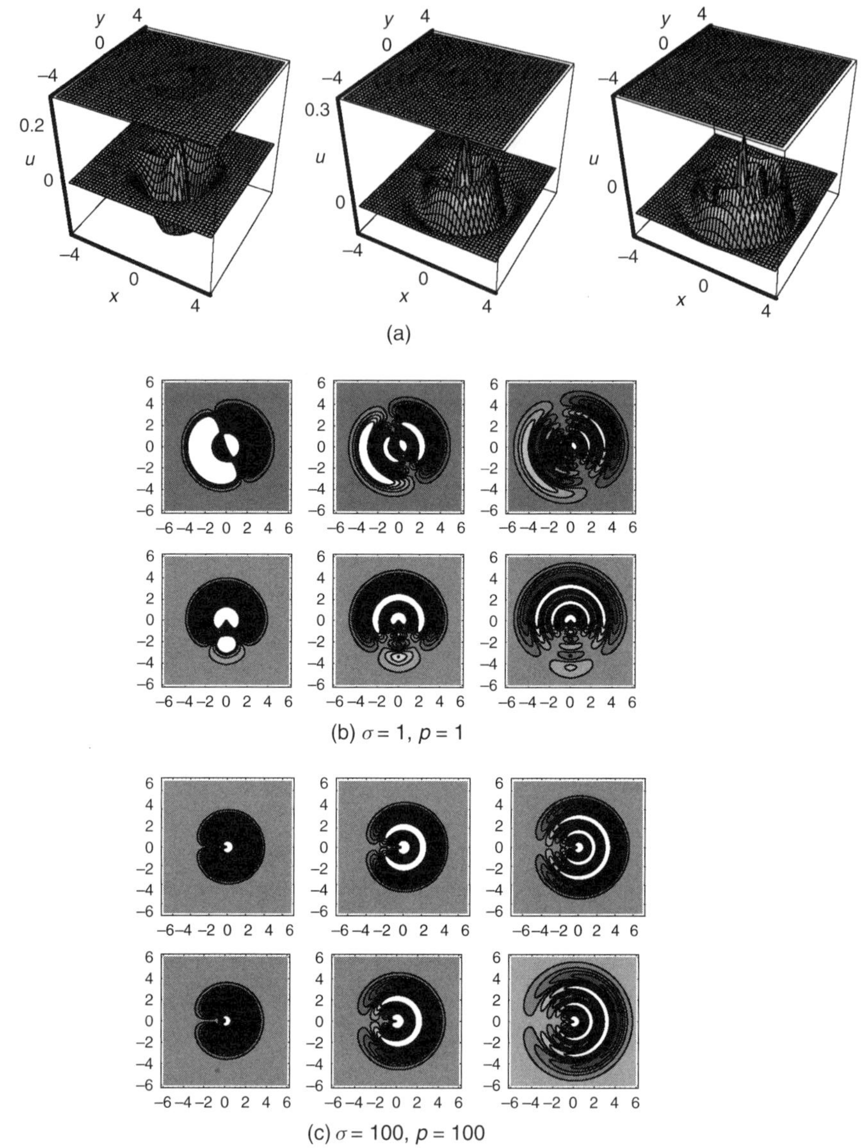

FIGURE 4.15 Modulational instability in the half-moon soliton family, for $m = 1/2$ and $q = 0$. (a) Field distribution in $z = 0$. (b and c) Top views at propagation distances $z = 10, 100$, from top to bottom. In (b) $\sigma = 1$, $P = 1$ and in (c) $\sigma = 100$, $P = 100$, respectively. $n = 1, 3, 5$, from left to right in each graph.

real axis, where it is null. The number of layers depends on n and there are $n + 1$ layers. At the initial stage ($z = 0$), the solitons display the half-moon shape [67], as the numerical solutions stem from the modulated WS modes. Along propagation (Fig. 4.15b) we observe that, at $z = 10$, the WS break up into two out-of-phase portions, each similar to the initial half-moon. At $z = 100$, these distributions become two irregular half-moon solitons. The larger the parameter n, the larger the half-moon WS deformation with propagation distance, indicating strong modulational instabilities. However, if larger values of σ and P are chosen, for example, $\sigma = 100$, $P = 100$, the propagation of half-moon solitons becomes more stable, although with slight diffraction, as shown in Figure 4.15c.

4.3.5.3 Symmetric Multipole Solitons When $m>0$ is an integer and $q = 0$, we obtain symmetric multipole solitons, including symmetric single- and multilayer necklace solitons. Once again, their initial distributions stem from the linear WS modes at $z = 0$. In propagation, they experience strong modulational instabilities. Figure 4.16 shows some examples. These soliton families have adjacent alternating bright and dark regions along the azimuth. For $n = 0$, the field at $z = 0$ along the radius has only one layer and the WS form symmetric single-layer necklace solitons. Increasing the propagation distance, these WS families become strongly asymmetric. The number of bright (or dark) spots decreases and the adjacent bright (or dark) spots overlap and merge along the azimuth, as at $z = 100$.

For $n>0$, we observe the symmetric multilayer necklace WS. At $z = 0$ they form adjacent alternating bright and dark regions along the radial direction, as well. There are $n + 1$ layers and the maximum field is located at the outside layer along the radius. As shown in Figure 4.16b and c, by increasing the propagation distance the symmetric multipole WS families experience symmetry-breaking instability and merging phenomena for larger m. For $m = 7$, for example, in the bottom rows, the number of bright/dark regions and their contrast reduce with propagation. They maintain $n = 1$ layers along the radius but become asymmetric multipole solitons, with a decreasing number of bright and dark regions. However, the stability of symmetric necklace WS greatly improves; that is, they propagate stably and without merging for longer distances, provided larger σ and P are selected. This is clearly seen in Figure 4.16c.

4.3.5.4 Asymmetric Multipole Solitons In the limit $q = 0$ and for $m>1$, a rational number, we observe asymmetric multipole soliton families. For m we pick half-integer values, because theoretical studies [66] suggest an interesting dynamical behavior there. These asymmetric multipole WS contain asymmetric single- and multilayer necklace solitons (see also Reference 15).

4.4 CONCLUSIONS

Spatial solitons in NLCs have stimulated broad interest. Soliton families in strongly NN media have also attracted attention in applied sciences. During the last few

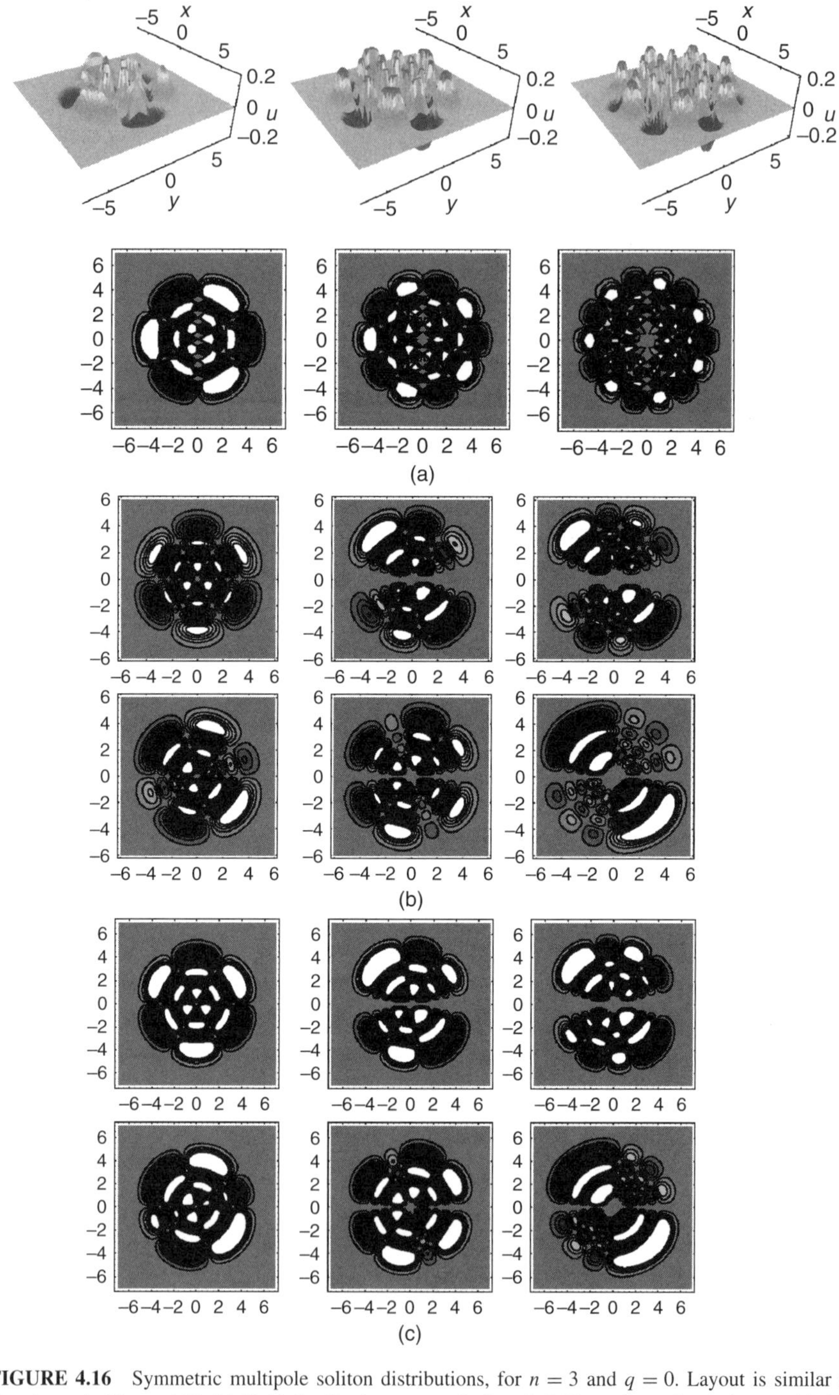

FIGURE 4.16 Symmetric multipole soliton distributions, for $n = 3$ and $q = 0$. Layout is similar to that shown in Figure 4.15. (a) Field distributions at $z = 0$. (b and c) Views from above, for $z = 10, 100$ from top to bottom. In (b) $\sigma = 1$, $P = 1$ and in (c) $\sigma = 100$, $P = 100$, respectively. In all the graphs $m = 3, 5, 7$ from left to right.

years, not only the existence of nonlocal spatial solitons was experimentally demonstrated but also various desirable features were identified and mathematically modeled. Spatial solitons involve a large number of problems, including the exact analytical solutions of the NLSE in various forms and dimensions. The corresponding theories have produced good agreement with experiments and displayed the potential of spatial solitons in applications. We expect the research on soliton families to develop further theoretically, experimentally, and in the quest for actual implementations.

In conclusion, in this chapter, we have provided a mathematical framework for the theoretical understanding of NL nonlocal localization phenomena in NLCs. We have introduced two techniques: the first uses the self-similar method, which is effective when considering soliton symmetries; the second is based on numerical methods. Starting from the general NLSE, in the limit of strong nonlocal nonlinearity, and employing self-similar approach, the evolution equations have been simplified to the highly nonlocal NLSE, yielding exact accessible soliton solutions in one and two dimensions. Numerical methods were employed to solve the general NLS equation, and the exact solutions were used as initial conditions. The numerics were based on the split-step Fourier method, also known as the beam-propagation method. We have further discussed the stability of soliton solutions by comparing analytical and numerical results. We have made an attempt to summarize the most relevant mathematical aspects concerning nonlocal spatial soliton families in NLCs; however, owing to the remarkable recent growth of activities in the field, a few important contributions could not be covered here. Some of those are addressed in other chapters of this book.

Acknowledgments

This work was funded by several agencies, including the National Science Foundation of Guangdong Province, under Grant No. 1015283001000000 (China), the Science Research Foundation of Shunde Polytechnic (China), and the Qatar National Research Foundation through the project NPRP 09-462-1-074 (Qatar).

REFERENCES

1. G. Assanto and M. Karpierz. Nematicons: Self-localised beams in nematic liquid crystals. *Liq. Cryst.*, 36(10):1161–1172, 2009.
2. G. Assanto, A. Fratalocchi, and M. Peccianti. Spatial solitons in nematic liquid crystals: from bulk to discrete. *Opt. Express*, 15(8):5248–5259, 2007.
3. G. Assanto, M. Peccianti, and C. Conti. Nematicons: Optical spatial solitons in nematic liquid crystals. *Opt. Photon. News*, 14(2):44–48, 2003.
4. C. Rotschild, O. Cohen, O. Manela, M. Segev, and T. Carmon. Solitons in nonlinear media with an infinite range of nonlocality: First observation of coherent elliptic solitons and of vortex-ring solitons. *Phys. Rev. Lett.*, 95:213904, 2005.
5. W. Krolikowski and O. Bang. Solitons in nonlocal nonlinear media: Exact solutions. *Phys. Rev. E*, 63:016610, 2001.

6. W. Krolikowski, O. Bang, J. J. Rasmussen, and J. Wyller. Modulational instability in nonlocal nonlinear Kerr media. *Phys. Rev. E*, 64:016612, 2001.
7. S. K. Srivatsa and G. S. Ranganath. New nonlinear optical processes in liquid crystals. *Opt. Commun.*, 180:346–359, 2000.
8. J. Henninot, M. Debailleul, F. Derrien, G. Abbate, and M. Warenghem. (2D + 1) Spatial optical solitons in dye doped liquid crystals. *Synt. Met.*, 124:9–13, 2001.
9. D. Deng, X. Zhao, Q. Guo, and S. Lan. Hermite-Gaussian breathers and solitons in strongly nonlocal nonlinear media. *J. Opt. Soc. Am. B*, 24:2537–2544, 2007.
10. E. A Sziklas and A. E. Siegman. Mode calculations in unstable resonators with flowing saturable gain. 2: Fast Fourier transform method. *Appl. Opt.*, 14:1874–1889, 1975.
11. B. Fornberg and G. B. Whitham. A numerical and theoretical study of certain nonlinear wave phenomena. *Philos. Trans. R. Soc. London, Ser. A*, 289:373–403, 1978.
12. W. P. Zhong and L. Yi. Two-dimensional Laguerre-Gaussian soliton family in strongly nonlocal nonlinear media. *Phys. Rev. A*, 75:061801, 2007.
13. M. Soljačić and M. Segev. Integer and fractional angular momentum borne on self-trapped necklace-ring beams. *Phys. Rev. Lett.*, 86:420–423, 2001.
14. A. S. Desyatnikov and Y. S. Kivshar. Rotating optical soliton clusters. *Phys. Rev. Lett.*, 88:053901, 2002.
15. W. P. Zhong, M. Belić, R. H. Xie, and G. Chen. Two-dimensional Whittaker solitons in nonlocal nonlinear media. *Phys. Rev. A*, 78:013826, 2008.
16. W. P. Zhong, M. Belić, R. H. Xie, and T. W Huang. Three-dimensional Bessel light bullets in self-focusing Kerr media. *Phys. Rev. A.*, 82:033834, 2010.
17. W. P. Zhong, Z. P. Yang, R. H. Xie, M. Belić, and G. Chen. Two-dimensional spatial solitons in nematic liquid crystals. *Commun. Theor. Phys.*, 51:324–330, 2009.
18. M. Belić and W. P. Zhong. Two-dimensional spatial solitons in highly nonlocal nonlinear media. *Eur. J. Phys. D*, 53:97–106, 2009.
19. B. A. Malomed, D. Mihalache, F. Wise, and L. Torner. Spatiotemporal optical solitons. *J. Opt. B: Quantum Semiclass. Opt.*, 7:53–72, 2005.
20. B. B. Baizakov, B. A. Malomed, and M. Salerno. Multidimensional solitons in periodic potentials. *Europhys. Lett.*, 63:642–648, 2003.
21. W. P. Zhong, L. Yi, R. H. Xie, M. Belić, and G. Chen. Robust three-dimensional spatial soliton clusters in strongly nonlocal media. *J. Phys. B: At. Mol. Opt. Phys.*, 41:025402, 2008.
22. W. P. Zhong and M. Belić. Kummer solitons in strongly nonlocal nonlinear media. *Phys. Lett. A*, 373:296–298, 2009.
23. W. P. Zhong and M. Belić. Three-dimensional optical vortex and necklace solitons in highly nonlocal nonlinear media. *Phys. Rev. A*, 79:023804, 2009.
24. A. W. Snyder and D. J. Mitchell. Accessible solitons. *Science*, 276:1538–1541, 1997.
25. D. Mihalache, D. Mazilu, F. Lederer, B. A. Malomed, Y. V. Kartashov, L. C. Crasovan, and L. Torner. Three-dimensional spatiotemporal optical solitons in nonlocal nonlinear media. *Phys. Rev. E*, 73:025601, 2006.
26. W. P. Zhong, M. Belić, R. H. Xie, T. W. Huang, and Y. Q. Lu. Three-dimensional spatiotemporal solitary waves in strongly nonlocal media. *Opt. Commun.*, 283:5213–5217, 2010.
27. I. B. Burgess, M. Peccianti, G. Assanto, and R. Morandotti. Accessible light bullets via synergetic nonlinearities. *Phys. Rev. Lett.*, 102:203903, 2009.

28. M. Peccianti, I. B. Burgess, G. Assanto, and R. Morandotti. Space-time bullet trains via modulation instability and nonlocal solitons. *Opt. Express*, 18(6):5934–5941, 2010.
29. Y. V. Kartashov, V. A. Vysloukh, and L. Torner. Stability of vortex solitons in thermal nonlinear media with cylindrical symmetry. *Opt. Express*, 15:9378–9384, 2007.
30. S. Skupin, M. Saffman, and W. Krolikowski. Nonlocal stabilization of nonlinear beams in a self-focusing atomic vapor. *Phys. Rev. Lett.*, 98:263902, 2007.
31. W. Krolikowski, M. Saffman, B. Luther-Davies, and C. Denz. Anomalous interaction of spatial solitons in photorefractive media. *Phys. Rev. Lett.*, 80:3240–3242, 1998.
32. G. Assanto and M. Peccianti. Spatial solitons in nematic liquid crystals. *IEEE J. Quantum Electron.*, 39(1):13–21, 2003.
33. F. Dalfovo, S. Giorgini, L. P. Pitaevskii, and S. Stringari. Theory of Bose-Einstein condensation in trapped gases. *Rev. Mod. Phys.*, 71:463–512, 1999.
34. V. M. Perez-Garcia, V. V. Konotop, and J. J. Garcia-Ripoll. Dynamics of quasi-collapse in nonlinear Schrödinger systems with nonlocal interactions. *Phys. Rev. E* 62:4300–4308, 2000.
35. W. P. Zhong, M. Belić, R. H. Xie, G. Chen, and Y. Q. Lu. Dynamically compressed bright and dark solitons in highly anisotropic Bose-Einstein condensates. *Eur. J. Phys. D*, 55:147–153, 2009.
36. W. P. Zhong, M. Belić, Y. Q. Lu, and T. W. Huang. Traveling and solitary wave solutions to the one-dimensional Gross-Pitaevskii equation. *Phys. Rev. E*, 81:016605, 2010.
37. V. I. Kruglov, A. C. Peacock, and J. D. Harvey. Exact self-similar solutions of the generalized nonlinear Schrödinger equation with distributed coefficients. *Phys. Rev. Lett.*, 90:113902, 2003.
38. V. I. Kruglov, D. Mechin, and J. D. Harvey. Self-similar solutions of the generalized Schrödinger equation with distributed coefficients. *Opt. Express*, 12:6198–6207, 2004.
39. M. E. Fermann, V. I. Kruglov, B. C. Thomsen, J. M. Dudley, and J. D. Harvey. Self-similar propagation and amplification of parabolic pulses in optical fibers. *Phys. Rev. Lett.*, 84:6010–6013, 2000.
40. S. Chen, H. Liu, S. Zhang, and L. Yi. Compression of Hermite–Gaussian pulses in an engineered optical fiber absorber with varying dispersion and nonlinearity. *Phys. Lett. A*, 353:493–496, 2006.
41. S. Chen, L. Yi, D. Guo, and P. Lu. Self-similar evolutions of parabolic, Hermite-Gaussian, and hybrid optical pulses: Universality and diversity. *Phys. Rev. E*, 72:016622, 2005.
42. S. Chen and L. Yi. Chirped self-similar solutions of a generalized nonlinear Schrödinger equation model. *Phys. Rev. E*, 71:016606, 2005.
43. S. A. Ponomarenko and G. P. Agrawal. Do solitonlike self-similar waves exist in nonlinear optical media? *Phys. Rev. Lett.*, 97:013901, 2006.
44. V. I. Karpman. *Non-linear Waves in Dispersive Media*. Pergamon, New York, 1975.
45. G. I. Barenblatt. *Scaling, Self-Similarity, and Intermediate Asymptotics*. Cambridge University Press, Cambridge, 1996.
46. F. Simoni. *Nonlinear Optical Properties of Liquid Crystals and PDLC*. World Scientific, Singapore, 1997.
47. M. Peccianti, C. Conti, and G. Assanto. Optical multisoliton generation in nematic liquid crystals. *Opt. Lett.*, 28:2231–2233, 2003.

48. C. Conti, M. Peccianti, and G. Assanto. Observation of optical spatial solitons in a highly nonlocal medium. *Phys. Rev. Lett.*, 92:113902, 2004.
49. C. Conti, M. Peccianti, and G. Assanto. Route to nonlocality and observation of accessible solitons. *Phys. Rev. Lett.*, 91:073901, 2003.
50. M. Warenghem, J. F. Henninot, and G. Abbate. Non linearly induced self waveguiding structure in dye doped nematic liquid crystals confined in capillaries. *Opt. Express*, 2:483–490, 1998.
51. A. S. Desyatnikov, D. Neshev, and Y. S. Kivshar. Multipole composite spatial solitons: theory and experiment. *J. Opt. Soc. Am. B*, 19:586–595, 2002.
52. J. Wyller, W. Krolikowski, O. Bang, and J. J. Rasmussen. Generic features of modulational instability in nonlocal Kerr media. *Phys. Rev. E*, 66:066615, 2002.
53. W. Krolikowski, O. Bang, N. I. Nikolov, D. Neshev, J. Wyller, J. J. Rasmussen, and D. Edmundson. Modulational instability, solitons and beam propagation in spatially nonlocal nonlinear media. *J. Phys. B: At. Mol. Opt. Phys.*, 6:288–294, 2004
54. W. Krolikowski, O. Bang, J. Wyller, and J. J. Rasmussen. Optical beams in nonlocal nonlinear media. *Acta Phys. Pol., A*, 103:133–147, 2003.
55. W. Krolikowski, G. Mccarthy, M. Saffman, O. Bang, J. Wyller, and J. J. Rasmussen. *Focus on Lasers and Electro-optics Research*, ed. F. Columbus. Nova Science, New York, 2005.
56. O. Bang, W. Krolikowski, J. Wyller, and J. J. Rasmussen. Collapse arrest and soliton stabilization in nonlocal nonlinear media. *Phys. Rev. E*, 66:046619, 2002.
57. M. Peccianti, C. Conti, G. Assanto, A. De Luca, and C. Umeton. Nonlocal optical propagation in nonlinear nematic liquid crystals. *J. Nonlin. Opt. Phys. Mater.*, 12:525–538, 2003.
58. E. T. Whittaker. An expression of certain known functions as generalised hypergeometric functions. *Bull. Am. Math. Soc.*, 10:125–134, 1904.
59. B. Yang, W. P. Zhong, and M. Belić, Self-similar Hermite-Gaussian Spatial solitons in two-dimensional nonlocal nonlinear media. *Commun. Theor. Phys.*, 53:937–942, 2010.
60. D. Briedis, D. E. Petersen, D. Edmundson, W. Krolikowski, and O. Bang. Ring vortex solitons in nonlocal nonlinear media. *Opt. Express*, 13:435–443, 2005.
61. E. T. Whittaker and G. N. Watson. *A Course in Modern Analysis*, 4th edn. Cambridge University Press, Cambridge, 1990.
62. S. Skupin, O. Bang, D. Edmundson, and W. Krolikowski. Stability of two-dimensional spatial solitons in nonlocal nonlinear media. *Phys. Rev. E*, 73:066603, 2006.
63. A. I. Yakimenko, V. M. Lashkin, and O. O. Prikhodko. Dynamics of two-dimensional coherent structures in nonlocal nonlinear media. *Phys. Rev. E*, 73:066605, 2006.
64. W. P. Zhong, M. Belić, T. W. Huang, and L. Y. Wang. Superpositions of Laguerre-Gaussian Beams in Strongly Nonlocal Left-handed Materials. *Commun. Theor. Phys.*, 53:749–754, 2010.
65. Z. H. Musslimani and J. Yang. Self-trapping of light in a two-dimensional photonic lattice. *J. Opt. Soc. Am. B*, 21:973–981, 2004.
66. M. V. Berry. Optical vortices evolving from helicoidal integer and fractional phase steps. *J. Opt. A: Pure Appl. Opt.* 6:259–268, 2004.
67. Y. J. He, B. A. Malomed, D. Mihalache, and H. Z Wang. Crescent vortex solitons in strongly nonlocal nonlinear media. *Phys. Rev. A*, 78:023824, 2008.

5

External Control of Nematicon Paths

Armando Piccardi, Alessandro Alberucci, and Gaetano Assanto

Nonlinear Optics and OptoElectronics Lab, University ROMA TRE, Rome, Italy

5.1 INTRODUCTION

The excitation of spatial solitons in nematic liquid crystals (NLC) relies on dipolar interactions between the elongated organic molecules and the electric field associated to propagating light beams [1, 2]. The NLC reorientational response to low and/or high frequency fields, together with other light-NLC interactions—for instance dye-mediated responses [3, 4]—can also be employed to spatially modulate the distribution of the molecular director. This makes NLC cells one of the most versatile environments to generate largely tunable refractive index profiles. In this chapter, we describe how nematicons propagate within inhomogeneous refractive index distributions and illustrate a few methods to modify their trajectories with the use of external fields.

Whatever is the nature of the director distortion, for example optical or electrical [5], a number of considerations hold valid in all cases. First of all, the NLC dielectric anisotropy affects the interaction, in both quantitative and qualitative ways, with respect to isotropic media: as beam energy travels in general forming a nonzero angle (called *walk-off*) with respect to the wave-vector [6], a beam can be steered by spatial changes in optic axis, even without altering the phase fronts. Thus, the beam trajectory can depend on walk-off variations, other than index gradient. Moreover, nonlocality affects the width and shape of the induced index gradient. In fact, strong anchoring at the sample boundaries set the distribution of elastic energy that can be stored in the nematic slab [7]: it can be demonstrated that, due to intermolecular forces, director variations cannot extend by less than the minimum cell dimension [8]. Even if restrictions are imposed on the system tunability, for typical cell sizes (50–100 μm), the Mauguin limit [7, 9] is fulfilled all over

Nematicons: Spatial Optical Solitons in Nematic Liquid Crystals, First Edition.
Edited by Gaetano Assanto.

the sample [10] thanks to the nonlocality, ensuring the reciprocal independence of ordinary and extraordinary components [11], and enhancing the robustness of reorientational solitons (excited only by the extraordinary wave) when interacting with regions defined by a high index contrast. Moreover, the nonlocal response allows the use of the approximation of geometric optics to soliton propagation: provided that the beam width is much smaller than the perturbation width, ray optics can be successfully employed to calculate the nematicon trajectory in an arbitrary index distribution. As pointed out by Mitchell et al. [12], in highly nonlocal media a self-induced refractive index well can be approximated by a parabolic profile; the latter permits to elucidate light self-trapping by the well-known physics of the harmonic quantum oscillator.

This chapter is organized as follows: Section 5.2 is devoted to the basic equations describing a nematicon moving in an arbitrary potential: the director distribution is derived from the usual reorientation equation, taking into account both electric and optical contributions; then a NonLinear Schrödinger Equation (NLSE) comprehensive of all characteristic parameters is derived. A particle-like model is also introduced, with the Ehrenfest theorem used to express the soliton position as a function of an equivalent force governing the trajectory. Afterward, in Section 5.3, we describe a few experimental observations on nematicon control: external beam(s) is (are) shown to be able to deviate the soliton through changes in material properties, for example optical reorientation due to *direct* dipolar interactions or light-induced modifications of a guest–host system, defining pins or interfaces. In Section 5.4, we illustrate various configurations of electrodes able to modify the director distribution (the director control via magnetic fields is formally equivalent to the latter case) in 2D or 3D geometries: electrically driven reorientation is analyzed with uniform or patterned electrodes, varying the soliton path through changes in linear walk-off. Finally, in Section 5.5, we summarize the effects of electrically defined interfaces.

Other than summarizing the main aspects of nematicon interactions with linear potentials, this chapter collects a few strategies for light manipulation: keeping in mind the self-induced waveguide character of spatial solitons, the control of their propagation indicates various possibilities for *nematicon-based signal routers and processors*.

5.2 BASIC EQUATIONS

In this chapter, we refer to a planar cell of thickness L (along x, in a Cartesian system) and infinitely extended along the other dimensions. Naming θ the angle between the molecular director and the z axis, the anchoring conditions set the director to lie in the yz plane, with a uniform distribution $\theta = \theta_0$ throughout the sample. An external control (either optical or electrical) can modify its distribution at rest, altering the refractive index environment where solitons can propagate. From the Euler–Lagrange equations, we can cast the conservation of angular momentum into a *reorientation equation*, which describes the NLC elastic response to electrical as well as optical excitations [5]:

$$K\nabla^2\theta + \frac{\epsilon_0\epsilon_a}{4}\sin(2\theta)\,|A|^2 + \frac{\epsilon_0\Delta\epsilon_{LF}}{2}\sin(2\theta)\,E^2 = 0, \tag{5.1}$$

where K is the scalar Frank's elastic constant [7] representing the intermolecular restoring forces that balance the torques externally induced by optical and/or electric control fields A and E, respectively. As the anchoring imposes boundary conditions at the cell interfaces $\left[\theta(0, y, z) = \theta(L, y, z) = \theta_0\right]$, for arbitrary distributions of the external fields (and thus director), it is $\theta = \theta(x, y, z)$.

Let us now consider a wave packet propagating in a director background, with its wave-vector distribution centered in $\mathbf{k}//\hat{z}$. It is convenient to deal with the magnetic field associated to the extraordinary (*e*-) polarized wave, namely $\mathbf{H} = H\hat{x}$ [13]; from the expression of the dielectric tensor $\boldsymbol{\epsilon} = \left[\epsilon_{xx}, 0, 0; 0, \epsilon_{yy}, \epsilon_{yz}; 0, \epsilon_{zy}, \epsilon_{zz}\right]$, it follows that the electric field components are $E_x = 0$ and $E_z = \left(\epsilon_{yz}/\epsilon_{zz}\right)E_y - \left(\mathrm{i}/\omega\epsilon_{zz}\right)(\partial H/\partial y)$. Therefore, beam propagation is described by

$$\frac{\partial H}{\partial z} + \frac{\epsilon_{yz}}{\epsilon_{zz}}\frac{\partial H}{\partial y} - \mathrm{i}\omega\left(\epsilon_{yy} - \frac{\epsilon_{yz}^2}{\epsilon_{zz}}\right)E_y - \frac{\mathrm{i}}{\omega\mu_0}\frac{\partial^2 E_y}{\partial x^2} = 0, \tag{5.2}$$

$$\frac{\partial E_y}{\partial z} + \frac{\epsilon_{yz}}{\epsilon_{zz}}\frac{\partial E_y}{\partial y} - \mathrm{i}\omega\mu_0 H - \frac{i}{\omega\epsilon_{zz}}\frac{\partial^2 H}{\partial y^2} = 0, \tag{5.3}$$

where $\omega = k_0(\mu_0\epsilon_0)^{-1/2}$ is the optical frequency and k_0 is the vacuum wave-vector. When considering—as in the rest of this chapter—beams with $\mathbf{k}//\hat{z}$, the extraordinary refractive index and the walk-off can be expressed as $n_e(\theta) = [(\epsilon_{yy} - \epsilon_{yz}^2/\epsilon_{zz})/\epsilon_0]^{1/2}$ and $\delta = \arctan(\epsilon_{yz}/\epsilon_{zz})$, respectively. Even if the latter quantities are pointwise functions of the field intensity, in typical experimental conditions their profiles are much wider than the beam width owing to nonlocality [14, 15]. This allows one to assume both of them to be constant all over the beam profile, their values taken on the beam axis, namely $n_e^{(b)}$ and $\delta^{(b)}$. If the magnetic field has the form $H = A\mathrm{e}^{\mathrm{i}k_0 n_e^{(b)} z}$, we can reduce Equations 5.2 and 5.3 to

$$2\mathrm{i}n_e^{(b)}k_0\left[\frac{\partial A}{\partial z} + \tan\delta^{(b)}\frac{\partial A}{\partial y}\right] + D_y\frac{\partial^2 A}{\partial y^2} + \frac{\partial^2 A}{\partial x^2} + k_0^2\Delta n_e^2 A = 0, \tag{5.4}$$

where A is the field in the slowly varying envelope approximation (SVEA) and $\Delta n_e^2 = n_e^2 - \left(n_e^{(b)}\right)^2$ is the change in the extraordinary index. There are two contributions to Δn_e^2: a linear portion that stems from the control field E and a nonlinear one that depends on the soliton field and responsible for self-confinement. Equation 5.4 is an NLSE that describes the evolution of an optical field in an anisotropic medium with an arbitrary index profile $[\Delta n_e^2 = \Delta n_e^2(x, y, z)]$, the latter fixed by Equation 5.1, including linear/nonlinear changes in walk-off via $\tan\delta^{(b)}$.

To calculate the soliton trajectory in the presence of a generic distribution of the dielectric tensor, we can consider the coordinate z as an equivalent time t and assume $\hbar = 2k_0 n_e^{(b)}$, $m = 2k_0^2 (n_e^{(b)})^2$, $\mathbf{A}_{\rm eq} = A_{{\rm eq}y}\hat{y}$, $-(e/c)A_{{\rm eq}y} = \tan\delta^{(b)}$, $eU_{\rm eq} = -k_0^2 \Delta n_e^2 - k_0^2 (n_e^{(b)})^2 \tan^2\delta^{(b)}$. The latter render nematicon propagation formally equivalent to the motion of a charged particle in an electromagnetic potential. The corresponding Hamiltonian is $H_{\rm eq} = (1/2m)[\mathbf{p}_{\rm eq} - (e/c)\,\mathbf{A}_{\rm eq}]^2 + eU_{\rm eq}$, where m is the particle mass, e is its charge, c is the speed of light in vacuum, $\mathbf{p}_{\rm eq}$ is the equivalent momentum (given by the slope of the trajectory), and $\mathbf{A}_{\rm eq}$ and $U_{\rm eq}$ are the effective vector and scalar potentials, respectively; the expression of the associated electric field is $\mathbf{E}_{\rm eq} = -\nabla_T U_{\rm eq} + \partial \mathbf{A}_{\rm eq}/\partial t$ ($\nabla_T = \hat{x}\partial/\partial x + \hat{y}\partial/\partial y$). In analogy to potential terms generating a force acting on charged particles, we can derive an *equivalent transverse force* that modifies the soliton trajectory:

$$\mathbf{F} = 2k_0^2 \left(n_e \frac{\partial n_e}{\partial x}\hat{x} + n_e \frac{\partial n_e}{\partial y}\hat{y} + \left(n_e^{(b)}\right)^2 \frac{\mathrm{d}\tan\delta^{(b)}}{\mathrm{d}z}\hat{y} \right), \tag{5.5}$$

where the first two terms represent the refractive index gradient (corresponding to the scalar potential $U_{\rm eq}$) and the last term comes from the longitudinal variations in walk-off as perceived by the beam (owing to the equivalent vector potential $\mathbf{A}_{\rm eq}$). The position of the soliton "center of mass" $\mathbf{r}_b = \int \mathbf{r}\,|A|^2\,\mathrm{d}\mathbf{r} / \int |A|^2\,\mathrm{d}\mathbf{r}$ can be found by applying the Ehrenfest theorem [16] to Equation 5.4

$$2k_0^2 \left(n_e^{(b)}\right)^2 \frac{\mathrm{d}^2 \mathbf{r}_b}{\mathrm{d}z^2} = \mathbf{F}|_{\mathbf{r}=\mathbf{r}_b}, \tag{5.6}$$

with the NLC highly nonlocal response accounted for on the right-hand side. Summarizing, Equation 5.4 models the propagation of a finite size light beam in an NLC cell accounting for anisotropy, walk-off, and nonlinear index changes. The beam trajectory can get modified by a nonhomogeneous index profile generated by external electrical and/or optical fields as described by Equation 5.1. Moreover, the high nonlocality allows adopting a particle-like model for soliton propagation, illustrating the trajectory of a self-confined beam within the ray-optics approximation; hence, once the linear index background is defined, we can predict the resulting path of a nematicon.

5.3 NEMATICON CONTROL WITH EXTERNAL LIGHT BEAMS

Let us address the effect of an inhomogeneous index profile produced by an external beam. The setup is sketched in Figure 5.1: a nematicon is excited by launching an extraordinary polarized beam with $\mathbf{k}//\hat{z}$, with a *control* beam focused on the sample and normal to the nematicon wave-vector, in the yz plane. In this section, we discuss the role of control beams with various shapes on the nematicon path.

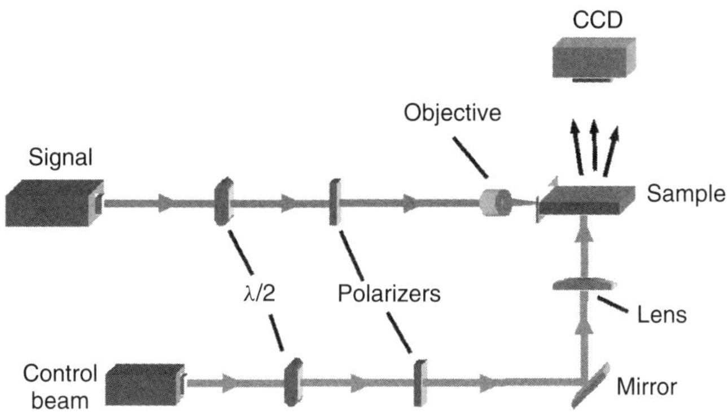

FIGURE 5.1 Typical setup used to optically induce a nonuniform index distribution.

5.3.1 Interaction with Circular Spots

Figure 5.2a shows the calculated refractive index perturbation due to a circular beam of size comparable with the thickness $L = 75$ μm of the sample. We recall that, given the boundary conditions in $x = 0$ and $x = L$ (planar anchoring), the director distribution depends also on x; however, when dealing with nematicons propagating in the middle of the cell, we can simply consider the (planar) contribution in $x = L/2$. A nematicon will be deflected by such an inhomogeneous potential: Figure 5.2b shows a self-confined near-infrared beam (1064 nm) probed

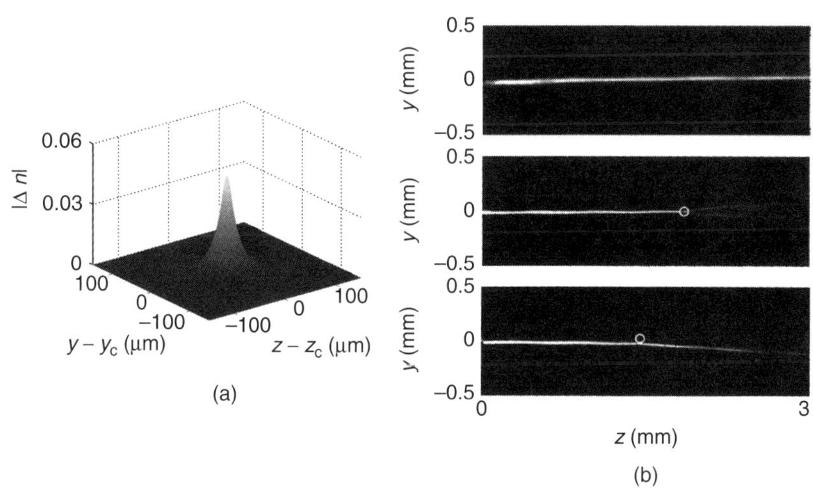

FIGURE 5.2 (a) Absolute value of the index profile induced in the plane $x = L/2$ by a circular control beam propagating along $\hat{x}$ and centered in (y_c, z_c), carrying a power of 32 mW and with a waist of 8 μm. (b) Effect of a circular defect (white circle): as the polarized control beam lowers the index, the soliton propagating with $\mathbf{k}//\hat{z}$ (top panel) is repelled away (center and bottom panels). The acquired photographs show a red probe signal guided by the near-infrared self-induced waveguide.

by a weak signal of different wavelength (633 nm) in a standard planar cell with the commercial mixture E7. The external (near infrared, in this case) control beam is focused in the yz plane with a waist of 8 μm, at intensity high enough to produce optical reorientation and linear polarization chosen in order to lower the refractive index "sensed" by the nematicon: a repulsive potential well acts on the nematicon and deviates its trajectory, with paths as shown in Figure 5.2b [17].

Figure 5.2b displays a few photographs of a red probe confined by the nematicon waveguide and illustrates the interaction of a nematicon with a circular index perturbation induced by a control spot focused to a waist of $w_c \approx 8$ μm (Figure 5.2a). If the latter is placed exactly along the soliton trajectory, the axisymmetric reduction in refractive index splits it into two identical lower power beams, thereby forming a Y-junction; both arms of the Y can remain self-localized or diffract after the interaction, depending on the initial power; the weak probe undergoes the same behavior. When the perturbation is shifted sideways in y, the system symmetry is broken and the soliton (as well as the probe) is "pushed" away in yz toward higher index regions.

The soliton deviation induced by the control beam is a function of both its position Δy with respect to the nematicon axis and its power P_c: considering the resulting soliton trajectories in a limited range of P_c (Fig. 5.3, with $P_c \leq 32$ mW to avoid detrimental thermal effects), a linear dependence exists between P_c and the corresponding steering angle α.

Besides the interaction, the control of soliton paths by way of external beams can be exploited toward simple all-optical logic functions. For instance, the control beam(s) can play the role of input variable(s) in a logic gate, associating a position in the plane yz to the logic output: switching from high to low output can be made to correspond to the displacement due to the control spot(s). Figure 5.4 shows the implementation of an AND gate: in this example, the two perturbing pins are generated via molecular photoisomerization of an azo-NLC [18, 19].

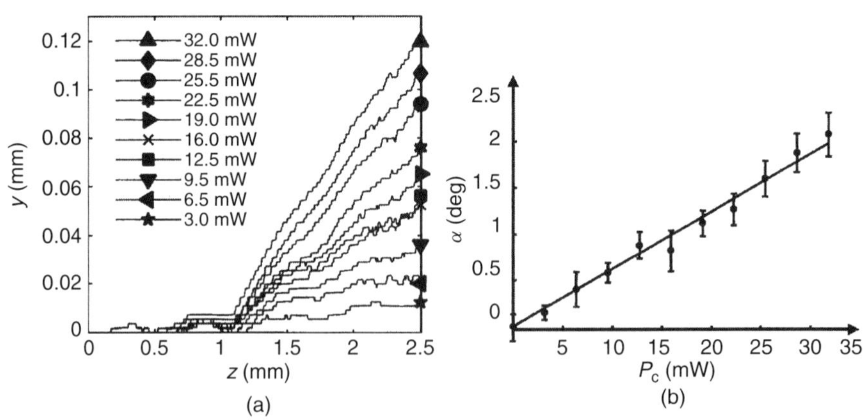

FIGURE 5.3 Soliton deflection for various control beam powers. (a) Nematicon trajectories and (b) angular steering for a control spot placed in $\Delta y = 9$ μm with respect to the unperturbed nematicon path.

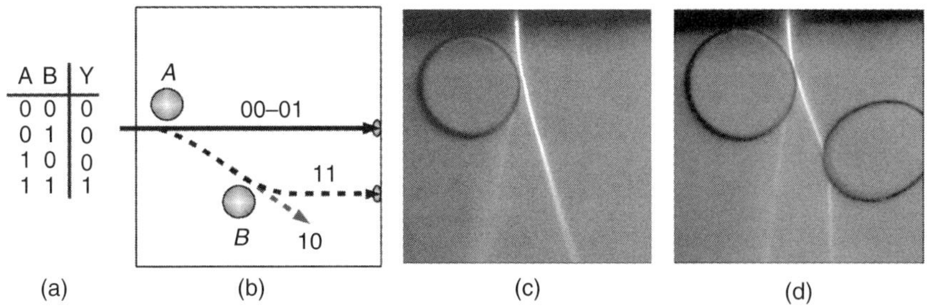

FIGURE 5.4 Nematicon-based AND logic gate. (a) Truth table. (b) Principle of operation: the position corresponding to a "true" output is reached just in the presence of both control spots (i.e., logic "11"). (c and d) Photographs corresponding to the logic inputs (c) "10" and (d) "11", respectively.

5.3.2 Dielectric Interfaces

Changing the shape of the perturbation, it is possible to access different regimes of interaction with nematicons; using an elongated control beam, a double-graded interface can be defined. A typical index distribution obtained in the middle of the cell is shown in Figure 5.5a for an attracting potential well.

In the experiments, we used a commercial NLC mixture 5CB doped with a small amount of *Methyl Red* dye (0.1% in weight) in a planar cell where only one of the glass–NLC interfaces was *rubbed* to provide anchoring, whereas the other one was left untreated. The exposure of this particular dye to visible light is known to modify its interaction energy with the surfaces: the consequent molecular adsorption/desorption [20] mediates a rearrangement of the NLC director around the dye molecules next to the boundaries. Owing to the dye dichroism in this guest–host system, the rearrangement is polarization dependent and determines a preferred NLC orientation, affected by intensity, exposure time, and impinging wavelength [21]. Although the rigorous physical explanation goes beyond our

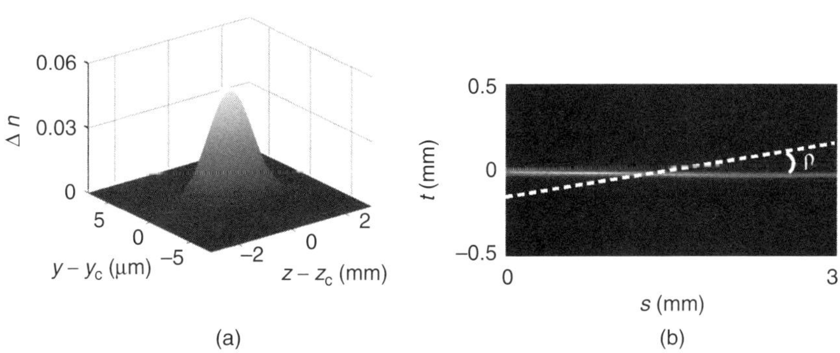

FIGURE 5.5 (a) Pictorial sketch of the index profile Δn generated by an elliptical control beam with long axis parallel to axis z. (b) Unperturbed soliton propagation. The dashed line indicates the long axis of the control spot, at an angle ρ with respect to the nematicon Poynting vector.

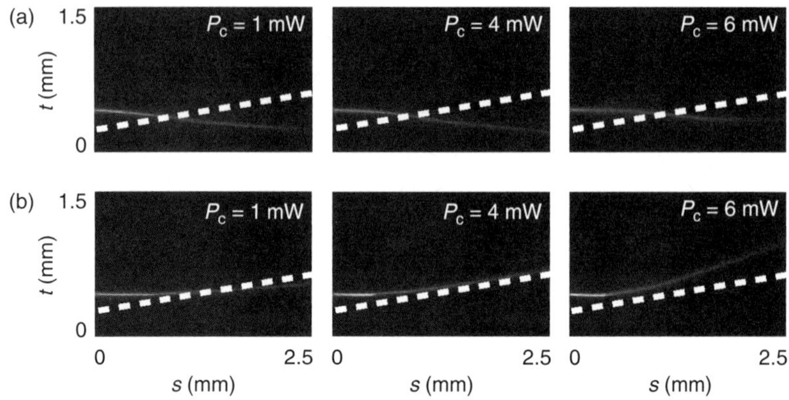

FIGURE 5.6 Optically induced double interface: (a) double refraction and (b) total internal reflection of a nematicon. The dashed lines indicate the long axis of the elliptical control beam.

scope, we can simply state that, illuminating the untreated surface, the control beam actually sets new anchoring conditions, thus affecting the director orientation in the whole thickness of the sample via intermolecular forces [22]. In the actual experimental conditions [using a cw (continuous wave) laser at $\lambda = 532$ nm and moderate intensities], the dye-mediated field–NLC interaction tends to reorient the molecular director perpendicularly to the polarization of the control field.

Figure 5.6 elucidates the effects of an elliptical beam with 5 μm and 3 mm long axes, impinging in the plane yz with its major axis at an angle $\rho \approx 9^\circ$ with respect to the nematicon Poynting vector [3] (see Figure 5.5b): when the control beam is polarized perpendicular to the soliton wave-vector, the index gradient is positive and the soliton undergoes double refraction as it crosses the two graded interfaces (Fig. 5.6a); the beam displacement is a nontrivial function of the control power P_c, the particular dye-NLC guest–host, the beam ellipticity, and the NLC anisotropy. For control light polarization along z, the index gradient is negative: at low P_c and moderate changes in director orientation, the nematicon goes through the double interface, whereas at higher P_c it undergoes total internal reflection (TIR) and is repelled by the interface (Fig. 5.6b). From refraction and reflection we estimated a maximum refractive index variation $\Delta n \approx 0.05$.

5.3.3 Comments

Although we illustrated three examples of *all-optical* control on nematicons, from pure optical reorientation to dye-enhanced field–NLC interactions, the trajectory of nematicons depends only on the resulting modulation of the director background, consistent with the behavior expected from the model discussed in Section 5.2. All the experimental observations confirm that self-confinement is preserved after propagation through a region with nonconstant refractive index: the soliton robustness to such collisions and the great NLC versatility to control the director distribution suggest the feasibility of all-optical signal readdressing based on self-trapped waves in highly nonlocal media.

5.4 VOLTAGE CONTROL OF NEMATICON WALK-OFF

Another means to induce changes in optic axis rely on external electric fields: to this extent Indium Tin Oxide (ITO)-transparent electrodes are often deposited on the inner glass–NLC interfaces of planar cells to apply a bias. Typical *rms* voltages are of a few volts, and the slow elastic response of NLC allows to employ frequencies down to 100 Hz (the upper limit depends on the equivalent capacitance). The high NLC birefringence corresponds to large walk-off angles, making soliton steering feasible with very large displacements; at the same time, electrode patterning permits to engineer the induced perturbation of director distribution, that is, refractive index for extraordinarily polarized light waves. Here, we describe various electrically defined potentials and their interaction with nematicons.

5.4.1 Out-of-Plane Steering of Nematicons

The simplest electrode geometry is one with a uniform thin film of conductor (ITO) deposited on both upper and lower glass–NLC interfaces. To maximize the steering, the surfaces are mechanically rubbed to enforce a planar alignment of the molecular director with $\theta_0 = 45^\circ$ with respect to the z axis [2]. As visible in Figure 5.7, the extraordinary polarization of an input beam launched with $\mathbf{k}//\hat{z}$ is parallel to y and the corresponding walk-off is $\delta \approx 7^\circ$ for the NLC mixture E7 at wavelength $\lambda =$ 1064 nm ($\Delta n \approx 0.2$). As expected, whereas the ordinary component is subject to the Freedericksz threshold [7] and propagates linearly at milliWatt powers, a nematicon can be excited by the extraordinary beam once sufficient power is injected.

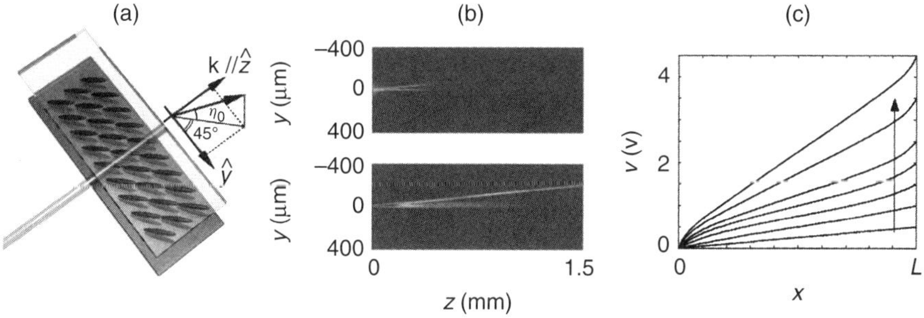

FIGURE 5.7 (a) Sketch of the sample. (b) Propagation of a mixed polarization beam in the absence of bias. At low power (top panel) both ordinary and extraordinary waves diffract, with corresponding Poynting vectors at an angle, owing to walk-off. At high power (bottom panel) the extraordinary wave is self-confined and the nematicon propagates at the linear walk-off angle with respect to z. (c) Bias induced electric potential inside the cell. The arrow indicates increasing voltages.

The application of a bias reorientates the molecular director toward the x axis in the whole cell: the elevation $\eta_0 = \eta_0(x)$ becomes a function of the applied voltage (Fig. 5.7), defining a new optic axis for each V. Consequently, the principal plane for the extraordinary eigenwave varies with voltage following the elevation angle. As the Poynting vector always lies in the plane defined by wave-vector and optic axis, it will be modified by the electric field as well, with the effect of precessing the nematicon path. As Figure 5.8 shows, when $V = 0V$, the walk-off is defined by the geometry, namely $\delta \approx 7^\circ$; after a nearly flat trend versus η_0, the walk-off gradually reduces for increasing η_0 (in turn a function of V) and eventually approaches 0° for $V \approx 3$V, corresponding to the director being parallel to the x axis ($\eta_0 \approx 90^\circ$). Owing to the initial θ_0, the corresponding nematicon "center of mass" moves in the xyz describing a semicone (Fig. 5.8a).

A direct comparison of the model with the top-view experimental observation is not possible, as the latter can only give information on the projection of the nematicon path on the yz plane. Naming ϕ the (voltage-dependent) angle formed by the extraordinary principal plane with axis y (corresponding to the angle between projection of η_0 on xy and axis y) and using geometric considerations, it is straightforward to find the *apparent* walk-off $\alpha = \arctan(\tan\delta\cos\phi)$, that is, the walk-off observable in the yz plane.

Figure 5.9a shows the experimental results for an extraordinarily polarized beam of waist 5 μm and power 5 mW; the measured apparent walk-off versus bias, interpolated with the previously calculated expression, demonstrates voltage-controlled soliton steering over the whole available angular interval: above the

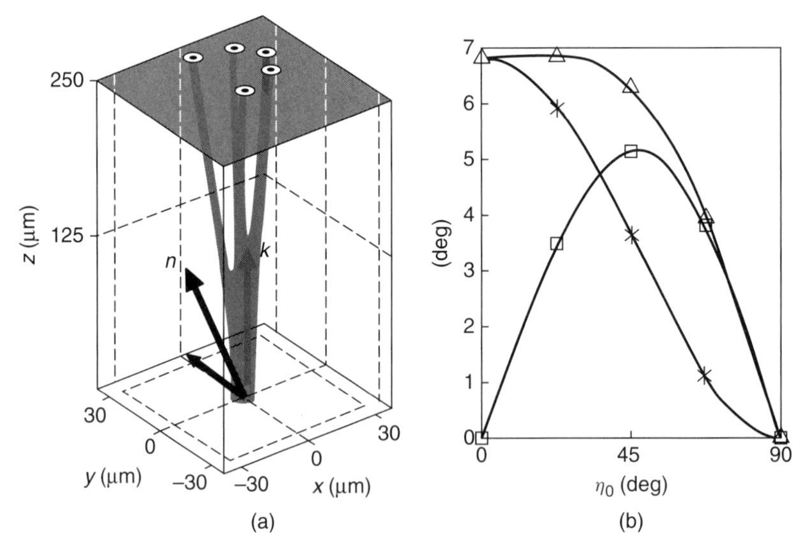

FIGURE 5.8 (a) Sketch of nematicon evolution versus applied voltage. Out-of-plane reorientation causes the trajectory to precess in a half cone. (b) Calculated walk-off (triangles) and its projection on xz (squares) and yz (asterisks); the latter is the apparent walk-off observable in experiments.

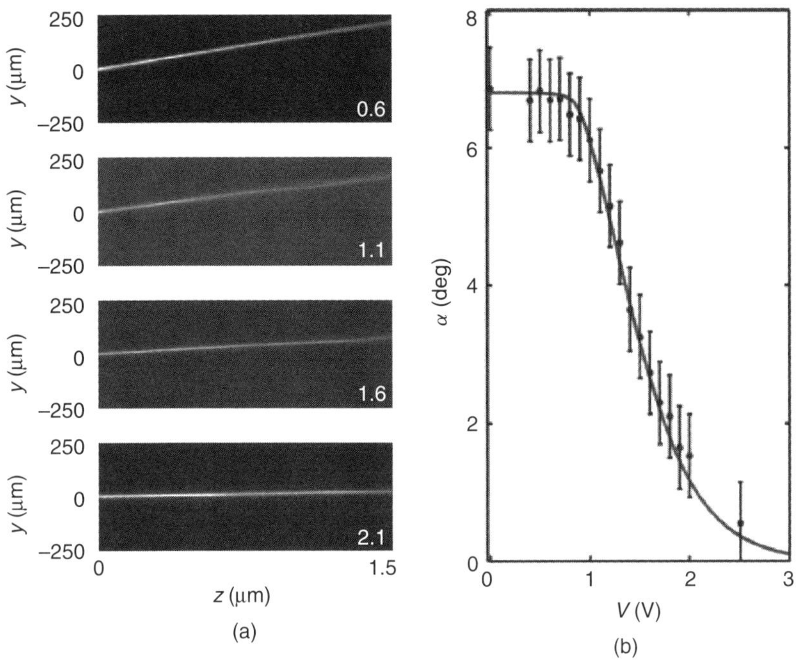

FIGURE 5.9 (a) Photographs of voltage-controlled soliton steering for $\lambda = 1064$ nm. (b) Experimental (points) and theoretical curve of the apparent walk-off versus applied voltage.

Freedericksz threshold ($V \approx 1$ V), the *apparent* walk-off monotonically decreases (see Figure 5.9b) until reorientation starts to saturate at high voltages ($V \approx 2.7$ V) and the nonlinearity becomes less effective, letting the light beam propagate without walk-off in a quasi-linear diffractive manner.

The nonplanar evolution of a nematicon is also determined by the confinement across x by way of the boundaries and the applied bias: as the soliton moves toward the upper (lower) interface $x = L$ ($x = 0$) owing to walk-off in xz, the refractive gradient introduced by V attracts the self-confined beam to the mid-plane $x = L/2$ where the index increase is maximum. Eventually, the self-trapped beam bounces across the cell as it propagates along z, with a periodic behavior depending on the elevation angle. The observation of the transverse dynamics of nematicons can be carried out by tilting the image-plane of the microscope objective by 45° (Fig. 5.10), revealing the voltage-tunable oscillations in nematicon trajectories [23].

5.4.2 In-Plane Steering of Nematicon

The geometry above emphasizes the role of walk-off and the possibility of skew trajectories: interactions of solitons with boundaries [24, 25] and the intrinsic 3D nature of this steering are hardly compatible with a planar technology. Alternative electrode geometries can overcome this problem and result in in-plane nematicon steering [26].

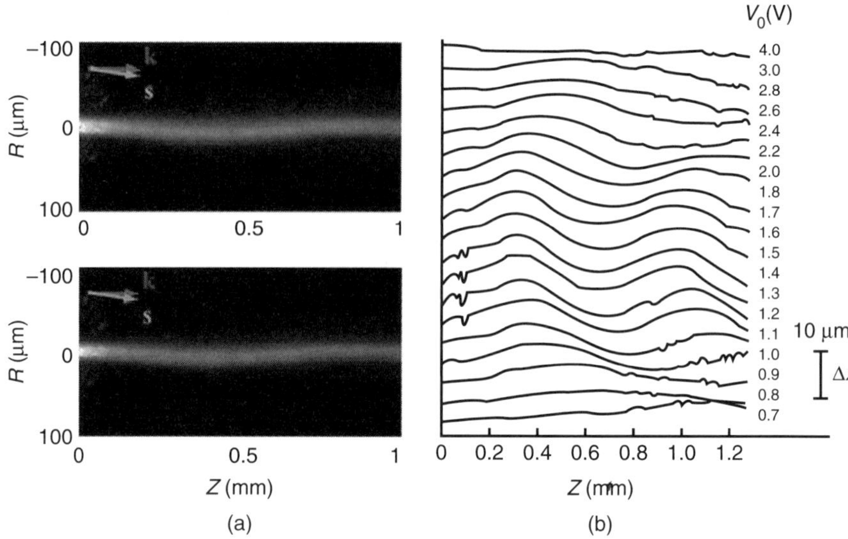

FIGURE 5.10 (a) Photographs of nematicon propagation in an observation plane rotated about 45° with respect to x, for two applied voltages. (b) Voltage-dependent nematicon trajectories.

Let us consider the sample sketched in Figure 5.11a–c: in a standard planar NLC cell, ITO electrodes are patterned on each interface (top and bottom) with a double interdigitated *comb* layout, each comb of period 2Λ and electrically connected to the same external bias, the teeth separated by Λ along z. By applying a low frequency voltage V between the electrodes, an alternating electric field E_{LF} is generated with period 2Λ: we can assume $E_{\mathrm{LF}} = 0$ in the NLC under (above) the electrodes and $E_{\mathrm{LF}} = V/(\Lambda - l)$ between them (l being the tooth width), with an exponential decay $\mathrm{e}^{-\kappa x}$ along the cell thickness, the coefficient κ modeling the coupling between the quasi-static field and the NLC molecules [27]. For $\Lambda \ll L$ and moderate voltages, the x-component of the field is averaged out by the nonlocal response and the dominant component is along z; hence, molecular reorientation essentially occurs in the yz plane. To model this geometry, we impose periodic boundary conditions to Equation 5.1, considering $|A| = 0$ and

$$|E_{\mathrm{LF}}| = \frac{V}{\Lambda - l}\mathrm{rect}_{\Lambda - l}\left[z - \Lambda - l/2\right]\mathrm{e}^{-\kappa x} \tag{5.7}$$

for $|x| < L/2$, with $\mathrm{rect}_d(z)$ equal to 1 for $z \in (-d/2\ d/2)$, 0.5 in $z = \pm d/2$ and 0 elsewhere. Numerical simulations of E7 director distribution in $0 < z < \Lambda$ for $L = 100\ \mu\mathrm{m}$, $\Lambda = 30\ \mu\mathrm{m}$, and $l = \Lambda/2$ confirm that disturbance in the director distribution do not penetrate appreciably the NLC slab: the applied electric field reorients the molecules near the electrodes, whereas the optic axis rotation around the mid-plane ($x = L/2$) is ensured by the elastic response. As a consequence, the distribution of θ is flat in $x = L/2$ and depends on the voltage [$\theta(L/2, y, z) = \theta_0(V)$,

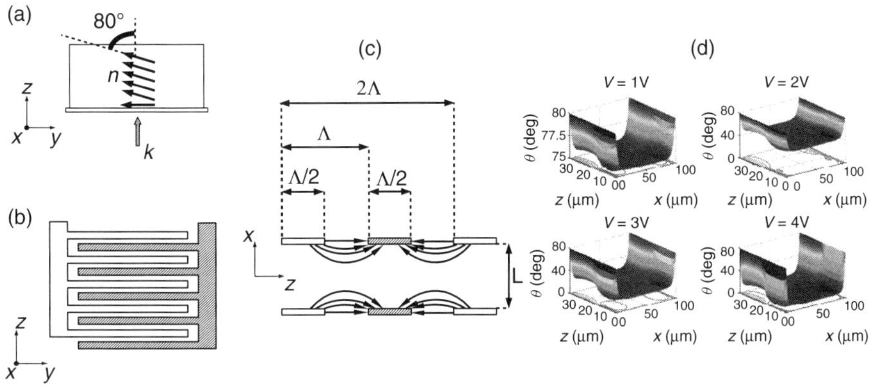

FIGURE 5.11 Details on the sample for in-plane soliton steering. (a) Top-view of the cell with indication of the initial director distribution. (b) Top and (c) side sketch of the electrode pattern: the arrows indicate the electric field lines. (d) Calculated director profile in the cell with interdigitated electrodes, for various applied voltages. In the middle of the cell the director distribution is flat.

see Fig. 5.11d]; in other words, the comb-patterned electrodes tune the uniform director distribution and the walk-off in the plane yz of the cell. When an extraordinarily (y)-polarized beam is launched in the sample, its propagation takes place in the principal plane yz, and the soliton trajectory is affected by variations in walk-off rather than in *apparent* walk-off as in the previous configuration.

Figure 5.12a shows photographs of a near-infrared beam of power $P = 2$ mW launched along z in a cell with interdigitated electrodes and $\theta_0 = 80^\circ$ at rest: for $V = 0$ V the Poynting vector of the extraordinary beam propagates at the walk-off $\delta(80^\circ) \approx 2^\circ$ with respect to z. Increasing the voltage, the nematicon trajectory is steered to higher angles up to $\delta_M \approx 6.2^\circ$ for $V = 1.75$ V (*rms* value), the maximum walk-off corresponding to $\theta_0 \approx 50^\circ$. Considering molecular reorientation in the principal plane $x = L/2$ due to an effective electric field along z, the equivalent $\theta_0 vs V$ reorientation equation is

$$K\frac{\partial^2\theta_0}{\partial z^2} - \frac{\Delta\epsilon_{\mathrm{LF}}\left|E_{\mathrm{eff}}\right|^2}{2}\sin\left(2\theta_0\right) = 0, \tag{5.8}$$

where, $\left|E_{\mathrm{eff}}\right| = \alpha(\Lambda)V$, with α being a fitting parameter depending on the electrode periodicity but not on the cell thickness. Figure 5.12b and c demonstrates the excellent agreement between calculated and measured walk-off and reorientation angle for $\alpha \approx 0.62$.

Noticeably, self-confinement occurs only in a limited range of voltages, because the reorientational response in Equation 5.4 depends on θ_0 (see Chapter 11): in this kind of cell, the nonlinearity is therefore tuned by bias, thus changing the amount of self-focusing for a given power [27, 28]. Figure 5.12d plots the beam waist after 10 Rayleigh lengths versus applied bias: nematicons are generated only when the voltage-tuned nonlinearity can compensate linear diffraction. Thus, besides being

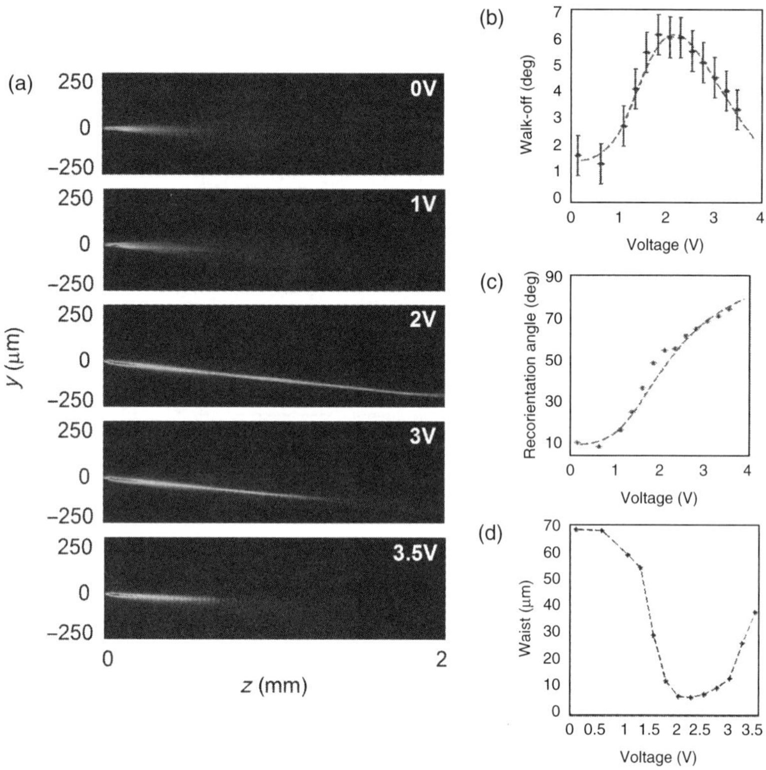

FIGURE 5.12 In-plane nematicon steering. (a) Photographs of a near-infrared ($\lambda = 1064$ nm) extraordinary beam for various cell biases. (b) Measured (dots) and calculated (dashed line) walk-off and (c) reorientation angle versus applied voltage. (d) Beam waist versus bias after propagation over 10 Rayleigh lengths. The θ dependence of the nonlinear response limits the range of voltages over which a nematicon is generated.

an ideal setting for in-plane beam steering, the geometry described can be used to adjust the core index of the nematicon-induced waveguide, allowing, for example, the nonlinear control (via soliton power) of the number of supported guided modes of a copolarized probe.

5.5 VOLTAGE-DEFINED INTERFACES

The previous examples of electrically driven nematicon steering were based on controlled changes of the optic axis—and thus walk-off—in a uniform birefringent medium. By properly patterning the electrodes, however, regions with different refractive indices can be realized into the same cell: this section deals with voltage-defined dielectric interfaces in NLC [29].

We consider a planar cell with a pattern of electrodes as sketched in Figure 5.13a, 100 μm thick and filled with E7. The top electrode is split into two regions across

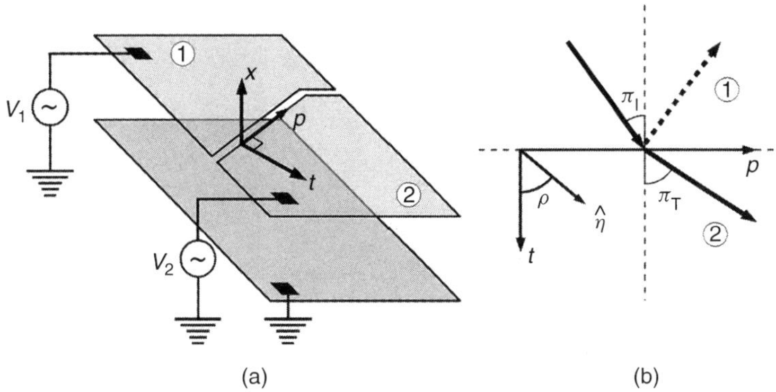

FIGURE 5.13 (a) Sketch of the sample with two top electrodes. (b) Geometry of the sample in the reference system of the interface.

a 100 μm wide gap, each region biased at a different voltage with respect to the uniform bottom electrode operating as ground terminal. The direction p along the gap forms an angle Θ with respect to z and the initial director distribution is uniform, with the optic axis at an angle ρ with the normal t to p (Figure 5.13b).

The application of in-phase voltages V_1 and V_2 to the two top electrodes induces distinct reorientations in the NLC under them, defining a graded interface in the region underneath the gap. The resulting director distribution can be calculated using a modified version of Equation 5.1, taking into account the elevation angle $\eta(V)$ and suitable boundary conditions.

An extraordinary wave interacting with this bias-controlled interface will perceive an index gradient, which is a function of $\Delta V = V_1 - V_2$. By launching an extraordinarily polarized beam of sufficient power across the interface (from first to second region), the trajectory of the resulting nematicon can be affected by the difference between the two corresponding extraordinary refractive indices n_{e1} and n_{e2}.

We stress that the two top electrodes allow to independently control the reorientation in each NLC region, i.e. the two indices, the nonlinearities, the walk-off angles, and the refractive gradient across the interface.

The ray-optics approximation is valid in this regime of propagation: the phenomenology is analogous to that of a plane wave impinging on an interface, thus Snell's law can be applied with corrections for anisotropy. With the letter Ψ we will indicate the angle measured on the plane yz, the quantities actually measured in experiments. For negative ΔV the refractive index in the first region is lower than that in the second one, and the angle of refraction Φ_T (with respect to t) is smaller than the angle of incidence Φ_I; conversely, when $V_1 > V_2$, it is $\Phi_T > \Phi_I$ until TIR can take place (for suitable Φ_I and ΔV). To maximize soliton steering, the initial director is set parallel to p ($\rho = 90^\circ$). In this case and with reference to Figure 5.14 nematicon impinges on the interface with $\Phi_I = 80^\circ$. For $V_2 = 0.7$ V and -0.8 V $< \Delta V < 0$ V (refraction, Fig. 5.14a), it undergoes a maximum angular deviation of 18°; conversely, for positive ΔV ($0V < \Delta V < 0.8V$) the nematicon

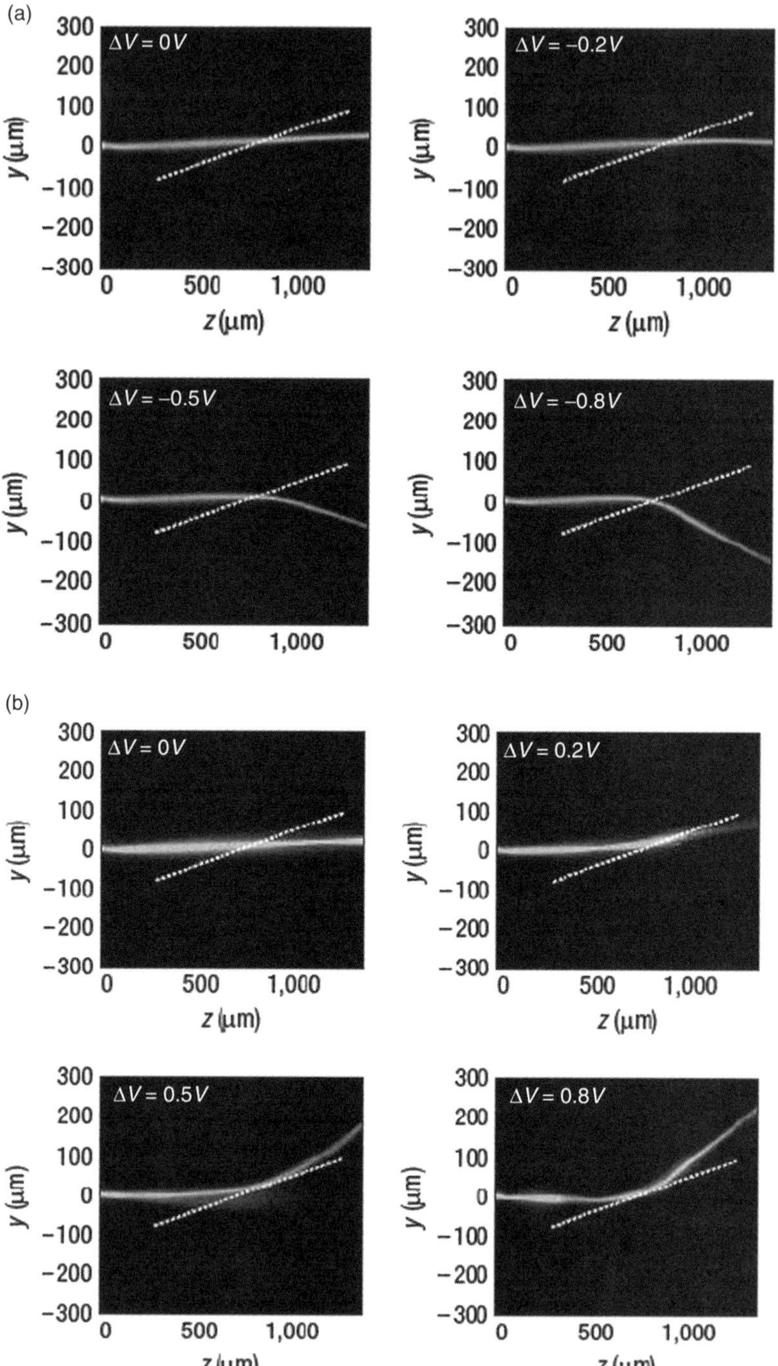

FIGURE 5.14 Nematicon interaction with a voltage-defined interface, for an incidence angle $\Phi = 80^\circ$. (a) For $\Delta V < 0$ V the nematicon undergoes refraction; (b) for $\Delta V > 0$ V and a suitable Δn TIR occurs.

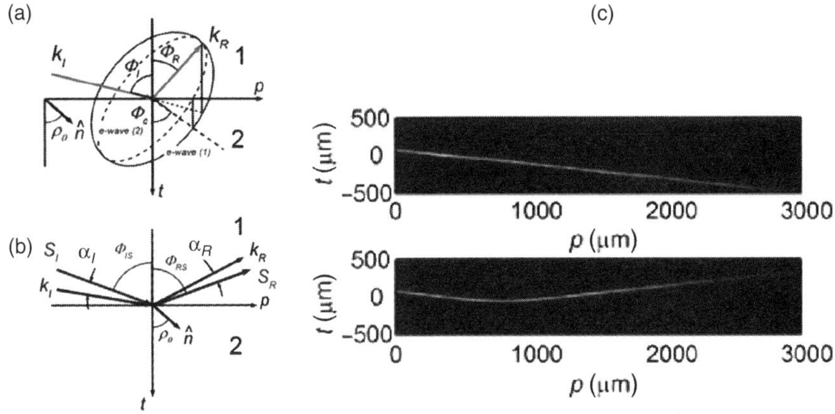

FIGURE 5.15 (a) Projection of the inverse surface of wave normals on the plane *tp*, for an extraordinary wave crossing the interface. Solid and dashed ellipses represent the indices in the first and second regions, respectively, the latter defining the critical angle Φ_C. If $\Phi_I < \Phi_C$, $\Phi_R \neq \Phi_I$ owing to anisotropy. (b) Geometric construction of the reflected wave and Poynting vectors. (c) Images of a soliton crossing the interface for $V_1 = V_2$ (top) and its TIR for $V_1 = 1.5$ and $V_2 = 0$ (bottom).

is reflected at the interface (Fig. 5.14b) with a resulting steering of 22°. In this sample, the overall electrically driven nematicon deviation reaches 40°, the largest electro-optical soliton steering ever reported.

Remarkably, as visible in the photographs, the adiabatic rotation of the director in the graded region allows for polarization adjustment of the extraordinary beam and conservation of self-confinement; as predicted by the particle-like model for solitons in nonlocal media (see Section 5.2), the nematicon survives the interaction with the interface even for large ΔV, when its transverse dynamics across x becomes appreciable due to variations in walk-off [11, 23].

Because soliton steering is measured by the angular deviation of the corresponding Poynting vector (subscript S), the angles Φ_I, Φ_T, and Φ_R are found by considering the measured apparent walk-off (see Section 5.4.1) $\alpha_I, \alpha_T, \alpha_R$, respectively. In the case of TIR, the above choice of ρ ensures that, considering the three dimensional nature of the reflection, the angles of the Poynting vectors of incident and reflected beams are always larger than those of the corresponding wave-vectors; moreover $\Phi_T = \Phi_I$ due to the symmetry and, therefore, the reflection is specular with regards to wave-vectors.

Conversely, a different ρ may cause nonspecular TIR owing to anisotropy [30]. The conservation of soliton wave-vectors can be visualized by employing the *inverse surface wave normals* (Fig. 5.15a) [6]. Conservation of wave-vector components along p requires a reflection angle larger than the incident one if the refractive index perceived by the impinging wave is larger than sensed by the reflected one (difference arising from the anisotropic nature of the medium); clearly, a necessary condition is ΔV positive and large enough to ensure TIR [30]. To illustrate this, we take $\rho = 50°$ and light impinging at an angle $\Phi_I \approx 80°$. In these conditions, the geometric construction leads to $\Phi_I = \Phi_{IS} + \alpha_I$ and $\Phi_R = \Phi_{RS} - \alpha_R$

(Fig. 5.15b). If we consider $V_1 = 1.5$V and $V_2 = 0$V, TIR is ensured; then, we can calculate $\alpha_I = 2.1^\circ$ and $\alpha_R = 2.4^\circ$, quantities required to link the Poynting vector directions with the corresponding wave-vectors. Experimentally, $\Phi_{IS} = 79^\circ$ and $\Phi_{RS} = 79.5^\circ$ are measured (Fig. 5.15c). The observable (in plane yz) nonspecularity of the Poynting vectors upon TIR is very slight owing to the opposite sign of the apparent walk-off [30]. In fact, from the measurements, we can estimate for the wave-vector $\Phi_I = 81.1^\circ$ and $\Phi_R = 76.6^\circ$, with a significant nonspecularity $\Delta\Phi = 4.5^\circ$, in quite good agreement with theoretical predictions [30].

5.6 CONCLUSIONS

Spatial solitons in NLC are robust and lend themselves to realizing a variety of routing schemes via their interactions with voltage or light defined dielectric perturbations. In this chapter, we reviewed a few selected examples of nematicon steering and processing using localized NLC defects as well as graded interfaces, in a variety of configurations, including uniform and patterned electrodes, external beams, and combinations of voltage and light. Although other noticeable examples of nematicon steering have been reported in the literature, including the use of air bubbles [31], the escape from a trapping potential [32, 33], nematicon–nematicon interactions [34–37], the use of light valves [11, 38, 39] and more [4, 40], a number of novel possibilities are still available using solitons in NLC and are likely to be exploited in the near future.

Acknowledgments

We are particularly grateful to Marco Peccianti, Malgosia Kaczmarek, Cesare Umeton, Francesco Simoni, and Claudio Conti for their essential contributions to the work reported hereby. A.A. thanks Regione Lazio for financial support.

REFERENCES

1. M. Peccianti, A. De Rossi, G. Assanto, A. De Luca, C. Umeton, and I. C. Khoo. Electrically assisted self-confinement and waveguiding in planar nematic liquid crystal cells. *Appl. Phys. Lett.*, 77(1):7–9, 2000.
2. M. Peccianti, C. Conti, G. Assanto, A. De Luca, and C. Umeton. Routing of anisotropic spatial solitons and modulational instability in nematic liquid crystals. *Nature*, 432:733, 2004.
3. A. Piccardi, G. Assanto, L. Lucchetti, and F. Simoni. All-optical steering of soliton waveguides in dye-doped liquid crystals. *Appl. Phys. Lett.*, 93:171104, 2008.
4. A. Piccardi, A. Alberucci, and G. Assanto. Self-turning self-confined light beams in guest-host media. *Phys. Rev. Lett.*, 104:213904, 2010.
5. I. C. Khoo. *Liquid Crystals: Physical Properties and Nonlinear Optical Phenomena*. Wiley Interscience, New York, 1995.
6. M. Born and E. Wolf. *Principles of Optics*. Pergamon Press, New York, 1975.

7. P. G. De Gennes and J. Prost. *The Physics of Liquid Crystals*. Oxford Science, New York, 1993.
8. A. Alberucci and G. Assanto. Propagation of optical spatial solitons in finite-size media: interplay between nonlocality and boundary conditions. *J. Opt. Soc. Am. B*, 24(9):2314–2320, 2007.
9. Ch. Mauguin. Sur les cristaux liquid de Lehmann. *Bull. Soc. Fr. Miner.*, 34:71–117, 1911.
10. A. Alberucci, M. Peccianti, G. Assanto, G. Coschignano, A. De Luca, and C. Umeton. Self-healing generation of spatial solitons in liquid crystals. *Opt. Lett.*, 30:1381–1383, 2005.
11. A. Alberucci, A. Piccardi, U. Bortolozzo, S. Residori, and G. Assanto. Nematicon all-optical control in liquid crystal light valves. *Opt. Lett.*, 35(3):390–392, 2010.
12. A. W. Snyder and D. J. Snyder. Accessible solitons. *Science*, 276:1538–1541, 1997.
13. A. Alberucci and G. Assanto. Nematicons beyond the perturbative regime. *Opt. Lett.*, 35(15):2520–2522, 2010.
14. C. Conti, M. Peccianti, and G. Assanto. Route to nonlocality and observation of accessible solitons. *Phys. Rev. Lett.*, 91:073901, 2003.
15. C. Conti, M. Peccianti, and G. Assanto. Observation of optical spatial solitons in a highly nonlocal medium. *Phys. Rev. Lett.*, 92:113902, 2004.
16. J. J. Sakurai. *Modern Quantum Mechanics*. Addison-Wesley, Reading, MA, 1994.
17. M. Peccianti A. Pasquazi, A. Alberucci and G. Assanto. Signal processing by opto-optical interactions between self-localized and free propagating beams in liquid crystals. *Appl. Phys. Lett.*, 87:261104, 2005.
18. N. Tabiryan, U. Hrozhyk, and S. Serak. Nonlinear refraction in photoinduced isotropic state of liquid crystalline azobenzenes. *Phys. Rev. Lett.*, 93:113901, 2004.
19. S. V. Serak, N. V. Tabiryan, M. Peccianti, and G. Assanto. Spatial soliton all-optical logic gates. *IEEE Photon. Technol. Lett.*, 18:1287, 2006.
20. E. Ouskova, Y. Reznikov, S. V. Shiyanovskii, L. Su, J. L. West, O. V. Kuksenok, O. Francescangeli, and F. Simoni. Photo-orientation of liquid crystals due to light-induced desorption and adsorption of dye molecules on an aligning surface. *Phys. Rev. E*, 64(5):051709, 2001.
21. E. Ouskova, D. Fedorenko, Y. Reznikov, S. V. Shiyanovskii, L. Su, J. L. West, O. V. Kuksenok, O. Francescangeli, and F. Simoni. Hidden photoalignment of liquid crystals in the isotropic phase. *Phys. Rev. E*, 63(2):021701, 2001.
22. L. Lucchetti, D. E. Lucchetta, O. Francescangeli, and F. Simoni. Sine: Surface induced nonlinear effects. *Mol. Cryst. Liq. Cryst.*, 375:1641–650, 2002.
23. M. Peccianti, A. Fratalocchi, and G. Assanto. Transverse dynamics of nematicons. *Opt. Express*, 12:6524, 2004.
24. B. Alfassi, C. Rotschild, O. Manela, M. Segev, and D. N. Christodoulides. Boundary force effects exerted on solitons in highly nonlocal nonlinear media. *Opt. Lett.*, 32:154, 2006.
25. A. Alberucci, M. Peccianti, and G. Assanto. Nonlinear bouncing of nonlocal spatial solitons at the boundaries. *Opt. Lett.*, 32(19):2795–2797, 2007.
26. A. Piccardi, M. Peccianti, G. Assanto, A. Dyadyusha, and M. Kaczmarek. Voltage-driven in-plane steering of nematicons. *Appl. Phys. Lett.*, 94(9):091106, 2009.

27. A. Alberucci, A. Piccardi, M. Peccianti, M. Kaczmarek, and G. Assanto. Propagation of spatial optical solitons in a dielectric with adjustable nonlinearity. *Phys. Rev. A*, 82(2):023806, 2010.

28. A. Piccardi, A. Alberucci, and G. Assanto. Power-dependent nematicon steering via walk-off. *J. Opt. Soc. Am. B*, 27:2398–2404, 2010.

29. M. Peccianti, A. Dyadyusha, M. Kaczmarek, and G. Assanto. Tunable refraction and reflection of self-confined light beams. *Nat. Phys.*, 2:737–742, 2006.

30. M. Peccianti, G. Assanto, A. Dyadyusha, and M. Kaczmarek. Non-specular total internal reflection of spatial solitons at the interface between highly birefringent media. *Phys. Rev. Lett.*, 98:113902, 2007.

31. Y. V. Izdebskaya, V. G. Shvedov, A. S. Desyatnikov, W. Krolikowski, and Y. Kivshar. Soliton bending and routing induced by interaction with curved surfaces in nematic liquid crystals. *Opt. Lett.*, 35(10):1692–1694, 2010.

32. M. Peccianti, A. Dyadyusha, M. Kaczmarek, and G. Assanto. Escaping solitons from a trapping potential. *Phys. Rev. Lett.*, 101(15):153902, 2008.

33. G. Assanto, A. A. Minzoni, M. Peccianti, and N. F. Smyth. Nematicons escaping a wide trapping potential: modulation theory. *Phys. Rev. A*, 79:033837, 2009.

34. M. Peccianti, K. A. Brzdakiewicz, and G. Assanto. Nonlocal spatial soliton interactions in bulk nematic liquid crystals. *Opt. Lett.*, 27(16):1460–1462, 2002.

35. M. Peccianti, C. Conti, G. Assanto, A. De Luca, and C. Umeton. All optical switching and logic gating with spatial solitons in liquid crystals. *Appl. Phys. Lett.*, 81(18):3335–3337, 2002.

36. M. Peccianti, C. Conti, G. Assanto, A. De Luca, and C. Umeton. Nonlocal optical propagation in nonlinear nematic liquid crystals. *J. Nonlin. Opt. Phys. Mater.*, 12(4):525–538, 2003.

37. Y. V. Izdebskaya, V. G. Shvedov, A. S. Desyatnikov, Y. Kivshar, W. Krolikowski, and G. Assanto. Incoherent interaction of nematicons in bias-free liquid crystal cells. *J. Eur. Opt. Soc.*, 5:10008, 2008.

38. A. Piccardi, U. Bortolozzo, S. Residori, and G. Assanto. Spatial solitons in liquid crystal light valves. *Opt. Lett.*, 34(6):737–739, 2009.

39. A. Piccardi, A. Alberucci, U. Bortolozzo, S. Residori, and G. Assanto. Readdressable interconnects with spatial soliton waveguides in liquid crystal light valves. *IEEE Photon. Technol. Lett.*, 22(10):694–696, 2010.

40. M. Peccianti and G. Assanto. Signal readdressing by steering of spatial solitons in bulk nematic liquid crystals. *Opt. Lett.*, 26(21):1690–1692, 2001.

6

Dynamics of Optical Solitons in Bias-Free Nematic Liquid Crystals

YANA V. IZDEBSKAYA, ANTON S. DESYATNIKOV, AND YURI S. KIVSHAR

Nonlinear Physics Centre, Research School of Physics and Engineering, The Australian National University, Canberra, ACT, Australia

6.1 SUMMARY

We present a review of our experimental work on spatial optical solitons in nematic liquid crystals (NLCs) in a planar cell without external fields. We investigate the nonlocal interaction of two mutually incoherent copropagating (CO) and counter-propagating (CP) bias-free *nematicons*; the analysis of the dynamics of CP solitons versus their relative separation and excitation shows the existence of stable vector solitons composed of two oppositely propagating nematicons. Efficient steering of nematicons is achieved by introducing defects in the liquid crystal layer: we demonstrate significant bending as well as large-angle total internal reflection of nematicons interacting with a curved interface between NLCs and air. Nematicons induce guiding channels capable of confining a weak signal: we identify experimentally the domains of nematicon power where higher order guided modes are supported. Finally, we present recent results on the astigmatic transformation of vortex beams in NLCs into spiraling dipole azimuthons accompanied by power-dependent charge-flipping of on-axis phase singularity.

6.2 INTRODUCTION

Spatial optical solitons have been studied extensively in diverse nonlinear media, in both one- and two-dimensional geometries [1]; they have significant potentials for signal processing, switching, and readdressing in the next generations of all-optical circuits. In this context, the giant optical nonlinearity arising from molecular

Nematicons: Spatial Optical Solitons in Nematic Liquid Crystals, First Edition.
Edited by Gaetano Assanto.

reorientation in NLCs has attracted significant attention [2, 3]. The reorientational nonlinearity allows generation of spatial optical solitons at relatively low optical powers, in the milliWatt region or below [4], and provides direct access to the study of fundamental aspects and applications of light interaction with self-assembling nonlinear soft matter. Both theoretical [5] and experimental [6] results have been reported for spatial optical solitons in NLC, also called *nematicons* [7]. The reorientational optical nonlinearity can be modulated by prealigning the NLC molecules with external electric field. Such bias allows for a versatile control over the orientation of the organic molecules at rest and, therefore, provides fine tuning of the strengths of both nonlinearity and nonlocality [8]. However, nematicons can also exist in simpler settings with unbiased NLC cells, where the molecule pretilt in the bulk of the planar cell is achieved by the anchoring conditions at the boundaries [9–13]. The nonlinear effects, induced by small tilts of the molecules by laser light, are sufficiently strong to observe self-focusing effects at milliwatt input powers, provided the input beam is extraordinarily polarized, that is, with electric field and molecular director coplanar in the plane (x, z), parallel to the cell interfaces [9]. Because of the optical anisotropy of the liquid crystalline molecules and the resulting birefringence of the nematic medium, light beams propagating in NLC walk-off the direction of their wave vectors (see also Chapter 1).

In this chapter, we present an overview of our recent experimental results on the generation and dynamics of spatial optical solitons in short planar NLC cells without external bias [14–18]. We investigate in detail the power-dependent nonlocal interaction of two identical mutually incoherent CO nematicons (Section 6.3) and CP nematicons (Section 6.4), demonstrating the existence of a class of *vector nematicons* consisting of two self-trapped beams propagating in opposite directions. Depending on the parameters of the setup we observe either stable stationary states or dynamic instabilities of two CP beams in the form of splitting and spatial entanglement. In Section 6.5, the interaction of spatial optical solitons with curved dielectric surfaces is studied in an NLC layer with injected air bubbles of various sizes; this interaction allows large-angle bending of nematicons as well as total internal reflection preserving the nondiffracting character of the self-trapped beams. In Section 6.6, we show that the waveguides induced by spatial solitons in unbiased NLCs can support various copolarized higher order guided modes. Section 6.7 presents our experimental observations of the so-called dipole azimuthons, robust higher order spatial solitons forming in nematics from an input vortex beam. Finally, Section 6.8 concludes this chapter.

6.3 FROM ONE TO TWO NEMATICONS

In the experiments, we use a short planar cell of length $L = 1.1$ mm (Fig. 6.1a) consisting of two polycarbonate slides spaced by 100 μm and filled with 6CHBT [19, 20], an NLC with relatively low absorption. The inner surfaces of the slides are unidirectionally rubbed in order to align the NLC molecular director $\mathbf{n}$ in the plane (x,z) at 45° with respect to the z axis. Such boundary conditions in (x,z) are

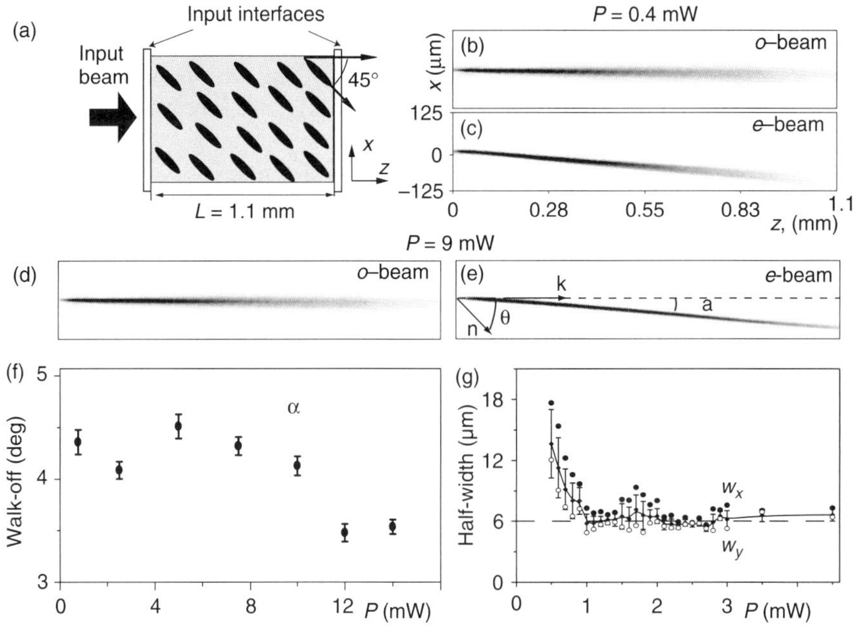

FIGURE 6.1 (a) Experimental setup: top view of the planar cell with ellipses indicating the orientation of the molecules in the plane (x,z). (b–e) Experimental images of the radiation scattered from ordinary (b, d) and extraordinary (c, e) beams in a bias-free NLC cell: linearly diffracting beams with $P = 0.4$ mW in (b, c) and $P = 9$ mW in (d) and self-trapped e-beam with $P = 9$ mW in (e). Here and below the darker areas indicate higher light intensity. (f) Walk-off angle α versus power, as defined in (e). (g) Nematicon HWHM in x and y directions, $w_{x,y}$,respectively, are shown by dots; mean radius and standard deviation of a circular fit of the half-peak intensity contour are shown by solid curve and error bars, respectively. Adapted from References 15 and 17.

analogous to the bulk prealignment in (y,z) by an external bias voltage [6]. Two additional glass slides are attached perpendicularly to the propagation axis z at two opposite sides of the cell in order to define air–NLC input/output interfaces, avoiding lensing and depolarization due to an NLC meniscus [6]. We use microscope objectives and CCD cameras to collect the light at the sample output and the light scattered above the cell along propagation.

First of all, we study the propagation of a single beam in such a cell for various excitations. Figure 6.1 shows typical top-view photos of (b,d) the ordinarily polarized o beam, with electric field parallel to y, and (c,e) the extraordinarily polarized e beam, with electric field parallel to x, launched with input powers $P = 0.4$ mW in (b,c) and $P = 9$ mW in (d,e). Clearly, the o beams in Figure 6.1b and 6.1d and the low power e beam in Figure 6.1c diffract, whereas the e beam in Figure 6.1e experiences self-focusing at power $P \geq 0.9$ mW and gives rise to a self-localized state, that is, a nematicon. Owing to the high birefringence $\Delta\epsilon \approx 0.15$ (refractive indices $n_e = 1.6718$ and $n_o = 1.5225$ at room temperature of 23°C at wavelength $\lambda = 532$ nm) in our geometry, the nematicon travels in the NLC along the direction of the Poynting vector with a significant spatial walk-off angle, $\alpha \approx 4.2°$, with respect to the z direction [15, 21]. The reduction of walk-off α with input power is plotted in Figure 6.1f.

Noteworthy, the nematicons are generated in a liquid uniaxial medium subject to slow dynamics [22] and instabilities [15, 23], where the birefringence-induced walk-off is power sensitive [15, 21] and nonlocality leads to beam breathing [24, 25]. In the absence of external bias, as in our configuration, these effects result in a bent nonuniform waveguide with fluctuations on a time scale smaller than the inverse maximum frame-rate of our camera (25 fps). Therefore, to quantitatively characterize the nematicon profile, we define the nematicon half-width at half-maximum (HWHM), $w_{x,y}$, as two half-sizes of a rectangle enclosing the contour line of the averaged intensity profile at the half-peak level. Figure 6.1g shows the experimental results at the cell output for a nematicon excited with the green laser at 532 nm. At low input power ($P < 0.9$ mW) the gradual reduction of beam size in Figure 6.1g shows its self-focusing, whereas at higher power $P > 0.9$ mW a nematicon is formed.

Next, we study the dynamics and interaction of two mutually incoherent—initially parallel—copropagating beams in NLC at various excitation levels. In order to generate two CO beams, we used a standard Mach–Zehnder arrangement. As discussed earlier [26], owing to the long-range character of the nonlocality the nematicons attract each other, more and more for increasing input powers. In most cases, this attraction is independent of the relative phase of the solitons, even when they are mutually coherent [26]. A sequence of images in Figure 6.2 shows the stationary trajectories of two initially parallel beams for different input powers. At low input power (0.5 mW) self-focusing is too weak to overcome diffraction, so the beams keep spreading as they propagate. By increasing the power to 2 mW we achieve stable propagation of solitons and their weak attraction. For higher excitations, the attraction is sufficiently strong to induce one (at $P \geq 3$ mW) or multiple ($P = 13.5$ mW) intersections in the nematicon trajectories.

6.4 COUNTER-PROPAGATING NEMATICONS

One of the simplest processes that leads, in nonlinear optics, to a variety of complex phenomena is the mutual interaction of two CP optical beams. Numerous concepts, such as phase conjugation, Bragg reflection by volume gratings, and wave-mixing, are based on this geometry. Nevertheless, a simple CP geometry can give rise to an extremely complicated and sometimes counterintuitive dynamic behavior, including mutual beam self-trapping and the formation of stationary states, as well as complex spatiotemporal instabilities [27].

Fundamental concepts motivated earlier studies of optical solitons created by two beams propagating in opposite directions, and it was shown that mutual self-trapping of two CP beams can lead to the formation of a novel type of *vector soliton* [28, 29], for both coherent and incoherent interactions. More detailed analyses [30, 31] revealed that CP solitons may display a variety of instabilities, accompanied by nontrivial temporal and spatial dynamics.

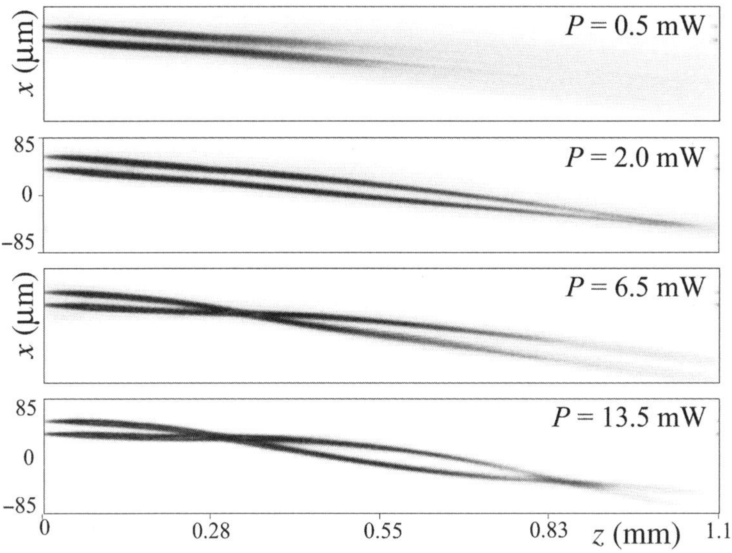

FIGURE 6.2 Experimentally recorded images of light scattered from two copropagating nematicons at various input powers P in each beam [14].

In order to investigate the propagation dynamics and interaction of CP solitons, we launched in our cell beams from opposite sides. When the two inputs propagate toward each other with a finite impact parameter r (the offset, see dashed lines in Figure 6.3A), we observe the formation of two nematicons and their bending in opposite transverse directions owing to walk-off. Because walk-off is power dependent (see also Chapter 11), the interaction of CP nematicons is parameterized by both the offset r and the input power P.

Typical experimental results on the interaction of CP solitons are presented in Figure 6.3. As expected from the single beam case in Figure 6.1, an increase in excitation $P > 0.9$ mW leads to self-focusing and the formation of nematicons. For a given offset r and small powers the beams do not interact, as shown in Figure 6.3A. Here and later, we distinguish the absence or the presence of interaction by blocking one of the beams and observing the other one that relaxes (in time) to its independent (unperturbed) trajectory. With decrease in the offset r (or increase in the input power P), the two CP beams still remain spatially separated but the attraction between them leads to additional bending, clearly visible in Figure 6.3B for $r = 130$ μm. The bound state of two CP nematicons is obtained for specific parameters r and P, as seen in Figure 6.3C. On the other hand, the sequence of Figure 6.3C and 6.3D for $r = 105$ μm shows how such bound state can be destroyed by an increase in power, which leads in Figure 6.3D to the development of a dynamic transverse instability of the two time-averaged CP beams in the form of spatial splitting and spatial entanglement.

The results of numerous experiments are summarized in Figure 6.4a with differently shaded regions for qualitatively different interaction scenarios, as described

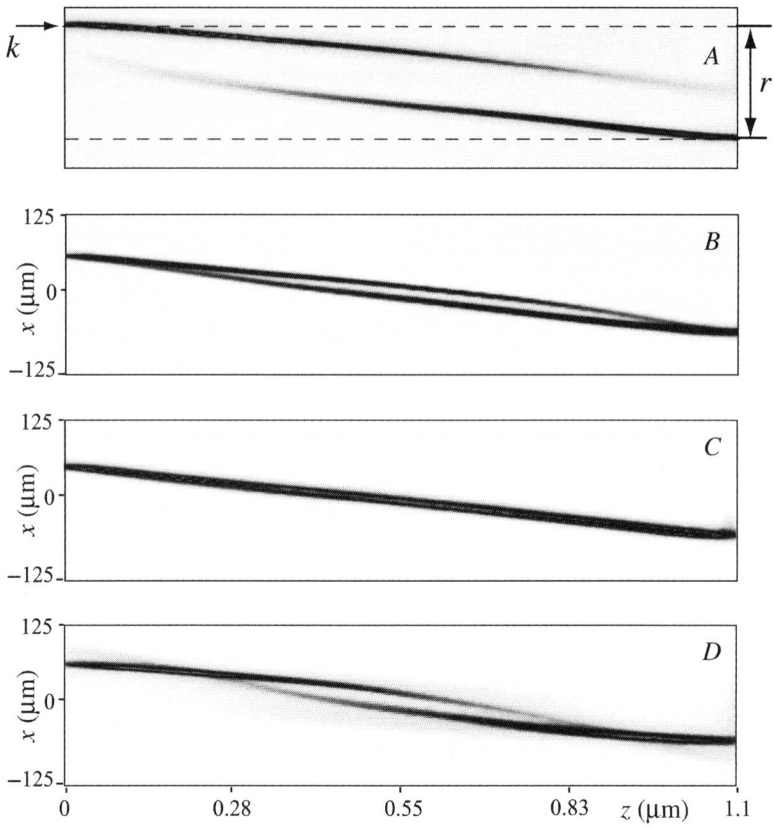

FIGURE 6.3 Experimental top-view snapshots of CP solitons in different interaction regimes [15]. (A) Separation $r = 180$ μm, no interaction; (B) $r = 130$ μm, soliton attraction; (C) $r = 105$ μm, bound state with a commonly induced waveguide; (D) $r = 105$ μm, instability. The input power in each beam is $P = 5$ mW in (A–C) and $P = 14$ mW in (D).

earlier. Importantly, in all our experiments, the beams are launched parallel to the optic axis z with equal powers P and, because the interaction is *phase independent*, the key parameters governing the dynamics are offset r and power P.

For powers below a threshold, no interaction is observed (blank area) and each nematicon trajectory does not bend when the CP soliton is launched (Fig. 6.3A). The threshold is offset dependent and it can be seen as the border of the light gray domain of attraction marked with black triangles in Figure 6.4a. By increasing the power we enter this domain of mutual attraction and observe the two trajectories bending toward each other while propagation remains stationary in time (Fig. 6.3B). Vector nematicons (Fig. 6.3C) are formed in the dark area with square markers, and the instability (Fig. 6.3D) is observed in a narrow region with circles in Figure 6.4a.

The domains in Figure 6.4a point out a counterintuitive feature, namely, an asymmetry with respect to the (dashed) line in $r = 105$ μm. This offset value corresponds to a “head-on” collision of the two CP solitons at small powers. Above

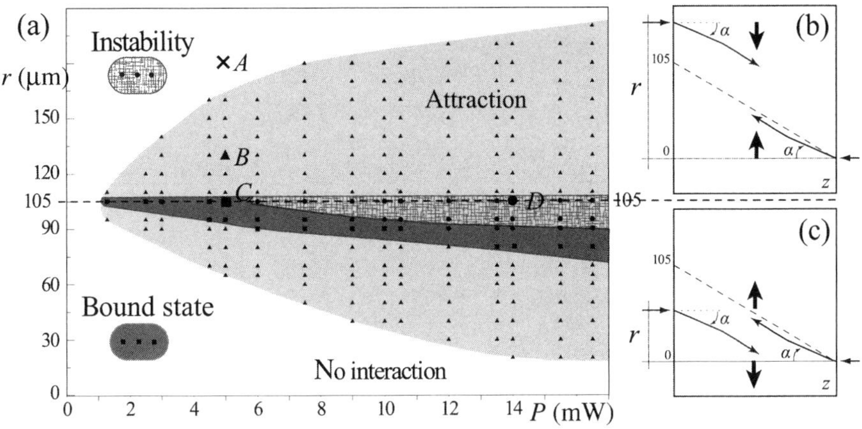

FIGURE 6.4 (a) Stability and dynamics diagram of the parameter domains (P,r) corresponding to different scenarios of CP soliton interaction [15]. Typical intensity snapshots at the points marked A–D are shown in Figure 6.3. (b and c) Sketches of two distinct regimes of interaction between forward and backward nematicons, separated by a critical value of the offset $r = 105$ μm. In both cases, the nonlinear attraction pulls the two CP solitons toward each other, either acting in the same (b) or in the opposite (c) direction with respect to the transverse walk-off indicated by straight vertical arrows.

and below this value a qualitative difference appears between the two configurations sketched in Figure 6.4b and 6.4c. For an offset $r > 105$ μm in Figure 6.4b, the walk-off of the two nematicons, indicated by curved arrows, acts in *the same* transverse direction of the mutual nonlinear attraction. Conversely, for $r < 105$ μm in Figure 6.4c, the two nematicons bend away from each other due to walk-off, counteracting mutual attraction. As a result, the bound states of CP nematicons with relatively straight trajectories appear at low powers only for offsets $r \simeq 105$ μm (see the dark area in Fig. 6.4a), because they require a lesser amount of curvature in order to merge into a joint waveguide. For $r \leq 105$ μm (such as that shown in Fig. 6.3C and Fig. 6.4a) vector CP nematicons require larger excitation powers.

6.5 INTERACTION OF NEMATICONS WITH CURVED SURFACES

Nematicons are excellent candidates for all-optical soliton control and signal routing. Efficient and tunable routing of nematicons can be achieved by reflection or refraction at the interface between two nonlinear media, each controlled by an independent external bias [32, 33] or by anchoring at the boundaries [34]. Similarly, localized nonlinear defects induced by an additional "control" light beam can be repulsive [13, 35] or attractive [13, 36], depending on the induced change (negative or positive) of the refractive index. However, if the boundary conditions are linear, for example, fixed by the anchoring of NLC molecules, the interface is repulsive [37, 38].

In this section, we report on the experimental demonstration of the strong interaction of nematicons with curved dielectric surfaces. To induce a curved

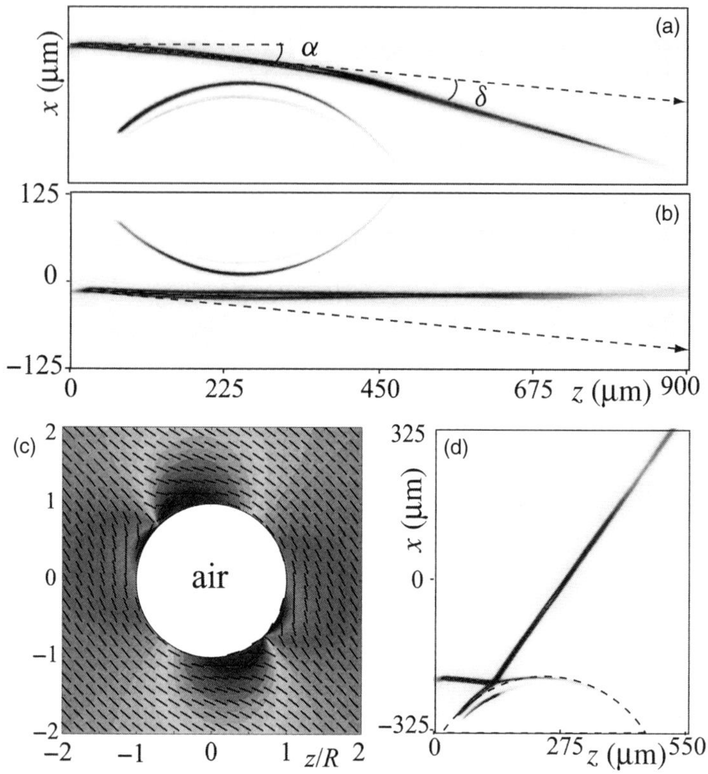

FIGURE 6.5 Experimental results on nematicon interaction with a curved dielectric surface [16]. (a and b) The interaction with the surface leads to bending of the nematicon trajectories on either side of the bubble surface, the latter shown with curved contour. The dashed arrows show the directions of unperturbed solitons propagating without bubble. (c) Calculated changes of molecular orientation (dashes) and refractive index (gray shade) induced by the cylindrical interface in NLC. (d) Total internal reflection of a nematicon from a curved dielectric surface; the dashed line shows the surface of the air bubble.

linear/nonlinear interface between air and NLC, we injected microscale air bubbles of different sizes into the layer. From the side of the cell, such extended defects have cylindrical shapes with generatrices orthogonal to the planar cell boundaries. In our experiments, we kept the input nematicon power $P = 3.5$ mW constant while varying the relative distance between soliton and defect.

The formation of a "free" nematicon, without air bubble, is accompanied by walk-off, which effectively tilts the straight-line trajectory by the angle α with respect to the axis z (see dashed arrows in Fig. 6.5a and b). In sharp contrast with the previous case, when the soliton approaches the bubble, its trajectory bends toward the surface, as is seen from gray-shaded intensity profiles in Figure 6.5a and b. Such bending indicates attraction by the surface. In good qualitative agreement, the numerically calculated changes of molecular orientation and refractive index induced by the bubble in Figure 6.5c show a significant increase of the latter (dark regions), or the induction of an attractive potential.

At the same time, the bending is quantitatively different for nematicons approaching the bubble from the top and the bottom (Fig. 6.5a and b). This apparent difference can be ascribed to the counteraction of two "forces," the walk-off and the attraction to the surface. Qualitatively, whereas attraction is in the direction of walk-off in Figure 6.5a and increases the nematicon inclination δ, it is opposite to and compensates the walk-off in Figure 6.5b.

Finally, when the target soliton trajectory crosses the bubble surface, as shown in Figure 6.5d, the soliton experiences total internal reflection, observed earlier in different settings [32]. Unlike the cases studied earlier, in our geometry, the nematicon steering may exceed 90°. Surprisingly, despite the very different propagation direction after reflection, the solitons propagate similar to rays of light, with hardly any visible differences in widths and profiles before and after collision with the curved surface.

6.6 MULTIMODE NEMATICON-INDUCED WAVEGUIDES

An attractive property of soliton-induced waveguides is their ability to guide weak signals of different wavelengths [6, 39, 40] as well as the possibility to reconfigure such waveguides by spatial steering. Another dimension in soliton-based optical switching is offered by the multimodal character of the self-induced waveguides, observed earlier with photorefractive solitons [39, 41]. In this regard, the *nematicon waveguides*, that is, those induced by spatial optical solitons in NLCs [6, 42], are of particular interest because of the long-range or highly nonlocal character of the reorientational nonlinearity. The transverse size of the nematicon-induced index perturbation can be up to one order of magnitude wider than the beam [24, 43, 44] and, therefore, such a waveguide is expected to be multimodal. Additional evidence of such multimodality is the existence of higher order nonlocal solitons [45–47] and incoherent solitons in photorefractive crystals [48] and nematics [49].

Here, we describe experimental observations of higher order modes guided by bias-free nematicons [17]. First, we study the dynamics of a weak extraordinary polarized signal (red, wavelength 632.8 nm) fundamental H_{00}[1] mode launched collinearly with the green nematicon ($\lambda = 532$ nm). Figure 6.6a and 6.6b shows top-view images of the extraordinary signal beam, resulting in either diffraction without nematicon (Fig. 6.6a) or guided-wave propagation by the presence of a nematicon (Fig. 6.6b). The power of the input signal (red) beam was 82 μW in all experiments.

Figure 6.6c–e shows experimental results at the cell output for a H_{00} mode. At low nematicon power ($P < 0.9$ mW), self-focusing reduces the waist in the green region and the corresponding size of the mode in the red region, until a

[1]We use the notation H_{mn} for Hermite-like modes with indices (m,n) because the proper distinction between hybrid EH and HE modes is not possible [50] in birefringent liquid crystals.

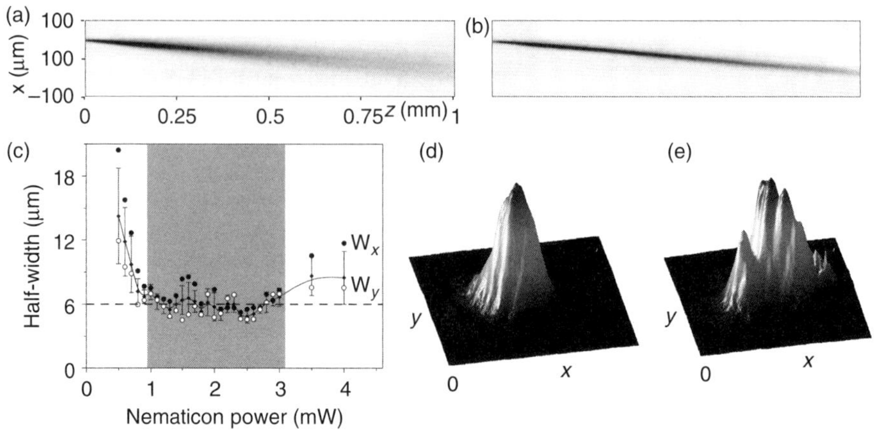

FIGURE 6.6 Fundamental H_{00} mode of the nematicon waveguide [17]. Experimental results for a weak extraordinarily polarized signal: (a) diffracting and (b) nematicon-guided beams, respectively; (c) HWHM $w_{x,y}$ (dots) of the H_{00} guided mode. The shaded area in (c) marks the region of stable guidance. (d and e) Averaged intensity distributions of the mode guided by nematicons of power (d) $P = 1.5\,\text{mW}$ and (e) $P = 4\,\text{mW}$, respectively.

nematicon is generated for $P > 0.9$ mW (Fig. 6.1g) and the red signal becomes guided (Fig. 6.6c). The shaded region in Figure 6.6c marks the existence of a stable H_{00} guided mode, with a typical averaged output transverse profile displayed in Figure 6.6d. Further increases in nematicon power P lead to a soliton waveguide supporting a larger number of guided modes, with a stronger mixing and an output exhibiting a multi–hump profile, as in Figure 6.6e. The latter transition for $P >$ 3 mW is accompanied by an increase in HWHM for the red signal, as apparent in Figure 6.6c, whereas the nematicon maintains its robust structure (Fig. 6.1g). Clearly, the nonlinear refractive index potential gets reshaped with nematicon excitation, inducing a multimoded channel waveguide at the output of which a complex pattern appears owing to superposition and mixing of guided modes [50].

To excite the first-order mode in the soliton-induced waveguide, we insert a thin glass plate in front of half of the signal beam and tilt it, introducing a π-phase jump and reproducing the phase profiles of H_{10} and H_{01} modes, respectively. The HWHM of both modes versus nematicon excitation (Fig. 6.7a–c) exhibits (shaded) regions of stable guidance, with averaged profiles shown in Figure 6.7b–d. We temporally average the output images of higher order modes as recorded during data acquisition (as discussed in Section 6.3; Fig. 6.1g). At low excitation powers, the strong diffraction of the signal beam is limited only in the y direction by the cell boundaries, thus $w_y < w_x$. At variance with soliton waveguides in photorefractive crystals, subject to a directional bias [41], both dipole-like modes coexist in the same interval of excitations, $1.2 < P[\text{mW}] < 2$, suggesting that, in this case, the role of the cell boundaries and the related index anisotropy [51] are minimal.

Using two glass plates, we generate the red signal mode H_{11}; see, for example, Figure 3e and f in Reference 17. The HWHM of this mode in both x and y

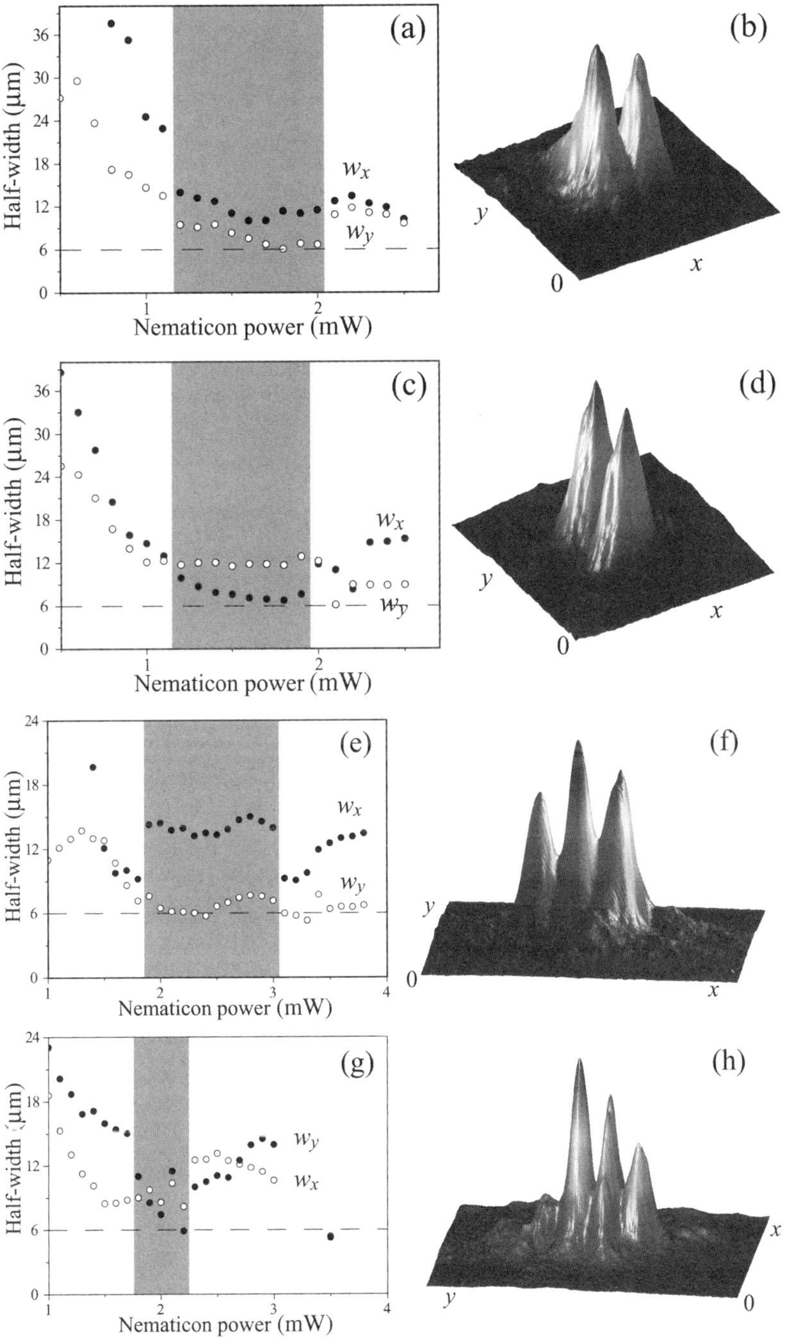

FIGURE 6.7 Experimental results [17] at the cell output for (a and b) H_{10}, (c and d) H_{01}, (e and f) H_{20}, and (g and h) H_{02} modes guided by a nematicon. (a,c,e,g) HWHM versus nematicon power P; (b,d,f,h) Averaged intensity outputs for $P = 2\,\text{mW}$ in a window 80.7×80.7 μm.

corresponds well to the sizes of the H_{10} and H_{01} modes above, ~12 μm, approximately twice wider than the nematicon. As expected, the stable domain for H_{11} is shifted toward higher powers with $1.4 < P[\text{mW}] < 2.3$ [17]. Finally, Figure 6.7 displays experimental results for the H_{20} and H_{02} modes, generated with two glass plates, as well. The H_{20} mode displays a relatively broad domain of stable existence, $1.8 < P[\text{mW}] < 3$. Its HWHM $w_x \simeq 15$ μm is nearly three times larger than the nematicon size. In sharp contrast, the vertically oriented H_{02} mode (Fig. 6.7h) is only quasi-stable in a narrow domain, $1.7 < P[\text{mW}] < 2.2$, due to the (anisotropic) role of the cell boundaries at $y = 0, d$ (where d is the cell thickness) with fixed director orientation and despite the fact that its total width, $2w_y < 24$ μm, is much smaller than $d = 100$ μm. It follows that the order of the supported guided modes is limited from above not only by the nematicon power and mode mixing process but also by the boundary-induced anisotropy [51]. Noteworthy, for H_{mn} with $(m, n) \leq 1$, the nematicon waveguides can be considered isotropic and circularly symmetric.

6.7 DIPOLE AZIMUTHONS AND CHARGE-FLIPPING

A link between fundamental optical spatial solitons [1] and doughnut-shaped vortices [52, 53] is provided by the existence of dynamic bound states of solitons in the form of rotating soliton clusters [54] and azimuthally modulated vortex solitons, or azimuthons [55, 56]. In this section, we report on the formation of dipole azimuthons in unbiased NLC, with nontrivial charge-flipping of on-axis phase dislocation [18].

A single-charged vortex beam is generated with a fork-type amplitude diffraction hologram from an extraordinarily polarized cw (continuous wave) laser beam of wavelength 800 nm. At low input power, $P < 0.9$ mW, the vortex beam uniformly diffracts without any noticeable self-action. As the power increases, $1 < P[\text{mW}] < 2.2$, the beam experiences self-focusing and its output size visibly reduces. In this regime, the initial radially symmetric vortex undergoes a drastic transformation, see for example, Figure 1 in Reference 18: the vortex "doughnut" breaks up into two beams with a dark core transformed into a tilted stripe, similar to vortex astigmatic transformations [57].

A further increase in input vortex power ($P > 2.3$ mW) leads to stronger narrowing of the output beam and to the formation of a self-trapped dipole-like beam. Figure 6.8 shows the longitudinal dynamics and the transverse output profiles of the dipole beam for three different input powers. The whole beam is tilted and bent due to walk-off and diffraction is suppressed. Note the dark line along the beam, splitting it into two lobes in Figure 6.8a–c. The corresponding output transverse patterns in Figure 6.8d–f contain two pronounced bright spots, clearly coupled together in a structure similar to a numerically simulated dipole azimuthon [58]. For different input powers the output dipole has a strongly varying elliptic shape tilted at different angles with respect to the boundaries. Importantly, this spatial

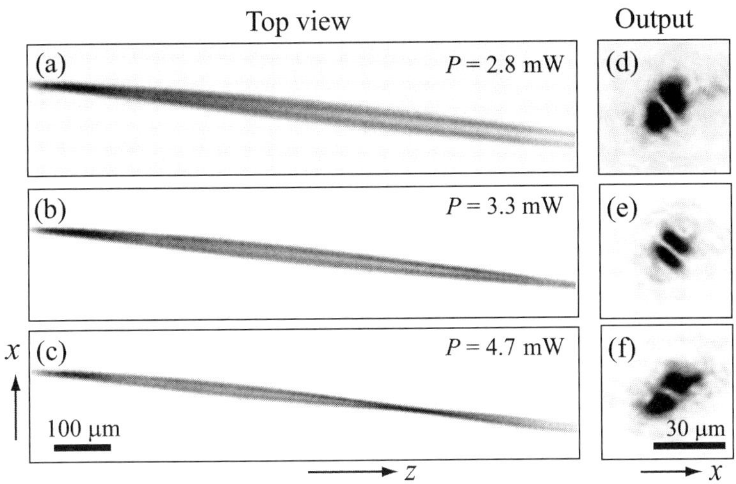

FIGURE 6.8 Generation of a dipole azimuthon, demonstrating power-controlled twist and breathing of the beam [18]. (a–c) Top views in the plane (x,z) of the beam with input charge $m = +1$ for three different input powers, as indicated. (d–f) Corresponding transverse intensity patterns at the cell output.

twist of the beam inside the cell is defined by the topological charge of the input vortex beam, in excellent agreement with theoretical predictions [58].

To analyze the singular phase structure of dipole azimuthons, we employ an interferometric technique: a tilted broad Gaussian beam interferes at an angle with the dipole at the cell output. The obtained interferograms in Figure 6.9 allow detecting the position of the phase dislocations by the characteristic presence of fork dislocations. Two opposite topological charges, $m = \pm 1$, are distinguished by the fork orientation, either down or up. Remarkably, we observe nontrivial topological reactions of the on-axis phase dislocation of nonlocal dipole azimuthons with the sudden appearance of a triplet of vortices. As a representative example, Figure 6.9 shows the output interferograms for the input vortex with topological charge $m = +1$. At small powers, the input phase dislocation is preserved (Fig. 6.9a, left). However, increasing power above a critical value, such as $P = 2.6$ mW in Figure 6.9b, we observe three spatially separated phase singularities in Figure 6.9a, middle. The central (on-axis) vortex has a topological charge $m_0 = -1$ opposite to the one in Figure 6.9a, left; two satellite vortices with $m_{1,2} = +1$ compensate this difference keeping the total charge unchanged, $m = m_0 + m_1 + m_2 = +1$. Increasing the input power results in spatial separation of the triplet and then attraction of vortices, followed by a second flipping of the central vortex. This time, the charge flips from $m_0 = -1$ to $m_0 = +1$ above $P \sim 3.3$ mW (see Fig. 6.9a, right). With further increase in power up to 3.7 mW, the charge-flipping process repeats again.

It is noteworthy that analogous experiments with an input vortex beam carrying charge $m = -1$ lead to an opposite direction for the spatial twist of the dipole (Fig. 6.8) accompanied by topological reactions of three vortices similar to Figure 6.9. This observation allows us to conclude that charge-flipping is a

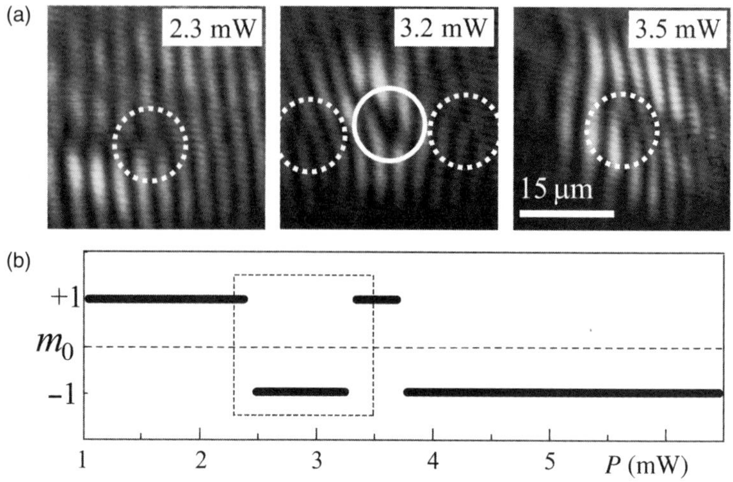

FIGURE 6.9 (a) Experimentally acquired interferograms with topological reactions of nonlocal dipole azimuthons. The circles indicate the positions of phase singularities with charges +1 (dash) and −1 (solid), respectively. (b) Observed topological charge m_0 of an on-axis singularity; the dashed square indicates the range of input powers in (a) [18].

robust and reproducible effect. The sudden splitting of a vortex line into three spatially separated lines or, conversely, the reconnection of three vortex lines into one, is similar to the diagrams of pitchfork bifurcations (supercritical and subcritical, respectively) with transition driven by optical power. The numerical simulations of the initially deformed vortex beam in isotropic nonlocal media reveal a similar dynamic behavior of vortex lines, that is, the formation of pitchfork topological reactions [59] during propagation, for example, Figure 4 in Reference 18. The apparent reason for such complex dynamics is the anisotropic deformation of the nematicon-induced waveguide, similar to astigmatic mode conversion [57], which also facilities the quasi-periodic transformations of nonlocal solitons [47, 60, 61].

6.8 CONCLUSIONS

We have presented an overview of experimental studies of a rich variety of phenomena with spatial optical solitons in an NLC-filled planar cell without external bias. The reorientational nonlinearity is strong enough to support stable nematicons as well as their robust interactions. We have investigated the nonlocal interactions of CO and CP nematicons and observed the formation of vector nematicons with two components propagating in opposite directions. We have experimentally determined the domains of two key parameters, transverse separation of CP beams r and their excitation power P, where the vector nematicon remains stable or exhibits dynamic instabilities. Furthermore, we have experimentally studied the interaction of nematicons with curved dielectric surfaces and demonstrated that, by varying the impact parameter, the soliton beam can be efficiently steered by exploiting its attractive interaction with the surface. We have experimentally observed total internal reflection of soliton beams from an interface between air and liquid crystals, with nematicon deviations exceeding 90°. Each nematicon induces an actual waveguide: we reported on the experimental observation of higher order modes guided by soliton-induced waveguides. We found that the nematicon waveguides operate in a bounded power region specific to each guided mode. Below this region, the guided beams diffract; above this region, mode mixing and coupling give rise to unstable outputs. Finally, we have generated self-trapped vortices in NLCs and observed their transformations into spiraling dipole azimuthons. We have observed nonlinearity-induced charge-flipping of the central phase dislocation as splitting of the on-axis phase singularity into three vortex lines, similar to pitchfork bifurcations. Our findings contribute toward a better understanding of nonlinear optical processes in anisotropic media and the dynamics of self-trapped laser beams in nonlocal media, and they can be potentially useful for all-optical processing of information in futuristic soliton-based optical switches and interconnects.

Acknowledgments

The authors acknowledge financial support from the Australian Research Council, and they are grateful for essential contributions of their collaborators, G. Assanto, V. Shvedov, W. Krolikowski, and M. Belić.

REFERENCES

1. Y. S. Kivshar and G. P. Agrawal. *Optical Solitons: From Fibers to Photonic Crystals*. Academic Press, San Diego, CA, 2003.
2. N. V. Tabirian, A. V. Sukhov, and B. Y. Zel'dovich. Orientational optical nonlinearity of liquid crystals. *Mol. Cryst. Liq. Cryst*., 136(1):1–139, 1986.
3. I. C. Khoo. *Liquid crystals: Physical Properties and Nonlinear Optical Phenomena*. Wiley, New York, 1995.
4. J. Beeckman, K. Neyts, X. Hutsebaut, C. Cambournac, and M. Haelterman. Simulations and experiments on self-focusing conditions in nematic liquid-crystal planar cells. *Opt. Express*, 12(6):1011–1018, 2004.
5. D. W. McLaughlin, D. J. Muraki, and M. J. Shelley. Self-focussed optical structures in a nematic liquid crystal. *Physica D*, 97(4):471–497, 1996.
6. M. Peccianti, G. Assanto, A. De Luca, C. Umeton, and I. C. Khoo. Electrically assisted self-confinement and waveguiding in planar nematic liquid crystal cells. *Appl. Phys. Lett*., 77:7–9, 2000.
7. G. Assanto, M. Peccianti, and C. Conti. Nematicons: Optical spatial solitons in nematic liquid crystals. *Opt. Photon. News*, 14:44–48, 2003.
8. M. Peccianti, C. Conti, and G. Assanto. Interplay between nonlocality and nonlinearity in nematic liquid crystals. *Opt. Lett*., 30:415–417, 2005.

9. M. Peccianti, C. Conti, G. Assanto, A. De Luca, and C. Umeton. Highly anisotropic spatial solitons, modulational instability and their routing in liquid crystals. *Nature*, 432:733–737, 2004.
10. A. Alberucci, M. Peccianti, G. Assanto, G. Coschignano, A. De Luca, and C. Umeton. Self-healing generation of spatial solitons in liquid crystals. *Opt. Lett.*, 30:1381–1383, 2005.
11. A. Fratalocchi, A. Piccardi, M. Peccianti, and G. Assanto. Nonlinearly controlled angular momentum of soliton clusters. *Opt. Lett.*, 32:1447–1449, 2007.
12. A. Fratalocchi, A. Piccardi, M. Peccianti, and G. Assanto. Nonlinear management of the angular momentum of soliton clusters: theory and experiments. *Phys. Rev. A*, 75:063835, 2007.
13. A. Piccardi, G. Assanto, L. Lucchetti, and F. Simoni. All-optical steering of soliton waveguides in dye-doped liquid crystals. *Appl. Phys. Lett.*, 93:171104, 2008.
14. Y. V. Izdebskaya, V. Shvedov, A. S. Desyatnikov, W. Krolikowski, G. Assanto, and Y. S. Kivshar. Incoherent interaction of nematicons in bias-free liquid-crystal cells. *J. Eur. Opt. Soc.-Rapid*, 5:10008, 2010.
15. Y. V. Izdebskaya, V. G. Shvedov, A. S. Desyatnikov, W. Z. Krolikowski, M. Belic, G. Assanto, and Y. S. Kivshar. Counterpropagating nematicons in bias-free liquid crystals. *Opt. Express*, 18(4):3258–3263, 2010.
16. Y. V. Izdebskaya, V. G. Shvedov, A. S. Desyatnikov, W. Z. Krolikowski, and Y. S. Kivshar. Soliton bending and routing induced by interaction with curved surfaces in nematic liquid crystals. *Opt. Lett.*, 35:1692–1694, 2010.
17. Y. V. Izdebskaya, A. S. Desyatnikov, G. Assanto, and Y. S. Kivshar. Multimode nematicon waveguides. *Opt. Lett.*, 36:184–186, 2011.
18. Y. V. Izdebskaya, A. S. Desyatnikov, G. Assanto, and Y. S. Kivshar. Dipole azimuthons and charge-flipping in nematic liquid crystals. *Opt. Express*, 19(22):21457–21466, 2011.
19. J. Baran, Z. Raszewski, R. Dabrowski, J. Kedzierski, and J. Rutkowska. Some physical properties of mesogenic 4-(trans-4′-n-Alkylcyclohexyl) isothiocyanatobenzenes. *Mol. Cryst. Liq. Cryst.*, 123:237–245, 1985.
20. R. Dabrowski, J. Dziaduszek, and T. Szczucinski. Mesomorphic characteristics of some new homologous series with the isothiocyanato terminal group. *Mol. Cryst. Liq. Cryst.*, 124:241–257, 1985.
21. A. Piccardi, A. Alberucci, and G. Assanto. Soliton self-deflection via power-dependent walk-off. *Appl. Phys. Lett.*, 96:061105, 2010.
22. J. Beeckman, K. Neyts, X. Hutsebaut, C. Cambournac, and M. Haelterman. Time dependence of soliton formation in planar cells of nematic liquid crystals. *IEEE J. Quantum Electron.*, 41:735–740, 2005.
23. C. Conti, M. Peccianti, and G. Assanto. Complex dynamics and configurational entropy of spatial optical solitons in nonlocal media. *Opt. Lett.*, 31:2030–2032, 2006.
24. C. Conti, M. Peccianti, and G. Assanto. Observation of optical spatial solitons in a highly nonlocal medium. *Phys. Rev. Lett.*, 92:113902, 2004.
25. M. Peccianti, A. Fratalocchi, and G. Assanto. Transverse dynamics of nematicons. *Opt. Express*, 12:6524–6529, 2004.
26. M. Peccianti, K. A. Brzdiakiewicz, and G. Assanto. Nonlocal spatial soliton interactions in nematic liquid crystals. *Opt. Lett.*, 27:1460–1462, 2002.

27. M. S. Petrović, M. R. Belić, C. Denz, and Y. S. Kivshar. Counterpropagating optical beams and solitons. *Laser Photon. Rev.*, 5(2):214–233, 2011.

28. M. Haelterman, A. P. Sheppard, and A. W. Snyder. Bimodal counterpropagating spatial solitary waves. *Opt. Commun.*, 103:145–152, 1993.

29. O. Cohen, R. Uzdin, T. Carmon, J. W. Fleischer, M. Segev, and S. Odulov. Collisions between optical spatial solitons propagating in opposite directions. *Phys. Rev. Lett.*, 89:133901, 2002.

30. M. Belić, P. Jander, A. Strinić, A. Desyatnikov, and C. Denz. Self-trapped bidirectional waveguides in a saturable photorefractive medium. *Phys. Rev. E*, 68:025601, 2003.

31. P. Jander, J. Schröder, C. Denz, M. Petrović, and M. Belić. Dynamic instability of self-induced bidirectional waveguides in photorefractive media. *Opt. Lett.*, 30:750–752, 2005.

32. M. Peccianti, A. Dyadyusha, M. Kaczmarek, and G. Assanto. Tunable refraction and reflection of self-confined light beams. *Nat. Phys.*, 2:737–742, 2006.

33. J. Beeckman, K. Neyts, and M. Haelterman. Patterned electrode steering of nematicons. *J. Opt. A: Pure Appl. Opt.*, 8:214–220, 2006.

34. M. Warenghem, J. F. Blach, and J. F. Henninot. Measuring and monitoring optically induced thermal or orientational non-locality in nematic liquid crystal. *Mol. Cryst. Liq. Cryst.*, 454:297–314, 2006.

35. S. Serak, N. V. Tabiryan, M. Peccianti, and G. Assanto. Spatial soliton all-optical logic gates. *IEEE Phot. Tech. Lett.*, 18(12):1287–1289, 2006.

36. A. Alberucci, A. Piccardi, U. Bortolozzo, S. Residori, and G. Assanto. Nematicon all-optical control in liquid crystal light valves. *Opt. Lett.*, 35:390–392, 2010.

37. A. Alberucci, M. Peccianti, and G. Assanto. Nonlinear bouncing of nonlocal spatial solitons at the boundaries. *Opt. Lett.*, 32:2795–2797, 2007.

38. A. Alberucci, G. Assanto, D. Buccoliero, A. S. Desyatnikov, T. R. Marchant, and N. F. Smyth. Modulation analysis of boundary-induced motion of optical solitary waves in a nematic liquid crystal. *Phys. Rev. A*, 79:043816, 2009.

39. M.-F. Shih, M. Segev, and G. Salamo. Circular waveguides induced by two-dimensional bright steady-state photorefractive spatial screening solitons. *Opt. Lett.*, 21:931–933, 1996.

40. J. Petter and C. Denz. Guiding and dividing waves with photorefractive solitons. *Opt. Commun.*, 188:55–61, 2001.

41. J. Petter, C. Denz, A. Stepken, and F. Kaiser. Anisotropic waveguides induced by photorefractive (2+1)D solitons. *J. Opt. Soc. Am. B*, 19(5):1145–1149, 2002.

42. M. Peccianti and G. Assanto. Signal readdressing by steering of spatial solitons in bulk nematic liquid crystals. *Opt. Lett.*, 26:1690–1692, 2001.

43. X. Hutsebaut, C. Cambournac, M. Haelterman, J. Beeckman, and K. Neyts. Measurement of the self-induced waveguide of a solitonlike optical beam in a nematic liquid crystal. *J. Opt. Soc. Am. B*, 22(7):1424–1431, 2005.

44. J.-F. Henninot, J.-F. Blach, and M. Warenghem. Experimental study of the nonlocality of spatial optical solitons excited in nematic liquid crystal. *J. Opt. A: Pure Appl. Opt.*, 9(1):20–25, 2007.

45. X. Hutsebaut, C. Cambournac, M. Haelterman, A. Adamski, and K. Neyts. Single-component higher-order mode solitons in nematic liquid crystals. *Opt. Commun.*, 233:211–217, 2004.

46. J. Beeckman, K. Neyts, P. J. Vanbrabant, R. James, and F. A. Fernandez. Finding exact spatial soliton profiles in nematic liquid crystals. *Opt. Express*, 18(4):3311–3321, 2010.
47. D. Buccoliero, A. S. Desyatnikov, W. Z. Krolikowski, and Y. S. Kivshar. Laguerre and Hermite soliton clusters in nonlocal nonlinear media. *Phys. Rev. Lett.*, 98:053901, 2007.
48. M. Mitchell and M. Segev. Self-trapping of incoherent white light. *Nature*, 387(6636):880–883, 1997.
49. M. Peccianti and G. Assanto. Incoherent Spatial Solitary Waves in Nematic Liquid Crystals. *Opt. Lett.*, 26:1791–1793, 2001.
50. A. M. Snyder and J. D. Love. *Optical Waveguide Theory*. Chapmen and Hall, London, 1983.
51. C. Conti, M. Peccianti, and G. Assanto. Spatial solitons and modulational instability in the presence of large birefringence: the case of highly nonlocal liquid crystals. *Phys. Rev. E*, 72:066614, 2005.
52. J. F. Nye and M. V. Berry. Dislocations in wave trains. *Proc. R. Soc. London, Ser. A*, 336:165–190, 1974.
53. A. S. Desyatnikov, Y. S. Kivshar, and L. Torner. Optical vortices and vortex solitons. *Prog. Opt.*, 47:291–391 (Ed. E. Wolf, Elsevier, 2005).
54. M. Soljacic and M. Segev. Integer and fractional angular momentum borne on self-trapped necklace-ring beams. *Phys. Rev. Lett.*, 86:420–423, 2001.
55. A. S. Desyatnikov, A. A. Sukhorukov, and Y. S. Kivshar. Azimuthons: Spatially modulated vortex solitons. *Phys. Rev. Lett.*, 95:203904, 2005.
56. S. Lopez-Aguayo, A. S. Desyatnikov, and Y. S. Kivshar. Azimuthons in nonlocal nonlinear media. *Opt. Express*, 14:7903–7908, 2006.
57. E. Abramochkin and V. Volstnikov. Beam transformations and nontransformed beams. *Opt. Commun.*, 83:123–135, 1991.
58. S. Lopez-Aguayo, A. S. Desyatnikov, Y. S. Kivshar, S. Skupin, W. Krolikowski, and O. Bang. Stable rotating dipole solitons in nonlocal optical media. *Opt. Lett.*, 31:1100–1102, 2006.
59. G. Molina-Terriza, J. Recolons, and L. Torner. The curious arithmetic of optical vortices. *Opt. Lett.*, 25:1135–1137, 2000.
60. D. Buccoliero and A. S. Desyatnikov. Quasi-periodic transformations of nonlocal spatial solitons. *Opt. Express*, 17(12):9608–9613, 2009.
61. D. Buccoliero, A. S. Desyatnikov, W. Krolikowski, and Y. S. Kivshar. Boundary effects on the dynamics of higher-order optical spatial solitons in nonlocal thermal media. *J. Opt. A: Pure Appl. Opt.*, 11(9):094014, 2009.

7

Interaction of Nematicons and Nematicon Clusters

CATHERINE GARCÍA-REIMBERT AND ANTONMARIA A. MINZONI

Department of Mathematics and Mechanics, IIMAS, Fenomenos Nonlineales y Mecánica, Universidad Nacional Autónoma de México, Mexico D.F., Mexico

NOEL F. SMYTH

School of Mathematics and Maxwell Institute for Mathematical Sciences, University of Edinburgh, Edinburgh, Scotland, United Kingdom

7.1 INTRODUCTION

The theory for modeling and evolution of a single solitary wave in a nematic liquid crystal, a nematicon, has been described in Chapter 3. One of the key concepts discussed in that chapter was that of the nonlocal response of the nematic medium. This nonlocal response means that the perturbation due to the optical beam on the nematic extends far beyond the waist of the optical beam itself. Although an important effect for a single nematicon, it stops the catastrophic collapse that occurs in the case of a local two-dimensional optical solitary wave [1], it becomes vital when the interaction of multiple nematicons is studied. The nonlocal perturbation of the nematic means that two optical beams can interact through the nematic even though they appear to be well separated. Indeed, the interaction between two or more nematicons will be found to have a close connection with the action at a distance of Newtonian gravitation [2].

A primary motivation for the study of the interaction of nematicons is that this interaction can be used as the basis for all-optical routing and logic devices based on nematic liquid crystals [3–7]. The fundamental concept is that the trajectory of a signal beam can be altered by the presence of a control beam, or control beams, resulting in the signal beam being routed to a given output. This control is both through the relative physical positions of the control beams and their powers. The detailed understanding of the interaction of nematicons, particularly their interaction

Nematicons: Spatial Optical Solitons in Nematic Liquid Crystals, First Edition.
Edited by Gaetano Assanto.

via the nematic, is therefore vital for the development of all-optical devices based on nonlinear beams in nematic liquid crystals. These potential applications, as well as intrinsic interest, have motivated a great deal of study of the interaction of nematicons propagating in the same direction [8–13] and counterpropagating nematicons [14–16].

All proposed logic devices are based on the adjustable routing of a signal beam from a given input to a range of possible outputs at the other end of the liquid crystal cell. The advantage of a liquid crystal as the intermediary medium is that there are no fixed "circuits," unlike a wire-based device, so that any path through the cell is possible. This reconfigurable flexibility of a liquid-crystal-based logic device is potentially a great advantage. There are two basic mechanisms for optically controlling the trajectory of a signal beam in a liquid crystal cell. The first involves the interaction of two nematicon beams [3, 6, 8–10, 12]. The strength of the interaction between two or more nematicons depends on both their relative separation and their optical power, with stronger interaction occurring at closer separation and higher power. In the case of solitary waves for local nonlinear Schrödinger-type (NLS-type) equations

$$\mathrm{i}\frac{\partial u}{\partial z} + \frac{1}{2}\nabla^2 u + F(|u|^2)u = 0, \tag{7.1}$$

the interaction between solitary waves depends not only on their separation and power but also on their relative phase [1]. If the solitary waves are in phase, then they attract and if they are π out of phase, they repel [1]. For other phase differences, the interaction is more complicated [1]. In the case of nonlocal media, such as liquid crystals, there is the added effect of attraction due to the medium-mediated mutual interaction. This nonlocal interaction can be sufficient to overcome the repulsion of out of phase solitary waves [8–12, 17]. When the solitary waves have angular momentum, stable rotating bound vector solitary waves can result [9–12, 17]. Therefore, in experimental situations, the phase difference between the nematicons plays no role in the switching properties of the proposed logic devices. There are a number of ways in which switching can be realized. The output point of a signal beam can be controlled by the absence or presence of a control beam. Multiple beams can be routed to different outputs, depending on the presence or absence of other beams and the strength of the mutual interaction. In this manner, elementary logic can be performed whereby the output point of a given signal depends on the properties of other signals.

A second type of beam interaction scenario involves nematicons interacting with localized beams propagating in a direction orthogonal to the nematicon [5, 4, 7]. In this case, these orthogonal beams form a localized refractive index change, which alters the nematicon trajectory when it passes close to the beam. This is similar to the familiar refraction of light, except in this case the refractive index change has a nonlinear dependence on the beam power. Again, the absence or presence of the control beam can switch the nematicon between different outputs. An additional benefit of using control beams is that it is easy to introduce multiple beams in a cell, resulting in complicated nematicon trajectories. By actually passing a nematicon through the orthogonal control beam, it is possible to split the nematicon, resulting in a Y junction [5].

This discussion of potential applications of nematicons in all-optical logic and signal processing devices shows the importance of a detailed understanding of the interaction of multiple nematicons. This topic is discussed in detail in the following sections.

7.2 GRAVITATION OF NEMATICONS

The simplest configuration in which to consider the interaction of nematicons is that of so-called two-color nematicons. In this configuration, the two nematicons are based on polarized beams of two different wavelengths (colors). The mathematical analysis of the incoherent interaction of two nematicons is much easier than that of coherent interaction. This is because in the case of coherent interaction the nematicons interact directly [17], whereas in the case of incoherent interaction they only interact through the nematic. This results in far fewer terms in the resulting modulation equations and a far clearer picture of the interaction. The detailed setup of the liquid crystal cell is as for a single nematicon and is illustrated in Figure 7.1. A static electric field is applied in the x direction to overcome the Freédericksz threshold, so that in the absence of light the optical director is pretilted at an angle $\hat{\theta}$ to the z direction. The angle θ will then measure the perturbation of the director angle from this pretilt due to the optical beams. In nondimensional form, the equations governing the propagation of the two color beams in the cell are

$$i\frac{\partial u}{\partial z} + \frac{1}{2}D_u\nabla^2 u + 2A_u u\theta = 0, \tag{7.2}$$

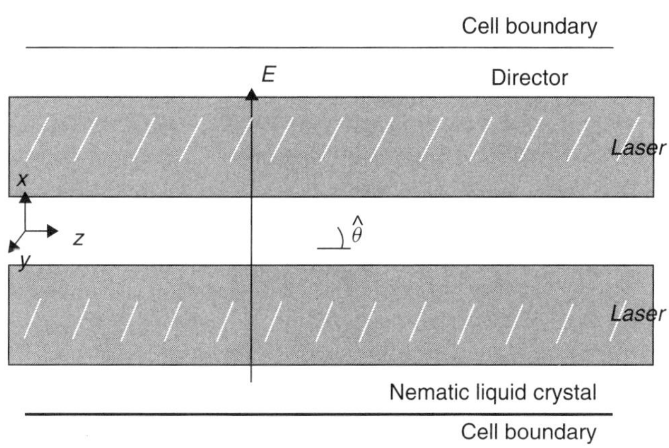

FIGURE 7.1 Liquid crystal cell with two polarized light beams of different colors. $\hat{\theta}$ is the pretilt angle due to the electric field E in the x direction. (*Source*: Reproduced with permission from Figure 7.1 in Reference 18.)

$$\mathrm{i}\frac{\partial v}{\partial z} + \frac{1}{2}D_v\nabla^2 v + 2A_v v\theta = 0, \tag{7.3}$$

$$\nu\nabla^2\theta - 2q\theta = -2A_u|u|^2 - 2A_v|v|^2, \tag{7.4}$$

see Alberucci et al. [19]. Here, u and v are the slowly varying envelopes of the electric fields of the two beams. The Laplacian ∇^2 is in the (x,y) plane. The coefficients D_u and D_v are the diffraction coefficients for the two colors and A_u and A_v are the coupling coefficients between the light and the nematic for the two colors. The parameter ν measures the elasticity of the nematic and q is related to the square of the static electric field that pretilts the nematic [20–22]. These two-color nematicon equations have the Lagrangian formulation

$$L = \mathrm{i}\left(u^*u_z - uu_z^*\right) - D_u|\nabla u|^2 + 4A_u\theta|u|^2 + i\left(v^*v_z - vv_z^*\right) - D_v|\nabla v|^2$$
$$+ 4A_v\theta|v|^2 - \nu|\nabla\theta|^2 - 2q\theta^2, \tag{7.5}$$

where the superscript * denotes the complex conjugate. The nematicon equations (Eqs 7.2–7.4) show that the nematicons interact only via the director (Eq. 7.4) and not via direct beam on beam interaction [17]. The incoherent interaction of nematicons of the same color is also governed by the system (Eqs 7.2–7.4), with $D_u = D_v$ and $A_u = A_v$.

The evolution of the two-color nematicons, governed by Equations 7.2–7.4, will be analyzed using the averaged Lagrangian method discussed in Chapter 3. To this end, the profiles of the two nematicons will be assumed to be given by the trial functions

$$\begin{aligned} u &= a_u f(\chi_u/w_u)\,\mathrm{e}^{\mathrm{i}\psi_u} + \mathrm{i}g_u\mathrm{e}^{\mathrm{i}\psi_u}, \\ v &= a_v f(\chi_v/w_v)\,\mathrm{e}^{\mathrm{i}\psi_v} + \mathrm{i}g_v\mathrm{e}^{\mathrm{i}\psi_v}, \\ \theta &= \alpha_u(f(\chi_u/\beta_u))^2 + \alpha_v(f(\chi_v/\beta_v))^2, \end{aligned} \tag{7.6}$$

where

$$\begin{aligned} \chi_u &= \sqrt{\left(x-\xi_u\right)^2 + \left(y-\eta_u\right)^2}, \\ \chi_v &= \sqrt{\left(x-\xi_v\right)^2 + \left(y-\eta_v\right)^2}, \\ \psi_u &= \sigma_u + U_u\left(x-\xi_u\right) + V_u\left(y-\eta_u\right), \\ \psi_v &= \sigma_v + U_v\left(x-\xi_v\right) + V_v\left(x-\eta_v\right). \end{aligned} \tag{7.7}$$

At this stage, the actual detailed functional form $f(\zeta)$ of the nematicons will not be specified. Indeed, in many interaction scenarios involving nematicon interactions in liquid crystals, the trajectories of the nematicons are essentially independent of the transverse profile of the nematicon [2, 18, 23, 24]. In the present chapter, both hyperbolic secant $f(\zeta) = \operatorname{sech} \zeta$ and Gaussian $f(\zeta) = \exp(-\zeta^2)$ profiles are used as examples.

Substituting the trial functions (Eq. 7.6) into the Lagrangian (Eq. 7.5) and averaging by integrating in x and y from $-\infty$ to ∞ yield, as explained in detail in Chapter 3, the averaged Lagrangian

$$\begin{aligned}\mathcal{L} = &-2\left(I_2 a_u^2 w_u^2 + \Lambda_u g_u^2\right)\left(\sigma_u' - U_u \xi_u' - V_u \eta_u'\right) - 2I_1 a_u w_u^2 g_u' + 2I_1 g_u w_u^2 a_u' \\ &+ 4I_1 a_u g_u w_u w_u' - 2\left(I_2 a_v^2 w_v^2 + \Lambda_v g_v^2\right)\left(\sigma_v' - U_v \xi_v' - V_v \eta_v'\right) \\ &- 2I_1 a_v w_v^2 g_v' + 2I_1 g_v w_v^2 a_v' + 4I_1 a_v g_v w_v w_v' - D_u I_{22} a_u^2 - D_v I_{22} a_v^2 \\ &- D_u\left(I_2 a_u^2 w_u^2 + \Lambda_u g_u^2\right)\left(U_u^2 + V_u^2\right) - D_v\left(I_2 a_v^2 w_v^2 + \Lambda_v g_v^2\right)\left(U_v^2 + V_v^2\right) \\ &- 4\nu I_{42}\alpha_u^2 - 4\nu I_{42}\alpha_v^2 - 2q I_4 \alpha_u^2 \beta_u^2 - 2q I_4 \alpha_v^2 \beta_v^2 + \mathcal{L}_I. \end{aligned} \tag{7.8}$$

The interaction component of the averaged Lagrangian is

$$\begin{aligned}\mathcal{L}_I = &\, 2A_u A^2 B^2 a_u^2 w_u^2 \left[\alpha_u \beta_u^2 Q_1^{-1} + \alpha_v \beta_v^2 Q_2^{-1} \mathrm{e}^{-\gamma_1}\right] \\ &+ 2A_v A^2 B^2 a_v^2 w_v^2 \left[\alpha_v \beta_v^2 Q_3^{-1} + \alpha_u \beta_u^2 Q_4^{-1} \mathrm{e}^{-\gamma_2}\right] \\ &- 4\nu \alpha_u \alpha_v \beta_u^2 \beta_v^2 Q_5^{-2}\left[1 - \gamma_3\right]\mathrm{e}^{-\gamma_3} - 2q A^2 \alpha_u \alpha_v \beta_u^2 \beta_v^2 Q_5^{-1} \mathrm{e}^{-\gamma_3}. \end{aligned} \tag{7.9}$$

Here,

$$\begin{aligned} &Q_1 = A^2\beta_u^2 + B^2 w_u^2, \quad Q_2 = A^2\beta_v^2 + B^2 w_u^2, \quad Q_3 = A^2\beta_v^2 + B^2 w_v^2, \\ &Q_4 = A^2\beta_u^2 + B^2 w_v^2, \quad Q_5 = \beta_u^2 + \beta_v^2. \\ &\gamma_1 = \frac{\rho^2}{Q_2}, \quad \gamma_2 = \frac{\rho^2}{Q_4}, \quad \gamma_3 = \frac{\rho^2}{A^2 Q_5}, \\ &\rho^2 = \left(\xi_u - \xi_v\right)^2 + \left(\eta_u - \eta_v\right)^2. \end{aligned} \tag{7.10}$$

The areas, modulo 2π, of the shelves of radiation under the nematicons are

$$\Lambda_u = \frac{1}{2}R_u^2 \quad \text{and} \quad \Lambda_v = \frac{1}{2}R_v^2. \tag{7.11}$$

The variable ρ is the distance between the two nematicons and will be found to be analogous to the distance between masses in Newtonian gravitation. Taking variations of the averaged Lagrangian (Eq. 7.8) results in the modulation equations

$$\frac{\mathrm{d}}{\mathrm{d}z}\left[I_2 a_u^2 w_u^2 + \Lambda_u g_u^2\right] = 0, \tag{7.12}$$

$$I_1 \frac{\mathrm{d}}{\mathrm{d}z}\left(a_u w_u^2\right) = \Lambda_u g_u \left[\sigma_u' - \frac{1}{2}D_u U_u^2 - \frac{1}{2}D_u V_u^2\right], \tag{7.13}$$

$$\frac{d\xi_u}{dz} = D_u U_u, \quad \frac{d\eta_u}{dz} = D_u V_u, \tag{7.14}$$

$$I_1 \frac{dg_u}{dz} = \frac{1}{2} D_u I_{22} a_u w_u^{-2} - A_u A^2 B^4 a_u w_u^2 \Big[\alpha_u \beta_u^2 Q_1^{-2} + \alpha_v \beta_v^2 Q_2^{-2} e^{-\gamma_1} - \alpha_v \beta_v^2 \rho^2 Q_2^{-3} e^{-\gamma_1}\Big], \tag{7.15}$$

$$I_2 \left[\frac{d\sigma_u}{dz} - \frac{1}{2} D_u U_u^2 - \frac{1}{2} D_u V_u^2\right] = -D_u I_{22} w_u^{-2} + A_u A^2 B^2 \Big[\alpha_u \beta_u^2 \left(A^2 \beta_u^2 + 2B^2 w_u^2\right) Q_1^{-2} + \alpha_v \beta_v^2 \left(A^2 \beta_v^2 + 2B^2 w_u^2\right) Q_2^{-2} e^{-\gamma_1} - \alpha_v w_u^2 \beta_v^2 \rho^2 Q_2^{-3} e^{-\gamma_1}\Big], \tag{7.16}$$

$$\frac{d}{dz}\left[\left(I_2 a_u^2 w_u^2 + \Lambda_u g_u^2\right) U_u\right] = -2A^2 B^2 (\xi_u - \xi_v) \Big[A_u \alpha_v a_u^2 w_u^2 \beta_v^2 Q_2^{-2} e^{-\gamma_1} + A_v \alpha_u a_v^2 w_v^2 \beta_u^2 Q_4^{-2} e^{-\gamma_2}\Big] + 2q\alpha_u \alpha_v \beta_u^2 \beta_v^2 \left(\xi_u - \xi_v\right) Q_5^{-2} e^{-\gamma_3} + 4\nu A^{-2} \alpha_u \alpha_v \beta_u^2 \beta_v^2 \left(\xi_u - \xi_v\right) Q_5^{-3} \Big[1 - \rho^2 Q_5^{-1}\Big] e^{-\gamma_3}, \tag{7.17}$$

$$\frac{d}{dz}\left[\left(I_2 a_u^2 w_u^2 + \Lambda_u g_u^2\right) V_u\right] = -2A^2 B^2 (\eta_u - \eta_v) \Big[A_u \alpha_v a_u^2 w_u^2 \beta_v^2 Q_2^{-2} e^{-\gamma_1} + A_v \alpha_u a_v^2 w_v^2 \beta_u^2 Q_4^{-2} e^{-\gamma_2}\Big] + 2q\alpha_u \alpha_v \beta_u^2 \beta_v^2 \left(\eta_u - \eta_v\right) Q_5^{-2} e^{-\gamma_3} + 4\nu A^{-2} \alpha_u \alpha_v \beta_u^2 \beta_v^2 \left(\eta_u - \eta_v\right) Q_5^{-3} \Big[1 - \rho^2 Q_5^{-1}\Big] e^{-\gamma_3}, \tag{7.18}$$

plus the algebraic equations

$$2\left(2\nu I_{42} + q I_4 \beta_u^2\right) \alpha_u = A^2 B^2 \beta_u^2 \left(A_u a_u^2 w_u^2 Q_1^{-1} + A_v a_v^2 w_v^2 Q_4^{-1} e^{-\gamma_2}\right) - q A^2 \alpha_v \beta_u^2 \beta_v^2 Q_5^{-1} e^{-\gamma_3} - 2\nu \alpha_v \beta_u^2 \beta_v^2 Q_5^{-2} \left(1 - A^{-2} \rho^2 Q_5^{-1}\right) e^{-\gamma_3}, \tag{7.19}$$

$$q I_4 \alpha_u = A^2 B^4 \left(A_u a_u^2 w_u^4 Q_1^{-2} + A_v a_v^2 w_v^4 Q_4^{-2} e^{-\gamma_2}\right) + A_v A^4 B^2 a_v^2 w_v^2 \beta_u^2 \rho^2 Q_4^{-3} e^{-\gamma_3} - 2\nu \alpha_v \beta_v^2 Q_5^{-3} \Big[\beta_v^2 - \beta_u^2 - \left(\beta_v^2 - 3\beta_u^2\right) \rho^2 A^{-2} Q_5^{-1} - A^{-4} \beta_u^2 \rho^4 Q_5^{-2}\Big] e^{-\gamma_3} - q\alpha_v \beta_v^2 Q_5^{-2} \left(A^2 \beta_v^2 + \beta_u^2 \rho^2 Q_5^{-1}\right) e^{-\gamma_3}, \tag{7.20}$$

together with symmetric equations in the v color, for the evolution of the parameters of the two-color nematicons. Equation 7.12 is the equation of conservation of mass in the sense of scale invariances of the Lagrangian (Eq. 7.5) [25]. Although it physically corresponds to conservation of mass in the application of the NLS equation to water waves, in the present optical context it corresponds to conservation of power, or equivalently, photon number.

The modulation Equations 7.17 and 7.18 are x and y momentum equations for the color u. The equations for total momentum conservation in the x and y directions can then be obtained by adding the momentum Equations 7.17 and 7.18 to their v color counterparts, to give total x momentum conservation as

$$\frac{d}{dz}\left[\left(I_2 a_u^2 w_u^2 + \Lambda_u g_u^2\right) U_u + \left(I_2 a_v^2 w_v^2 + \Lambda_v g_v^2\right) U_v\right] = 0 \tag{7.21}$$

and total y momentum conservation as

$$\frac{d}{dz}\left[\left(I_2 a_u^2 w_u^2 + \Lambda_u g_u^2\right) V_u + \left(I_2 a_v^2 w_v^2 + \Lambda_v g_v^2\right) V_v\right] = 0. \tag{7.22}$$

That these are, in fact, x and y momentum conservation equations can be verified by the application of Nöther's theorem to the Lagrangian (Eq. 7.5) based on invariances to shifts in the space directions x and y [26].

The modulation Equations 7.12–7.20 and their symmetric v counterparts can be put in a simpler form using Kepler coordinates. Let us define the vector position of the u color nematicon by $\boldsymbol{\xi}_u = (\xi_u, \eta_u)$ and that of the v color nematicon by $\boldsymbol{\xi}_v = (\xi_v, \eta_v)$. The relative displacement of the nematicons, defined by Equation 7.10, in vector form, is

$$\boldsymbol{\rho} = \boldsymbol{\xi}_u - \boldsymbol{\xi}_v. \tag{7.23}$$

The "mass" conservation equation (Eq. 7.12) shows that the "masses" of the u and v color nematicons can be defined to be

$$M_u = I_2 a_u^2 w_u^2 + \Lambda_u g_u^2 \quad \text{and} \quad M_v = I_2 a_v^2 w_v^2 + \Lambda_v g_v^2. \tag{7.24}$$

The center of "mass" of the two nematicons can then be defined to be

$$\boldsymbol{R} = \frac{D_v M_u \boldsymbol{\xi}_u + D_u M_v \boldsymbol{\xi}_v}{D_v M_u + D_u M_v}. \tag{7.25}$$

It should be noted that the masses of the nematicons have been weighted by the diffraction coefficients in this definition of the center of mass. These weightings, other than being a mere analogy with Newtonian gravitation, reflect the different linear optical propagation of the two beams. Relative to the center of mass (Eq. 7.25), polar coordinates will be defined based on ρ and the polar angle denoted by ϕ, similar to the case of the Kepler problem. In a similar manner to the gravitational problem, the equations for the positions of the two-color nematicons can then be shown to be the conservation of angular momentum equation

$$\rho^2 \frac{d\phi}{dz} = L, \tag{7.26}$$

where L is the constant angular momentum, and the radial equations

$$\frac{\mathrm{d}\boldsymbol{R}}{\mathrm{d}z} = \boldsymbol{0} \tag{7.27}$$

and

$$\frac{\mathrm{d}^2\rho}{\mathrm{d}z^2} - L^2\rho^{-3} = -\frac{D_v M_u + D_u M_v}{M_u M_v}\frac{\partial \Phi}{\partial \rho}. \tag{7.28}$$

The potential Φ, which can be found from the averaged Lagrangian (Eq. 7.8) using a point transformation to calculate the corresponding averaged Hamiltonian, is

$$\begin{aligned}\Phi = &-2A_u A^2 B^2 \alpha_v a_u^2 w_u^2 \beta_v^2 Q_2^{-1} \mathrm{e}^{-\gamma_1} - 2A_v A^2 B^2 \alpha_u a_v^2 w_v^2 \beta_u^2 Q_4^{-1} \mathrm{e}^{-\gamma_2} \\ &+ 4\nu\alpha_u\alpha_v\beta_u^2\beta_v^2 Q_5^{-2}\left(1 - A^{-2}\rho^2 Q_5^{-1}\right)\mathrm{e}^{-\gamma_3} \\ &+ 2qA^2\alpha_u\alpha_v\beta_u^2\beta_v^2 Q_5^{-1}\mathrm{e}^{-\gamma_3}.\end{aligned} \tag{7.29}$$

These dynamical equations for the positions of the nematicons are the same as those for two masses in Newtonian gravitation [27], except that the potential is not the inverse separation potential of Newton's law of gravitation. In addition, the potential (Eq. 7.29) is nonmonotonic and so there exist multiple steady states for the separation of the nematicons [28], unlike the unique steady state for gravitation. The center of mass equation (Eq. 7.27) is just a statement of conservation of linear momentum.

The gravitational forms (Eqs 7.26–7.28) of the modulation equations for the two-color nematicons have an obvious extension to n-color interacting nematicons [29].

7.3 IN-PLANE INTERACTION OF TWO-COLOR NEMATICONS

Let us consider the interaction of two-color nematicons in the nonlocal limit, so that the nonlocality parameter ν is large. Typical experimental values of ν are $O(100)$ [30]. As the nematicons propagate over a large z distance they evolve to a steady state by shedding diffractive radiation. Hence, loss to shed diffractive radiation must be included in the modulation equations of Section 7.2 [22, 31, 32]. The largest effect on the nematicons of this loss to shed diffractive radiation is the loss of mass, or optical power in this context [22, 32, 33]. The momentum and energy shed in diffractive radiation is of higher order than the shed mass [33]. The mass conservation equation (Eq. 7.12) and its v color counterpart will then need to be modified to incorporate this loss. Furthermore, as the shelves of low wave number radiation under the nematicons must link to the shed radiation, Equation 7.15 for g_u and the equation for its v color counterpart g_v must also be modified. The details of this modification are given in Kath and Smyth [33], García-Reimbert et al. [34], Minzoni et al. [32], and in Chapter 3 and so will not be repeated here. The final result is that loss terms are added to the mass Equation 7.12 and Equation 7.15 for g_u, so that they become

$$\frac{\mathrm{d}}{\mathrm{d}z}\left(I_2 a_u^2 w_u^2 + \Lambda_u g_u^2\right) = -2D_u \delta_u \tilde{\Lambda}_u \kappa_u^2, \tag{7.30}$$

and

$$I_1 \frac{\mathrm{d}g_u}{\mathrm{d}z} = \frac{1}{2} D_u I_{22} a_u w_u^{-2} - A_u A^2 B^4 a_u w_u^2 \Big[\alpha_u \beta_u^2 Q_1^{-2} + \alpha_v \beta_v^2 Q_2^{-2} \mathrm{e}^{-\gamma_1} - \alpha_v \beta_v^2 \rho^2 Q_2^{-3} \mathrm{e}^{-\gamma_1}\Big] - 2D_u I_1 \delta_u g_u. \tag{7.31}$$

In these equations, the loss coefficient δ_u is

$$\delta_u = \frac{-\sqrt{2\pi}}{32\mathrm{e}\kappa_u \tilde{\Lambda}_u} \int_0^z \pi\kappa_u(z') \ln((z-z')/\tilde{\Lambda}_u) \left[\left\{\left[\frac{1}{2}\ln((z-z')/\tilde{\Lambda}_u)\right]^2 + \frac{3\pi^2}{4}\right\}^2 + \pi^2\left[\ln((z-z')/\tilde{\Lambda}_u)\right]^2\right]^{-1} \frac{\mathrm{d}z'}{(z-z')}. \tag{7.32}$$

Finally,

$$\kappa_u^2 = \frac{1}{\tilde{\Lambda}_u}\left[I_2 a_u^2 w_u^2 - I_2 \hat{a}_u^2 \hat{w}_u^2 + \Lambda_u g_u^2\right], \tag{7.33}$$

where

$$\tilde{\Lambda}_u = \frac{1}{2}\left(7\beta_{u1/2}\right)^2, \quad \beta_{u1/2} = \beta_u \mathrm{sech}^{-1}\left(1/\sqrt{2}\right). \tag{7.34}$$

The mass equation in the v color and the equation for g_v are modified in a similar manner to incorporate loss. In the expression (Eq. 7.33) for κ_u the ^ superscript denotes fixed point values.

The final steady nematicon state is found from the equation for total energy conservation for the nematicon equations (Eqs 7.2–7.4). This energy conservation equation is most easily found using Nöther's theorem based on the invariance of the Lagrangian (Eq. 7.5) to shifts in z [26]. In this manner, we find that the conserved energy density is

$$H = D_u|\nabla u|^2 - 4A_u\theta|u|^2 + D_v|\nabla v|^2 - 4A_v\theta|v|^2 + \nu|\nabla\theta|^2 + 2q\theta^2. \tag{7.35}$$

Averaging this conserved energy density by integrating in x and y from $-\infty$ to ∞ yields the energy conservation modulation equation

$$\begin{aligned}\frac{d\mathcal{H}}{dz} = \frac{d}{dz}\Bigg\{ &- 2A_u A^2 B^2 \alpha_v a_u^2 w_u^2 \beta_v^2 Q_2^{-1} e^{-\gamma_1} - 2A_v A^2 B^2 \alpha_u a_v^2 w_v^2 \beta_u^2 e^{-\gamma_2} Q_4^{-1} \\ &+ 4\nu \alpha_u \alpha_v \beta_u^2 \beta_v^2 \left(1 - \rho^2 Q_5^{-1}\right) Q_5^{-2} e^{-\gamma_3} + 2q A^2 \alpha_u \alpha_v \beta_u^2 \beta_v^2 Q_5^{-1} e^{-\gamma_3} \\ &+ I_{22} D_u a_u^2 + D_u \left(I_2 a_u^2 w_u^2 + \Lambda_u g_u^2\right)\left(U_u^2 + V_u^2\right) - \frac{2A_u A^2 B^2 \alpha_u a_u^2 w_u^2 \beta_u^2}{\left(A^2 \beta_u^2 + B^2 w_u^2\right)} \\ &+ 4\nu I_{42} \alpha_u^2 + 2q I_4 \alpha_u^2 \beta_u^2 + I_{22} D_v a_v^2 + D_v \left(I_2 a_v^2 w_v^2 + \Lambda_v g_v^2\right)\left(U_v^2 + V_v^2\right) \\ &- \frac{2A_v A^2 B^2 \alpha_v a_v^2 w_v^2 \beta_v^2}{\left(A^2 \beta_v^2 + B^2 w_v^2\right)} + 4\nu I_{42} \alpha_v^2 + 2q I_4 \alpha_v^2 \beta_v^2 \Bigg\} = 0. \end{aligned} \tag{7.36}$$

The modulation equation (Eq. 7.15), together with its v color counterpart, give the fixed point relations between $\hat{a}_u$ and $\hat{w}_u$ and $\hat{a}_v$ and $\hat{w}_v$. With these relations, the energy conservation equation (Eq. 7.36) then gives the fixed point amplitudes and widths from the known initial beams.

Figure 7.2 shows a comparison of the amplitudes and positions of the nematicons as given by the full numerical solution of the nematicon equations (Eqs 7.2–7.4) and the solution of the modulation equations with loss to diffractive radiation for a typical set of initial conditions. For this set of initial conditions the nematicons are initially moving in the same direction. The nematicons oscillate about each other following a mean trajectory as they evolve. This mean trajectory is given by conservation of linear momentum and is the center of mass position (Eq. 7.25), as is discussed later. There is excellent agreement in the positions of the nematicons, with a slight disagreement in the periods of the position oscillations for larger values of z. The agreement between the amplitudes is not so good. The numerical amplitude evolution shows much more complicated behavior than that given by the modulation equations, which shows smooth oscillations that decay in amplitude owing to radiation loss. As the amplitude-width evolutions of the nematicons form nonlinear oscillators, the differences in the amplitudes of the oscillations as given by the numerical and modulation solutions translate into period differences.

Figure 7.3 shows similar comparisons between the full numerical and modulation solutions for an initial condition for which the nematicons initially move in opposite directions. In this case the total linear momentum is less, so that the mean speed of the nematicons is not as large. The position comparison is again excellent, with a slight period difference for large z. The numerical amplitude again shows more complicated behavior and is more complicated than the evolution shown in Figure 7.3a. The reason for this more complicated amplitude evolution as shown by the full numerical solution is discussed in the following text.

Figure 7.4 shows the profile $|u|$ of the u color beam at a fixed value of z for various values of the nonlocality parameter ν. It can be seen that the interaction

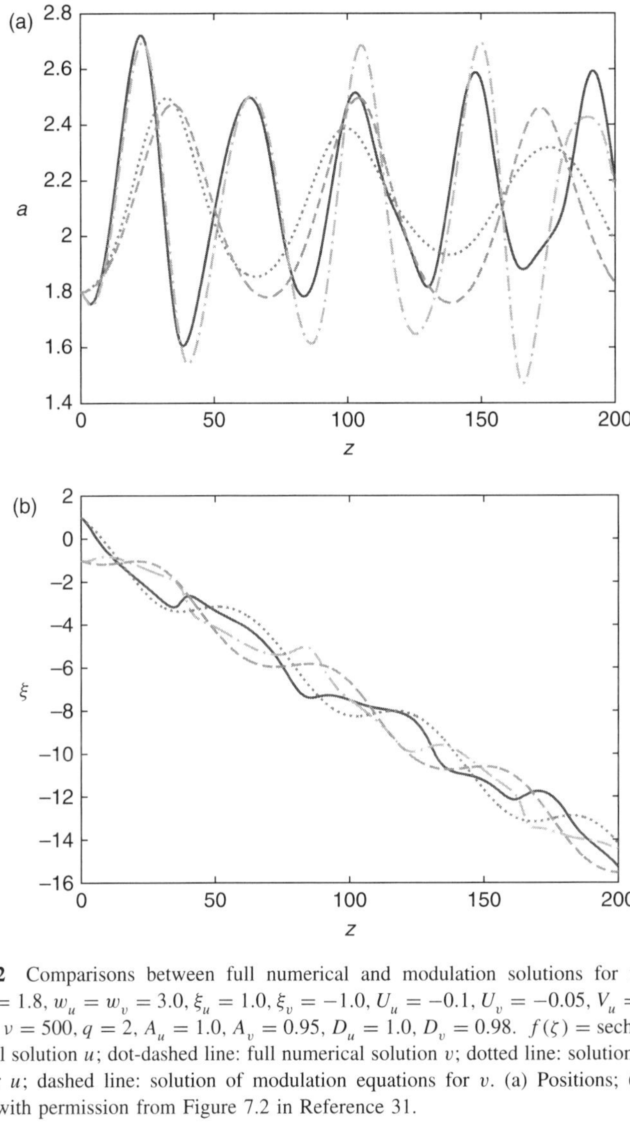

FIGURE 7.2 Comparisons between full numerical and modulation solutions for the initial values $a_u = a_v = 1.8$, $w_u = w_v = 3.0$, $\xi_u = 1.0$, $\xi_v = -1.0$, $U_u = -0.1$, $U_v = -0.05$, $V_u = V_v = 0$, $\eta_u = \eta_v = 0$ with $\nu = 500$, $q = 2$, $A_u = 1.0$, $A_v = 0.95$, $D_u = 1.0$, $D_v = 0.98$. $f(\zeta) = \operatorname{sech} \zeta$. Solid line: full numerical solution u; dot-dashed line: full numerical solution v; dotted line: solution of modulation equations for u; dashed line: solution of modulation equations for v. (a) Positions; (b) amplitudes. Reproduced with permission from Figure 7.2 in Reference 31.

of the beam with the beam in the v color has resulted in a significant distortion of its profile, particularly for values of ν in the experimental range $O(100)$. In addition, the shelves of radiation extending out from the beam show distortion. As the nonlocality parameter ν increases, the beam distortion decreases. The reason for this distortion of the beam shape is clear. The beam is being accelerated by its interaction with the v color beam and does not undergo symmetric interaction over its profile. This introduces asymmetry in the interaction, which deforms the profile. These changes in profile are, of course, not taken into account by the fixed

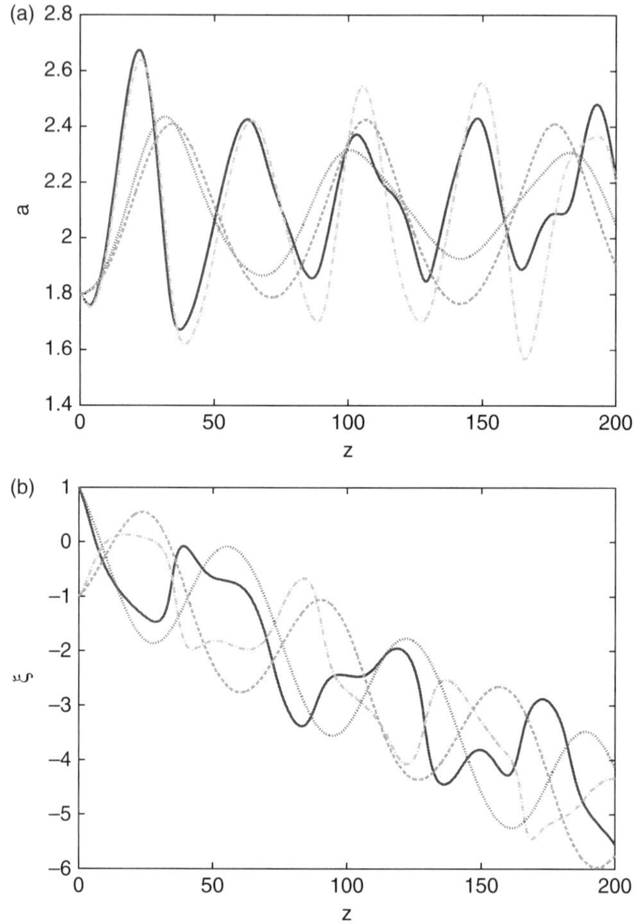

FIGURE 7.3 Comparisons between full numerical and modulation solutions for the initial values $a_u = a_v = 1.8$, $w_u = w_v = 3.0$, $\xi_u = 1.0$, $\xi_v = -1.0$, $U_u = -0.1$, $U_v = 0.05$, $V_u = V_v = 0$, and $\eta_u = \eta_v = 0$, with $\nu = 500$, $q = 2$, $A_u = 1.0$, $A_v = 0.95$, $D_u = 1.0$, $D_v = 0.98$. $f(\zeta) = \text{sech}\,\zeta$. Solid line: full numerical solution u; dot-dashed line: full numerical solution v; dotted line: solution of modulation equations for u; dashed line: solution of modulation equations for v. (a) Positions; (b) amplitudes.

trial functions (Eq. 7.6). Indeed, it is not obvious how to introduce realistic beam distortions into the trial functions. The beam positions are not greatly affected by the beam distortions as the trajectories are determined by the linear momentum of the beams, which depends on the total mass (optical power), an integral quantity not sensitive to the details of the beam profile. Fixed trial functions can then give good results for the beam trajectories.

The Kepler-like modulation equations of Section 7.2 predict that the nematicons will orbit about each other, much as for two gravitating bodies. However, when loss to shed diffractive radiation is included, the nematicons can spiral into each

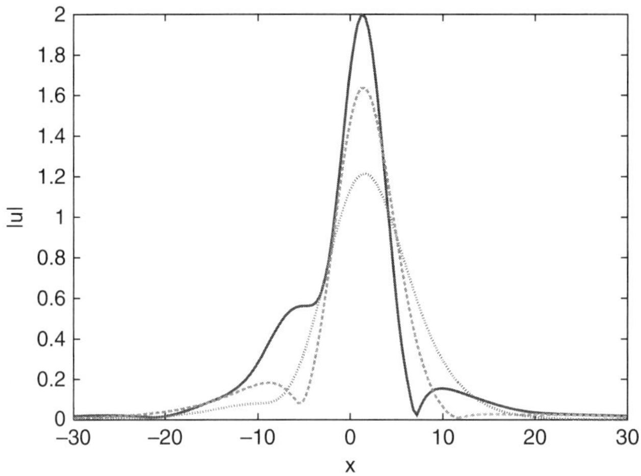

FIGURE 7.4 Full numerical solution for $|u|$ at $z = 100$ and $y = 0$ for the initial values $a_u = a_v = 1.2$, $w_u = w_v = 4.0$, $\xi_u = 1.0$, $\xi_v = -1.0$, $U_u = 0.1$, $U_v = -0.05$, $V_u = V_v = 0$, and $\eta_u = \eta_v = 0$, with $q = 2$, $A_u = 1.0$, $A_v = 0.95$, $D_u = 1.0$, $D_v = 0.98$. $f(\zeta) = \text{sech}\,\zeta$. Solid line: solution for $\nu = 250$; dashed line: solution for $\nu = 500$; and dotted line: solution for $\nu = 1000$. (*Source*: Reproduced with permission from Figure 7.3 in Reference 31.)

other and form a bound state propagating at the same position. For nematicons propagating in the plane with no angular momentum, this shedding of radiation will always result in a bound state as long as their initial linear momentum difference is not too large. When the nematicons have angular momentum, the situation is more complicated. As the nematicons shed radiation they lose angular momentum. There exists a threshold angular momentum at which there is a cutoff in angular momentum shed to radiation in the form of spiral waves [29]. This cutoff means that a stable bound state of orbiting two-color nematicons can form even though there is loss to diffractive radiation.

Let us then consider in-plane nematicon interaction, in the x plane say, resulting in a bound state of copropagating nematicons. If linear momentum loss to shed diffractive radiation is neglected, the x linear momentum conservation equation (Eq. 7.21) and the trajectory equation (Eq. 7.14) and their v color counterparts show that the position of this copropagating bound state is given by

$$\hat{\xi}' = \hat{\xi}'_u = \hat{\xi}'_v = \frac{D_u D_v M_0}{I_2 \left(D_u a_{v0}^2 w_{v0}^2 + D_v a_{u0}^2 w_{u0}^2 \right)}, \tag{7.37}$$

where the initial x linear momentum is

$$M_0 = I_2 \left(a_{u0}^2 w_{u0}^2 U_{u0} + a_{v0}^2 w_{v0}^2 U_{v0} \right). \tag{7.38}$$

Here, the 0 subscripts refer to initial values of the quantities. Figure 7.5 shows a comparison between $\hat{\xi}'$ as given by the full numerical solution of the two-color

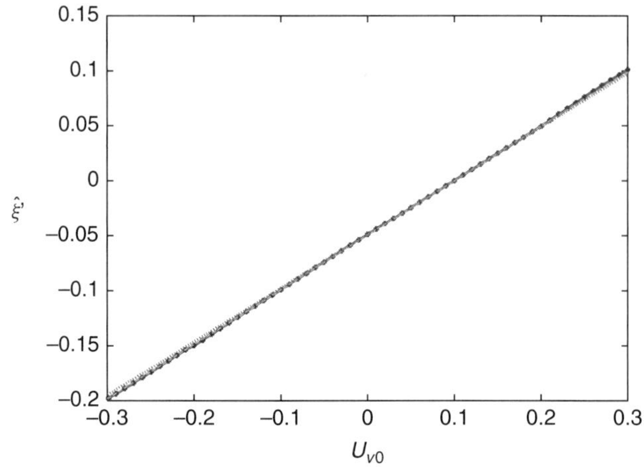

FIGURE 7.5 Steady value $\hat{\xi}'$ as a function of U_{v0} for the initial values $a_u = a_v = 1.8$, $w_u = w_v = 3.0$, $\xi_u = 1.0$, $\xi_v = -1.0$, and $U_u = 0.1$ with $\nu = 500$, $q = 2$, $A_u = 1.0$, $A_v = 0.95$, $D_u = 1.0$, $D_v = 0.98$. $f(\zeta) = \text{sech}\ \zeta$. Crosses: full numerical solution; dashed line: solution of modulation equations; dotted line: momentum conservation result. (*Source*: Reproduced with permission from Figure 7.4 in Reference 31.)

nematicon equations (Eqs 7.2–7.4), the solution of the modulation equations with radiative loss, and the momentum conservation result (Eq. 7.37). It can be seen that there is excellent agreement between the numerical and modulation solutions. Furthermore, the momentum conservation result (Eq. 7.37), which ignores momentum shed in diffractive radiation, gives an excellent prediction for $\hat{\xi}'$. This excellent agreement shows that in the nonlocal limit there is little momentum shed in diffractive radiation, unlike in the local limit [18]. This conforms with the general trend that as the nonlocality ν increases, the rate of loss to diffractive radiation decreases and it takes longer for nematicons to evolve to the steady state.

7.4 MULTIDIMENSIONAL CLUSTERS

The analogy of nematicon motion with the gravitational Kepler problem, explored in Section 7.2, suggests that other configurations known from classical mechanics could have equivalents in nematicon motion in liquid crystals, examples being the Lagrange triangle solution and figure of 8 solution from Newtonian gravitation [35] and cluster solutions, which include large bodies surrounded by smaller bodies rotating in a synchronized manner, similar to that which occurs with Saturn and its rings. It needs to be emphasized that the experimental verification of such configurations may not be easy, at least for current liquid crystal types, owing to scattering and other losses in the liquid crystal, which become more important when several beams impinge on the same cell. However, it is of interest to examine the similarities and differences between the nematicon case and the equivalent

Newtonian gravitation configurations. The differences are due to the nonmonotonic nature of the interaction potential between nematic beams, as shown by the potential (Eq. 7.29), and, to a lesser extent, by the deformable, nonrigid nature of nematicons, in contrast to gravitating masses. In particular, the diffractive radiation shed by the nematic beams as they move about a common center can cause the ultimate merging or collapse of the multinematicon cluster. In this section, we describe some of the analysis of Simon [35] for the gravitational Lagrange triangular three-body solution and its extension to three interacting nematicons and perform a numerical examination of the effect of a vortex on a cluster of two rotating nematicons.

Let us begin by considering the analog of the Lagrange triangle solution of Newtonian gravitation [35] for three nematicons of three different colors (wavelengths) [29]. In the gravitational case, the homogeneity of the potential, together with its monotonicity, results in rigidly rotating masses with the masses at the vertices of an equilateral triangle. Moreover, in the limit in which one of the masses is much larger than the other two, this gravitation triangle configuration is linearly stable [35]. This linear stability becomes apparent when the triangle configuration is treated as two decoupled two-body Kepler problems for which the small masses rotate around the large one, keeping the triangle configuration. A natural analogy is then the evolution of three nematicons of different colors (wavelengths). These evolve according to the equations [19, 29]

$$\mathrm{i}\frac{\partial u_j}{\partial z} + \frac{1}{2}D_j\nabla^2 u_j + A_j\theta u_j = 0, \quad j = 1, 2, 3, \tag{7.39}$$

$$\nu\nabla^2\theta - 2q\theta = -2\sum_{j=1}^{3} A_j|u_j|^2. \tag{7.40}$$

Here, D_j and A_j are the diffraction coefficients and coupling coefficients for the three colors, respectively. Let us consider initial conditions of the form

$$u_j = a_{j0} \;\operatorname{sech} \zeta_{j0}\, \mathrm{e}^{\mathrm{i}U_{j0}(x-\xi_{j0})+\mathrm{i}V_{j0}(y-\eta_{j0})}, \quad j = 1, 2, 3, \tag{7.41}$$

where

$$\zeta_{j0} = \sqrt{(x-\xi_{j0})^2 + (y-\eta_{j0})^2}/w_{j0}. \tag{7.42}$$

The peaks (ξ_j, η_j) of the nematicons are placed at the vertices of a triangle. The velocities (U_j, V_j) are chosen to be approximately orthogonal to the position vectors of the peaks from the center of the triangle to ensure rotation about this center.

Figure 7.6 shows an example of an unstable three-color nematicon cluster. This example shows a typical instability whereby two of the nematicons lock and the third one escapes. The two locked nematicons orbit each other as for the two-color-coupled nematicons of Section 7.3.

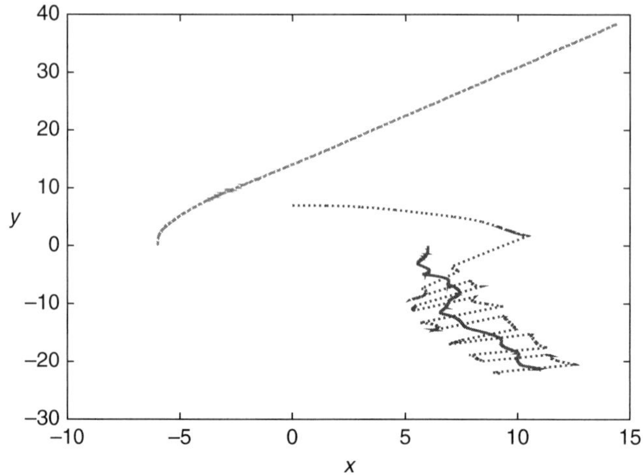

FIGURE 7.6 Numerical positions (x,y) of three nematicons up to $z = 200$. u_1: solid line; u_2: dashed line; u_3: dotted line. The initial conditions are $a_1 = 1.8$, $w_1 = 3.0$, $a_2 = 1.3$, $w_2 = 3.0$, $a_3 = 1.3$, $w_3 = 3.0$, $\xi_1 = 6.0$, $\eta_1 = 0.0$, $\xi_2 = -6.0$, $\eta_2 = 0.0$, $\xi_3 = 0.0$, $\eta_3 = 7.0$, $U_1 = 0.0$, $V_1 = -0.2$, $U_2 = 0.0$, $V_2 = 0.2$, $U_3 = 0.3$, $V_3 = 0.0$ with $q = 10$, $\nu = 50$, $D_1 = 1.0$, $D_2 = 0.95$, $D_3 = 0.98$, $A_1 = 1.0$, $A_2 = 0.9$ and $A_3 = 0.95$.

To gain an understanding of the dynamics of three-color nematicon clusters, let us develop a modulation theory based on the trial functions

$$u_j = a_j \operatorname{sech}(\zeta_j)\, e^{i\sigma_j + iU_j(x-\xi_j) + iV_j(y-\eta_j)}, \quad j = 1, 2, 3,$$

$$\theta = \sum_{j=1}^{3} \alpha_j \operatorname{sech}^2 \kappa_j \tag{7.43}$$

for the nematicons and the director response [32, 33]. Here,

$$\zeta_j = \sqrt{(x-\xi_j)^2 + (y-\eta_j)^2}/w_j, \quad \kappa_j = \sqrt{(x-\xi_j)^2 + (y-\eta_j)^2}/\beta_j \tag{7.44}$$

and all the parameters in the trial functions depend on the distance z down the cell. The three-color nematicon equations (Eqs 7.39 and 7.40) have the Lagrangian

$$L = \sum_{j=1}^{3} \left[i\left(u_j^* u_{jz} - u_j u_{jz}^*\right) - D_j |\nabla u_j|^2 + 4A_j |u_j|^2 \theta \right] - \nu |\nabla \theta|^2 - 2q\theta^2. \tag{7.45}$$

Substituting the trial functions (Eq. 7.43) into this Lagrangian and integrating x and y over the whole plane, $-\infty < x,\, y < \infty$, that is averaging, result in the averaged

Lagrangian

$$\mathcal{L} = \sum_{i=1}^{3} \Bigg[-2I_2 a_i^2 w_i^2 \left(\sigma_i' - U_i \xi_i' - V_i \eta_i'\right) - D_i I a_i^2 - D_i I_2 a_i^2 w_i^2 \left(U_i^2 + V_i^2\right)$$
$$- 4\nu I_{42}\alpha_i^2 - 2q I_4 \alpha_i^2 \beta_i^2 + \frac{2A^2 B^2 A_i \alpha_i a_i^2 w_i^2 \beta_i^2}{A^2 \beta_i^2 + B^2 w_i^2} \Bigg] - \sum_{j=1}^{3} \sum_{i=1, i\neq j}^{3} \Phi_{ij}(r_{ij}) \tag{7.46}$$

with the various integrals I_i and I_{ij} given in Appendix 7.A. From the potential (Eq. 7.29), the interaction potential Φ_{ij} of the nematicons is

$$\begin{aligned} \Phi_{ij}(r_{ij}) = &-2A^2 B^2 A_i \alpha_j a_i^2 w_i^2 \beta_j^2 Q_{1ij}^{-1} \mathrm{e}^{-r_{ij}^2/Q_{1ij}} \\ &- 2A^2 B^2 A_j \alpha_i a_j^2 w_j^2 \beta_i^2 Q_{1ji}^{-1} \mathrm{e}^{-r_{ij}^2/Q_{1ji}} \\ &+ 4\nu \alpha_i \alpha_j \beta_i^2 \beta_j^2 Q_{2ij}^{-2} \left(1 - A^{-2} r_{ij}^2 Q_{2ij}^{-1}\right) \mathrm{e}^{-r_{ij}^2/(A^2 Q_{2ij})} \\ &+ 2q A^2 \alpha_i \alpha_j \beta_i^2 \beta_j^2 Q_{2ij}^{-1} \mathrm{e}^{-r_{ij}^2/(A^2 Q_{2ij})}. \end{aligned} \tag{7.47}$$

The separation of the nematicons is

$$r_{ij}^2 = \left(\xi_i - \xi_j\right)^2 + \left(\eta_i - \eta_j\right)^2. \tag{7.48}$$

Finally, the quantities Q_{1ij} and Q_{2ij} appearing in the averaged Lagrangian are

$$Q_{1ij} = A^2 \beta_j^2 + B^2 w_i^2 \quad \text{and} \quad Q_{2ij} = \beta_i^2 + \beta_j^2. \tag{7.49}$$

Taking variations of the averaged Lagrangian (Eq. 7.46) with respect to the parameters gives the equations governing the evolution and positions of the nematicons. As in Section 7.2, the equations for the positions $\boldsymbol{\xi}_i = (\xi_i, \eta_i)$ of the nematicons can be put in the classical mechanics form

$$m_i \frac{\mathrm{d}^2 \boldsymbol{\xi}_i}{\mathrm{d}z^2} = - \sum_{j=1, j\neq i}^{3} \nabla \Phi_{ij}(r_{ij}), \tag{7.50}$$

where the mass, or optical power, of a nematicon is

$$m_i = \frac{I_2 a_i^2 w_i^2}{D_i}. \tag{7.51}$$

However, to derive the three-color Lagrange solution and study its stability, it is more convenient to use the coordinate system sketched in Figure 7.7 [35]. In this coordinate system, the motion of the center of mass of the nematicons is decoupled

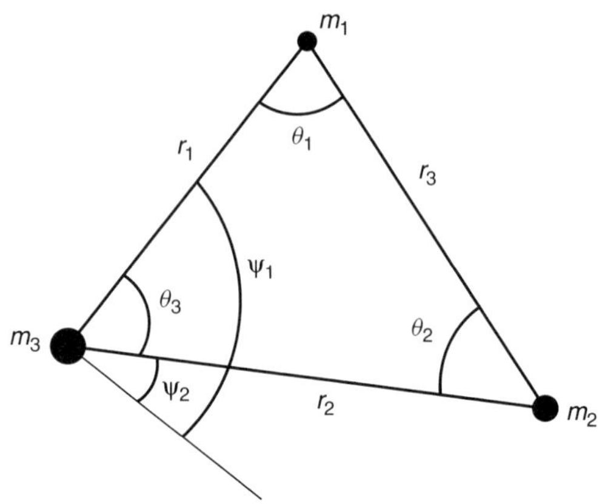

FIGURE 7.7 Coordinate system for the three-color nematicon solution. (*Source*: Reproduced with permission from Figure 7.1 in Reference 29.)

from their relative motion. In this coordinate system, equations (Eq. 7.50) for the positions of the nematicons become

$$m_1\left(\ddot{r}_1 - r_1\dot{\psi}_1^2 + \frac{\Phi'_{13}}{m_3} + \frac{\Phi'_{23}\cos\theta_3}{m_3}\right) = -\Phi'_{13} - \Phi'_{12}\cos\theta_1, \tag{7.52}$$

$$m_1\left(r_1\ddot{\psi}_1 + 2\dot{r}_1\dot{\psi}_1 - \frac{\Phi'_{23}\sin\theta_3}{m_3}\right) = -\Phi'_{12}\sin\theta_1, \tag{7.53}$$

$$m_2\left(\ddot{r}_2 - r_2\dot{\psi}_2^2 + \frac{\Phi'_{23}}{m_3} + \frac{\Phi'_{13}\cos\theta_3}{m_3}\right) = -\Phi'_{23} - \Phi'_{21}\cos\theta_2, \tag{7.54}$$

$$m_2\left(r_2\ddot{\psi}_2 + 2\dot{r}_2\dot{\psi}_2 + \frac{\Phi'_{13}\sin\theta_3}{m_3}\right) = \Phi'_{12}\sin\theta_2. \tag{7.55}$$

The rigidly rotating Lagrange triangle solution then has $\psi_2 = \omega z$, so that $\psi_1 = \omega z + \theta_3$, and constant values for the radii r_1 and r_2. The angles θ_1, θ_2, and θ_3 are then determined from Equations 7.52–7.55. At the steady state, Equations 7.53 and 7.55 for the angles ψ_1 and ψ_2 give

$$\Phi'_{23}\frac{\sin\theta_3}{m_3} = \Phi'_{12}\frac{\sin\theta_1}{m_1}. \tag{7.56}$$

On using the sine rule

$$\frac{\sin\theta_3}{r_3} = \frac{\sin\theta_1}{r_2} \tag{7.57}$$

this becomes

$$\frac{r_3}{r_2} = \frac{m_3}{m_1}\frac{\Phi'_{12}}{\Phi'_{23}}. \tag{7.58}$$

By symmetry, we then have that the three-color Lagrange nematicon solution is

$$\frac{r_1}{r_3} = \frac{m_2}{m_3}\frac{\Phi'_{13}}{\Phi'_{12}}, \quad \frac{r_2}{r_1} = \frac{m_1}{m_2}\frac{\Phi'_{23}}{\Phi'_{13}}, \quad \frac{r_3}{r_2} = \frac{m_3}{m_1}\frac{\Phi'_{12}}{\Phi'_{23}}. \tag{7.59}$$

This gives the lengths of the sides of the triangle. Finally, the radial equations (Eqs 7.52 and 7.54) give the angular velocity of the rotating triangle as

$$\omega^2 = \frac{\Phi'_{13}\left(m_1 + m_2 + m_3\right)}{r_1 m_1 m_3}. \tag{7.60}$$

In experimental situations the nonlocality parameter ν is large, $\nu = O(100)$ [20, 36]. Let us therefore examine the three-color Lagrange solution in the limit of $\nu \gg 1$. In this limit the potential (Eq. 7.47) can be approximated by

$$\Phi_{ij}(r_{ij}) = 4\nu P_{ij}\left(Q_{2ij}^{-1} + \frac{qA^2}{2\nu} - \frac{r_{ij}^2}{A^2 Q_{2ij}^2}\right) \mathrm{e}^{-r_{ij}^2/(A^2 Q_{2ij})}, \tag{7.61}$$

with

$$P_{ij} = \alpha_i \alpha_j \beta_i^2 \beta_j^2 Q_{2ij}^{-1}. \tag{7.62}$$

The first difference between the Lagrange solution for gravitation and for three-color nematicons is that the three-color potential (Eq. 7.61) is not monotonic as it is for Newtonian gravitation. Furthermore, the shape of the potential depends on the masses, whereas in the gravitation case the masses only scale the potential. It is then apparent that the three-color Lagrange solution will not be an equilateral triangle, as this uniform shape depends on the masses just scaling the potential. To see what possible shapes for the Lagrange three-color solution can occur, let us take advantage of the nonmonotonicity of the potential and construct a solution in the form of an elongated isosceles triangle. This particular solution is tractable using the large nonlocality limit (Eq. 7.61) of the potential. To this end, let us take two equal masses (optical powers) $m_2 = m_3$ and further assume that $m_2 \gg m_1$. The triangle is then elongated with $\theta_1 \sim 0, \theta_2 = \theta_3 \approx \pi/2$ and $r_3 = r_1$ to leading order. From the general triangle solution (Eqs 7.59 and 7.60), we then obtain

$$r_1\omega^2 = \frac{2}{m_1}\Phi'_{13}(r_1) \quad \text{and} \quad r_2\omega^2 = \frac{2}{m_2}\Phi'_{23}(r_2). \tag{7.63}$$

The r_3 in Equation 7.59 is then automatically satisfied by symmetry. Now because r_2 is assumed to be the short side and r_1 is assumed to be the long side of the triangle, we choose the large root of the first of Equation 7.63 and the small root of the second of Equation 7.63. We therefore see that an elongated isosceles triangle solution is possible for three nematicons. This solution is not possible for the monotone gravitational potential, but is natural for the nonmonotone nematicon potential.

Let us now study the linearized stability of the Lagrange three-color solution. Full details of this stability analysis can be found in Assanto et al. [29]. It is sufficient to note that the stability of the Lagrange three-color solution stems from the different slopes of the potential at the equilibrium positions of the nematicons. This again shows that the isosceles triangle shape is essential for stability when the potential is nonmonotone.

Unlike gravitating masses, nematicons are not rigid, but deform during their evolution and shed diffractive radiation. In Chapter 3 and in Section 7.3 it was shown how diffractive radiation generated from the shelf of low wave number radiation under an evolving nematicon provides a mechanism for mass loss, which in turn provides a damping that stabilizes nematicons. However, it is also known that the shedding of angular momentum in the form of a spiral wave can produce the collapse of a rotating coherent structure [37]. In light of this, let us now examine the effect of the spiral waves shed by the rotating three-color Lagrange structure.

To model the diffractive radiation shed by the jth-color nematicon, let us consider the linearized electric field equation (Eq. 7.39)

$$\mathrm{i}\frac{\partial u_j}{\partial z} + \frac{D_j}{2}\nabla^2 u_j = 0. \tag{7.64}$$

The linearized equation is appropriate as the radiation has small amplitude compared with the nematicon. In polar coordinates (ρ, φ) relative to the center of mass of the cluster, each nematicon moves in a circle of radius R_j with an angular position $\zeta_j(z)$. In effect, the nematicon acts as a source for the radiation equation (Eq. 7.64). Let us take the shelf under the jth nematicon to be of the form $g_j(\rho, \varphi - \zeta_j(z))$. The angular extent of the flat shelf is taken to be λ_j and the radial extent is taken as $R_j - \epsilon \le \rho \le R_j + \epsilon$. The detailed functional form of g_j is not needed, but can be determined as detailed in Chapter 3 and in Section 7.3 using conservation of mass. We then need to solve the radiation equation (Eq. 7.64) together with the boundary condition

$$u_j(R_j + \epsilon) = g_j(R_j + \epsilon, \varphi - \zeta_j(z)) = h_j(\varphi - \zeta_j(z)). \tag{7.65}$$

Let us obtain the solution of the radiation equation (Eq. 7.64) using a geometric optics expansion. The validity of this geometric optics expansion is based on the relatively rapid rotation of the triangle. We therefore expand in the form

$$h_j(\varphi - \zeta_j(z)) = \sum_{n=-\infty}^{\infty} h_{jn} \mathrm{e}^{\mathrm{i}n(\varphi - \zeta_j(z))}. \tag{7.66}$$

We then construct the solution for u_j mode by mode in the geometric optics expansion form

$$u_j(\rho, \varphi, z) = \sum_{n=-\infty}^{\infty} A_n \mathrm{e}^{\mathrm{i}nS_j(\rho,\varphi,z)}. \tag{7.67}$$

Substituting this geometric optics expansion into the radiation equation (Eq. 7.64) gives at first order the eikonal equation

$$\frac{\partial S_j}{\partial z} = -\frac{D_j}{2}\left[\left(\frac{\partial S_j}{\partial \rho}\right)^2 + \frac{1}{\rho^2}\left(\frac{\partial S_j}{\partial \varphi}\right)^2\right]. \tag{7.68}$$

Higher order terms give transport equations for the A_n, but these will not be needed here. The boundary condition for the phase function S is

$$S_j(R_j + \epsilon, \varphi, z) = \varphi - \zeta_j(z) \tag{7.69}$$

on using the boundary condition (Eq. 7.65). A separation of variables solution for S can hence be found as

$$S_j = F_j(\rho) + \varphi - \zeta_j(z). \tag{7.70}$$

This solution gives the phase of the spiral wave shed by the nematicon.

Substitution of the separation of variables solution (Eq. 7.70) into the eikonal equation (Eq. 7.68) gives the equation for F_j

$$\frac{D_j}{2}\left(\frac{\mathrm{d}F_j}{\mathrm{d}\rho}\right)^2 = \frac{\mathrm{d}\zeta_j}{\mathrm{d}z} - \frac{D_j}{2\rho^2} \tag{7.71}$$

with the boundary condition $F_j(R_j + \epsilon) = 0$. It should be noted that as $F_j = F_j(\rho)$ and $\zeta_j = \zeta_j(z)$, this equation is inconsistent. The resolution of this is that the angular velocity ζ_j is a slowly varying function of z, and so can be taken as constant to the order considered here. We then have that the phase of the outgoing spiral wave is determined using the positive square root of Equation 7.71 as

$$\frac{\mathrm{d}F_j}{\mathrm{d}\rho} = \sqrt{\frac{2}{D_j}}\left[\frac{\mathrm{d}\zeta_j}{\mathrm{d}z} - \frac{D_j}{2\rho^2}\right]^{1/2}. \tag{7.72}$$

An asymptotic solution of this equation can be easily obtained for large ρ. Using this asymptotic form, the phase S_j of the spiral wave is

$$S_j(\rho, \varphi, z) \sim \varphi - \zeta_j(z) - \rho\sqrt{\dot{\zeta}_j} \tag{7.73}$$

for large ρ, with the dot referring to derivatives with respect to z. It is clear from Equation 7.72 that for a spiral wave to exist the local rotation speed $\dot{\zeta}_j$ must satisfy

$$\dot{\zeta}_j > \frac{D_j}{2(R_j + \epsilon)^2}. \tag{7.74}$$

If this relation is not satisfied the spiral waves are cut off. We then see that only rapidly rotating triangles shed spiral waves.

We shall now show how this cutoff behavior prevents the Lagrange cluster from collapsing. To do this, we shall calculate the loss of angular momentum from the jth nematicon and show that its angular momentum settles to an equilibrium value. The total angular momentum l_j of the jth-color nematicon is

$$l_j = \mathrm{Im} \int_{-\infty}^{\infty} \int_{-\infty}^{\infty} (\mathbf{r} \times \nabla u_j)\, u_j^* \, dx \, dy. \tag{7.75}$$

The loss of angular momentum at the edge of the shelf of radiation under the nematicon is [33, 34]

$$\frac{dl_j}{dz} = -\frac{R_j}{2} \int_0^{2\pi} \left(u_j \frac{\partial u_j^*}{\partial \rho} + u_j^* \frac{\partial u_j}{\partial \rho} \right) \Bigg|_{\rho = R_j + \epsilon} d\varphi. \tag{7.76}$$

To evaluate the integral in this loss expression we need the geometric optics solution for the radiation u_j evaluated at the edge of the shelf. The boundary condition (Eq. 7.65) gives $A_n = h_{jn}$. Using this relation, the loss expression (Eq. 7.76) becomes

$$\frac{dl_j}{dz} = -\pi R_j \left[\dot{\zeta}_j - \frac{D_j}{2(R_j + \epsilon)^2} \right]^{1/2} \left(\frac{2}{D_j} \right)^{1/2} \sum_{n=1}^{\infty} |h_{jn}|^2. \tag{7.77}$$

As the mass of the nematicon is Equation 7.51,

$$\dot{\zeta}_j = \frac{D_j l_j}{I_2 R_j a_j^2 w_j^2} \tag{7.78}$$

on using the classical mechanics expression for angular momentum. We then have a closed equation for the angular momentum l_j.

Equation 7.77 shows that, initially, the nematicons slow due to the loss of angular momentum. As they slow, the spiral waves are eventually cut off owing to the cutoff condition (Eq. 7.74). After this point, the angular momentum of the coherent Lagrange triangle is conserved and the structure is stable.

The earlier analysis shows that isosceles triangle Lagrange solutions are stable and produce no diffractive radiation after they settle to the stable motion state. These theoretical predictions were tested by numerically solving the three-color nematicon equations (Eqs 7.39 and 7.40) with an equilateral triangle initial condition. Figure 7.8 shows a typical example, which has astronomical equivalents [38]. It can be seen that the structure undergoes dramatic reshaping as it rotates and translates, the translation being due to the initial linear momentum of the center of mass. It is further seen that the structure is reshaping itself into an isosceles triangle, confirming our stability analysis. More detailed views of the positions of the three nematicons in the Lagrange structure is shown in Figure 7.9. These detailed

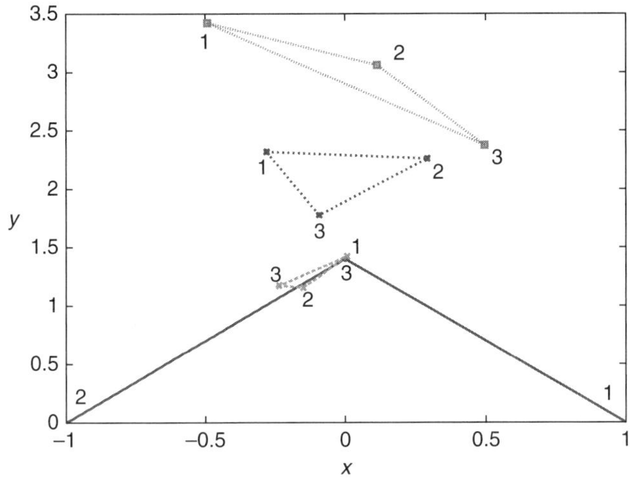

FIGURE 7.8 Position of nematicon peaks. $z = 0$ (solid line); $z = 100$ (cross); $z = 200$ (asterisk); $z = 300$ (square). 1 refers to u_1, 2 refers to u_2, and 3 refers to u_3. Initial conditions are $a_1 = 1.8$, $w_1 = 3.0$, $a_2 = 1.8$, $w_2 = 3.0$, $a_3 = 1.44$, $w_3 = 3.0$, $\xi_1 = 1.0$, $\eta_1 = 0.0$, $\xi_2 = -1.0$, $\eta_2 = 0.0$, $\xi_3 = 0.0$, $\eta_3 = 1.4$, $U_1 = 0.0$, $V_1 = -0.03$, $U_2 = 0.0$, $V_2 = 0.03$, $U_3 = 0.038$, and $V_3 = 0.0$, with $q = 10$, $\nu = 500$, $D_1 = 1.0$, $D_2 = 0.95$, $D_3 = 0.98$, $A_1 = 1.0$, $A_2 = 0.9$, and $A_3 = 0.95$. (*Source*: Reproduced with permission from Figure 7.1 in Reference 29.)

x and y positions confirm that the nematicons are orbiting about each other, with a general translation of the structure in the y direction.

A detailed comparison between the full numerical solution and the solution of the modulation equations is shown in Figure 7.10. The agreement between the numerical and modulation solutions is surprisingly good, given the drastic simplifications made to derive the modulation equations. The modulation solution reproduces the broad features of the motion, with the quantitative differences due to the reshaping of the nematicons, as discussed in Section 7.3. The assumptions used to derive the modulation equations (Eq. 7.50) were even more restrictive compared to those discussed in Section 7.3 as it was assumed that the amplitude and width of the nematicons were fixed.

These numerical and modulation results suggest that the rotating Lagrange isosceles triangle solution is strongly attractive. We therefore expect that three color clusters with angular momentum will reshape and radiate diffractive radiation, to eventually settle into an isosceles triangle configuration.

7.5 VORTEX CLUSTER INTERACTIONS

Another analog with Newtonian gravitation, which could be of interest, is that of Saturn's rings interacting with the small moons, called shepherding moons, which occur in the vicinity of the rings, or even orbit within the rings. The nematicon

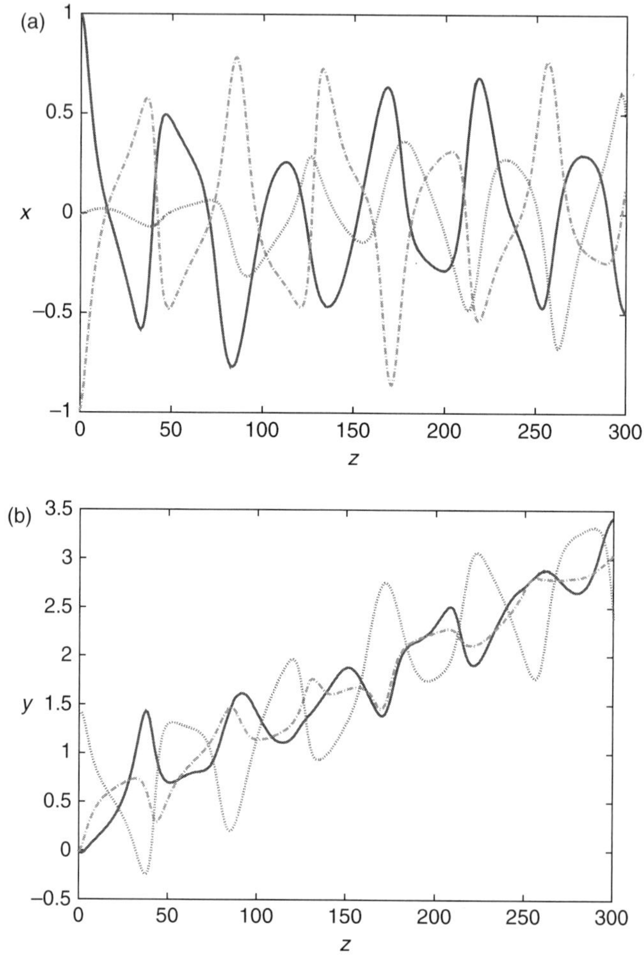

FIGURE 7.9 Numerical positions of three nematicons as a function of z: (a) x position and (b) y position. u_1: solid line; u_2: dot-dashed line; u_3: dotted line. The initial conditions are $a_1 = 1.8$, $w_1 = 3.0$, $a_2 = 1.8$, $w_2 = 3.0$, $a_3 = 1.44$, $w_3 = 3.0$, $\xi_1 = 1.0$, $\eta_1 = 0.0$, $\xi_2 = -1.0$, $\eta_2 = 0.0$, $\xi_3 = 0.0$, $\eta_3 = 1.4$, $U_1 = 0.0$, $V_1 = -0.03$, $U_2 = 0.0$, $V_2 = 0.03$, $U_3 = 0.038$, and $V_3 = 0.0$, with $q = 10$, $\nu = 500$, $D_1 = 1.0$, $D_2 = 0.95$, $D_3 = 0.98$, $A_1 = 1.0$, $A_2 = 0.9$, and $A_3 = 0.95$. (*Source*: Reproduced with permission from Figure 7.2 in Reference 29.)

analog of this gravitational interaction is a central vortex in one color interacting with nematicon beams. For simplicity, let us take a vortex interacting with two nematicon beams. There is the possibility of a strong distortion of the stable vortex due to a resonance between the angular motion of the dipole formed by the nematicons and the normal modes of the vortex. This is the same mechanism proposed by Maxwell [39] in his continuum theory of the rings of Saturn. This theory predicted that sustained tidal-like waves could be produced by the moons orbiting around the rings.

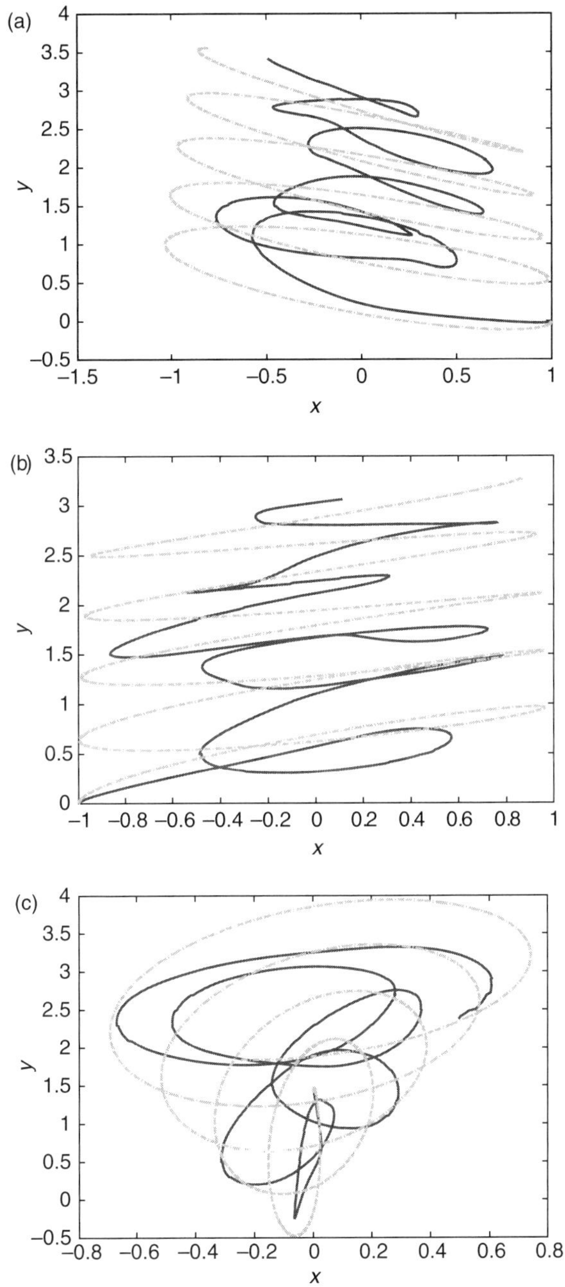

FIGURE 7.10 Comparison of nematicon position between the numerical (solid line) and modulation solutions (dot-dashed line). Initial conditions are $a_1 = 1.8$, $w_1 = 3.0$, $a_2 = 1.8$, $w_2 = 3.0$, $a_3 = 1.44$, $w_3 = 3.0$, $\xi_1 = 1.0$, $\eta_1 = 0.0$, $\xi_2 = -1.0$, $\eta_2 = 0.0$, $\xi_3 = 0.0$, $\eta_3 = 1.4$, $U_1 = 0.0$, $V_1 = -0.03$, $U_2 = 0.0$, $V_2 = 0.03$, $U_3 = 0.038$, and $V_3 = 0.0$, with $q = 10$, $\nu = 500$, $D_1 = 1.0$, $D_2 = 0.95$, $D_3 = 0.98$, $A_1 = 1.0$, $A_2 = 0.9$, and $A_3 = 0.95$. (*Source*: Reproduced with permission from Figure 7.4 in Reference [29].)

To test this Saturn-like configuration for the three-color nematicon equations (Eqs 7.39 and 7.40), these equations were numerically solved for the initial conditions

$$u_j = a_{j0}\text{sech}\ \zeta_{j0}\ \mathrm{e}^{\mathrm{i}U_{j0}(x-\xi_{j0})+\mathrm{i}V_{j0}(y-\eta_{j0})}, \quad j=1,2, \quad u_3 = a_{30}\rho \mathrm{e}^{-\rho/w_{30}}\mathrm{e}^{\mathrm{i}\varphi} \tag{7.79}$$

with

$$\zeta_{j0} = \sqrt{(x-\xi_{j0})^2 + (y-\eta_{j0})^2}/w_{j0}, \quad j=1,2. \tag{7.80}$$

Here, (ρ, φ) are polar coordinates in the (x,y) plane.

Figure 7.11 shows the trajectories of the two nematicons u_1 and u_2 up to $z = 200$. It can be seen that, whereas the individual trajectories are complicated, the two nematicons are orbiting about the origin, which is the position of the center of the vortex u_3. The two nematicons u_1 and u_2 orbit at smaller radii than their initial values. The reason for this can be seen from the initial values at $z = 0$ and numerical solutions at $z = 200$ for u_1, u_2, and u_3 shown in Figure 7.12. The vortex has evolved to have a smaller radius and the two nematicons have moved to be located on the ring on which the vortex has its maximum as this is the center of the waveguide induced by the vortex. The two nematicons have also distorted the symmetric shape of the vortex. Although the modulation equations governing nonlinear beams

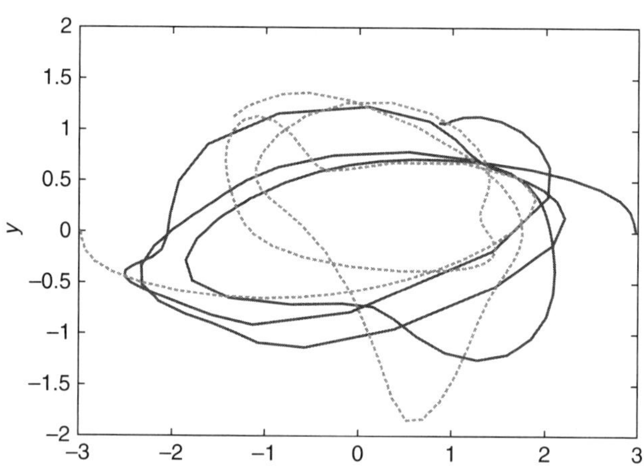

FIGURE 7.11 Positions of u_1 color (solid line) and u_2 color (dashed line) as given by full numerical solution (Eqs. 7.39 and 7.40) for initial conditions (Eq. 7.79). Parameter values are $a_{10} = 1.5, w_{10} = 3.0, a_{20} = 1.5, w_{20} = 3.0, a_{30} = 1.5, w_{30} = 3.0, \xi_{10} = 3.0, \eta_{10} = 0.0, \xi_{20} = -3.0, \eta_{20} = 0.0, U_{10} = 0.0, V_{10} = 0.05, U_{20} = 0.0$, and $V_{20} = -0.05$, with $q = 10, \nu = 500, D_1 = 1.0, D_2 = 0.95, D_3 = 0.98, A_1 = 1.0, A_2 = 0.9$, and $A_3 = 0.95$.

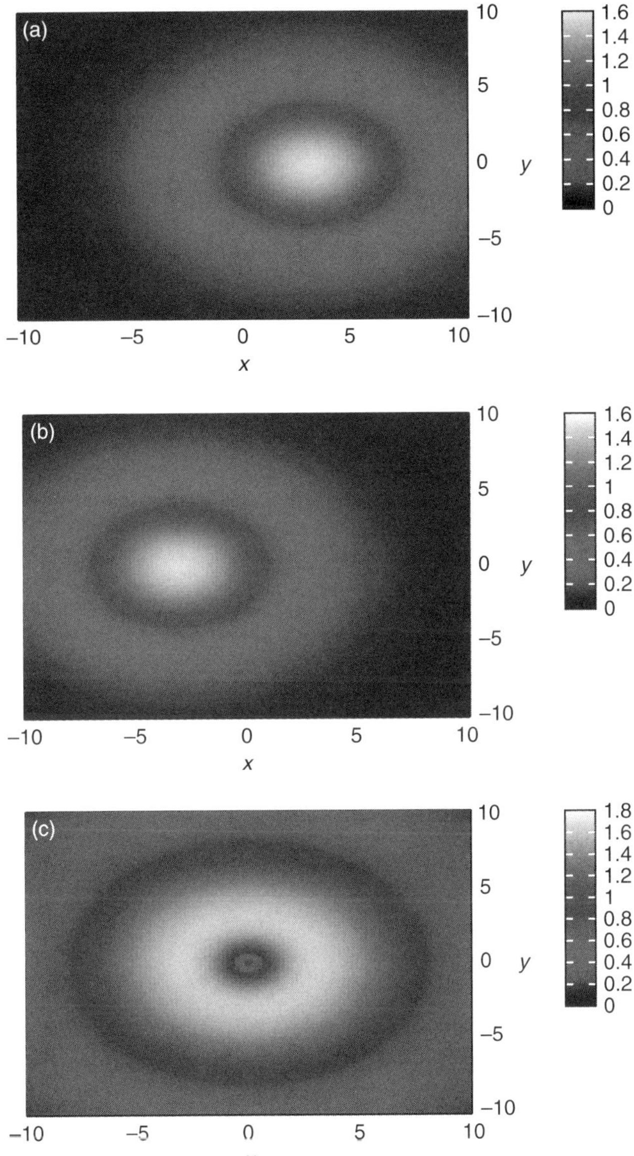

FIGURE 7.12 Numerical solution of Equations 7.39 and 7.40 for initial conditions (Eq. 7.79). (a) $|u_1|$ at $z = 0$, (b) $|u_2|$ at $z = 0$, (c) $|u_3|$ at $z = 0$, (d) $|u_1|$ at $z = 200$, (e) $|u_2|$ at $z = 200$, (f) $|u_3|$ at $z = 200$. Parameter values are $a_{10} = 1.5$, $w_{10} = 3.0$, $a_{20} = 1.5$, $w_{20} = 3.0$, $a_{30} = 1.5$, $w_{30} = 3.0$, $\xi_{10} = 3.0$, $\eta_{10} = 0.0$, $\xi_{20} = -3.0$, $\eta_{20} = 0.0$, $U_{10} = 0.0$, $V_{10} = 0.05$, $U_{20} = 0.0$, and $V_{20} = -0.05$, with $q = 10$, $\nu = 500$, $D_1 = 1.0$, $D_2 = 0.95$, $D_3 = 0.98$, $A_1 = 1.0$, $A_2 = 0.9$, and $A_3 = 0.95$. *(Continued)*

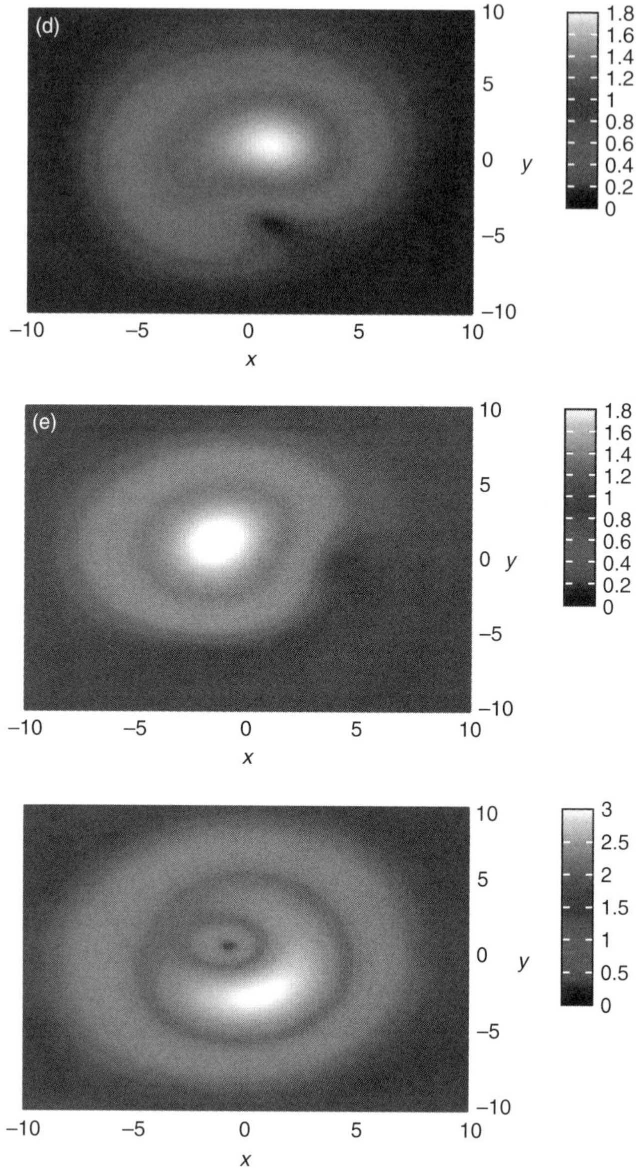

FIGURE 7.12 *(Continued)*

in liquid crystals are of the same form as those governing gravitating masses, nonlinear beams are not rigid masses and their distortion introduces dynamical effects that are not present in gravitation. To include the distortion of the vortex using modulation theory, we would need to include the evolution of its internal modes, as studied in Chapter 15, and obtain a wave-type equation for the evolution of its width forced by the motion of the nematicon beams. The comparison of the results of this extended modulation theory with Maxwell's theory for the waves generated in Saturn's rings would be an interesting extension of the theory of this chapter.

Recently, other types of gravitational clusters with periodic structures have been investigated and their solution forms derived [40]. It would be of interest to study the nematicon equivalents of these gravitational clusters and determine the influence of the nonmonotonic nematicon potential (7.29) on their existence and stability.

7.6 CONCLUSIONS

The modulation theory developed in Chapter 3 has been used to investigate the interaction of two and more nematicons, solitary waves in a nematic liquid crystal. The amplitude and width evolution of the nematicons essentially decouples from their position and velocity evolution. On totally decoupling the amplitude and width evolution from the position and velocity evolution, the modulation equations describing the position and velocity evolution of the nematicons are the same as those governing the interaction of masses in Newtonian gravitation. However, the interaction potential is more complicated than the inverse separation potential of Newtonian gravitation. The nematicon potential gives two orbits for two interacting nematicons, the one with the larger radius being stable and the other being unstable. Under the modulation theory approximation, the evolution of interacting nematicons is then governed by the same basic principles as is the case with gravitating masses, these being conservation of total momentum and energy. The principal difference between the interaction of nematicons and gravitating masses is that interacting nematicons shed diffractive radiation in order to evolve to a steady state. Normally, gravitating bodies do not shed mass in order to reach stable orbits. Another difference between gravitating masses and interacting nematicons is that nematicons need a minimum mass, otherwise they break up into diffractive radiation. Furthermore, above a critical mass, they will break up into multiple nematicons. The close analogy between gravitating masses and interacting nematicons means that classical solutions from Newtonian gravitation can be carried over to the latter field, for instance, the Lagrange three-body solution discussed here.

The analogy between interacting nematicons and gravitating masses can be pushed further. In addition to solitary wave beams (nematicons), the nematicon equations possess stable vortex solutions. Such vortices are analogous to many interacting masses, for example, the rings of Saturn and the other gas giants of the solar system or the asteroid belt. Nematicons can interact with such a vortex in a manner similar to the shepherding moons of Saturn. These analogies between interacting nematicons and gravitating masses can be further utilized to gain insight into and understanding of the interaction and evolution of nematicons.

APPENDIX: INTEGRALS

The integrals I_i and $I_{i,j}$ arising in the modulation equations are

$$I_1 = \int_0^\infty \zeta f(\zeta)\, d\zeta, \quad I_2 = \int_0^\infty \zeta f^2(\zeta)\, d\zeta,$$
$$I_{22} = \int_0^\infty \zeta \left[\frac{df}{d\zeta}\right]^2 d\zeta, \quad I_{x32} = \int_0^\infty \zeta^3 f^2(\zeta)\, d\zeta,$$
$$I_{42} = \frac{1}{4}\int_0^\infty \zeta \left[\frac{d}{d\zeta} f^2(\zeta)\right]^2 d\zeta, \quad I_4 = \int_0^\infty \zeta f^4(\zeta)\, d\zeta. \tag{7.A.1}$$

For a sech beam profile $f(\zeta) = \text{sech}\ \zeta$

$$I_1 = 2C, \quad I_2 = \ln 2, \quad I_{22} = \frac{1}{3}\ln 2 + \frac{1}{6}, \quad I_{x32} = 1.352314016\ldots,$$
$$I_{42} = \frac{2}{15}\ln 2 + \frac{1}{60}, \quad I_4 = \frac{2}{3}\ln 2 - \frac{1}{6}. \tag{7.A.2}$$

Here, C is the Catalan constant, $C = 0.915965594\ldots$ [41]. For a Gaussian beam profile $f(\zeta) = \exp(-\zeta^2)$

$$I_1 = \frac{1}{2}, \quad I_2 = \frac{1}{4}, \quad I_{22} = \frac{1}{2}, \quad I_{x32} = \frac{1}{8}, \quad I_{42} = \frac{1}{8}, \quad I_4 = \frac{1}{8}. \tag{7.A.3}$$

The constants A and B arising in the modulation equations are

$$A = \frac{I_2\sqrt{2}}{\sqrt{I_{x32}}} \quad \text{and} \quad B = \sqrt{2I_2}. \tag{7.A.4}$$

Acknowledgments

This research was supported by the Royal Society of London under Grant No. JP090179.

REFERENCES

1. Y. S. Kivshar and G. Agrawal. *Optical Solitons: From Fibers to Photonic Crystals*. Academic Press, San Diego, CA, 2003.
2. G. Assanto, N. F. Smyth, and A. L. Worthy. Two colour, nonlocal vector solitary waves with angular momentum in nematic liquid crystals. *Phys. Rev. A*, 78:013832, 2008.
3. M. Peccianti, C. Conti, G. Assanto, A. De Luca, and C. Umeton. All-optical switching and logic gating with spatial solitons in liquid crystals. *Appl. Phys. Lett.*, 81:3335–3337, 2002.
4. S. V. Serak, N. V. Tabiryan, M. Peccianti, and G. Assanto. Spatial soliton all-optical logic gates. *IEEE Photon. Technol. Lett.*, 18:1287–1289, 2006.

5. A. Pasquazi, A. Alberucci, M. Peccianti, and G. Assanto. Signal processing by opto-optical interactions between self-localized and free propagating beams in liquid crystals. *Appl. Phys. Lett.*, 87:261104, 2005.
6. G. Assanto and M. Peccianti. Routing light at will. *J. Nonl. Opt. Phys. Mater.*, 16:37–48, 2007.
7. A. Alberucci, A. Piccardi, U. Bortolozzo, S. Residori, and G. Assanto. Nematicon all-optical control in liquid crystal light valves. *Opt. Lett.*, 35:390–392, 2010.
8. M. Peccianti, K. A. Brzdakiewicz, and G. Assanto. Nonlocal spatial soliton interaction in nematic liquid crystals. *Opt. Lett.*, 27:1460–1462, 2002.
9. A. Fratalocchi, A. Piccardi, M. Peccianti, and G. Assanto. Nonlinear management of the angular momentum of soliton clusters: Theory and experiments. *Phys. Rev. A*, 75:063835, 2007.
10. A. Fratalocchi, M. Peccianti, C. Conti, and G. Assanto. Spiraling and cyclic dynamics of nematicons. *Mol. Cryst. Liq. Cryst.*, 421:197–207, 2004.
11. S. Lopez-Aguayo, A. S. Desyatnikov, Y. S. Kivshar, S. Skupin, W. Krolikowski, and O. Bang. Stable rotating dipole solitons in nonlocal optical media. *Opt. Lett.*, 31:1100–1102, 2006.
12. D. Buccoliero, S. Lopez-Aguayo, S. Skupin, A. S. Desyatnikov, O. Bang, W. Krowlikowski, and Y. S. Kivshar. Spiraling solitons and multipole localized modes in nonlocal nonlinear media. *Physica B*, 394:351–356, 2007.
13. G. Assanto, M. Peccianti, K. A. Brzdkiewicz, A. De Luca, and C. Umeton. Nonlinear wave propagation and spatial solitons in nematic liquid crystals. *J. Nonlin. Opt. Phys. Mater.*, 12:123–134, 2003.
14. J. F. Henninot, M. Debailleul, and M. Warenghem. Tunable non-locality of thermal non-linearity in dye doped nematic liquid crystal. *Mol. Cryst. Liq. Cryst.*, 375:631–640, 2002.
15. J. F. Henninot, J. F. Blach, and M. Warenghem. Experimental study of the nonlocality of spatial optical solitons excited in nematic liquid crystal. *J. Opt. A: Pure Appl. Opt.*, 9:20–25, 2007.
16. Y. V. Izdebskaya, V. G. Shvedov, A. S. Desyatnikov, W. Z. Krolikowski, M. Belic, G. Assanto, and Y. S. Kivshar. Counterpropagating nematicons in bias-free liquid crystals. *Opt. Express*, 18:3258–3263, 2010.
17. C. García-Reimbert, A. A. Minzoni, T. R. Marchant, N. F. Smyth, and A. L. Worthy. Dipole soliton formation in a nematic liquid crystal in the nonlocal limit. *Physica D*, 237:1088–1102, 2008.
18. B. D. Skuse and N. F. Smyth. Two-colour vector soliton interactions in nematic liquid crystals in the local response regime. *Phys. Rev. A*, 77:013817, 2008.
19. A. Alberucci, M. Peccianti, G. Assanto, A. Dyadyusha, and M. Kaczmarek. Two-color vector solitons in nonlocal media. *Phys. Rev. Lett.*, 97:153903, 2006.
20. C. Conti, M. Peccianti, and G. Assanto. Route to nonlocality and observation of accessible solitons. *Phys. Rev. Lett.*, 91:073901, 2003.
21. C. Conti, M. Peccianti, and G. Assanto. Observation of optical spatial solitons in a highly nonlocal medium. *Phys. Rev. Lett.*, 92:113902, 2004.
22. C. García-Reimbert, A. A. Minzoni, N. F. Smyth, and A. L. Worthy. Large-amplitude nematicon propagation in a liquid crystal with local response. *J. Opt. Soc. Am. B*, 23:2551–2558, 2006.

23. G. Assanto, B. D. Skuse, and N. F. Smyth. Optical path control of spatial optical solitary waves in dye-doped nematic liquid crystals. *Photon. Lett. Poland*, 1:154–156, 2009.
24. G. Assanto, B. D. Skuse, and N. F. Smyth. Solitary wave propagation and steering through light-induced refractive potentials. *Phys. Rev. A*, 81:063811, 2010.
25. D. J. Kaup and A. C. Newell. Solitons as particles, oscillators, and in slowly changing media: a singular perturbation theory. *Proc. R. Soc. London, Ser. A*, 361:413–446, 1978.
26. I. M. Gelfand and S. V. Fomin. *Calculus of Variations*. Prentice-Hall, Englewood Cliffs, NJ, 1963.
27. H. Goldstein. *Classical Mechanics*. Addison-Wesley, Reading, MA, 1981.
28. G. Assanto, A. A. Minzoni, and N. F. Smyth. Light self-localization in nematic liquid crystals: modelling solitons in nonlocal reorientational media. *J. Nonlin. Opt. Phys. Mater.*, 18:657–691, 2009.
29. G. Assanto, C. García-Reimbert, A. A. Minzoni, N. F. Smyth, and A. L. Worthy. Lagrange solution for three wavelength solitary wave clusters in nematic liquid crystals. *Physica D*, 240:1213–1219, 2011.
30. G. Assanto, A. A. Minzoni, M. Peccianti, and N. F. Smyth. Optical solitary waves escaping a wide trapping potential in nematic liquid crystals: modulation theory. *Phys. Rev. A*, 79:033837, 2009.
31. B. D. Skuse and N. F. Smyth. Interaction of two colour solitary waves in a liquid crystal in the nonlocal regime. *Phys. Rev. A*, 79:063806, 2009.
32. A. A. Minzoni, N. F. Smyth, and A. L. Worthy. Modulation solutions for nematicon propagation in non-local liquid crystals. *J. Opt. Soc. Am. B*, 24:1549–1556, 2007.
33. W. L. Kath and N. F. Smyth. Soliton evolution and radiation loss for the nonlinear Schrödinger equation. *Phys. Rev. E*, 51:1484–1492, 1995.
34. C. García-Reimbert, A. A. Minzoni, and N. F. Smyth. Spatial soliton evolution in nematic liquid crystals in the nonlinear local regime. *J. Opt. Soc. Am. B*, 23:294–301, 2006.
35. K. R. Simon. *Mechanics*, 2nd edn. Addison Wesley, Reading, 1960.
36. G. Assanto, M. Peccianti, and C. Conti. Optical spatial solitons in nematic liquid crystals: Nematicons. *Opt. Phot. News*, 14:45–48, 2003.
37. A. A. Minzoni, N. F. Smyth, and A. L. Worthy. Pulse evolution for a two dimensional Sine-Gordon equation. *Physica D*, 159:101–123, 2001.
38. J. A. Carter, D. C. Fabrycky, D. Ragozzine, M. J. Holman, S. N. Quinn, D. W. Latham, L. A. Buchhave, J. Van Cleve, W. D. Cochran, M. T. Cote, M. Endl, E. B. Ford, M. R. Haas, J. M. Jenkins, D. G. Koch, J. Li, J. J. Lissauer, P. J. MacQueen, C. K. Middour, J. A. Orosz, J. F. Rowe, J. H. Steffen, and W. F. Welsh. KOI-126: A triply eclipsing hierarchical triple with two low-mass stars. *Science*, 120 1274, 2011.
39. J. C. Maxwell. On theories of the constitution of Saturn's rings. *Proc. R. Soc. Edinb.*, 4:99–101, 1862.
40. C. G. Azpetia. Aplicación del grado ortogonal en sistemas Hamiltonianos. Tesis Doctoral Universidad Nacional Autónoma de México, 2010.
41. M. Abramowitz and I. A. Stegun. *Handbook of Mathematical Functions with Formulas, Graphs and Mathematical Tables*. Dover Publications, New York, 1972.

8

Nematicons in Light Valves

Stefania Residori and Umberto Bortolozzo
INLN, Université de Nice-Sophia Antipolis, CNRS, Valbonne, France

Armando Piccardi, Alessandro Alberucci, and Gaetano Assanto
Nonlinear Optics and OptoElectronics Lab, University ROMA TRE and CNISM, Rome, Italy

8.1 INTRODUCTION

Solitons are ubiquitous in nature [1] and appear in such different fields as fluid dynamics, plasma physics, acoustics, magnetohydrodynamics, quantum electrodynamics, and Bose–Einstein condensates, just to cite a few. In optics, bright spatial solitons have been identified as nonlinear solutions of the propagation equation when light travels in materials with a dielectric constant increasing with the intensity of light [2]. Optical spatial solitons are self-trapped beams able to create their own dielectric waveguide and propagate in it without diffractive spreading. A wealth of theoretical and experimental work has been devoted to optical solitons, not only for their fundamental interest [3] but also for their potential applications in photonics, waveguiding, and communications [4].

Nowadays, the ability to optically control soliton dynamics, their paths and trajectories, has attracted considerable attention in the perspective of implementing all-optical and parallel architectures in optical interconnects. This chapter discusses optical spatial solitons in liquid crystal (LC) light valves. In these systems, optical spatial solitons arise because of the molecular reorientational nonlinearity of LC, whereas the optical control is provided by a photosensitive layer of the cell.

The first part of this chapter, Section 8.2, is a brief introduction to the reorientational optical nonlinearity of LC and soliton formation in nematics. Section 8.3 provides a general presentation of the liquid crystal light valve (LCLV), with a description of the cell structure and its working principle. Spontaneous pattern formation in the presence of a feedback mirror is also briefly described

Nematicons: Spatial Optical Solitons in Nematic Liquid Crystals, First Edition.
Edited by Gaetano Assanto.

as an example of transverse configuration. Section 8.4 describes a light valve optimized for the excitation and propagation of spatial solitons in the longitudinal direction of the LC layer, together with the presentation of stable nematicons and of ways of changing their walk-off. Then, Section 8.5 deals with the propagation of solitons in the presence of local and optically induced *defects* of the molecular orientation. Experiments are discussed together with a theoretical model supporting the observations. Section 8.6 demonstrates how the previously exposed effects can be effectively employed to realize logic gates by optically addressing nematicons. Finally, Section 8.7 concludes and illustrates the perspectives for future work.

8.2 REORIENTATIONAL KERR EFFECT AND SOLITON FORMATION IN NEMATIC LIQUID CRYSTALS

LC are strongly anisotropic materials composed of elongated organic molecules. Their main feature is that there is a temperature range within which they constitute mesophases, that is, they form phases with properties intermediate between solids and liquids. Different LC phases exist, exhibiting higher or lower degrees of long-range order [5]. The optical properties of LC are very appealing, because of their large birefringence, polarization selectivity, and transparency over the full visible and near-infrared ranges [6].

Nematic LC, the most commonly used especially in liquid crystal displays (LCDs), are characterized by a preferential orientation of the long axes of the molecules. Therefore, in the nematic phase, all molecules are—in average—aligned along a preferential direction, called the nematic director $\vec{n}$. Because of their anisotropy, LC molecules have different polarizabilities along their long and short axes, with dielectric anisotropy $\Delta\varepsilon = \varepsilon_{||} - \varepsilon_{\perp}$, $\varepsilon_{||}$ and $\varepsilon_{\perp}$ being the dielectric constants parallel and orthogonal to $\vec{n}$, respectively. When an electric field, or a voltage, is applied across the nematic layer, induced dipole moments arise because of such anisotropy and the molecules tend to reorient and become aligned with the direction of the applied field. This process is called *Fréedericksz transition* and requires the applied field to be high enough to overcome the elastic restoring forces of the LC layer, usually a few volts in common nematic cells [5, 6].

Moreover, a nematic LC layer behaves as a strongly birefringent material, characterized by different refractive indices for light beams polarized parallel or orthogonal to $\vec{n}$, that is, the extraordinary and ordinary indices, respectively. Typical values of the extraordinary and ordinary indices of a nematic LC (e.g., E7) are $n_e = 1.7$ and $n_o = 1.5$, respectively, that is, birefringences as large as $\Delta n = n_e - n_o = 0.2$. On applying an electric field, molecular reorientation can lead to a change of the optic axis of the nematic layer as a whole. This allows to control the refractive index and/or birefringence and/or walk-off in the LC layer as a function of the applied voltage. Based on this principle, LCDs exploit changes in transmissivity of a twisted nematic cell placed in between a couple of crossed polarizers [7].

8.2.1 Optically Induced Reorientational Nonlinearity

It is also possible to drive the (re)orientation of the LC molecules by using directly the electric field of a light beam impinging on the nematic layer. When the polarization of the incoming beam is such that the projection of the electric field on the nematic director is zero (ordinary wave) we usually refer to it as the *optical Fréedericksz transition* (OFT) [8]. In this case, there is a threshold for molecular reorientation to occur and, indeed, OFT requires intensities as high as a few hundred kiloWatts per square centimeter. At powers below the Fréedericksz threshold, molecular reorientation occurs only if the beam polarization has a nonzero projection on the nematic director, that is, reorientation can be triggered only by an extraordinary wave component.

When, due to the action of a light beam, molecules reorient, they change direction toward the polarization of the input field if the high frequency dielectric anisotropy, $\Delta\varepsilon_{\mathrm{hf}} = n_{\mathrm{e}}^2 - n_{\mathrm{o}}^2$, is positive [6]. Therefore, optically induced reorientation leads to a change of the refractive index from its lower to higher values, that is, from ordinary to extraordinary index in OFT [9]. The amount of reorientation, therefore, of refractive index change is proportional to the input beam intensity, which provides a self-focusing Kerr-like effect [10]. Focusing Kerr nonlinearities can, in general, be exploited to compensate beam spreading due to diffraction, allowing to generate bright spatial solitons [2–4]. As we see in the following sections, the reorientational effect in nematic LC has been proved to be an efficient mechanism to create spatial optical solitons.

8.2.2 Spatial Solitons in Nematic Liquid Crystals

Optical spatial solitons result from the balance of diffraction and self-focusing in nonlinear media; their transverse index profile is able to confine the soliton as well as other signals [3, 4]. In nematic LC, solitons arise from the balance of diffraction with the self-focusing molecular reorientation under the action of an extraordinarily polarized field.

Pioneering experiments on optical solitons in LC were reported for a laser beam propagating in a capillary tube filled with nematics [11]. It was shown that a beam propagating inside a nematic layer undergoes a strong self-focusing effect followed by filamentation, or soliton formation, when the light intensity increases. Since then, stable and controllable spatial solitons in nematics, also called nematicons, have been obtained in planar cell geometries [12]. It was shown that nematicons can be controlled by changing their route by modifying their walk-off and/or the perceived refractive index, thanks to the application of an external low-frequency electric field (Chapter 5). Nematicons have, thus, been proved to be excellent candidates for signal readdressing and processing in reconfigurable circuits [13, 14].

Modulational instability regimes were also studied in similar setups [15, 16], showing the spontaneous filamentation of initially uniform extended beams on propagation in a nematic layer. Recently, a new mechanism for spontaneous soliton formation has been devised and experimentally proved, which is based on the establishment of optical wave turbulence, characterized by low nonlinearity and spontaneous soliton formation at very long distances inside a nematic planar cell after the propagation of an extended (and with initial phase noise) beam [17].

The theoretical framework for soliton formation and beam filamentation in nematics is very general and includes two coupled differential equations, one describing the noninstantaneous and nonlocal (diffusive) response of the LC and the other accounting for the beam propagation, namely (see Chapter 1)

$$K\,\nabla_{\perp}^{2}\theta - \frac{\varepsilon_0 \Delta\varepsilon}{2L^2} V_{\mathrm{LC}}^{2}\theta + \frac{\varepsilon_0}{4}(n_{\mathrm{e}}^{2} - n_{\mathrm{o}}^{2})|A|^2 = 0,$$
$$i\frac{\partial A}{\partial z} + \frac{1}{2k_0}\nabla_{\perp}^{2}A + \frac{k_0(n_e^2 - n_o^2)}{2}\theta A = 0, \qquad (8.1)$$

where θ is the molecular orientation angle in the LC layer, in the linear approximation and for θ around $\theta_0 = \pi/4$; K is the LC elastic constant, taken equal for splay, twist, and bend (one constant approximation); $\Delta\varepsilon$ is the LC dielectric anisotropy; A is the amplitude of the optical field; k_0 is the optical wave number; and ε_0 is the vacuum permittivity.

In the past decade, a high degree of control, such as routing and steering of nematicons, has been achieved by varying the voltage applied across the nematic slab (Chapter 5) [12–14]. However, an all-optical control of spatial solitons is highly desirable in order to optimize the routing/steering process, to achieve high levels of parallelism, to realize flexible multiplexing configurations, and to allow interconnects to be reconfigured in real-time. In view of implementing such all-optical control protocols, preliminary attempts were carried out with intense external beams, also using azo-dye-doped LC and surface effects [18–20]. More recently, a finely controllable setup has been realized by using an LCLV in place of a standard planar cell [21]. Stable nematicons were created in the light valve and the photo-controlled transduction of light into effective voltage across the LC layer, due to the valve operation, allowed for a flexible and all-optical control of the nematicon dynamics.

8.3 LIQUID CRYSTAL LIGHT VALVES

LCLV are optically addressable spatial light modulators, able to impose a phase modulation on the wave front of a readout beam, whereas the control/input signal is provided by another optical beam [22]. The general structure of an LCLV comprises two components, the photoreceptor and the electro-optic material, sometimes separated by a dielectric mirror [23]. Historically, the photoreceptor was a photoconductor, such as selenium or cadmium sulfide [24], amorphous silicon [25] or GaAs [26]. Therefore, in general, LCLV are hybrid structures, combining organic layers—typically made of nematic LC—with solid-state photosensitive substrates. The input beam activates the photoreceptor, which produces a corresponding charge field on the electro-optic material. The readout light is modulated in its double pass through the electro-optic element in a retroreflective scheme.

More recently, photorefractive LCLV have been realized by combining a nematic LC layer with a thin monocrystalline $Bi_{12}SiO_{20}$ (BSO) crystal [27]. The BSO, well known for its photorefractive properties [28], was chosen for its large photoconductivity and transparency in the visible range together with the possibility of having large monocrystalline samples with good optical quality and uniform dark resistance. The maximum photoconductivity of the BSO is in the blue-green region of the spectrum. The photorefractive LCLV is a novel type of spatial light modulator that works in transmissive configurations, where input and readout beams may in general coincide. We see, in the following discussion, how this type of LCLV can be optimized for the generation and all-optical control of nematicons.

8.3.1 Cell Structure and Working Principle

A typical BSO-made LCLV is sketched in Figure 8.1. A thin layer of nematic LC is sandwiched between a glass plate and a wall made of the photorefractive BSO crystal, cut in the shape of a thin plate (1 mm thickness, 20×30 mm^2 lateral size). Before assembling the cell, the inner surfaces of both the BSO and the glass plate are coated with polyvinyl alcohol polymer and mechanically rubbed to obtain the planar alignment of the LC (nematic director $\vec{n}$ parallel to the confining walls). After this treatment, the gap between the two walls, of typical thickness $L = 14\mu$m, is filled with the LC. Transparent electrodes, made of indium tin oxide (ITO) are deposited on the outer surface of the BSO and the inner surface of the glass plate: they allow applying an external bias V_0 across the cell.

When an electric field is applied across the nematic layer, all the molecules tend to reorient in such a way that they become aligned with the direction of

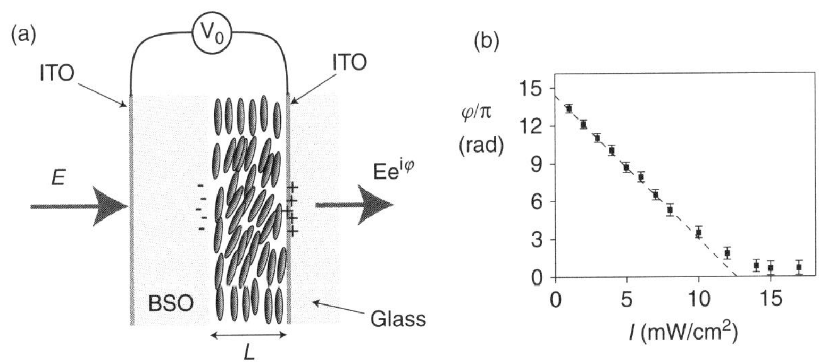

FIGURE 8.1 The liquid crystal light valve. (a) Sketch of the cell structure: the thick wall (left side) is the photoconductive BSO crystal; when a light beam impinges on this wall free charges are photogenerated and lead to an increase of the effective voltage across the liquid crystal layer, hence, to molecular reorientation. At the exit of the valve, the beam acquires a phase shift φ, which is a function of the applied voltage V_0 and of the intensity I of the beam itself. (b) Typical phase retardation measured after the LCLV versus the incoming beam intensity. The light wavelength is 532 nm, and the applied bias is 20 V root mean square at a frequency of 2 kHz.

the applied field. This implies a change in the optic axis of the LC layer; hence, an incoming light field experiences a large refractive index change. Typically, the voltage applied is AC, with a root mean square rms value from 2 V to 20 V and a frequency from 50 Hz to 20 kHz. When a light beam impinges on the LCLV, because of the photoconductivity of the BSO, photogeneration of charges occurs at its surface; hence, the effective voltage across the LC layer increases locally, according to the illumination. As a consequence, the LC molecules reorient and, due to their birefringence, the light beam acquires a phase shift φ at the exit of the valve, which is a function of the applied voltage and of the beam intensity itself.

A typical response function of the LCLV is shown in Figure 8.1b, where the phase shift φ acquired by the light beam passing through the valve is plotted against the input beam intensity I. The response is linear up to intensities of the order of 10 mW/cm^2. Then, it saturates to a value that corresponds to the maximum birefringence, $\Delta n \approx 0.2$, when all the LC molecules are aligned along the direction of the applied field. In between, that is, from the initial planar alignment to the final orthogonal alignment, the average orientation angle of the LC molecules varies from 0 to $\pi/2$, leading to a phase shift φ that, in the linear part of the LCLV response, changes from 0 to several π. In this range of parameters, the LCLV behaves as a Kerr-like medium, providing an effective change of the refractive index proportional to the input light intensity. Correspondingly, the LCLV nonlinear coefficient is as large as $n_2 \simeq -6$ cm^2/ W, the minus sign accounting for the defocusing character of the nonlinearity (the refractive index changes from n_e to n_o when increasing the voltage or the light intensity) [29].

The response time of the LCLV is dictated by the time τ_{LC} required for the collective motion of the LC molecules to establish over the whole thickness of the nematic layer. This is given by [5]

$$\tau_{LC} = \gamma L^2/K, \tag{8.2}$$

where γ is the LC rotational viscosity and K is the LC elastic constant (taken equal for splay, twist, and bend deformations). For $L = 14\mu$m and typical values of the LC constants, τ_{LC} is of the order of 100 ms. Note that, owing to the slow relaxation time, the LC molecules cannot follow the oscillation of the applied AC field. Instead, they perform a static reorientation, in which they reach an equilibrium position fixed by the rms value of the voltage. Because of the complex impedances of the different layers comprising the LCLV, there is an optimal frequency range within which the contrast of the voltage transferred to the LC with and without illumination is maximum [27, 29, 30].

Finally, the nonlinear mechanism in the LCLV is different from optically induced reorientation, even though both deal with orientation of the LC molecules. Indeed, whereas in the LCLV nonlinearity the molecules are driven by the low-frequency external bias and light intervenes through the photoconductive transducer, in the second case molecules are directly oriented by the electric field of the incoming light beam. This second mechanism requires much higher intensity for obtaining an equivalent distortion of the nematic layer. In the LCLV, the two effects combine for the control of spatial soliton propagation: nematicons can be excited via optically induced reorientation, whereas the LCLV nonlinearity allows local and dynamical addressing of the refractive index.

8.3.2 Optical Addressing in Transverse Configurations

Transverse configurations have been used to address the LCLV in various nonlinear optical experiments. In particular, spontaneous pattern formation in the presence of an optical feedback and by using a retroreflective LCLV have been largely studied in the past, showing different types of spatial structuring of the optical wave front and various spatiotemporal dynamical regimes [26,31–33]. For the same type of retroreflective LCLV, the control of optical localized structures [34] and the control of front propagation dynamics [35] have been efficiently implemented by illuminating the photoconductive side of the valve with suitable spatially modulated optical potentials.

More recently, the BSO-made transmissive LCLV has allowed more simplified schemes for optical feedback experiments [36]. Figure 8.2a shows the typical setup for spontaneous pattern formation under transverse optical feedback. In the linear part of its response, the LCLV can be assimilated to a Kerr medium, providing a phase change proportional to the input intensity (Fig. 8.1b). In the presence of a feedback mirror placed at a distance $d/2$ with respect to the LCLV, any initial amplitude perturbation is converted into phase modulation by the nonlinear response of the LCLV, whereas the free propagation from the valve to the mirror reconverts phase into amplitude modulation [31]. If the light intensity is increased above a certain threshold, an instability takes place and makes the initially uniform wave front become spatially structured, as shown in Figure 8.2b. The spatial period of the structure, $\Lambda = \sqrt{2\lambda d}$, is a geometric mean of the total free propagation length d and of the light wavelength λ. For a laser wavelength in the visible, it is typically of the order of a few hundred micrometers for d of the order of a few centimeters and can be easily changed by varying the distance between the valve and the mirror.

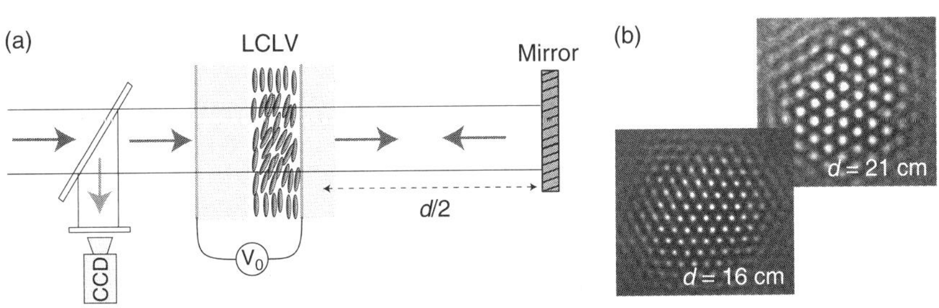

FIGURE 8.2 Spontaneous pattern formation in an LCLV with single mirror feedback. (a) Experimental setup: $d/2$ is the distance between the LCLV and the feedback mirror and d is the total free propagation length (back and forth). (b) Hexagonal pattern (near-field image) obtained for different free propagation distances.

The transmissive configuration of the photorefractive light valve also offers the possibility of performing wave mixing experiments [29, 37]. The beam coupling takes place over a large working area, and extra optical/electric control can be achieved by addressing the photoconductive layer. Two-wave mixing experiments have led to signal beam amplification [38] and the realization of nonlinear optical cavities [39, 40] as well as to slow-light experiments [41].

Finally, it is worth mentioning the possibility of addressing the BSO with arbitrary distributions or optical potentials. For instance, controlled distributions of disorder have been used to illuminate the photoconductive side of the valve, inducing corresponding distributions of defects in the molecular orientation; the latter has provided experimental evidence of speckle instability under the simultaneous presence of disorder and nonlinearity [42].

8.4 SPATIAL SOLITONS IN LIGHT VALVES

For the propagation of stable spatial solitons we have prepared specific BSO-made light valves aimed at favoring the self-guided propagation in a longitudinal dimension of the LC layer [21]. To this end, we have employed large sample thicknesses and oblique planar anchoring at the surfaces. The first condition facilitates the in-coupling of the beam in the LC layer, whereas the second one is used to favor the optically induced reorientation.

8.4.1 Stable Nematicons: Self-Guided Propagation in the Longitudinal Direction

A schematic drawing of the soliton-optimized LCLV is shown in Figure 8.3. It consists of a glass slide and a photoconductor slab of BSO, both coated with a layer of ITO for the application of an external voltage bias and held parallel to one another by 50 μm Mylar spacers to obtain the necessary thickness. An additional glass slide is sealed on one side of the LCLV (see Figure 8.3a) and provides the interface for launching the input beam, thus preventing depolarization and meniscus effects at the entrance of the LC layer. All the internal surfaces are coated with a polyvinyl alcohol polymer and mechanically rubbed to provide molecular anchoring and alignment in the nematic phase. As shown in Figure 8.3b, the rubbing is parallel to y on the input facet to optimize the coupling of the extraordinary wave in the bulk NLC; in the bulk φ_0 is the angle of the initial $\vec{n}$ with $\hat{z}$, being $\vec{n} \cdot \hat{x} = 0$ as well. Once assembled, the LCLV is filled with the nematic LC E48, with $n_{\|} = 1.7536$ and $n_{\perp} = 1.5230$ at the employed wavelength. Figure 8.3c sketches the typical experimental configuration for soliton excitation. Typically it is $\varphi_0 = \pi/4$, in order to maximize the optical nonlinearity, and δ_0 is the soliton walk-off.

A TEM_{00} mode from a He–Ne laser, wavelength 632.8 nm, is focused onto the cell entrance at $x = L/2$ with a waist of about 6 μm and wave vector $\vec{k}||\hat{z}$. We use an optical microscope and a camera to image the beam evolution in the y–z plane, collecting the photons scattered by the LC out of the plane of propagation. In order

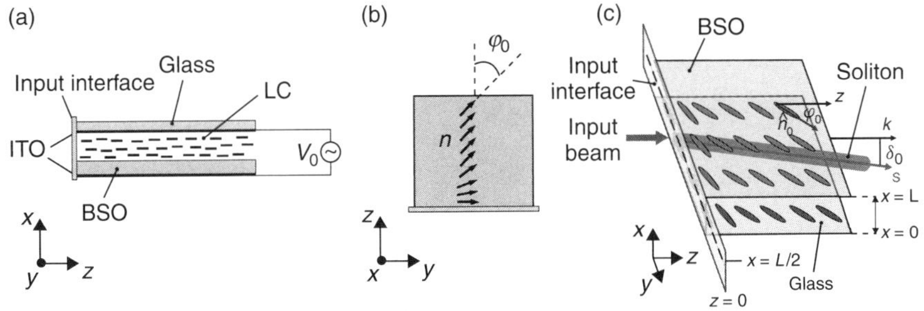

FIGURE 8.3 (a) Side view of the LCLV: the liquid crystal E48 is sandwiched between a BSO photoconductive slab and a glass slide, spaced by $L = 50$ μm. (b) Distribution of the molecular director $\vec{n}$ in the cell midplane (top view). Following a (smooth) transition layer near the input facet, the angle between $\vec{n}$ and the z axis is φ_0. (c) Sketch of the LCLV and experimental configuration for soliton excitation: the input beam $\lambda = 632.8$ nm impinges on $x = L/2$ with wave vector $\vec{k}$ normal to xy. $\vec{s}$ is the Poynting vector for $V_0 = 0$ V, and δ_0 is the corresponding walk-off.

to characterize the LCLV as a suitable environment for nematicon propagation, the applied voltage is first set to $V_0 = 0$ V. Optically induced reorientation occurs for the extraordinary beam when the intensity is sufficiently high. Sample results are shown in Figure 8.4. When the beam is polarized parallel to x (ordinary wave), we observe diffraction independent of excitation, as shown in Figure 8.4a. Conversely, when the beam is polarized parallel to y (extraordinary wave), self-focusing compensates diffraction and gives rise to spatial solitons for powers greater than or equal to 1.5 mW, as can be seen in Figure 8.4d. In Figure 8.4b a mixed polarization state (electric field linearly polarized at $\pi/4$ with respect to both x and y) and $P = 2.5$ mW yields a diffracting beam for the ordinary component and a nematicon for the extraordinary component. In Figure 8.4c, the extraordinary polarized beam diffracts at low power, $P < 1$ mW.

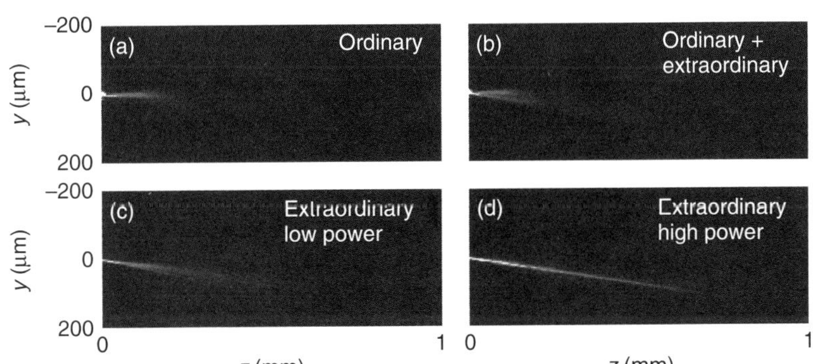

FIGURE 8.4 Beam evolution in the y–z plane of the LCLV. (a) Ordinary wave input, E_x and $P =$ 2 mW. (b) Mixed polarization input with electric field linearly polarized at $\pi/4$ with respect to both x and y, $P = 2.5$ mW. (c) Extraordinary polarization E_y for $P < 1$ mW (low power). (d) An extraordinary wave input E_y for $P = 2$ mW leads to a soliton.

Owing to the large birefringence of E48, with $\Delta n \sim 0.23$, solitons propagate with a Poynting vector $\vec{s}$ at a large walk-off angle $\delta_0 \sim 7^\circ$ with respect to their wave vector $\vec{k}$. The Poynting vectors corresponding to ordinary and extraordinary beam excitations are visible in Figure 8.4b. Once excited, nematicons propagate as self-confined beams over distances exceeding 15 Rayleigh lengths, that is, as long as their power remains high enough in spite of the propagation (scattering) losses.

8.4.2 Tuning the Soliton Walk-Off

The soliton walk-off in the $y-z$ plane can be changed by the application of a voltage V_0 across the LC thickness, as previously done in standard LC cells (Chapter 5) [12]. In the LCLV, we apply a low-frequency voltage of variable (rms) amplitude across the cell and record the corresponding soliton paths [21]. The results are shown in Figure 8.5. Figure 8.5a shows the nematicon propagation for various applied voltages. Correspondingly, Figure 8.5b plots the measured soliton walk-off. For voltages above 2 V (corresponding to the Freedericksz threshold for reorientation), the apparent walk-off (i.e., the observable angle between $\vec{s}$ and $\vec{k}$) reduces, approaching 0° for a bias $V_0 = 6$ V such that the director $\vec{n}$ becomes nearly parallel to x and no further molecular reorientation takes place. Thereby, when δ approaches zero the beam loses self-confinement because the nonlinearity fades away, as apparent from the bottom photograph in Figure 8.5a. We stress how the curve in Figure 8.5b follows the qualitative behavior of standard LC samples, but differs for the voltage scale, owing to the equivalent resistance of the BSO layer [12].

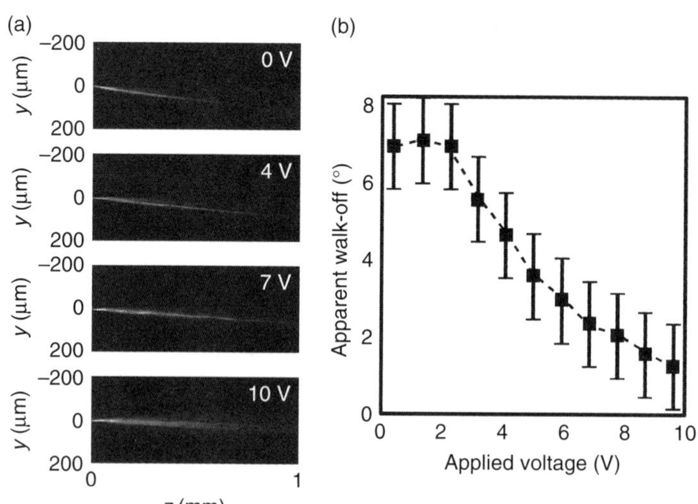

FIGURE 8.5 Controlling nematicon walk-off versus applied voltage in a biased LCLV. (a) Soliton propagation for various applied biases. (b) Apparent walk-off in the $y-z$ plane versus applied voltage.

Other than via the applied voltage, the nematicon walk-off in the LCLV can be optically controlled by uniformly illuminating the BSO, thanks to its photoconductive properties. For a fixed bias and under uniform external illumination with an expanded laser beam, the LC layer where a nematicon propagates can be tuned in orientation. To illuminate the LCLV, we use a solid-state laser operating at 532 nm, well within the spectral range where the BSO has maximum photoconductivity. Then, we launch a 2 mW nematicon at 632.8 nm as described earlier and keep the bias fixed at $V_0 = 3.5$ V and frequency $f = 80$ Hz while varying the external beam intensity. Figure 8.6 shows the measured change in propagation angle (apparent walk-off) versus the external control intensity I_{ext}. We can see that the walk-off is efficiently controlled by the external light; hence, the soliton is angularly steered, in analogy to what is observed in Figure 8.5 but now by way of an all-optical control.

Another important feature of the LCLV is the response time introduced by both LC and photoconductive layers when the external electric field is applied: whereas the former, dictated by the time of LC molecules to be reoriented (typically in 100 ms, Chapter 13), does not allow a frequency below 10 Hz, the latter imposes an upper bound, determined by the BSO–LC equivalent electrical impedance. This cutoff frequency can be observed in Figure 8.7, showing the apparent walk-off under voltage bias and illumination versus frequency: clearly, no change in nematicon trajectory occurs above 500 Hz.

To establish a direct comparison between electrical and optical control of the walk-off, we resort to a simple model describing the photoconductor response in the linear approximation [33]. In this case, the effective voltage applied to the LC layer can be expressed as

$$V_{LC} = \Gamma V_0 + \alpha I_{\text{ext}}, \tag{8.3}$$

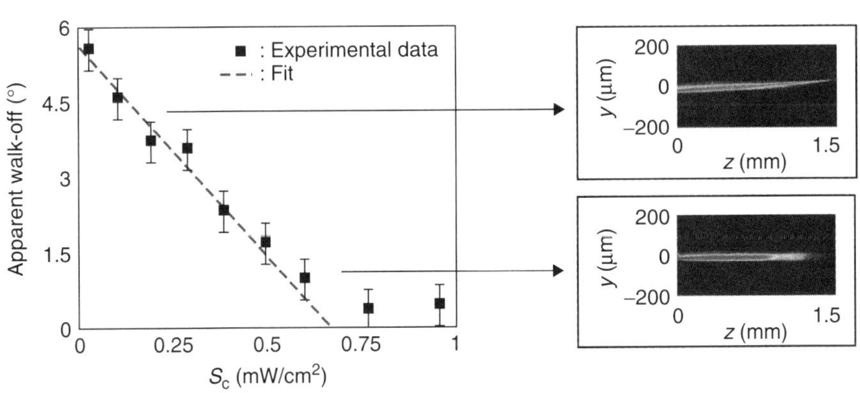

FIGURE 8.6 Control of nematicon path by uniform illumination of the BSO layer. The wavelength of the external beam is 532 nm, and the valve is biased at $V_0 = 3.5$ V, frequency $f = 80$ Hz. The walk-off angle is plotted versus the intensity I_{ext} of the illuminating beam. Top and bottom insets show the 2 mW soliton trajectories for low and high intensity illumination of the BSO, respectively.

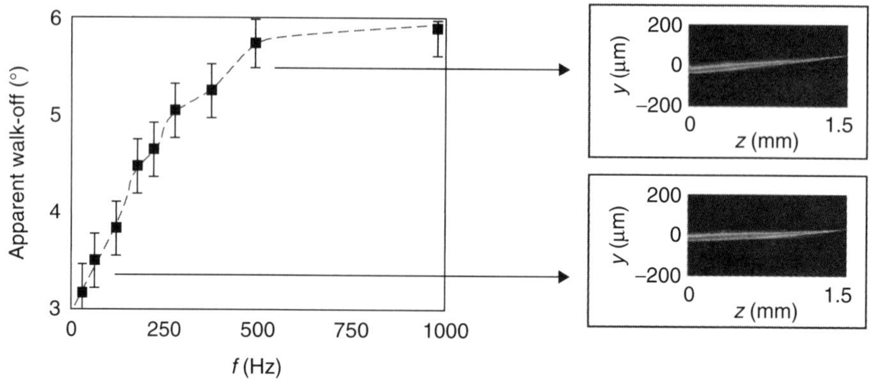

FIGURE 8.7 Frequency dependence on controlling nematicon walk-off. The LCLV, with external illumination and an applied voltage of rms amplitude 3.5 V, acts as a low pass filter. The insets show 2 mW soliton trajectories for $f = 75$ Hz (bottom) and $f = 500$ Hz (top); above 500 Hz the LCLV no longer modifies the soliton trajectory.

where Γ accounts for the impedances of the dielectric layers and α is a phenomenological parameter describing the linear response of the photoconductor in the LCLV. The values $\Gamma = 0.6$ and $\alpha = 3 \cdot 10^3$ Vcm^2/W apply to our valve; hence, as the external intensity I_{ext} varies from *0* to 1 mW/cm^2 in a fixed bias LCLV, the latter operates as if the applied voltage V_0 increases from *0* to 5 V (i.e., increases in V_{LC} from *0* to 3V). For a fixed bias $V_0 = 3.5$ V and increasing the illumination as earlier, conversely, the walk-off varies linearly with an equivalent voltage dynamics from $V_0 = 3.5$ to $V_0 = 8.5$ V (i.e., V_{LC} from 2.1 to 5.1 V). Linearity breaks down above $I_{\text{ext}} = 0.7$ mW/cm^2, when saturation of the nematic response (director parallel to the quasistatic electric field) comes into play.

8.5 SOLITON PROPAGATION IN 3D ANISOTROPIC MEDIA: MODEL AND EXPERIMENT

The photoconductive transduction of intensity into applied voltage across the LC layer allows for selectively addressing local areas of the LC underneath the BSO. Then, local control of the soliton trajectory can be implemented by using finite size external beams rather than uniform illumination of the BSO [43, 44]. To this purpose, a small spot laser beam at 532 nm can be employed to induce lenslike director/index perturbations along the path of nematicons, as sketched in Figure 8.8a. The solitons propagating in the vicinity of a light-induced defect in the molecular order can sense a change in refractive index and a local variation of walk-off, bending their trajectories along paths depending on the distance from the perturbation, its size, and its shape (see Figure 8.8b–c).

The propagation of nematicons near externally induced "defects" along a local longitudinal coordinate z can be modeled—in the most general case—by a set of

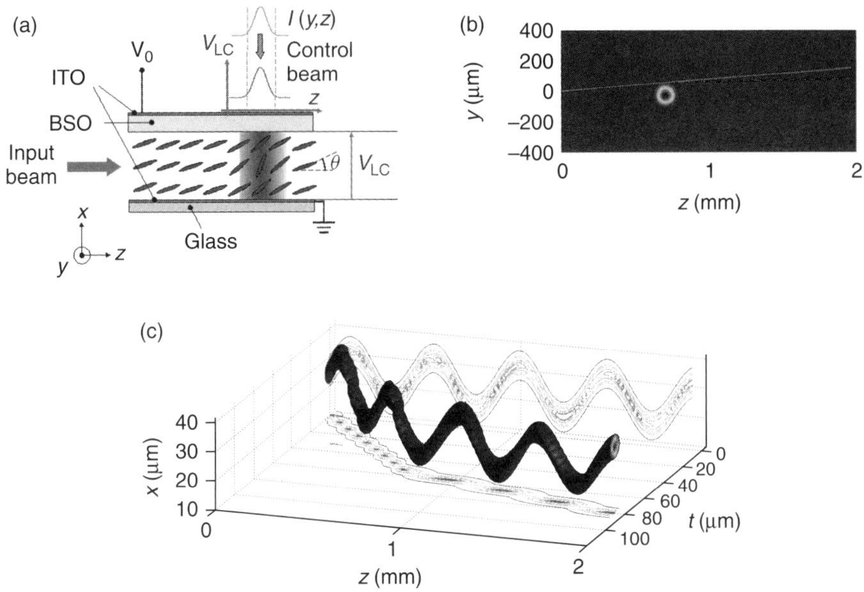

FIGURE 8.8 (a) Sketch of nematicon control by local external illumination: the top profile is the external illumination, and the bottom profile is the corresponding voltage change. Ellipses represent the distribution of the LC molecular director, at the angle θ with respect to z, in the illuminated area. (b) Measured (bottom solid line) and calculated (dashed) nematicon trajectory in the proximity of a round optically induced perturbation. The top solid line refers to the unperturbed case. (c) Computed 3D graph of the soliton path around the light-driven defect, in the presence of walk-off and a voltage-induced potential in xz.

coupled partial differential equations, including the Poisson equation governing the electrostatic potential, the reorientational equations governing the director distribution (determined by two angles that fix the local orientation of the director), and finally, the Maxwell equations ruling light evolution. The solution of this complete model is quite heavy, but some drastic simplifications can be applied in order to enlighten the fundamental concepts and, at the same time, to save computational power. In order to determine the molecular director distribution in the presence of a voltage and external control beam, we have to compute the two angles θ and φ that the director makes, respectively, with z in the $x-z$ plane and with z in the $y-z$ plane throughout the bulk LC (Fig. 8.9) when a soliton propagates in the LCLV [45]. By considering all the various contributions, we have $\theta = \theta_{Vc} + \theta_{NL}$ and $\varphi = \varphi_{Vc} + \varphi_{NL}$, where $\theta_{Vc} \equiv \theta_V + \theta_c$ and $\varphi_{Vc} \equiv \varphi_V + \varphi_c$ are the background values due to the bias, θ_V, φ_V, and to the control beam, θ_c, φ_c, respectively, and $\theta_{NL}, \varphi_{NL}$ are the nonlinear (reorientational) perturbations induced by the spatial soliton. The soliton-induced perturbations at typical experimental powers (<3 mW) are negligible with respect to θ_{Vc} and φ_{Vc}; hence, we can safely take $\theta \simeq \theta_{Vc}$ and $\varphi \simeq \varphi_{Vc}$, that is, the soliton does not appreciably modify the defect when passing nearby. Moreover, by assuming an infinitely extended cell in the $y-z$ plane, we can

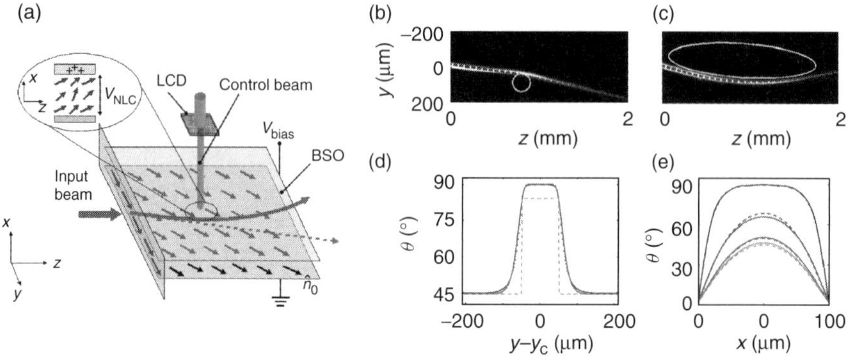

FIGURE 8.9 (a) Sketch of the LCLV showing the perturbation induced by an LCD-shaped control spot. (b) Induced circular defect at $y_c = 110$ μm, $z_c = 800$ μm with radius 50 μm. (c) As in (b) but with an elliptical defect (axes are 1550 and 200 μm long) located in $y_c = -30$ μm, $z_c = 900$ μm. The white contours represent the induced profiles; dashed curves are the calculated soliton trajectories. (d) Defect cross section along y in $x = L/2$, due to the control beam (dashed line) in (b): 3D computed (dotted line) and approximated (solid line) responses in reorientation θ. (e) Defect cross section along x due to the spot in (b): calculated 3D (solid lines) and approximated profiles (dashed lines) versus x at various distances from the spot axis: $y - y_c = 0$, 60, 80, 100 μm from top to bottom.

also take $\theta_V = \theta_V(x)$ and $\varphi_V = \varphi_0$ throughout the LC volume, φ_0 fixed by anchoring. The low-frequency electric field effectively acting on the LC is increased by the control-beam-induced charges via the associated decrease in BSO resistivity, with behavior roughly dictated by Equation 8.3. We do not account for variations of the intensity I_{ext} along x, as the control beam can be considered collimated inside the BSO layer. We also assume that the low-frequency electric field components lying in the y–z plane are negligible with respect to the x-component; hence, only θ_{Vc} changes under the action of V_{LC}. This is physically plausible because owing to the geometry and the dielectric constant of BSO larger than the LC ($\varepsilon_{BSO} = 56$, $\varepsilon_{||} = 19.5$) [28], the field lines in yz concentrate in the BSO slab. In the single elastic constant approximation, the reorientation equation in 3D takes the form

$$\nabla^2\theta_{Vc} + \gamma \sin(2\theta_{Vc})\frac{V_{LC}^2}{L^2} = 0, \tag{8.4}$$

where $\gamma = \varepsilon_0 \Delta\varepsilon/(2K)$. As said earlier, at standard intensities the torque due to direct coupling between the control beam and the LC molecules can be neglected. We can, thus, solve Equation 8.4 with the proper boundary conditions to determine the director profile around the defect [46].

Once the optic axis distribution in the LC volume for a given control spot or defect is known, we only have to determine light propagation in such index landscape. In particular, we are interested in computing the soliton trajectory for adiabatic changes in the dielectric tensor and in the limit of high nonlocality, consistent with the experimental conditions. Under them, the evolution of the beam envelope A can be found via the scalar equation

$$2in_e k_0 \left(\frac{\partial A}{\partial z_l} + \tan \delta_l \frac{\partial A}{\partial y_l} \right) + \frac{\partial^2 A}{\partial x_l^2} + D_y \frac{\partial^2 A}{\partial y_l^2} + k_0^2 (n_e^2 - n_o^2) \sin \left[2(\Psi - \delta_l) \right] \psi A = 0, \tag{8.5}$$

where Ψ and ψ are the local angle between wave vector and director in the absence of a soliton and its nonlinear variation, respectively, and D_y is the diffraction coefficient along y. The reference system $x_l y_l z_l$ is a moving frame rotating with the beam such that z_l and y_l are always parallel to the local wave vector and electric extraordinary displacement vector, respectively. Equation 8.5 holds valid in the paraxial approximation and for a director distribution slowly varying with respect to the intensity width (Chapter 5). We are now able to compute the soliton trajectory: explicitly, after defining the position of the soliton center of mass $\langle \vec{r}_b \rangle (\vec{r}_b = x_l \hat{x}_l + y_l \hat{y}_l)$ as

$$\langle \vec{r}_b \rangle = \frac{\iint \vec{r}_b I \, \mathrm{d}x_l \, \mathrm{d}y_l}{\iint I \, \mathrm{d}x_l \, \mathrm{d}y_l}, \tag{8.6}$$

and applying the Ehrenfest theorem to Equation 8.5, we find

$$\frac{\mathrm{d}^2 \langle \vec{r}_b \rangle}{\mathrm{d}z_l^2} = \vec{F} \,|_{\vec{r}_b = \langle \vec{r}_b \rangle}, \tag{8.7}$$

with the force $\vec{F}$, normal to the local wave vector, given by

$$\vec{F} = \frac{1}{\left(n_e^{(b)}\right)^2} \left[\left(n_e \frac{\partial n_e}{\partial x_l} \right) \hat{x}_l + \left(n_e \frac{\partial n_e}{\partial y_l} \right) \hat{y}_l + \left(\left(n_e^{(b)}\right)^2 \frac{\partial \tan \delta_l}{\partial z_l} \right) \hat{y}_l \right]. \tag{8.8}$$

The superscript "(b)" indicates the extraordinary refractive index computed at the beam peak. The first two terms on the right-hand side of Equation 8.8 are due to index gradients as in isotropic media; the third term is directed along y_l and stems from anisotropy: it accounts for longitudinal changes in walk-off.

Equation 8.7 predicts a soliton path independent of waist. Moreover, in our experimental conditions the transverse force always attracts the soliton toward the axis of the defect induced by the control beam, as the extraordinary refractive index is larger where V_{LC} is higher. In the absence of external illumination, the soliton evolves with sinusoidal oscillations in xz because of the joint action of walk-off and of the index well created by the bias. In the plane y–z, the soliton appears to propagate straight, with slope $\tan(\delta_{yz})$, $\tan(\delta_{yz})$ being the apparent walk-off [12].

An example of measured and calculated nematicon bending in the light valve near a light-induced lenslike perturbation is shown in Figure 8.8b, with an excellent agreement between model and experiment. Figure 8.8c displays the three-dimensional trajectory of a nematicon under the influence of the transverse electric potential (bias) and the defect. Owing to both walk-off out of the propagation plane y–z and refractive index perturbation, the soliton propagates oscillating up and down in the valve while bending its trajectory; hence, it moves in a complex 3D path distorted by both bias and illumination [43].

8.5.1 Optical Control of Nematicon Trajectories

Rather than a small beam, arbitrary shape and size perturbations can be applied by means of an external spatial light modulator (or LCD). To realize a 3D configuration of soliton control, the profile of the control spot(s) is adjusted by using a programmable LCD as shown in Figure 8.9a. Examples of curved soliton paths due to circular and elliptically shaped spots are shown in Figure 8.9b–c for $V_0 = 5$ V, $I_{ext} = 100$ μ W/cm^2, and 2 mW solitons launched with wave vector $\vec{k}||\hat{z}$ and Poynting vector at walk-off $\delta_0 = 7°$ with respect to the axis z. Noteworthy, flat-top beams are employed in order to maximize the light-induced index gradient for a fixed difference between maximum and minimum indices.

The computed trajectories are in excellent agreement with the experimental data, as visible in Figure 8.9b–c, dashed lines. For a control spot with small curvature (Fig. 8.9b), the soliton senses a pointlike perturbation and is deflected toward its center. For lesser curvature (Fig. 8.9c) the soliton is able to follow the profile edge, as the effective force due to the index gradient is weaker. Finally, Figure 8.9d–e shows the director distribution θ_{Vc} along y (x) when a circular control beam shines the photoconductor, with voltage V_0 and control intensity I_{ext} so as to saturate the director reorientation and maximize the gradient in dielectric tensor. The flat-top shape of the control beam (dashed lines) is smoothed in the LC because of the nonlocal response, with length of the transition zone fixed by the width of the reorientational response (Chapter 11).

8.6 SOLITON GATING AND SWITCHING BY EXTERNAL BEAMS

The optical potential previously superposed on the nematicon can be dynamically modified by acting on the programmable LCD in order to implement the desired soliton operations [47]. By using the model introduced earlier and the LCD spatial modulator, we can design and demonstrate a number of low power optically driven routers and logic gates, with logic input represented by the control spots and signal carried by solitons.

Figure 8.10 shows an experimental implementation of an XNOR gate, with the soliton output y-position equal to the unperturbed one y_0 (i.e., in the absence of control beams) corresponding to a high (true) logic state, equivalent to a low (false) state if the soliton is located elsewhere. As demonstrated by the evolution in yz (Fig. 8.10a) and by the acquired transverse profiles (Fig. 8.10b), the evolution without perturbation (00) and with both control inputs (11) leads the soliton to the same output y_0: this is achieved by a proper choice in the positions of the two control beams (power and shape are equal) such that the deflection from the

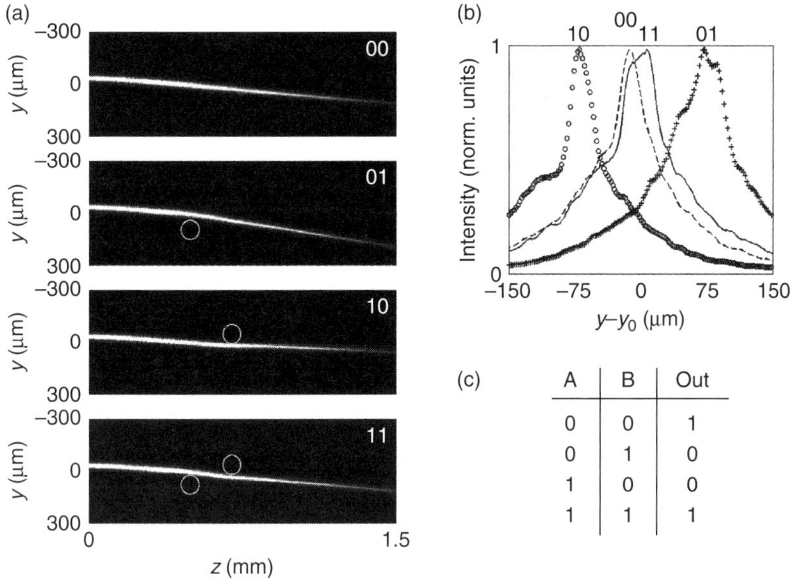

A	B	Out
0	0	1
0	1	0
1	0	0
1	1	1

FIGURE 8.10 XNOR logic gate. (a) Soliton evolution in yz for various configurations of control spots (bits) as indicated by the lettering. Circles indicate the spots in $(y, z) = (100, 500)$ μm and $(y, z) = (0, 700)$ μm for binary inputs 01 and 10. (b) Nematicon profiles in $z = 1.5$ mm, with corresponding inputs. (c) Truth table of an XNOR.

rightmost defect exactly compensates the deviation due to the leftmost control beam.

Having experimentally demonstrated all-optical logic gating in LCLV, our next step is a half adder (HA), that is, the building block of any digital processor. Figure 8.11 demonstrates the realization of a binary all-optical HA in LCLV. In fact, when only one control beam is turned on, the soliton output position remains the same no matter which input is on; the output position is labeled S (Fig. 8.11b,c) and gives the Sum bit (in this case a high logic state). Consistent with the previous definition, S is dark (low logic state) when the two control beams are both turned on or off. The output position when both control beams are on can be taken as the carry bit (labeled C in Figure 8.11b–c): for every other pair of logic inputs, C remains low. It is noteworthy that such configuration is quite versatile: in fact, different logic gates can be realized by taking each output position as the unique output state. The geometry in Figure 8.11 acts as an AND, XOR, and NOR when the output position corresponds to C, S, and L, respectively. Finally, Figure 8.11c compares acquired and calculated soliton trajectories, with good agreement.

Having demonstrated the all-optical logic gates in LCLV, we are interested in realizing other fundamental devices for signal processing: routers. To realize a digital two-bit router we employ two circular control beams of equal size and shape, and associate their presence or absence with true (high) or false (low) input bits, respectively. Every combination (out of $2^2 = 4$) of Boolean inputs is able to

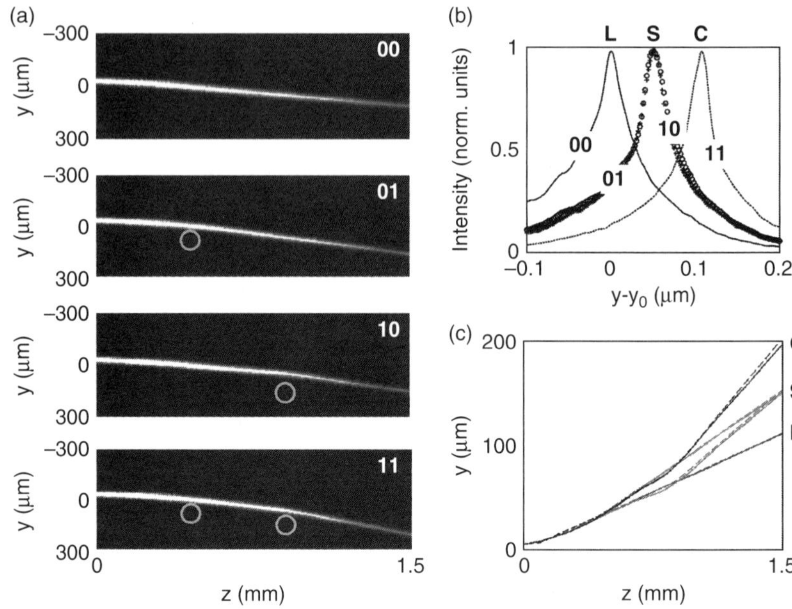

FIGURE 8.11 Realization of a half adder. (a) Soliton propagation for various defect configurations, as indicated by the Boolean legends. (b) Transverse profiles in $z = 1.5$ mm corresponding to various logic outputs. y_0 is the transverse position of the unperturbed nematicon (inputs 00). (c) Soliton trajectories in yz: data (solid lines) and simulations (dashed lines); labels C, S, and L refer to carry, sum, and low, respectively.

reroute a spatial soliton (and the signal traveling within it) to a different output at a given z. The operation of this 1×4 spatial demultiplexer is illustrated in Figure 8.12a, the transverse profiles shown in Figure 8.12b yield an effective cross-talk of about −5 dB, while the solid (dashed) lines in Figure 8.12c are the acquired (calculated) trajectories.

Taking advantage of soliton deflection versus shape of control spots, we can use elliptical spots to increase the deviations and realize a three-bits router, as illustrated in Figure 8.13. By choosing a suitable assortment of elliptical spots at given positions, the spatial soliton can be steered over a wider transverse interval of 400 μm, thus implementing a 1×8 demultiplexer controlled by combinations (out of $2^3 = 8$) of three binary inputs. In this case, the measured cross-talk of about −2.5 dB is due to the smaller distance between different control spots (bits) with respect to the 1×4 demultiplexer mentioned earlier.

Besides digital functions, with the LCLV one can build simple analog signal processors: we use the control spots to modulate nematicon–nematicon interaction (Chapter 1) in order to realize X and Y junctions. As shown in Figure 8.14a–d, two nematicons (each of them of power of $P = 2$ mW to ensure self-confinement) are launched with parallel wave vectors $\vec{k} \uparrow\uparrow \hat{z}$ in the plane y–z, but displaced along y of $d \approx 100$ μm. The input separation is larger than the cell thickness, that is, the nonlocality range: hence, the index profiles of the two solitons do not overlap in the

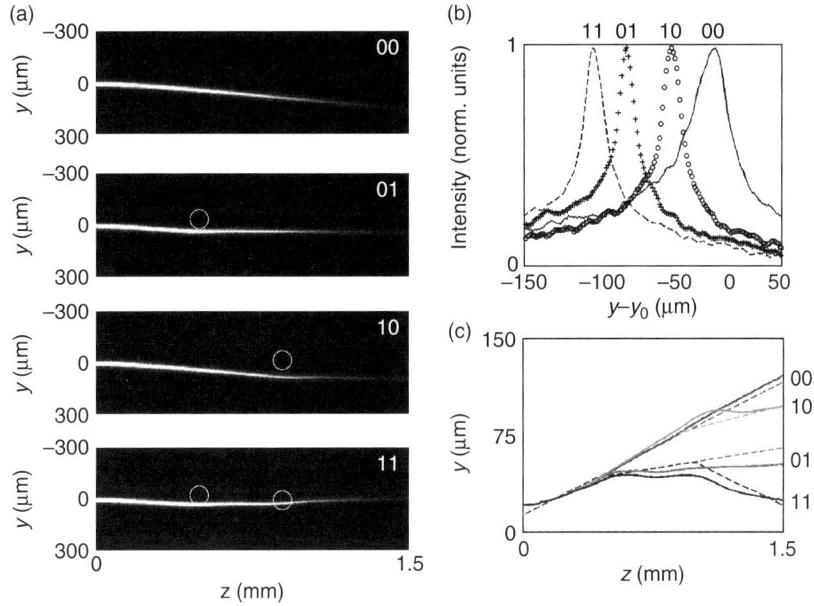

FIGURE 8.12 Two bits 1 × 4 router. (a) Photographs of soliton trajectories controlled by two control spots (circles). (b) Output profiles (z = 1.5 mm) and (c) nematicon trajectories for various inputs (Boolean legends). In (b), solid and dashed lines refer to experiments and simulations, respectively. y_0 corresponds to the logic input 00.

absence of control spots, preventing soliton–soliton interactions [48]. Conversely, when a "defect" is turned on between the two solitons, the beams are attracted to the defect and to one another, thereby interact. By varying curvature and position of the defect, the two beams can undergo either strong attraction—eventually crossing and forming an X junction (Fig. 8.14b,d)—, or weak interaction—merging at the output (Fig. 8.14c,e) and forming a Y junction [44, 47].

8.7 CONCLUSIONS AND PERSPECTIVES

In conclusion, we have shown that 3D control of nematicon trajectories can be realized through external beam addressing/multiplexing in LCLV. Nematicons have been shown to be robust and predictable solutions of the propagation in specifically designed light valves, with theoretical models confirming and supporting the observations. Gating and switching of solitons has been implemented by using arbitrary shape, position, and size of the control beams by means of a commercially available spatial light modulator.

New scenarios are foreseen in LCLV via the interplay of transverse patterns and propagating solitons [32].

Finally, we like to stress the unique wealth of possibilities offered by LCLV for implementing all-optical controlled schemes of light propagation and localization.

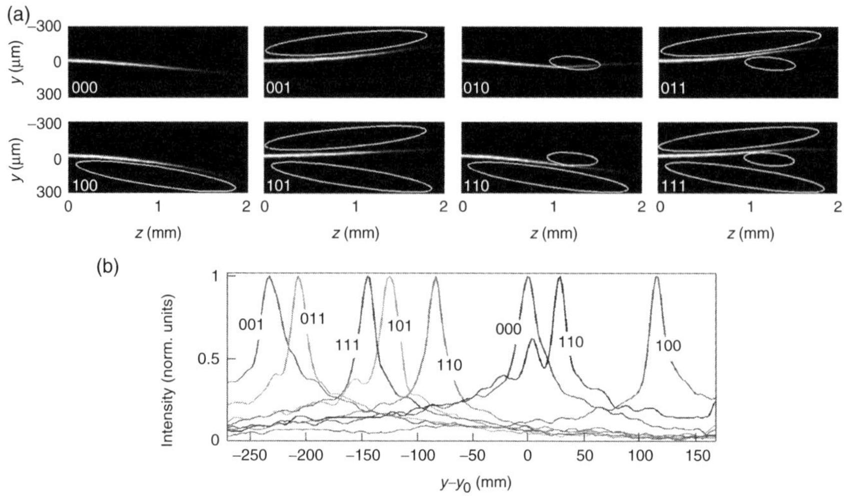

FIGURE 8.13 Three-bit 1 × 8 digital router controlled by elliptical external beams. (a) Photographs of soliton paths produced by the logic inputs in the legends. The closed lines indicate location and shape of the control spots. (b) Output ($z = 2$ mm) transverse profiles for various Boolean inputs. y_0 corresponds to the logic input 000.

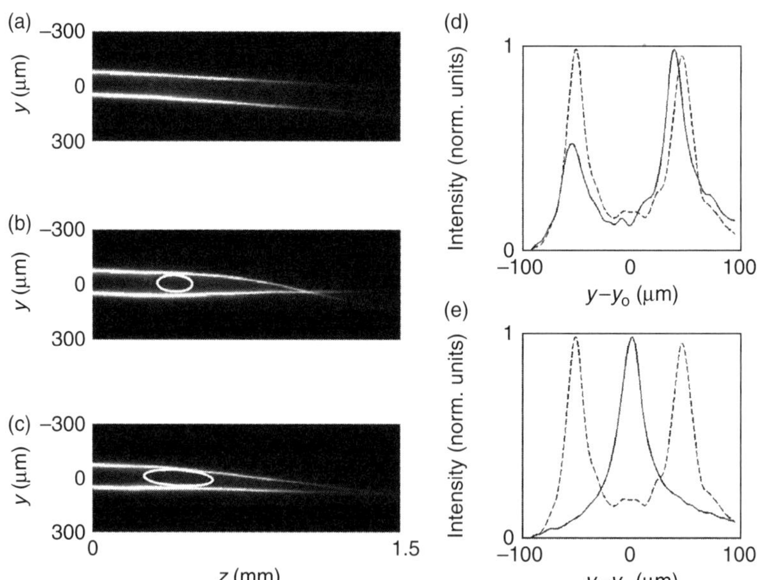

FIGURE 8.14 Implementation of an analog optical processor. (a) Two parallel solitons separated by $d \approx 100$ μm do not interact. By inserting a "defect" between them, we can modulate their interaction, implementing (b) an X junction when (d) the outputs are exchanged with respect to the inputs, or (c) a Y junction when (e) the beams merge into a single output.

Acknowledgments

S.R. and U.B. acknowledge the financial support of the ANR international program, Project ANR-2010-INTB-402-02, "COLORS". A.A. thanks Regione Lazio for financial support.

REFERENCES

1. A. C. Newell. *Solitons in Mathematics and Physics*. SIAM, Philadelphia, PA, 1985.
2. R. Y. Chiao, E. Garmire, and C. H. Townes. Self-trapping of optical beams. *Phys. Rev. Lett.*, 13:479–482, 1964.
3. Y. S. Kivshar and G. P. Agrawal. *Optical Solitons*. Academic Press, San Diego, CA, 2003.
4. G. I. Stegeman and M. Segev. Optical spatial solitons and their interactions: universality and diversity. *Science*, 286:1518–1523, 1999.
5. P. G. De Gennes and J. Prost. *The Physics of Liquid Crystals*, 2nd edn. Oxford Science Publications/Clarendon Press, Oxford, 1993.
6. I. C. Khoo. *Liquid Crystals: Physical Properties and Nonlinear Optical Phenomena*, 2nd edn. Wiley Interscience, New York, 2007.
7. P. Yeh and C. Gu. *Optics of Liquid Crystal Displays*. Wiley, New York, 1999.
8. N. V. Tabiryan, A. V. Sukhov, and B. Zeldovich. Orientational optical nonlinearity of liquid crystals. *Mol. Cryst. Liq. Cryst.*, 136:1–139, 1986.
9. I. C. Khoo. Nonlinear optics of liquid crystalline materials. *Phys. Rep.*, 47:221–267, 2009.
10. J. V. Moloney and A. C. Newell. *Nonlinear Optics*. Addison-Wesley, Redwood City, CA, 1991.
11. E. Braun, L. P. Faucheux, and A. Libchaber. Strong self-focusing in nematic liquid crystals. *Phys. Rev. A*, 48:611–622, 1993.
12. M. Peccianti, C. Conti, G. Assanto, A. De Luca, and C. Umeton. Routing of anisotropic spatial solitons and modulational instability in nematic liquid crystals. *Nature*, 432:733–737, 2004.
13. M. Peccianti and G. Assanto. Nematic liquid crystals: a suitable medium for self-confinement of coherent and incoherent light. *Phys. Rev. E*, 65:035603, 2002.
14. M. Peccianti, A. Dyadyusha, M. Kaczmarek, and G. Assanto. Tunable refraction and reflection of self-confined light beams. *Nat. Phys.*, 2:737–742, 2006.
15. M. Peccianti, C. Conti, and G. Assanto. Optical modulational instability in a nonlocal medium. *Phys. Rev. E*, 68:025602, 2003.
16. C. Conti, M. Peccianti, and G. Assanto. Complex dynamics and configurational entropy of spatial optical solitons in nonlocal media. *Opt. Lett.*, 31:2030–2032, 2006.
17. U. Bortolozzo, J. Laurie, S. Nazarenko, and S. Residori. Optical wave turbulence and the condensation of light. *J. Opt. Soc. Am. B*, 26:2280–2284, 2009.
18. A. Pasquazi, A. Alberucci, M. Peccianti, and G. Assanto. Signal processing by opto-optical interactions between self-localized and free propagating beams in liquid crystals. *Appl. Phys. Lett.*, 87:261104, 2005.

19. S. V. Serak, N. V. Tabiryan, M. Peccianti, and G. Assanto. Spatial soliton all-optical logic gates. *IEEE Photon. Technol. Lett.*, 18:1287–1289, 2006.
20. A. Piccardi, G. Assanto, L. Lucchetti, and F. Simoni. All-optical steering of soliton waveguides in dye-doped liquid crystals. *Appl. Phys. Lett.*, 93:171104, 2008.
21. A. Piccardi, U. Bortolozzo, S. Residori, and G. Assanto. Spatial solitons in liquid-crystal light valves. *Opt. Lett.*, 34:737–739, 2009.
22. U. Efron and G. Liverscu. *Spatial Light Modulator Technology: Materials, Devices and Applications*. Dekker, New York, 1995.
23. D. Armitage, J. I. Thackara, and W. D. Eades. Photoaddressed liquid crystal spatial light modulators. *Appl. Opt.*, 28:4763–4771, 1989.
24. J. Grinberg, A. Jacobson, W. P. Bleha, and L. Miller. A new real-time non-coherent to coherent light image converter - The hybrid field effect liquid crystal light valve. *Opt. Eng.*, 14:217–225, 1975.
25. P. R. Ashley and J. H. Davis. Amorphous silicon photoconductor in a liquid crystal spatial light modulator. *Appl. Opt.*, 26:241–246, 1978.
26. S. A. Akhmanov, M. A. Vorontsov, and Y. Y. Ivanov. Large-scale transverse nonlinear interactions in laser beams; new types of nonlinear waves; onset of optical turbulence. *JETP Lett.*, 47:707–711, 1988.
27. P. Aubourg, J. P. Huignard, M. Hareng, and R. A. Mullen. Liquid crystal light valve using bulk monocrystalline $Bi_{12}SiO_{20}$ as the photoconductive material. *Appl. Opt.*, 21:3706–3712, 1982.
28. P. Günter and J. P. Huignard. *Photorefractive Materials and their Applications 1*. Springer, Berlin, 2006.
29. U. Bortolozzo, S. Residori, and J. P. Huignard. Beam coupling in photorefractive liquid crystal light valves. *J. Phys. D: Appl. Phys.*, 41:24007, 2008.
30. A. Brignon, I. Bongrand, B. Loiseaux, and J. P. Huignard. Signal-beam amplification by two-wave mixing in a liquid-crystal light valve. *Opt. Lett.*, 22:1855–1857, 1997.
31. S. Residori. Patterns, fronts and structures in a liquid-crystal-light-valve with optical feedback. *Phys. Rep.*, 416:201–274, 2005.
32. O. Descalzi, M. G. Clerc, S. Residori, and G. Assanto Eds. Localized States in Physics: Solitons and Patterns. Springer-Verlag, Berlin, Heidelberg, 2011.
33. M. G. Clerc, A. Petrossian, and S. Residori. Bouncing localized structures in a liquid-crystal light-valve experiment. *Phys. Rev. E*, 71:015205, 2005.
34. U. Bortolozzo and S. Residori. Storage of localized structure matrices in nematic liquid crystals. *Phys. Rev. Lett.*, 96:037801, 2006.
35. F. Haudin, R. G. Elías, R. G. Rojas, U. Bortolozzo, M. G. Clerc, and S. Residori. Driven front propagation in 1D spatially periodic media. *Phys. Rev. Lett.*, 103:128003, 2009.
36. U. Bortolozzo, S. Residori, A. Petrosyan, and J. P. Huignard. Pattern formation and direct measurement of the spatial resolution in a photorefractive liquid crystal light valve. *Opt. Commun.*, 263:317–321, 2006.
37. U. Bortolozzo, S. Residori, and J. P. Huignard. Slow-light through nonlinear wave-mixing in liquid crystal light-valves. *C. R. Physique*, 10:938–948, 2009.
38. U. Bortolozzo, S. Residori, and J. P. Huignard. Enhancement of the two-wave-mixing gain in a stack of thin nonlinear media by use of the Talbot effect. *Opt. Lett.*, 31:2166–2168, 2006.

39. U. Bortolozzo, A. Montina, F. T. Arecchi, J. P. Huignard, and S. Residori. Complex dynamics of a unidirectional optical oscillator based on a liquid-crystal gain medium. *Phys. Rev. Lett.*, 99:023901, 2007.

40. A. Montina, U. Bortolozzo, S. Residori, and F. T. Arecchi. Non-Gaussian statistics and extreme waves in a nonlinear optical cavity. *Phys. Rev. Lett.*, 103:173901, 2009.

41. S. Residori, U. Bortolozzo, and J. P. Huignard. Slow and fast light in liquid crystal light valves. *Phys. Rev. Lett.*, 100:203603, 2008.

42. U. Bortolozzo, S. Residori, and P. Sebbah. Experimental observation of speckle instability in Kerr random media. *Phys. Rev. Lett.*, 106:103903, 2011.

43. G. Assanto, A. Piccardi, A. Alberucci, S. Residori, and U. Bortolozzo. Liquid crystal light valves: a versatile platform for Nematicons. *Phot. Lett. Poland*, 1:151–153, 2009.

44. A. Alberucci, A. Piccardi, U. Bortolozzo, S. Residori, and G. Assanto. Readdressable interconnects with spatial soliton waveguides in liquid crystal light valves. *IEEE Photon. Technol. Lett.*, 22:694–696, 2010.

45. A. Alberucci, A. Piccardi, U. Bortolozzo, S. Residori, and G. Assanto. Nematicon all-optical control in liquid crystal light valves. *Opt. Lett.*, 35:390–392, 2010.

46. G. Assanto, A. Piccardi, A. Alberucci, S. Residori, and U. Bortolozzo. Nematicons in liquid crystal light valves. *Mol. Cryst. Liq. Cryst.*, 527:98–108, 2010.

47. A. Piccardi, A. Alberucci, U. Bortolozzo, S. Residori, and G. Assanto. Soliton gating and switching in liquid crystal light valve. *Appl. Phys. Lett.*, 96:071104, 2010.

48. M. Kwasny, A. Piccardi, A. Alberucci, M. Peccianti, M. Kaczmarek, M. Karpierz, and G. Assanto. Nematicon-nematicon interactions in a medium with tunable nonlinearity and fixed nonlocality. *Opt. Lett.*, 36(13): 2566–2568, 2011.

9

Propagation of Light Confined via Thermo-Optical Effect in Nematic Liquid Crystals

MARC WARENGHEM, JEAN-FRANCOIS BLACH, AND JEAN-FRANCOIS HENNINOT

Unité de Catalyse et de Chimie du Solide, Faculté des Sciences, Université d'Artois, Lens, France

9.1 INTRODUCTION

Every material is temperature sensitive, and its physical parameters are temperature dependent. The optical properties are not an exception; hence, the index of refraction depends on temperature. In terms of applications, this could generally be regarded as a drawback, as a temperature change affects the index, and thus the propagation and, eventually, the operation of the device. Fortunately, for most materials, such dependence is quite weak. The best way to characterize it is the thermal coefficient of refraction $\frac{dn}{dT}$. For most manufactured glasses, this figure is around $\pm 10^{-6}$ K^{-1}, a few orders of magnitude larger than that for organic liquids (5×10^{-4} K^{-1} for acetone) and a few more for liquid crystals (10^{-2} K^{-1} for pentyl cyanobiphenyl). To give an insight of what such a figure means, let us take a textbook example: the minimum deflection of a beam by a prism. The change of that deflection is proportional to the change of the prism index of refraction. An increase of 10 K in the temperature of a prism (equilateral section) made of glass with a $dn/dT = 10^{-6}$ K^{-1} results in a variation of the beam minimum deflection of $4 \times 10^{-4\circ}$; for a $\frac{dn}{dT}$ four orders of magnitude larger [e.g., 10^{-2} K^{-1} in nematic liquid crystal (NLC)], this variation would be of 4°, which is no longer insignificant. As usual in optics, this thermal effect would be detectable if the change in phase shift or equivalently in optical path (index $\times$ propagation length) would be large enough. For instance, although the atmosphere index is very close to 1 and has a very small $\frac{dn}{dT}$, atmospheric thermal fluctuations or mirage can be

Nematicons: Spatial Optical Solitons in Nematic Liquid Crystals, First Edition.
Edited by Gaetano Assanto.

clearly visible! The thermal effect occurs whatever the origin of the temperature gradient. In other words, any temperature gradient induced by a beam would affect its own propagation, as for other nonlinearities. The effects associated with the thermal nonlinearity will be perceivable if the thermal coefficient of refraction or/and the induced temperature gradient is large enough. To observe these effects, the experimental setup is, in principle, simple: the index gradient is generated by local heating because of absorption of a nonuniform intense beam. In this chapter, we explore this capability: a beam with a nonuniform intensity (Gaussian) propagates through a medium with a large thermal coefficient of refraction, namely, an NLC. Before going into this topic, let us make some general and essential comments. The first two deal with the semantics of the expressions "nonlinear optics" and "optical soliton," and the third contains more physics.

1. Herein the considered nonlinearity does not involve the third-order dielectric susceptibility and somehow differs from conventional nonlinear optics. However, several well-known nonlinear effects have been observed, the basic ones being self-focusing, which occur as $\frac{dn}{dT}$ is positive, and self-defocusing, as this derivative is negative. Self-phase-modulation and self-trapping of beams have been observed as well.
2. The expression "self-trapping" sounds more appropriate than "spatial optical soliton" in the sense that absorption is inherent to the involved process.
3. Finally, a comment on the physical aspect of this specific thermo-optical effect: the process that drives this nonlinearity is a diffusion process and is therefore nonlocal. The thermally induced index change spreads out of the beam location, symmetrically with respect to the source, contrary to the photorefractive effect in which the index distribution can be translated only one way from the source. Moreover, it involves heat flow, which has no limits except the device itself: the range of nonlocality is higher than in nonlocal effect involving mass or charge displacement, such as in photorefractive media. The response times are different as well.

Although this chapter deals with effects observed in NLC, it should be noticed that thermo-optic effects—lensing, self-focusing, and defocusing of high-power laser beams—have been observed in other media (e.g., in gas) and modeled long ago [1–4]. Restricting our attention to NLC media, we have to consider two different geometries: thin and thick samples. In other words, does the beam propagate within the medium over a short or long distance? In the case of a thin sample, the beam crosses the sample and experiences classical nonlinear effects, such as self-phase-modulation and bistability [5] or some lens effect (self-focusing), which has been studied quite early [6, 7] and used—for instance—to characterize the material [8]. These two geometries—thin and thick samples—can be addressed from another point of view: the beam travels either across (thin samples) or along (thick samples) the NLC boundary: in other words, not only the propagation distance but also the nematic elasticity is different. However, as this chapter deals with the thermal effect and not with the orientational effect, we do not consider this aspect:

it is discussed in the other chapters of this book. We thus limit our attention to beams propagating a long distance through an NLC and how the large dependence of the index of refraction on temperature coupled to a nonuniform intensity of the beam affects its propagation. In Section 9.2, we report experimental observations on nonlinear propagation, and in Section 9.3, we focus on the methods proposed to analyze and characterize the effects and, among others, the nonlocality. We then compare, in Section 9.4, the thermal response with the orientational one: do they act with the same strength? Finally, we explore the potential applications of the thermal dependence (Section 9.5).

9.2 FIRST OBSERVATION IN NLC

Generally speaking, three conditions are required to observe the temperature-induced confined propagation of a beam. They are sketched in Figure 9.1.

First, the optical intensity should be spatially nonuniform. Second, the medium in which the light propagates should be slightly absorbing: the regions where the optical field is large absorb more and thus heat more than those where there is no optical field. A thermal equilibrium will be reached, which should ensure a stable temperature gradient. The third condition to be fulfilled concerns the medium as well: it should have an index that depends on temperature to transform the temperature gradient into an index gradient large enough to act in a detectable manner on the beam. As already mentioned in the introduction, NLC has one of the largest thermal coefficients, which can be either positive (focusing medium) or negative (defocusing medium).

In the geometries explored, the first condition is usually achieved by using a Gaussian beam, which is easy to handle, control, and characterize before and after nonlinear propagation. The second condition—to render the medium slightly

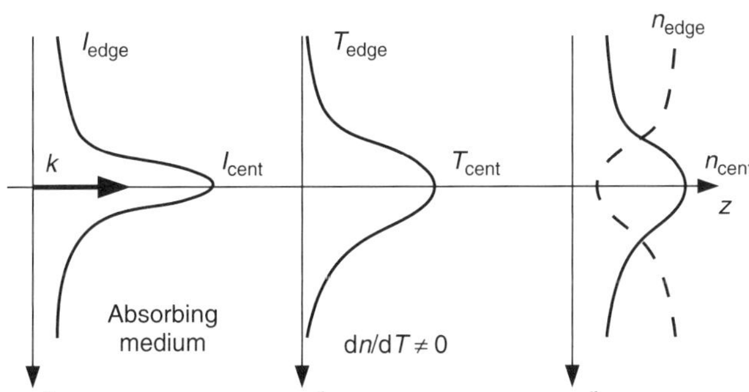

FIGURE 9.1 Principle of thermal nonlinear propagation: the beam intensity should be nonuniform, it should propagate in an absorbing medium, and the thermal coefficient of refraction dn/dT should be large. The nonuniform intensity generates a hotter region (on-axis of propagation in this sketch), which in turn generates an index gradient (focusing for $dn/dT > 0$, solid line; defocusing for $dn/dT < 0$ dashed line). *Source:* Reprinted from Reference 9. Copyright 2008, with permission from the Optical Society of America.

absorbing—is fulfilled by adding a very small amount of a dye to the liquid crystal. A concentration of a few 0/00 is enough to induce the required absorption while preserving the nematic nature of the mixture. The temperature of the phase transition is also slightly modified, the indices of refraction and their temperature dependence being practically unchanged. To fulfill the third condition, the choice of liquid crystals is almost obvious, because of the very large temperature dependence of their indices. However, there are several possible geometries. Indeed, while dealing with liquid crystals we deal with anisotropic media; hence, the two principal indices of refraction do not behave in the same way with temperature. Let us first recollect this feature. As shown in Figure 9.2, the extraordinary index always decreases with temperature; on the contrary, the ordinary index may increase (which is the case in most NLCs), or have a minimum.

It is beyond the scope of this chapter to detail the physical reasons for such behavior: the reader may refer to the appropriate bibliography [10–14]. Whatever the chosen liquid crystal, the polarization of the beam becomes an essential parameter to be controlled in the experiments. The simplest case consists in propagating a linearly polarized beam, exciting one of the two eigenmodes of the birefringent material. The extraordinary component experiences defocusing, whereas the ordinary component focuses, provided that the ordinary index increases with temperature (Fig. 9.2a). The geometries that clearly separate the two linear eigen polarizations are shown in Figure 9.3. Any intermediate situation might be explored, although more tricky to analyze and use.

Most of the experiments on thermally induced nonlinear propagation have been performed using a mixture of pure 5CB (Merck) and anthraquinone dye (AQ1) at low concentration (1/00, w/w). The real part of the index of refraction was measured for that mixture, using a total internal reflection (TIR) setup and was found to be practically identical to that of the pure material ($n_o = 1.53$ at 633 nm).

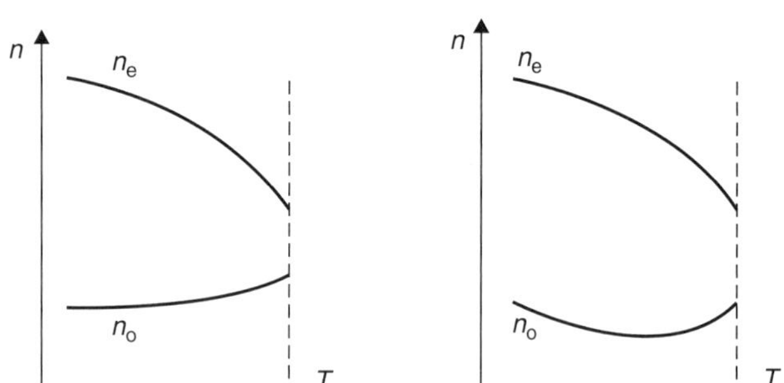

FIGURE 9.2 Temperature dependence of refractive index in nematic liquid crystals. The extraordinary index is always decreasing with temperature, whereas the ordinary one can be either increasing [(a); e.g., 5CB, Merck], or going through a minimum [(b); e.g., PCH & mixture of PCH5 + PCH7, Hoffmann-La Roche [10, 11]). T_{NI} is the clearing point, the temperature above which the material turns to be a standard isotropic liquid.

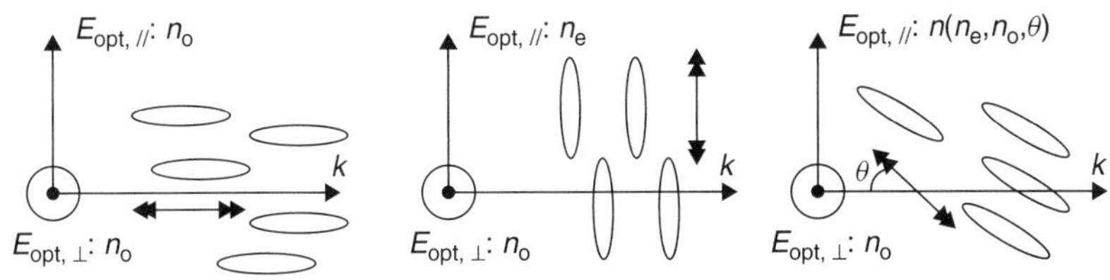

FIGURE 9.3 Geometries that allow exciting separately the two eigenmodes (ordinary and extraordinary waves) in NLC. The ellipses stand for the NLC director—averaged molecular alignment. The optic axis is marked with a double black arrow. The index of refraction is indicated for each optical field component.

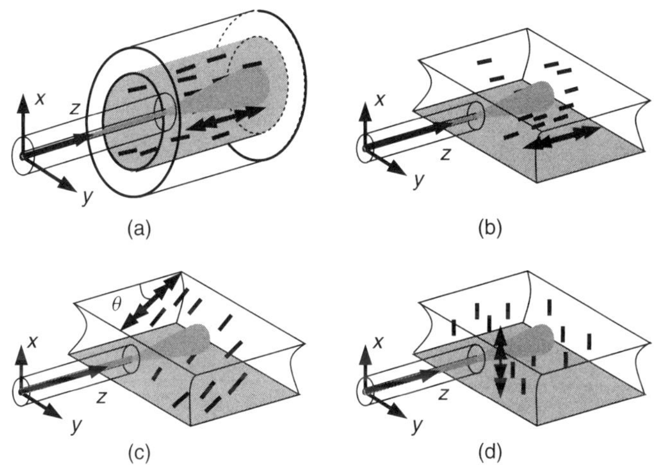

FIGURE 9.4 The four explored geometries. The geometries sketched in (a) and (b) are equivalent: only the ordinary index is involved. The NLC alignment is parallel to the plane yz in both cells (a) and (b). The optic axis is tilted away from the direction of propagation z in cell (c). The optic axis is parallel to the x-direction in cell (d) (homeotropic alignment). The ordinary index is involved for an input beam polarization parallel to y in all cells. *Source:* Reprinted from Reference 15. Copyright 2001, with permission from Elsevier.

Its temperature dependence looks like the case plotted in Figure 9.2a: the thermal coefficient of refraction is always positive and, provided that the correct polarization is used, the beam would be self-focused. The thermally induced nonlinear propagation has been tested under the geometries illustrated in Figure 9.3. The mixture is confined either in a capillary tube (inner diameter: 250 μm) or in a planar cell (thickness around 125 μm). The alignment of the nematic is obtained in both cases by a conventional technique (rubbed polymer coating), and is checked by visual inspection with an optical microscope. The beam under study (CW laser beam at 633 nm or 532 nm) was injected via an optical fiber directly inserted in the NLC sample. The fiber was either simply cleaved (beam waist: 3 μm) or tapered (beam waist:10 μm). The setup geometries are shown in Figure 9.4.

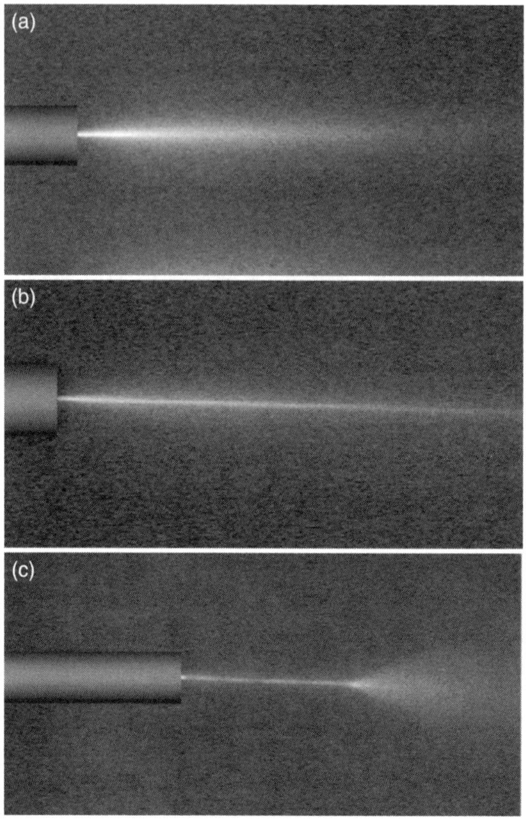

FIGURE 9.5 Sequence of events observed as the input power is increased. (a) Normal diffraction, (b) self-trapped beam, and (c) isotropic tube. *Source:* Reprinted from Reference 18. Copyright 2004, with permission from the American Institute of Physics.

The first observation of long-distance thermally induced nonlinear beam propagation in NLC was reported in 1998 [16]. Schematically, as the input beam power is increased from 0 to around 6 mW, two regimes are observed. The first regime starts at around 4 mW: the beam self-focuses and remains collimated (equivalent to a spatial optical soliton) practically up to 6 mW. This is the input power at which the second regime is observed: a well-collimated beam, with a length that linearly and reversibly depends on the input power. As the input power is increased, the length increases; as the input power is decreased, the length decreases. These two regimes are shown in Figure 9.5. Further studies were undertaken to better understand these results. Varying the dye concentration [16] or the birefringence [17] does not change the sequence of events described earlier, it just changes the optical powers at the onset of the regimes. The capillary geometry (Fig. 9.4a) or the equivalent cell geometry (Fig. 9.4b) yield the same sequence, although these two geometries have a different boundary and thus a different heat flow [15]. The

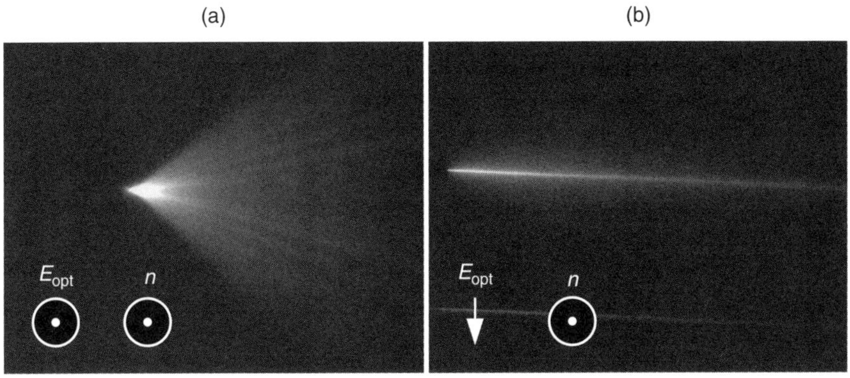

FIGURE 9.6 Cell with a homeotropic alignment (geometry of Figure 9.4d). The optic axis ***n*** is perpendicular to the plane of the figure. (a) The optical beam excites the extraordinary component and is defocused. (b) The optical beam excites the ordinary component and is collimated (the input waist of the beam is 3 μm, and the propagation distance is about 1 mm). *Source:* Reprinted from Reference 15. Copyright 2001, with permission from Elsevier.

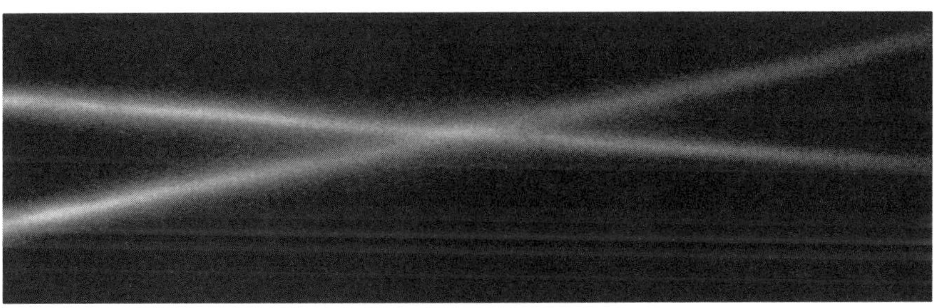

FIGURE 9.7 Cell with a homeotropic alignment (geometry of Figure 9.4d). The optic axis is perpendicular to the plane of the figure. For both beams, the polarization is in the plane of the figure and excites the ordinary component. The beams are collimated and keep propagating after crossing their paths.

geometry in Figure 9.4c, which is polarization dependent, was checked as well [15]. As expected for large θ angle (Fig. 9.15), the extraordinary beam is clearly defocused and the ordinary one is collimated, as shown in Figure 9.6.

The low power regime was identified as a solitonlike regime, the second regime as an isotropic tube. These results are detailed in the next section. Let us mention two results that bring us to the conclusion that we are dealing with a solitonlike behavior. First, it was checked that two solitons crossing their paths do not change their propagation: this is shown in Figure 9.7 [15]. Second, the set of two coupled equations that rule the process were numerically solved and showed that a solitonlike regime can be excited by a purely thermal effect [19].

To summarize these observations, as the input optical power is increased, the beam propagates normally (usual divergence) up to some value, say P_s, and then it starts to self-focus and is collimated to form a quasisoliton. At a slightly larger input power, say P_i, a phase transition occurs; the light is trapped in a tube of a nematic material that has turned to the isotropic phase. The length of that tube reversibly depends on the input optical power. The values of powers P_s and P_i depend on the actual sample and geometry. This scenario is the result of several experiments that were undertaken to point out the main parameters driving the process: they are detailed in the next section.

9.3 CHARACTERIZATION AND NONLOCALITY MEASUREMENT

The observations mentioned earlier give rise to several questions: is the beam really well collimated (a soliton)? At which input power does the phase transition really start? A related issue is can the temperature gradient—in other words, the nonlocality—be controlled? How do two solitons behave when they are copropagate and close to each other? To answer these questions, several experiments were conducted, aiming at better characterizing the nonlinear propagation. To check how good the collimation is, the most direct approach consists in performing an image analysis: it is straightforward to get the intensity profile of a cross section of the beam, at different distances from the beam input. As shown in Figure 9.8, the radial distribution of the intensity is unchanged when measured at two different locations, 400 μm away from each other [20]. Besides the collimation confirmation, this image analysis gives the order of magnitude for the radius of the self-written waveguide; however, what is obtained this way is just an integrated value of the scattered light and should be considered with caution.

Another way to check the collimation relies on interferometry. The cell is placed in a Mach–Zehnder interferometer, as shown in Figure 9.9 [21]. The fringe pattern reveals the disturbance created by the temperature gradient. Such

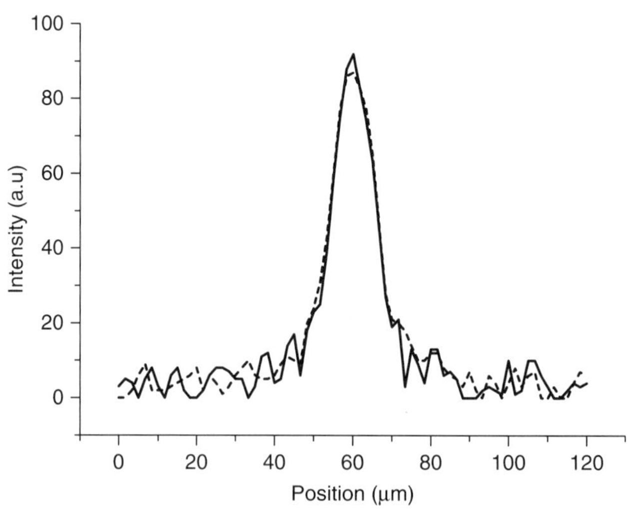

FIGURE 9.8 Radial intensity profiles obtained from Figure 9.5b at two positions 400 μm away from one another (dashed line: entrance; solid line after 400 μm of propagation). *Source:* Reprinted from Reference 19. Copyright 2000, with permission from the American Institute of Physics.

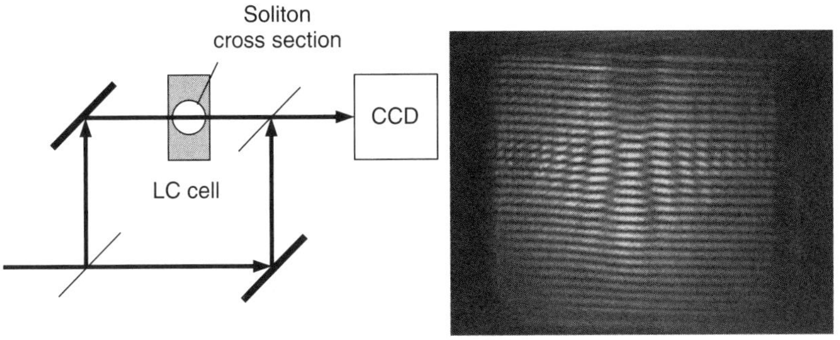

FIGURE 9.9 Mach–Zehnder type apparatus to measure index distribution across the NLC cell. The soliton is launched perpendicular to the drawing. The photo shows an example of what is captured by the CCD camera. Such an approach can be used for any index change, either thermal or orientational. *Source:* Reprinted from Reference 21. Copyright 2005, with permission from SPIE.

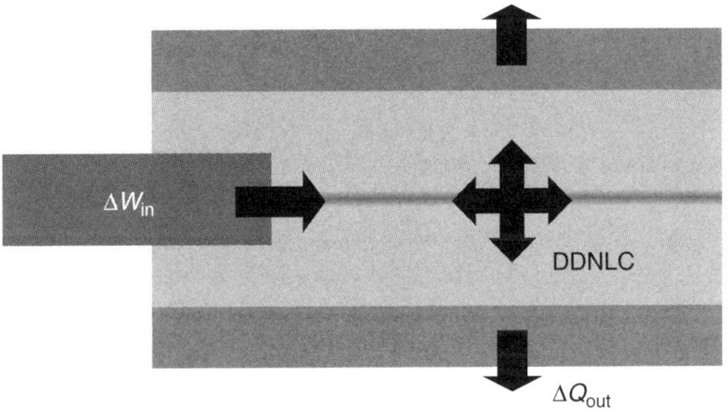

FIGURE 9.10 The input beam energy W_{in} is partially transformed into heat, which is expelled out of the sample in every direction. In the case of dye-doped NLC, the heat flow is anisotropic.

an interferometric technique can be used whatever the origin of the index change, thermal or orientational. Indeed, a numerical method was developed to extract the index change from such a fringe pattern [22].

Although these two methods to analyze the collimated regime are integrated methods (integrated scattered light for the image analysis and integrated phase shift for the interferometric technique), the integration is the same for an image analysis performed at the beginning of the propagation or away from it. It is therefore appropriate to state that the beam is collimated as the intensity profile is found to be constant along the propagation direction. The second question that was addressed aims at clarifying the onset of the phase transition. This issue deals more with heat transfer than with optics. However, it turns out that the understanding of this point has a direct consequence on the optics of the problem. It allows for a better control of the induced temperature, that is, index gradient. In other words, it allows controlling the nonlocality. Let us sketch the process. As shown in Figure 9.10, the beam continuously brings some energy into the medium, which

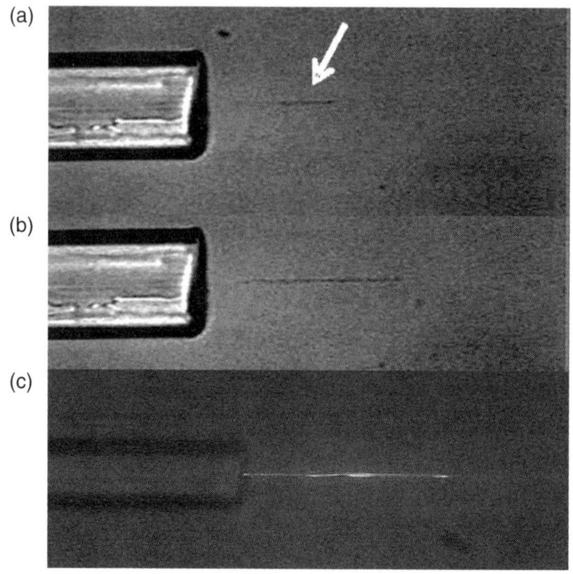

FIGURE 9.11 The onset of phase transition. As the optical input power increases, the phase transition starts a few micrometers away from the input fiber (a: black line with white arrow). Then the isotropic tube enlarges symmetrically from that point (b). As the isotropic phase reaches the fiber, the tube extends to its end (c). *Source:* Reprinted from Reference 18. Copyright 2004, with permission from the American Institute of Physics.

absorbs and locally heats up. This heat is expelled in every direction. The induced temperature distribution will depend on the different parameters of the sample such as heat capacity of the various components, input energy, and the overall geometry that rules the heat exchange with the surrounding. Accounting for the fact that the heat diffusion process is quite slow, the temperature, and hence the index distribution, will be robust with respect to beam intensity fluctuations faster than the diffusion characteristic time. Therefore, by using a chopped laser beam with the right characteristic times, it should be possible to control the input energy.

Using a laser pulse of 60 μs duration with a 500 μs repetition rate, slowly increasing the total power, it was possible to control the heat exchange and spot out the onset of the phase transition, as shown in Figure 9.11 [18].

This scenario is confirmed by the numerical calculation of the temperature distribution in the cell [18]. The hottest point in the cell, that is, where the phase transition will most likely start, is found to be on the propagation axis (which is obvious) but a bit away from the fiber source, as shown in Figure 9.12. This is in agreement with the experimental result: 150 μm calculated and 180 μm measured in Figure 9.11 (top). This is a consequence of the fact that the fiber heat conductivity is higher than in NLC. The heat transfer is symmetrically distributed in the transverse direction, whereas the fiber breaks up the symmetry in the longitudinal direction. This result is consistent with the one found for another geometry [23].

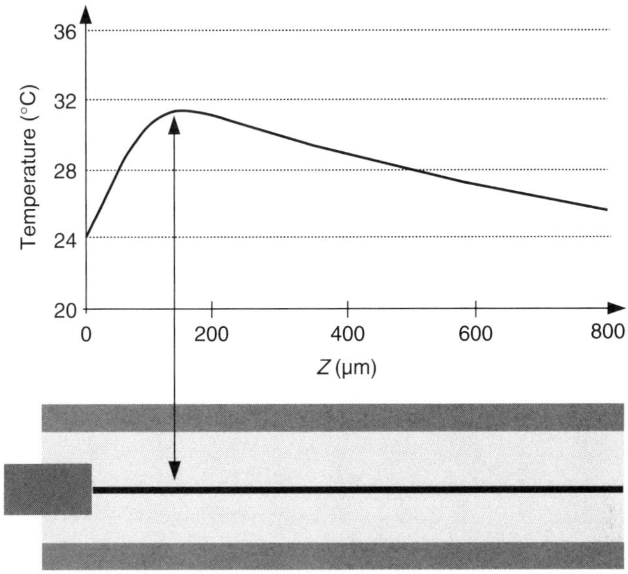

FIGURE 9.12 On-axis temperature distribution, obtained from finite element calculation in a cell as sketched at the bottom. The hottest point is found to be at 150 μm (black arrow). *Source:* Reprinted from Reference 18. Copyright 2004, with permission from the American Institute of Physics.

The method of controlling the input energy by chopping the input beam can be used to control the temperature gradient in the self-collimated regime. In other words, it allows to control the nonlocality, which in our case is directly linked to the temperature gradient size as compared to the one of the input beam. Moreover, the optical molecular reorientation being a slow process, it is possible to control the nonlocality induced by this process as well, provided the repetition and duration of the pulses are not too small compared to the elastic response time that drives the reorientational process.

As the nonlocality can be somehow controlled, can it be measured? In the system described herein, the nonlocality can be characterized by comparing the nonlinear effect, namely, the index distribution, with the cause, namely, the beam width. Assuming a soliton regime, the beam width (W_b, soliton waist) and the induced index profile (W_n) remain constant along the propagation direction. The ratio W_n/W_b may be a measure of the nonlocality. One way to get an estimate of this figure consists in measuring the beam waist from the profile, as shown in Figure 9.8, and the index profile from a fringe pattern such as the one shown in Figure 9.9. In the reported experiments, the ratio is estimated to be around 2–3. Actually, although the approach sounds simple, this figure should be considered very cautiously. As already mentioned, the obtained intensity profile is an integration of the scattered light, whereas the index profile is an integration of the phase shift across the cell thickness. Thus the obtained waists are not the actual values and, more importantly, the integrations are quite different in nature for both measurements; hence, strictly speaking, they cannot be compared.

Another way to get an estimate of the induced index gradient consists in modeling the experiment.

The nonlinear propagation is modeled by two partial differential equations: one for the propagation, which is easily reduced to the scalar Helmholtz (Eq. 9.1) and the other for the heat diffusion process, which is a classical Fourier process (Eq. 9.2). These two equations are coupled via the temperature dependence of the ordinary index of refraction (Eq. 9.3). In terms of the scalar field Ψ, the propagation constant β, the wave number k, the ordinary index of refraction $n(T)$, and the thermal constants, absorption α and thermal diffusion κ, of the material, these equations read

$$\Delta_{\perp}\Psi + [k^2 n^2(T) - \beta^2]\Psi = 0, \tag{9.1}$$

$$\Delta_{\perp}T = -\frac{\alpha_{\perp}}{\kappa_{\perp}}\Psi\Psi^*, \tag{9.2}$$

$$n(T) = a + b\mathrm{e}^{cT}. \tag{9.3}$$

The coefficients a, b, and c in the thermal dependence of n (Eq. 9.3) were obtained from experimental values in 5CB [24]. The set of equations was numerically solved, yielding an index profile of around 15 μm in size, close to the one measured from interferometry [19]. The method that yields a good estimation of the ratio $W_{\mathrm{n}}/W_{\mathrm{b}}$ consists in looking at the interaction between two self-waveguides. The geometry is illustrated in Figure 9.13.

Two fibers launch two counter-propagating solitons. The fibers can be displaced in order to vary the distance between the two beams and the two fibers (d and z, Fig. 9.13). A camera is set above the cell, which captures the scattered light. One soliton is launched first (from fiber A, Fig. 9.13): it travels straight (dashed line, Fig. 9.13). Then the second one is launched: it travels through the index gradient created by the first soliton and creates an extra index gradient that disturbs the first soliton trajectory: both solitons are bent, and the measurement of the deflection

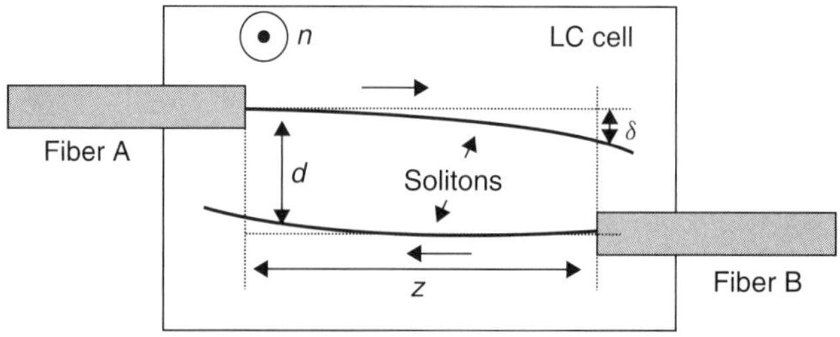

FIGURE 9.13 Geometry of the experiment of mutual deflection of solitons used to measure the index distribution. The optic axis $\boldsymbol{n}$ is perpendicular to the figure; the polarization of both beams is parallel to the plane of the figure.

(solid lines, δ, Fig. 9.13) is the signature of this mutual influence. The deflection was measured using image analysis, together with the acquired beam profiles (Fig. 9.14). Assuming that a soliton behaves like a ray, the trajectory is a matter of the eikonal equation (Eq. 9.4) that describes the path of a light beam in an inhomogeneous medium characterized by index $n(r)$:

$$\frac{\mathrm{d}}{\mathrm{d}s}\left(n\frac{\mathrm{d}r}{\mathrm{d}s}\right) = \nabla n. \tag{9.4}$$

From experimental observations, the paraxial approximation applies and Equation 9.4 reduces to

$$\frac{\mathrm{d}n}{\mathrm{d}z}\frac{\mathrm{d}r}{\mathrm{d}z} = \frac{\mathrm{d}n}{\mathrm{d}r}. \tag{9.5}$$

From the knowledge of the temperature dependence of the ordinary index and an estimate of the temperature decrease along the propagation, $\frac{\mathrm{d}n}{\mathrm{d}z}$ can be evaluated over the total length of propagation. The experimental evaluation of $\frac{\mathrm{d}r}{\mathrm{d}z}$ is derived from the measurement of δ for different distances d between the fibers (Fig. 9.13) and leads to the determination of the radial index variation $\frac{\mathrm{d}n}{\mathrm{d}r}$ [25].

Solving this equation, it was possible to retrieve the index profile [25]. The ratio W_n/W_b is reported to be around 3. The actual value of this figure is most likely

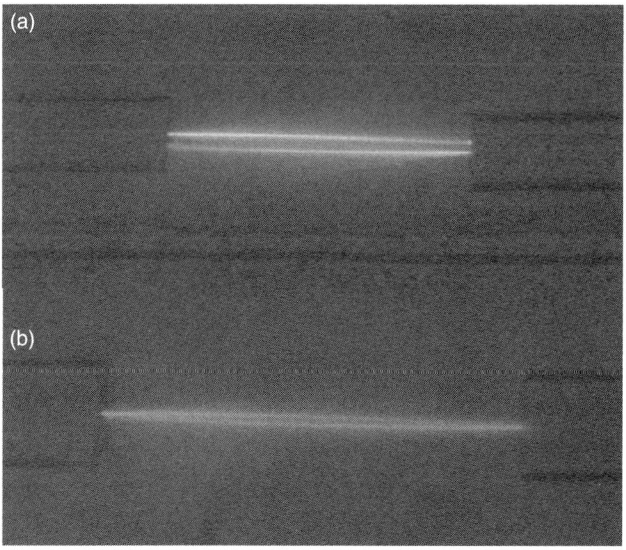

FIGURE 9.14 Example of mutual interaction of two counter-propagating solitons, for a separation between fibers of 30 μm (a) and 10 μm (b). *Source:* Reprinted from Reference 21. Copyright 2005, with permission from SPIE.

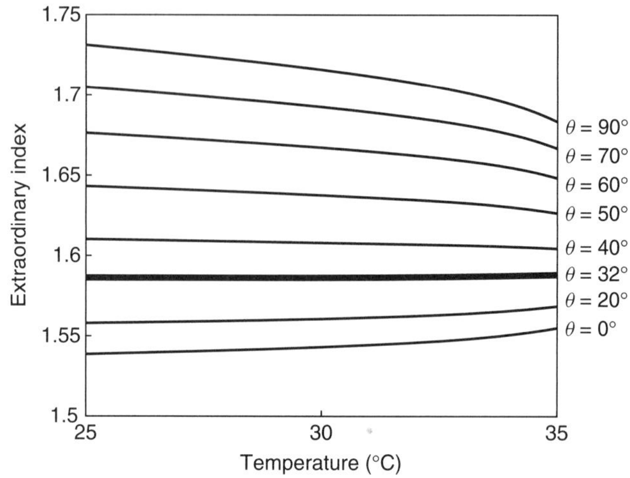

FIGURE 9.15 Temperature dependence of the extraordinary index for several values of the angle θ between the optic axis and the direction of propagation. For $\theta = 32^\circ$, the index is practically independent on T (calculated using values of n_e and n_o from Reference 24). *Source:* Reprinted from Reference 9. Copyright 2008, with permission from the Optical Society of America.

larger as the found beam waist is, once again, an integration; conversely, the index profile is correct, although not very accurate because it is calculated from rough measurements.

These methods to characterize the soliton are not process dependent; this means that they can be used for a soliton generated by the reorientation of the NLC (nematicon) [26]. To fully characterize the nonlinear propagation, it would be interesting to compare the two processes that may induce a soliton in a liquid crystalline material, that is, thermal and orientational. The next section addresses this issue.

9.4 THERMAL VERSUS ORIENTATIONAL SELF-WAVEGUIDES

As it was pointed out in Section 9.2, the extraordinary index always decreases with temperature. In an experiment dealing with nematicons, it is the extraordinary component that is involved and a temperature change would play a defocusing role, counteracting the orientational focusing effect. This is why the NLC used in these experiments is carefully chosen (most of the times E7) to avoid such counteracting effect. On the contrary, it might be interesting to compare both effects: is it possible to excite simultaneously the two kinds of solitons, and which one is stronger? At first glance, the coexistence of both solitons looks unlikely; however, it was possible to observe it and, in addition, to show that the orientational effect is much stronger than the thermal one, that is, it requires a lower input beam intensity to be excited [9]. The geometry used to obtain those two solitons is the one in Figure 9.4c. The polarization of the input beam is linear and tilted with respect to the plane of the figure: within the NLC, both ordinary and extraordinary components are excited.

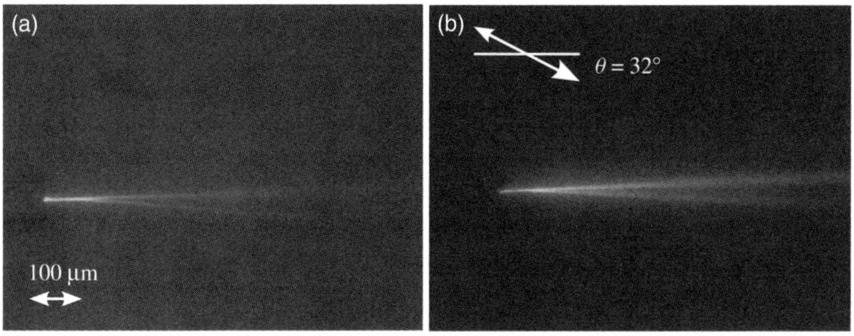

FIGURE 9.16 Coexistence of "thermal" and "orientational" solitons [(a) $P = 2.4$ mW], free propagation [(b), $P = 1.5$ mW] (the input waist of the beam is 3 μm and the distance of propagation is about 1 mm). *Source:* Reprinted from Reference 9. Copyright 2008, with permission from the Optical Society of America.

The intensity of each component depends on the angle of the input polarization, which can be varied. In such a scheme, the actual extraordinary index n is a well-known function of n_e, n_o, and θ, the angle between the optic axis and the direction of propagation

$$n = \frac{n_\mathrm{e} n_\mathrm{o}}{\left[n_\mathrm{e}^2 \cos^2(\theta) + n_o^2 \sin^2(\theta)\right]^{1/2}}. \tag{9.6}$$

Accounting for the opposite behavior of each of the principal indices n_e and n_o, one expects some value for the tilt angle θ at which the temperature dependence for the resulting extraordinary index is strongly reduced. Indeed, for the employed NLC (5CB), the resulting extraordinary index is practically insensitive to temperature for an angle of 32°, as shown in Figure 9.15.

As shown in Figure 9.16, obtained using 5CB slightly doped with the dye AQ1, in the geometry 9.4c, with an angle θ set to 32°, the coexistence of an ordinary "thermal" soliton with an "orientational" extraordinary soliton is possible. Regarding the tolerance on the angle θ: how far from the calculated value $\theta = 32°$ can the two solitons still coexist? Surprisingly, the coexistence was observed over a large range of tilt angles. However, looking more carefully at the role of each of the contributions, thermal and orientational, this experimental result is in agreement with a very simple picture, as developed in Reference 9 and briefly summarized later. The index gradient for the extraordinary component is generated by a change in the tilt angle, due to optical reorientation, and by a change of temperature, due to local absorption. In a direction transverse to propagation, the gradient is expressed as

$$\frac{\mathrm{d}n}{\mathrm{d}r} = \frac{\partial n}{\partial \theta}\frac{\mathrm{d}\theta}{\mathrm{d}r} + \frac{\partial n}{\partial T}\frac{\mathrm{d}T}{\mathrm{d}r}. \tag{9.7}$$

This can be understood as a temperature gradient weighed by a thermal portion $\frac{\partial n}{\partial T}$ added to an orientational gradient $\frac{\mathrm{d}\theta}{\mathrm{d}r}$, weighed by an orientational portion $\frac{\partial n}{\partial \theta}$. The dependence of n on T is known for the material 5CB [24], and the dependence of n on the angle θ is given by Equation 9.6. It is therefore possible to calculate each of these two weights and to look for an overall positive index gradient for different values of the ratio between orientational and thermal gradients. It turns out that the coexistence of the two solitons was observed for tilt angles as large as 60°; this means that, after numerical calculation, although the temperature gradient is around 100 times larger than the orientational one, the orientational weight is larger than the "thermal" weight and, on the whole, gives rise to a positive index gradient for most tilt angles. In other words, the orientational effect is much stronger than the thermal one.

9.5 APPLICATIONS

Any application based on a temperature-dependent effect would be not too easy to implement, as it would require a severe external temperature control to ensure that the exploited effect not be perturbed by external parameters. However, at least two potential applications can be mentioned for thermal optical solitons. The first one is based on the bent waveguide, whereas the second can be used for fluorescence recovery.

9.5.1 Bent Waveguide

We developed an experiment to test the transmission properties of this confining structure. Using the same setup as the one in Figure 9.14, we launched two isotropic channels in opposite directions by injecting laser beams ($\lambda = 532$ nm) in two optical fibers, slightly tilted with respect to each other. The distance between the fibers along the z-axis was in the order of 400 μm. As stated earlier, it is possible to adjust the length of the isotropic channels by carefully adjusting the input power: this was done for each source in order to join the two channels (Fig. 9.17).

A low power He–Ne beam (probe) was injected in one fiber and detected at the output of the other one. We could thus check whether the bent structure could guide light. Using micropositioners, the angle between the fibers was precisely adjusted. By varying this angle, we checked the maximum bending of the isotropic channel that still allowed for the transmission of the probe. As the angle was too large, the probe beam "leaked" out of the first channel and was no longer guided to the second fiber. The limit was imposed by the index variation induced by the nematic/isotropic phase transition. We experimentally found that the probe signal is transmitted through the bent structure for a maximum angle of 7°. A rough calculation, based on total internal refraction considerations, shows that this limit angle yields an index variation of about 10^{-2}, consistent with the known values for the considered material ($T = 35.3^\circ$C, $\lambda = 632.8$ nm, $n_o = 1.548, n_e = 1.585$).

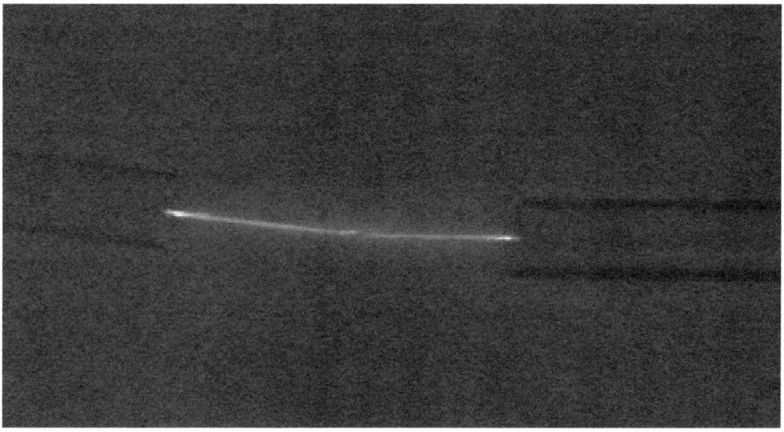

FIGURE 9.17 Isotropic bent channel obtained by joining two counter-propagating isotropic tubes.

9.5.2 Fluorescence Recovery

Up to now, spatial optical solitons have been mostly studied for potential applications in steering and routing of light beams [27, 28]. Recently, we proposed another use based on the enhancement of fluorescence recovery. The fluorescence signal is often weak and, as a result, its enhancement becomes a real challenge. When a spatial soliton propagates in an active medium, the corresponding luminescence will be partly trapped within the self-induced waveguide and driven to its output. We demonstrated the feasibility of such concept by comparing the collected fluorescence signal of a dye mixed in an NLC host, excited either by a freely expanding beam or by a spatial soliton. The setup is similar to the one we used for soliton interactions (Fig. 9.18).

Two single-mode fibers, facing each other, are introduced in the bulk of an NLC cell. A CW pump beam ($\lambda = 532$ nm) emitted from the left fiber propagates either freely or collimated (soliton), depending on the input power. The right fiber

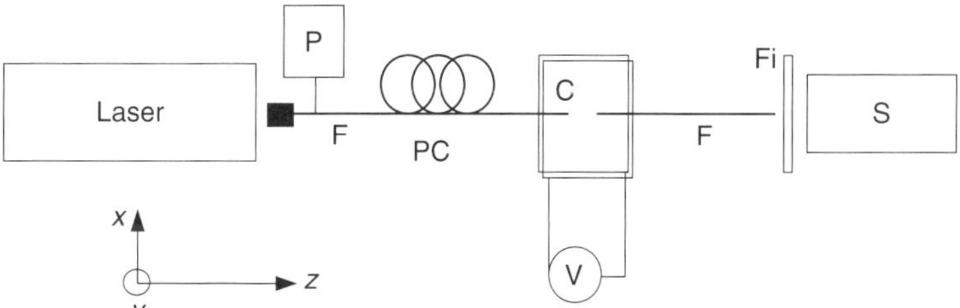

FIGURE 9.18 Experimental setup (P, power meter; PC, polarization controller; F, fibers; C, LC cell; Fi, notch filter; S, spectrometer; V, voltage generator). *Source:* Reprinted from Reference 29. Copyright 2010, with permission from the American Institute of Physics.

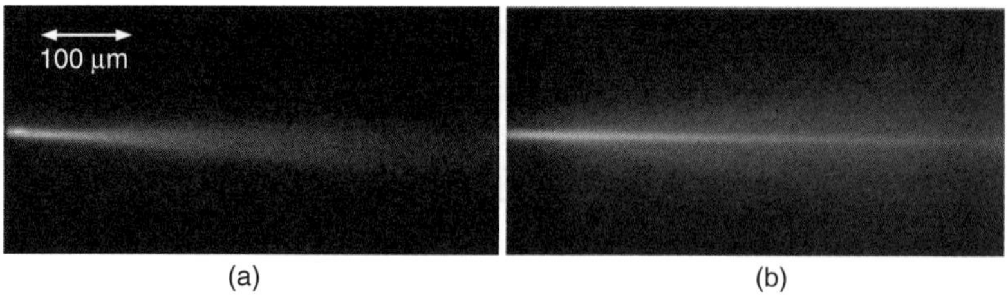

FIGURE 9.19 Laser beam propagating across the liquid crystal cell. (a) P_{in} = 1.5 mW (free propagation), (b) P_{in} = 3.5 mW (soliton). *Source:* Reprinted from Reference 29. Copyright 2010, with permission from the American Institute of Physics.

collects and transmits the light to a spectrometer after passing through a notch filter, to eliminate the pump. The material under study is a mixture of NLC 5CB and Quinizarin (0.1% in weight). This low concentration was chosen in order to obtain a weak signal of fluorescence and thus to test the sensitivity of the method. The collected spectra of fluorescence between 400 and 1000 nm are compared in both cases. The two regimes of propagation are shown in Figure 9.19: a normally diverging beam (Fig. 9.19a) and a soliton (Fig. 9.19b). The transverse size of the soliton was estimated by image analysis as 5.6 ± 0.5 μm, which corresponds to a diffraction length z_R of around 200 μm.

The first experiment consisted in increasing the pump power from 0.5 mW up to 5 mW, maintaining constant the interfiber distance ($d = 500$ μm). The soliton regime occurred above 3.5 mW. Examples of spectra recorded in the free regime and in the soliton regime are shown in Figure 9.20. The collected intensity is four times larger in the case of a soliton than in the linear case, taking into account the pump increase.

The evolution of the maximum intensity of fluorescence is plotted against input power in Figure 9.21. For low pump powers, the maximum intensity of the fluorescence increases linearly with pump power. At around 2.2 mW, thermal self-focusing starts and the collected intensity no longer varies linearly with the input power. For larger pump power, the soliton regime starts and the signal increases linearly once again, with a higher slope than before. For input pump powers higher than 5 mW, the liquid crystal heats too much and, when approaching the isotropic phase, becomes highly unstable. Changing the distance d between the fibers, from 125 to 500 μm, we observed that the fluorescence signal stayed quite constant for the soliton regime, whereas it decreased, as expected, in the free propagation case [29]. By this result, we demonstrated the possibility of using spatial solitons as a local probe of fluorescence in liquid suspensions.

9.6 CONCLUSIONS

Besides the very well-known display applications, NLCs should be considered as a very good test material to investigate complex electromagnetic wave propagation. In this chapter, we focused on the strong temperature dependence of refractive

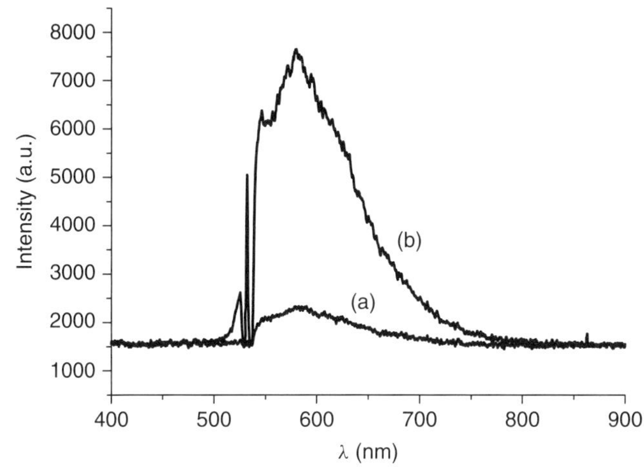

FIGURE 9.20 Fluorescence spectrum for (a) $P_{in} = 1.5$ mW (free beam) and (b) $P_{in} = 3.7$ mW (soliton), $d = 500$ μm. *Source:* Reprinted from Reference 29. Copyright 2010, with permission from the American Institute of Physics.

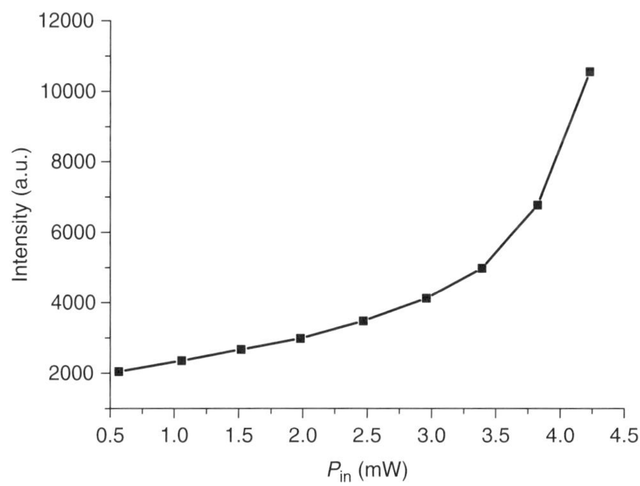

FIGURE 9.21 Evolution of the maximum intensity of the fluorescence spectrum ($\lambda = 590$ nm) versus input pump power for a constant fiber distance ($d = 500$ μm). The soliton regime starts around 3.5 mW, the signal increases quite sharply. *Source:* Reprinted from Reference 29. Copyright 2010, with permission from the American Institute of Physics.

indices, yielding to nonlinear propagation, especially self-waveguiding or quasispatial optical solitons. Different experiments were set up to characterize such confined propagation, the essential parameter being the induced index profile. In terms of applications, thermo-optical solitons are good candidates not only for optical routing but also for locally probing optical properties of material. Their strong nonlocality, associated with the diffusive properties of the thermo-optic effect, gives them good stability when they are coupled to optical fibers. The waveguide properties of the solitons were also exploited to enhance the recovery of a fluorescence signal from a mixture of dye and liquid crystals. Further investigations could be carried out in this direction, for example, concerning the luminescence of colloids of nanoparticles in NLC samples.

REFERENCES

1. J. P. Gordon, R. C. C. Leite, R. S. Moore, S. P. S. Porto, and J. R. Whinnery. Long transient effects in lasers with inserted liquid samples. *J. Appl. Phys.*, 36(1):3–8, 1965.
2. S. A. Akhmanov, D. P. Krindach, A. P. Sukhorukov, and R. V. Khokhlov. Nonlinear defocusing of lasers beams. *JETP Lett.*, 6(2):38–42, 1967.
3. S. A. Akhmanov, D. P. Krindach, A. V. Migulin, A. P. Sukhorukov, and R. V. Khokhlov. Thermal self actions of laser beams. *IEEE J. Quantum Electron.*, 4(10):568–575, 1968.
4. F. W. Perkins and E. J. Valeo. Thermal self-focusing of electromagnetic waves in plasmas. *Phys. Rev. Lett.*, 32(22):1234–1237, 1974.
5. P. Y. Wang, H. J. Zhang, and J. H. Dai. Laser-heating induced self-phase modulation, phase transition and bistability in nematic liquid crystals. *Opt. Lett.*, 13(6):479–481, 1988.
6. V. Volterra and E. Wiener-Avnear. CW thermal lens effect in thin layer of nematic liquid crystal. *Opt. Commun.*, 12(2):194–197, 1974.
7. V. Volterra and E. Wiener-Avnear. Laser induced isotropic holes in nematic liquid crystals. *Appl. Phys. A*, 6(2):257–259, 1975.
8. H. Ono, K. Takeda, and K. Fujiwara. Thermal lens produced in a nematic liquid crystal. *Appl. Spectrosc.*, 49(8):1189–1192, 1995.
9. M. Warenghem, J.-F. Blach, and J.-F. Henninot. Thermo-nematicon: an unnatural coexistence of solitons in liquid crystals? *J. Opt. Soc. Am. B*, 25(11):1882–1887, 2008.
10. J. Li, S. Gauzia, and S. T. Wu. High temperature-gradient refractive index liquid crystals. *Opt. Express*, 12(9):2002–2010, 2004.
11. M. Schadt and F. Muller. Class specific elastic, viscous, optical and dielectric properties of some nematic liquid crystals and correlation with their performances in twisted nematic displays. *Rev. Phys. Appl.*, 14:265–274, 1979.
12. R. Dabrowski, J. Dziaduszek, Z. Stolarz, and J. Kedzierski. Liquid crystalline material with low ordinary index. *J. Opt. Technol.*, 72(9):662–667, 2005.
13. I. C. Khoo and F. Simoni. *Physics of Liquid Crystalline Materials*. Gordon & Breach, London, pp. 365–392, 1991.
14. P. G. de Gennes, and J. Prost. *The Physics of Liquid Crystals*. 2nd edn. Oxford University Press, London, 1995.
15. J.-F. Henninot, M. Debailleul, F. Derrien, G. Abbate, and M. Warenghem. (2D+1) Spatial optical solitons in dye doped liquid crystals. *Synthetic Met.*, 124:9–13, 2001.
16. M. Warenghem, J.-F. Henninot, and G. Abbate. Non linearly induced self waveguiding structure in dye doped nematic liquid crystals confined in capillaries. *Opt. Express*, 2(12):483–490, 1998.

17. M. Warenghem, J.-F. Henninot, and G. Abbate. From Bulk Janossy effect to nonlinear self wave-guiding spatial soliton in dye-doped liquid crystal. *J. Nonlin. Opt. Phys.*, 8(3):341–360, 1999.

18. J.-F. Henninot, M. Debailleul, R. Asquini, A. d'Alessandro, and M. Warenghem. Self-waveguiding in an isotropic channel induced in dye-doped nematic liquid crystal and a bent self-waveguide. *J. Opt. A: Pure Appl. Opt.*, 6:315–323, 2004.

19. F. Derrien, J.-F. Henninot, M. Warenghem, and G. Abbate. Thermal (2D+1) spatial optical soliton in dye doped liquid crystal. *J. Opt. A: Pure Appl. Opt.*, 2:332–337, 2000.

20. M. Warenghem, J.-F. Henninot, F. Derrien, and G. Abbate. Thermal and orientational 2D+1 spatial optical solitons in dye doped liquid crystals. *Mol. Cryst. Liq. Cryst.*, 373:213–225, 2002.

21. M. Warenghem, J.-F. Henninot, and J.-F. Blach. Measurement, control and use of non-locality in some liquid crystal based devices. *Proc. SPIE*, 5947:27, 2005.

22. X. Hutsebaut, C. Cambournac, M. Haelterman, J. Beeckman, and K. Neyts. Measurement of the self-consistent waveguide of an accessible soliton. Nonlinear guided waves and their applications, Technical Digest, Toronto, paper TuC20, 2004.

23. H. Ono and Y. Harato. Characteristic of photothermal effects in guest-host liquid crystals by heat conduction analysis. *J. Opt. Soc. Am. B*, 16(12):2195–2201, 1999.

24. M. Warenghem and G. Joly. Liquid crystal refractive indices behavior versus wavelength and Temperature. *Mol. Cryst. Liq. Cryst.*, 207:205–218, 1991.

25. J.-F. Henninot, M. Debailleul, and M. Warenghem. Tunable non-locality of thermal non-linearity in dye doped nematic liquid crystal. *Mol. Cryst. Liq. Cryst.*, 375:631–640, 2002.

26. J.-F. Henninot, J.-F. Blach, and M. Warenghem. Experimental study of nonlocality of spatial optical soliton excited in nematic liquid crystal. *J. Opt. A: Pure Appl. Opt.*, 9:20–25, 2007.

27. J. Beeckman, K. Neyts, and M. Haelterman. Patterned electrode steering of nematicons. *J. Opt. A: Pure Appl. Opt.*, 8:214–220, 2006.

28. S. Serak, N. V. Tabiryan, M. Peccianti, and G. Assanto. Spatial soliton all-optical logic gates. *IEEE Photon. Technol. Lett.*, 18(12):1287–1289, 2006.

29. J.-F. Henninot, J.-F. Blach, and M. Warenghem. Enhancement of dye fluorescence recovery in nematic liquid crystals using a spatial optical soliton. *J. Appl. Phys.*, 107:113111, 2010.

10

Discrete Light Propagation in Arrays of Liquid Crystalline Waveguides

KATARZYNA A. RUTKOWSKA
Warsaw University of Technology, Warsaw, Poland
GAETANO ASSANTO
Nonlinear Optics and OptoElectronics Lab, University ROMA TRE, Rome, Italy
MIROSLAW A. KARPIERZ
Warsaw University of Technology, Warsaw, Poland

10.1 INTRODUCTION

In the past two decades, linear and nonlinear effects in discrete systems, in the form of photonic structures with a periodic modulation of refractive index (along one or more spatial directions), have become the subject of growing interest. The latter mainly stems from both the inherent physics and the variety of potential applications [1–3]. The unique properties of discrete structures originate from the discretization of light–matter interactions (both in linear and nonlinear regimes) taking place over a finite number of unit elements in a photonic lattice. The relevance of discrete systems has been supported by a wealth of materials and nonlinear mechanisms to be employed, including semiconductors [4–6], glasses [7, 8], polymers [9], photorefractives [10, 11], ferroelectrics [12], with a cubic and a quadratic nonlinear response, respectively. This has resulted in a wide range of novel and fascinating phenomena such as discrete blockers and routers [13], discrete diffraction and solitons [1, 14–21], gap solitons [22], discrete breathers [23], multiband solitons and their interactions [24, 25], linear and nonlinear Bloch oscillations (BO) [26–30], Landau–Zener (LZ) tunneling [30–33], and several other effects that have been successfully demonstrated. At the same time, some peculiar properties arising from lack of rotational symmetry in discrete systems have been pointed out toward all-optical switching and signal readdressing [9, 13, 15, 16, 34, 35].

Nematicons: Spatial Optical Solitons in Nematic Liquid Crystals, First Edition.
Edited by Gaetano Assanto.

Besides the solid-state materials mentioned earlier, a great deal of attention has been devoted to waveguide arrays in nematic liquid crystals (NLCs). The latter, together with those arrays fabricated in photorefractives [10, 11] and droplet arrays in Bose–Einstein condensates [36], represent an excellent example of systems with a tunable discreteness, that is, lattices where period, index contrast, and/or nonlinearity can be effectively adjusted. Moreover, NLCs, in addition to their peculiar optical properties (including huge birefringence and extended spectral transparency), have been proven to be excellent dielectrics for nonlinear optics. Their giant nonresonant reorientational nonlinearity results in low power requirements for spatial light localization, supporting spatial solitons in both one and two dimensions [37–42] as well as discrete solitons [42–47]. In addition, the significant electro-optic response of NLC allows for the creation of flexible and voltage-tunable architectures [43, 46–48], including those specifically designed for discrete optical phenomena. Using one-dimensional waveguide arrays in NLC, it is possible to observe the transition from discrete diffraction to spatial light localization, as well as all-optical signal switching and angular steering. As reviewed in this chapter, discrete diffraction and discrete solitons [42–47, 49], nonlinear angular steering [50], multiband vector breathers [51], LZ tunneling [52–54], and BO [55] have been investigated in voltage-tunable NLC-based discrete arrays for near-infrared (IR) light.

This chapter is organized as follows: after a brief generic description of discrete light propagation in one-dimensional photonic lattices (Section 10.2), we discuss the realization of such structures in NLC (Section 10.3); then we present various discrete optical phenomena in NLC waveguide arrays (Sections 10.4–10.8).

10.2 DISCRETE SYSTEMS

Stemming from the pioneering work by Somekh et al. on linear light tunneling in a one-dimensional (1D) optical lattice [56], by Jensen on the nonlinear directional coupler [57], and by Christodoulides et al. on infinite arrays of identical waveguides [17], intensive studies on linear and nonlinear discrete light propagation in 1D waveguide arrays have been pursued considering various materials and nonlinearities. A photonic structure with a periodically modulated refractive index (Fig. 10.1a) is equivalent to an array of parallel channel waveguides. Light propagation in such an array of waveguides can be modeled as a multicore directional coupler, representing the electromagnetic field as a superposition of modes guided in individual channels. An effect of light tunneling between neighboring waveguides is the characteristic propagation pattern, known as *discrete diffraction* and shown in Figure 10.1b. This pattern is typical of a linear regime, that is, at low optical powers, when the excitation increases in a self-focusing nonlinear medium, the change in (effective) refractive index and the induced wave vector mismatch between input and adjacent channels prevents resonant tunneling across the array, progressively isolating the excited channel(s) from the neighboring ones and leading to light trapping and propagation within the launch waveguide(s).

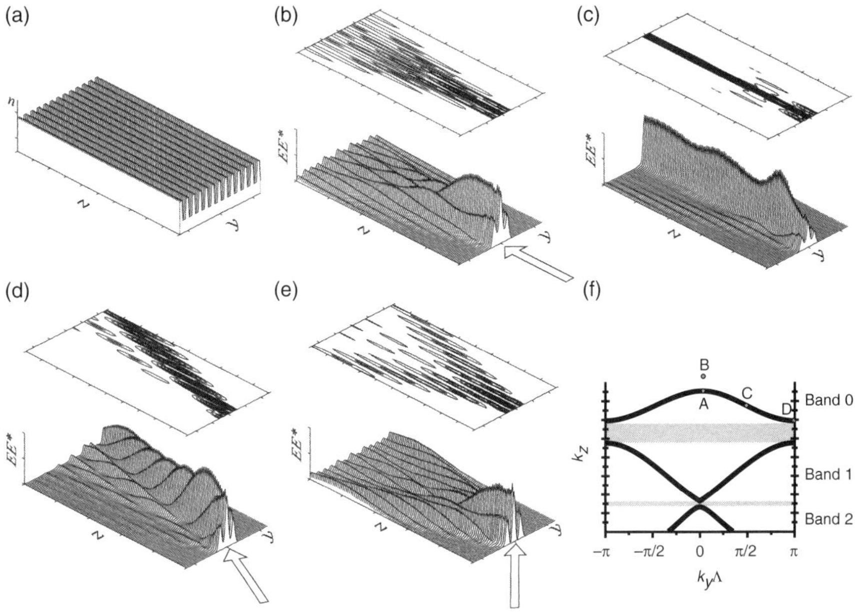

FIGURE 10.1 (a) Example of refractive index distribution in a one-dimensional waveguide array. (b) Discrete diffraction. (c) Discrete spatial soliton in the presence of a power-dependent refractive index (self-focusing). (d) Beam propagation when launched at the zero-diffraction angle. (e) Negative discrete diffraction. (f) Dispersion diagram for a discrete system, with the relation between the propagation constants k_z and $k_y\Lambda$, with k_y the transverse component of the wave vector (the Bloch wave number) and Λ the separation between waveguides (period of the refractive index modulation). Points A, B, C, and D correspond to the cases in panels (b), (c), (d), and (e), respectively.

Eventually, at high enough power, a discrete soliton, that is, a spatially localized wave, can propagate along the lattice without diffraction or spreading (Fig. 10.1c). The generation of such localized state can be viewed as resulting from the balance between discrete diffraction (due to waveguide-to-waveguide coupling) and the power-dependent phase mismatch between adjacent waveguides.

Owing to their periodicity in the transverse direction, discrete systems are characterized by a unique relation between the propagation constant k_z and the transverse component k_y of the wave vector. It is convenient to introduce the diffraction coefficient $D = -\partial^2 k_z/\partial k_y^2$, analogous to group velocity dispersion in dispersive media. The diffraction coefficient D can be zero in discrete systems at some excitation angles, that is, for some values of the transverse wave vector k_y: under such conditions, the input beam propagates across the array in a nearly nondiffracting manner, as displayed in Figure 10.1d [58]. This feature, combined with the generation of discrete solitons with zero transverse velocity at high powers, allows implementing all-optical signal steering. It is also possible to obtain negative diffraction (with $D<0$), as shown in Figure 10.1e. As in photonic crystals or in the case of electrons in periodic structures [59], a more accurate analysis requires the use of Bloch theory. The dispersion diagram (relating the propagation constant k_z to the Bloch wave number k_y) is composed of bands separated by gaps, wherein light propagation is forbidden (Fig. 10.1f). The study of modes in distinct bands together with the possibility of their interplay paves the way for the observation of gap solitons [22, 60], discrete breathers [23, 61], multiband vector solitons [24, 25, 51, 62, 63], and dynamic phenomena typical of quantum mechanical systems, including BO [26–30, 64] and LZ tunneling in "accelerated" lattices [30–33, 52–54].

10.3 WAVEGUIDE ARRAYS IN NEMATIC LIQUID CRYSTALS

The realization of a waveguide array in NLC is possible owing to their electro-optic response, which permits to transform an NLC layer into a 1D electromagnetic lattice. In order to introduce the required spatial distribution of the refractive index, a planar thin cell with a periodic set of fingered electrodes is used. Applying a voltage, the structure can be tuned from a planar waveguide supporting light confinement in one dimension (planar guiding in the NLC layer) to bidimensional confinement in a 1D array of identical weakly coupled channel waveguides. In such a way, a 1D photonic lattice is formed in NLC with refractive index modulation created and controlled by an external (low-frequency) electric field.

Figure 10.2 is a sketch of the structure. A thin (i.e., a few-micrometer thick) layer of the uniaxial NLC PCB (4′-*n*-pentyl-4-cyanobiphenyl) is sandwiched between two parallel glass plates, forming a standard planar waveguide with the NLC as the guiding layer. The thickness of the NLC layer is chosen in order to ensure transverse confinement along x in the absence of applied voltage (at the operating wavelength). The anchoring conditions imposed at the glass surfaces determine the planar alignment of the NLC within the cell, with the long axes of the molecules aligned along z, that is, along the direction of light propagation. The top glass plate is coated with a comb-shaped set of regularly spaced electrodes made of a 50 nm-thick transparent film of indium tin oxide (ITO), whereas the bottom one is uniformly covered by a grounded ITO film. The shape of the top electrode permits the application of a low-frequency (1 kHz) periodic voltage distribution [i.e., $V(y) = V(y + \Lambda)$] across the thickness d of the guiding NLC layer. It allows for the introduction of the spatially periodic refractive index modulation in NLC and leads to the formation of a one-dimensional dielectric lattice of constant period Λ. The lattice consists of identical channel waveguides providing light confinement in the transverse xy-plane and evanescently coupled to one another along y. In this chapter, we consider 50/50 duty cycle electrode patterns, that is, with equal finger widths and spacings. Samples with electrode periodicity ranging from 4 to 8 μm were tested experimentally, using NLC layer thicknesses up to a few micrometers in order to obtain evanescent-field coupling between channels (in the linear regime) and ensure a coupling length (millimeter scale) at the desired wavelength to be compatible with the experimental setup. This configuration provides considerable potentials for engineering the optical properties by tuning the sample parameters. As the confinement and the coupling strength of the bidimensional channel

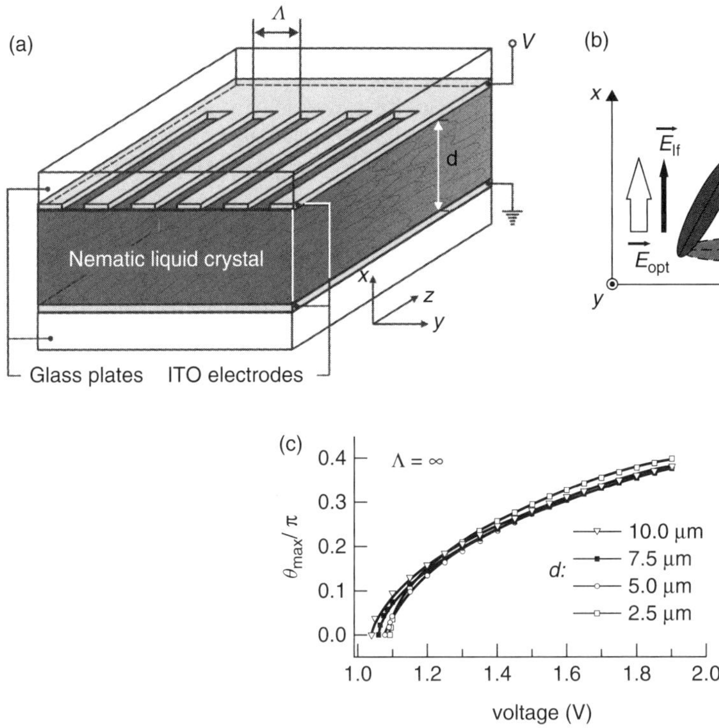

FIGURE 10.2 (a) Sketch of a one-dimensional NLC waveguide array with planar orientation. The top electrode is arranged in a periodic pattern of period Λ. (b) Sketch of the molecular reorientation (in the xz-plane) due to the electric field applied along x; the influence of low (E_{lf}) and optical frequency (E_{opt}) fields may allow for periodic modulation of refractive index and nonlinear optical effects, respectively. (c) Maximum reorientation versus applied bias for several cell thicknesses in a sample with uniform electrodes: reorientation starts after a threshold voltage, that is, the Freedericksz transition.

waveguides depend on the external bias, the main advantage of this structure is its voltage-controlled tunability, which, at variance with solid-state arrays, permits precise adjustments in both linear and nonlinear regime.

In order to obtain a specific periodic distribution of refractive index within the NLC layer, a low-frequency field E_{lf} is applied along x (Fig. 10.2b). In the ideal case of a planar configuration as considered here, the long axes of NLC molecules are uniformly oriented along z when no voltage is present. In NLCs with a positive low-frequency anisotropy, such as PCB with $\Delta\varepsilon_{lf} = 11.5$, the molecules tend to align their long axes (molecular director) along the external electric field (if the latter is above the Freedericksz threshold or in the presence of a small pretilt in xz), thereby changing the orientation angle θ (measured between z and the optic axis of uniaxial NLC) in xz and thus the optical properties of the NLC layer. Taking into account both electromagnetic and elastic forces and analyzing the total free energy [with term(s) depending on electric field(s) as well as terms describing the elastic contributions, as expressed through Frank's formalism], after its minimization, it is possible to obtain a differential (Euler–Lagrange) equation describing the relation between the orientation angle and the applied external field(s) (Chapter 1) [65, 66]. The low-frequency field distribution can be calculated from Maxwell's equations in the presence of the electric potential V. Through the fingered electrode, the periodic bias induces a spatial distribution of the effective refractive index for TM-like-polarized waves (i.e., with a dominant x-component of the electric field $E_{\rm opt}$, see Figure 10.2b), given by $n_{\rm eff} = n_{\rm o}n_{\rm e}/\sqrt{n_{\rm o}^2 \sin^2\theta + n_{\rm e}^2\cos^2\theta}$, with $n_{\rm o}$ and $n_{\rm e}$ being the ordinary and extraordinary NLC refractive indices, respectively.

The refractive indices of PCB [65–67] at room temperature and wavelength 1064 nm are 1.5158 (for ordinary polarization) and 1.6814 (for extraordinary polarization); they were used in all numerical simulations. At the same wavelength, the refractive index of BK7 glass is $n_{\rm g} = 1.5066$. In the single elastic constant approximation (i.e., equal for splay, bend, and twist deformations), the PCB elastic coefficient can be taken equal to $K = 3.2 \times 10^{-12}$ N. Figure 10.3a shows an example of electric field distribution within the 5 μm-thick NLC film biased by comb electrodes with 6-μm period. Figure 10.3c displays the resulting effective index distributions (for x-polarized light) in the xy-plane for three applied biases, whereas Figure 10.3b plots the results in the NLC midlayer, that is, for $x = 0$. The effective index increases more in the regions underneath the stripe electrodes, creating the channels and ensuring the lateral confinement of light. Moreover, Figure 10.3c demonstrates that a higher applied voltage induces a higher index modulation, provided the bias is low enough to avoid saturation effects; thus, the higher the bias the stronger is the light confinement.

Experimentally, the formation of a periodic lattice was tested with an optical polarizing (petrographic) microscope, placing the sample between crossed polarizers. Figure 10.3d shows typical results for the sample with $\Lambda = 8$ μm and $d = 6$ μm: at $V = 0$ V, no modulation in optical properties can be observed, whereas for biases above 0.7 V, a periodic modulation is impressed, with index contrast growing with bias.

The structure tunability, including waveguiding, can be achieved in four different ways: by changing the employed NLC and therefore its optical properties (including birefringence), by varying the thickness of the NLC layer d, by modifying the electrode geometry (i.e., the width of single fingers and their mutual separation) and, finally, by reorienting the optic axis via an external bias. After manufacturing the sample, the last approach is the only viable one. For example, the strength of discrete diffraction can be adjusted by the applied bias, as shown in Figure 10.4a: by increasing the voltage, the light transfer across adjacent channels diminishes because of higher and higher barriers separating the waveguides. In other words, light confinement gets stronger within each channel, causing a longer coupling length. However, saturation of reorientation tends to flatten the transverse index distribution when an excessive voltage is applied, weakening the spatial definition

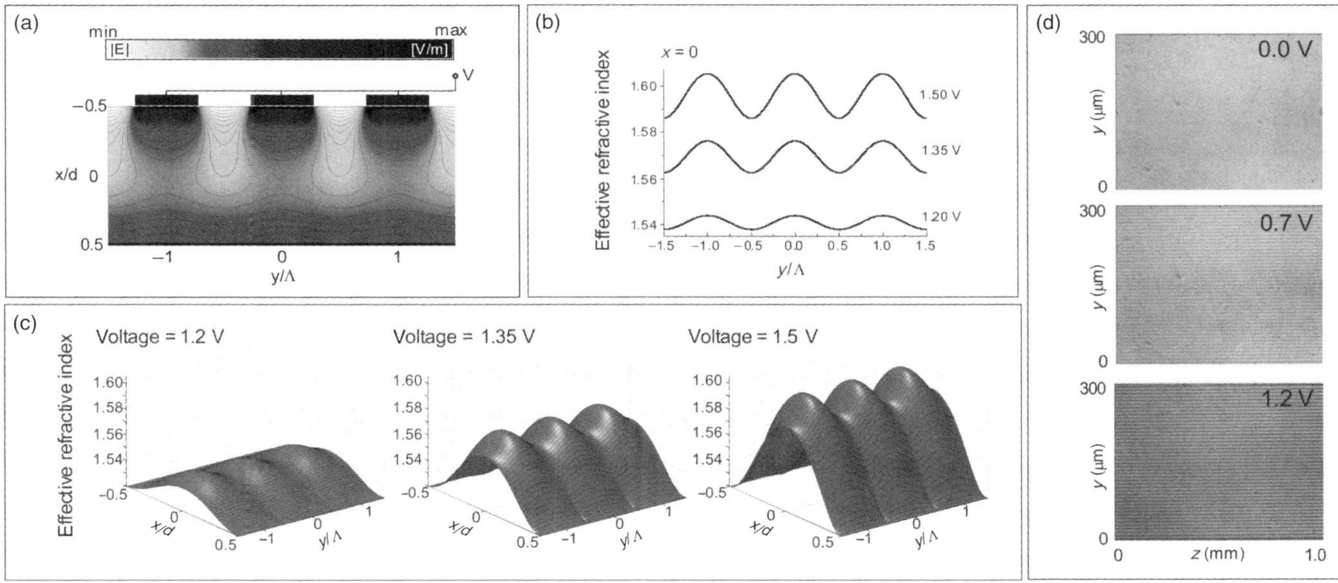

FIGURE 10.3 (a) Electric field distribution in a $d = 5$ µm NLC layer for a voltage $V = 1.35$ V applied to the periodic electrode ($\Lambda = 6$ µm). Effective index distribution (b) in the NLC midlayer and (c) in xy for three voltages. (d) Photographs of the NLC array (top view) placed under microscope in transmission mode through crossed polarizers. The refractive index modulation (contrast) clearly increases with bias.

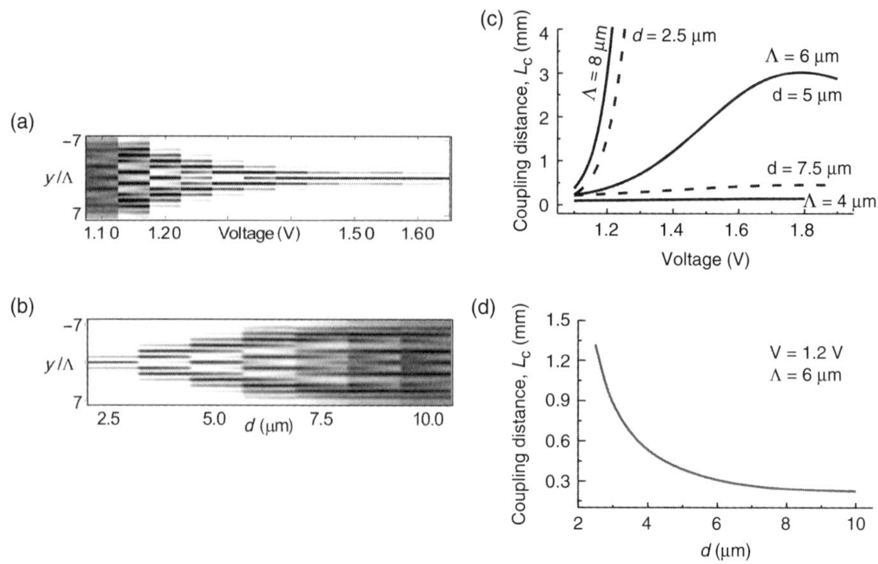

FIGURE 10.4 Coupling strength (at $\lambda = 1064$ nm) dependence on geometry and applied voltage. (a,b) Numerical simulations for a Gaussian beam with 2 μm initial waist propagating in the NLC array with 6 μm period (darker coloring corresponds to higher light intensity). Discrete diffraction for $d = 5$ μm (a) versus applied voltage in $z = 1$ mm and (b) versus cell thickness for $V = 1.2$ V. (c) Coupling distance versus bias for various geometric parameters of the array (at least one of the parameters is indicated next to each line. Dashed lines are used for $\Lambda = 5$ μm; and solid lines, for $d = 5$ μm). (d) Coupling distance versus NLC thickness for an electrode periodicity of 6 μm and an applied voltage $V = 1.2$ V.

of each guiding channel. We stress that, in the design of 1D waveguide arrays in NLC, a fundamental role is played by the degree of nonlocality, directly linked to d (Chapter 11) [68]. In fact, the intermolecular interactions are responsible for the diffusion of the induced orientation even in regions not under the electrodes (i.e., where the external electric field is negligible), smoothening the index distribution, and reducing the confinement; hence, in order to obtain narrow guides, it is necessary to employ a small thickness d. Figure 10.4b shows discrete diffraction versus layer thickness d for a fixed electrode pattern and bias.

As discrete diffraction relies on the coupling between neighboring channels, the optical properties of the array can be described by the coupling distance L_C, defined as the propagation distance after which the energy is transferred from one channel to the adjacent ones [69]. At any given wavelength, variations of the coupling distance can be computed with a two-dimensional mode solver. Figure 10.4c,d shows typical tuning curves: similar to one-dimensional coupled slab waveguides, the coupling distance depends exponentially on the refractive index contrast (here related to bias) and the spacing between waveguides. By reducing the NLC thickness, it is possible to better define the channel waveguides, resulting in increased coupling distances (Figure 10.4c,d). As expected, the latter can also be achieved by increasing the electrode periodicity (Fig. 10.4c). In actual devices, tunability is achieved by acting on the external voltage, as shown in Figure 10.4a,c. As previously discussed, for a fixed cell thickness, the coupling distance initially increases with voltage, until saturation becomes relevant and eventually produces a net decrease of coupling length.

Detailed numerical analyses have proved that an optimum ratio between electrode width and NLC thickness needs to be satisfied in order to maximize tunability in the range of applied biases. In actual experiments, several coupling distances have to be available because the observation window is significantly limited by scattering losses.

10.4 DISCRETE DIFFRACTION AND DISCRETE SOLITONS

Experiments were performed in 5- to 6- μm-thick NLC cells with comb periodicity of 6 μm. A beam from a Nd:YAG laser ($\lambda = 1064$ nm) was focused with a microscope objective to a spot of approximately 4 μm (in diameter) at the entrance of the sample. The beam propagation in the NLC layer was recorded by a high-resolution CCD camera imaging the photons scattered from the top of the sample in the x-direction. Figure 10.5a,b shows experimental results on light propagation for various input beam powers. At zero bias no light confinement occurs across y: the input beam diffracts in the transverse yz-plane (as in a homogeneous medium) (Fig. 10.5a); the beam diverges and progressively reduces its intensity, becoming hardly visible after a few Rayleigh ranges. Conversely, when the cell is biased, a periodic photonic lattice is formed and discrete diffraction takes place, with features depending on the applied voltage. Figure 10.5b shows discrete propagation of an input beam launched into a single channel for a voltage $V = 1.1$ V (sufficient to create an array of well-defined waveguides); owing to discrete diffraction, the energy is redistributed among the channels, as visible in Figure 10.5c graphing the transverse profiles of the beam in $z = 1.4$ mm (gray profiles are numerical simulations shown for comparison). In $z = 1.4$ mm, the excitation of power $P = 2$ mW in a single input waveguide is redistributed in 13 adjacent channels. It is to be noted that the divergence in this case is significantly lower than in the case of standard (continuous) diffraction in a uniform planar waveguide (i.e., no applied voltage), indicated by dashed lines. Figure 10.5b presents the nonlinear changes in discrete propagation for various input powers, with the corresponding transverse intensity profiles at $z = 1.4$ mm plotted in Figure 10.5c.

As anticipated, at high powers, the effective index of the input waveguide(s) is raised by light itself through reorientation. Such increase can counteract discrete diffraction via an induced phase mismatch between the excited waveguide(s) and the unexcited portion of the periodic structure. For sufficiently high intensities, the nonlinear response can overcome diffraction and support spatial self-localization, that is, self-confined light propagation in a channel subset of the array. The resulting discrete beam, with a transverse profile invariant along propagation, is called *discrete spatial soliton* [1–3]: discrete solitons may exist in the form of partially

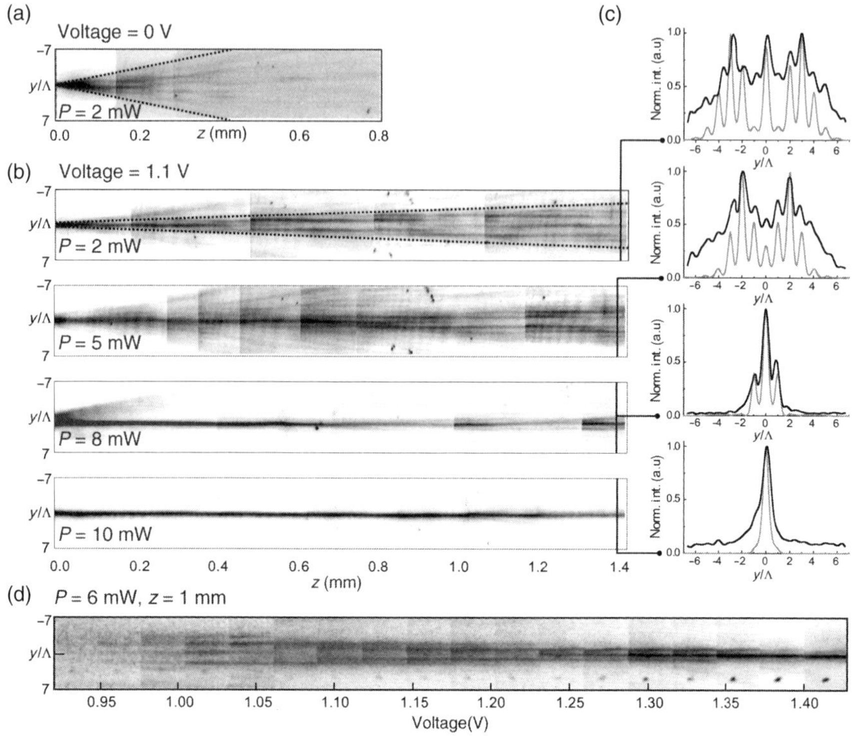

FIGURE 10.5 (a) Linear propagation (standard diffraction) in the homogeneous planar waveguide (zero voltage). (b) Discrete propagation ($V = 1.1$ V) for various input powers and a beam launched into a single channel. A discrete soliton forms at 10 mW. (c) Measured transverse intensity profiles (black lines) at $z = 1.4$ mm, corresponding to the cases in (b). The calculated profiles matching the experimental conditions are represented in gray. (d) Assembled photosequence versus applied voltage showing the diffraction of a 6 mW Nd:YAG beam in $z = 1$ mm.

delocalized or "wide" soliton with energy distributed among several channels, or in the form of localized or "narrow" solitons with all the energy confined in the input waveguide, as in the last panel of Figure 10.5b. Owing to the specific mechanism for light self-confinement in discrete systems, the power requirements for solitons are lower than in slab waveguides or bulk media. Specifically, discrete solitons in NLC result from the interplay between evanescent coupling, nonlinear response, and nonlocality (owing to intermolecular elastic forces). In this scenario, the giant nonlinear response, typical of molecular reorientation in NLC, allows observing discrete spatial solitons at a few milliWatts.

Besides self-effects controlled by light power, confinement can also be adjusted by bias. Figure 10.5d displays a sequence of photographs taken for an input power of 6 mW after a propagation length of 1 mm, showing transverse coupling and discrete propagation versus applied voltage: higher biases result in reduced angular beam divergence, with the coupling distance becoming longer. At low voltages, the beam energy distributes among several channels, whereas for voltages above 1.45 V, it is confined in a single waveguide.

10.5 OPTICAL MULTIBAND VECTOR BREATHERS

A multiband discrete breather originates from the superposition of (two or more) Floquet–Bloch (FB) modes of the lattice belonging to different bands of the dispersion diagram [59, 70]. Such modes are characterized by slightly different propagation constants. Analogous to vector solitons [71–74], discrete breathers stem from individual modes coupled via cross-phase modulation. The resulting two-component beam is capable of overcoming diffraction, even when constituent modes from each band (or gap) either diffract linearly or are localized into a discrete soliton [23]. In the case presented here, none of the individual modes is able to self-trap. The propagation constants of each mode as well as the breathing period can be effectively adjusted by tuning the array dispersion via the electro-optic reorientational response of NLC. With reference to the NLC array previously illustrated, it is possible to numerically compute the dispersion diagram of the linear eigenmodes versus applied bias by solving the differential equations governing molecular reorientation and then calculating the effective index for TM modes. To

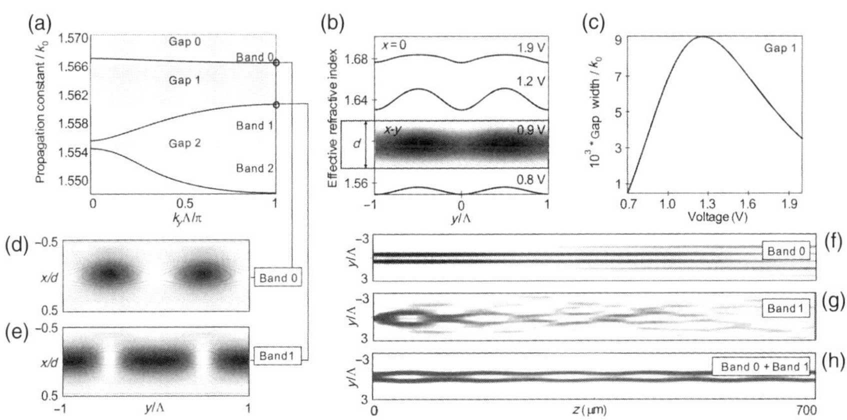

FIGURE 10.6 (a) Calculated bandgap diagram (i.e., dispersion relation) for an NLC waveguide array with $\Lambda = d = 6$ μm and for $V = 0.9$ V. (b) Effective index modulation at $x = 0$ (lines), as well as two-dimensional distribution (contour plot) at various voltages. (c) Calculated width of gap 1 versus bias. (d,e) Two-dimensional modal profiles (darker areas correspond to higher light intensity) for bands 0 and 1 and normalized Bloch wave vector $k_y\Lambda/\pi = 1$; the contour lines correspond to the refractive index distribution shown in Figure 10.3b. (f,g) Low power light propagation in the NLC array (for $V = 1$ V) computed with the beam propagation method (BPM) for a 0.2 mW Gaussian beam either (f) wide or (g) narrow injected between two adjacent waveguides. Modes belonging to (f) band 0 and (g) band 1 are excited, giving rise (h) to a vector-symmetric breather when superimposed.

this extent, the well-known FB analysis can be used [59]. Figure 10.6a displays the dispersion diagram calculated for the NLC array with $d = \Lambda = 6$ μm and an applied voltage $V = 0.9$ V: multiple (linear) FB bands (each corresponding to different FB modes) are separated by gaps (in gray), similar to solid-state physics. The full model based on FB theory yields discrete solitons belonging to the semi-infinite gap 0, as well as spatial gap solitons [75, 76] or FB solitons [23] involving two or more gaps. Figure 10.6d,e shows the intensity distribution of modes belonging to bands 0 and 1 for a transverse wave vector $k_y = \pi/\Lambda$. As in NLC arrays the effective index is easily modulated by the external bias (Fig. 10.6b), widths and location of the individual bands, and therefore the corresponding gaps, can be experimentally controlled and adjusted. Once again, the system is a tunable one-dimensional photonic lattice. Figure 10.6c shows the variations in the size of gap 1 versus bias. For voltages below 1.3 V, the effective index increases with bias and the gap widens; when the voltage exceeds 1.3 V, the nonlocal and saturating character of the reorientational response reduces the index modulation (Fig. 10.6b), with a reduction in gap width. A proper input superposition of FB linear modes can lead to light localization via cross-phase modulation, yielding periodic oscillations of the intensity distribution along propagation, that is, a multiband optical breather [51]. Figure 10.6f,h shows numerical simulations of the superposition of modes (which cannot self-trap when launched individually) from the two upper bands. FB modes belonging to band 0 (with intensity maxima in the channel regions) can be excited by a wide Gaussian beam launched between two channels (Fig. 10.6f). Conversely, modes of band 1 possess intensity maxima in between channels and can be excited by a narrow Gaussian beam centered between two waveguides (Fig. 10.6g). A symmetric breather, with spatially localized and periodically oscillating propagation (Fig. 10.6h) can be obtained when two Gaussian beams are injected together, combining their powers and interacting via cross-phase modulation.

In order to experimentally generate a symmetric breather, we used a narrow 1064-nm Gaussian beam (with a waist of about 5 μm) from a Nd:YAG laser. It was launched in the region between two neighboring channels to ensure the excitation of FB modes belonging to the first two bands. In such a configuration, a larger spatial overlap is obtained for the modes belonging to band 1 because of the chosen input profile. In fact, at low powers ($P = 0.2$ mW in Figure 10.7a) the diffraction pattern resembles the one in Figure 10.6g. As the power increases, light confinement can be enhanced via light-induced reorientation, thus making the excitation more effective toward modes from band 0 with respect to those from band 1. At high enough excitations (i.e., when the modes carry equal powers of 7 mW) a symmetric breather is obtained, with several oscillations clearly visible in Figure 10.7b; the on-axis light intensity (i.e., for $y = 0$) is displayed in Figure 10.7c. Linear propagation losses, mostly due to elastic Rayleigh scattering, are also apparent in the latter graph.

The breathing period dependence versus voltage was also measured, as in the sequence of photographs in Figure 10.7d. As the width of gap 1 can be effectively changed by bias (consistent with the variation of the effective index) as shown in Figure 10.6c, the oscillation period reaches its minimum for a voltage of about

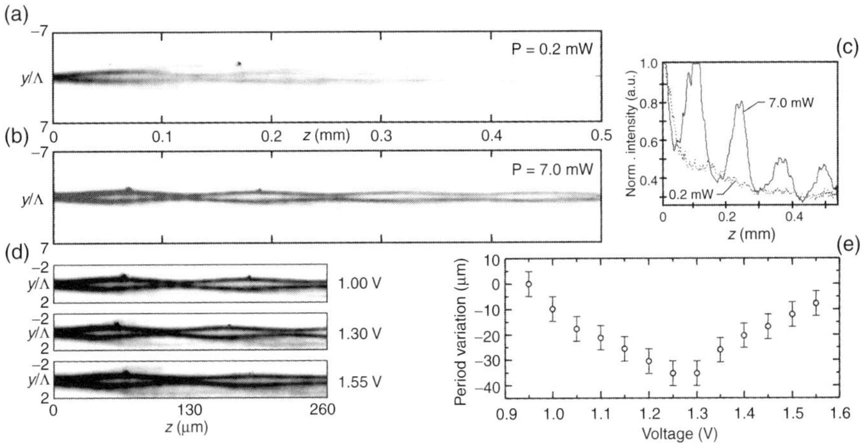

FIGURE 10.7 Experimental results on vector breathers in an NLC array with $\Lambda = 6$ μm and bias $V = 1$ V. (a) Low ($P = 0.2$ mW) and (b) high ($P = 7$ mW) power propagation of a Gaussian beam launched with waist 5 μm (along y) in between two channels. (c) On-axis light intensity distributions along propagation, corresponding to the cases in (a) and (b). (d) Photosequence showing the breather evolution for various applied voltages. (e) Measured variation in breathing period versus bias.

1.3 V, corresponding to the minimum width of gap 1. Above this value, the bandgap width reduces again, resulting in a longer period; such a trend is confirmed by experiments, as shown in Figure 10.7e.

10.6 NONLINEAR ANGULAR STEERING

Discrete systems can be employed for nonlinear beam steering because they lack rotational symmetry. At variance with continuous (isotropic) systems where all directions are equivalent, discrete arrays do not support solitary waves propagating aslant across the lattice. Whereas light beams can obliquely travel across a photonic array in the linear regime (with special directions defined by the minimum diffraction angle), discrete solitons can only propagate straight with zero transverse velocity: a discrete soliton travels along a waveguide channel nonlinearly detuned from the rest of the array.

Figure 10.8 demonstrates nonlinear beam steering in the NLC array with $\Lambda = 6$ μm, $d = 5 - 6$ μm and $V = 1.1$ V, launching a Gaussian beam of waist $w_y = 10$ μm at an angle of 1.90° with respect to z. At an input power of 3 mW, the tilted beam undergoes discrete diffraction; for $P = 11$ mW, light localizes in the input channel (Fig. 10.8a). The intensity cross sections in $z = 1.4$ mm illustrate an all-optically induced lateral shift of the beam (Fig. 10.8b). Nonlinear steering is also clearly visible in Figure 10.8c, displaying the output (i.e., for $z = 1.4$ mm) intensity distribution for various excitations in the range 3–11 mW. Such phenomenon has potential applications in all-optical switching, steering, and multiport routing [50, 77].

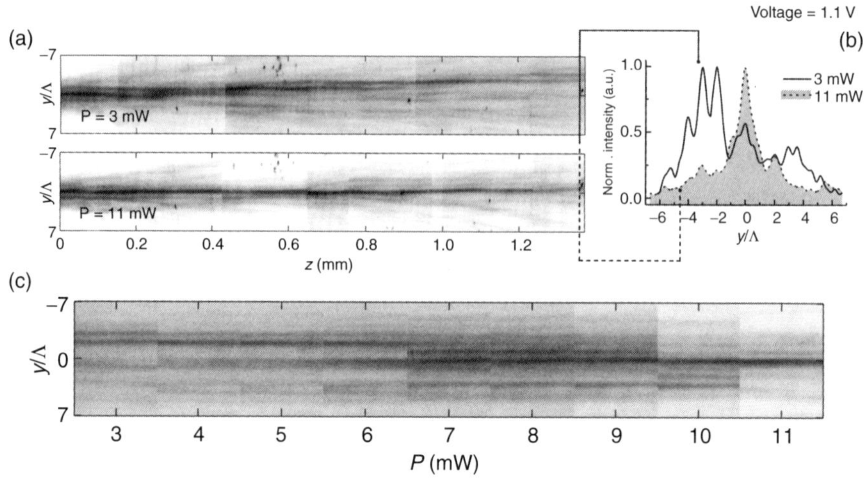

FIGURE 10.8 Experimental demonstration of nonlinear beam steering. (a) Minimum discrete diffraction of a 3 mW beam tilted by 1.90° and discrete soliton propagating straight for a power $P = 11$ mW. (b) Corresponding intensity profiles along y at $z = 1.4$ mm. (c) Assembled photosequence of discrete light propagation (at $z = 1.4$ mm) versus input power.

10.7 LANDAU–ZENER TUNNELING

LZ tunneling is another important topic related to optically discrete systems. LZ tunneling was introduced in the context of two-level quantum systems where the interaction (transition) between two different energy eigenstates can take place [78]; it may be generally associated with any quantized system accelerated by an external force [59]. In the nonadiabatic limit, the external perturbation makes the transition between different energy levels possible with a finite probability. In the case of optical lattices, light can tunnel between different FB bands when a refractive index gradient (acting as the perturbation that couples the eigenstates) is applied across the periodically modulated refractive index [31, 33]. In close analogy with semiconductor physics, a light beam, initially coupled to one single band, can transfer to another band if a nonadiabatic externally forced gradient is appreciable. The observable outcome is that the light intensity profile and/or the propagation direction will change due to tunneling. In the case discussed here, the gradient in refractive index responsible for the interband coupling is induced by light itself. Such all-optical acceleration in an NLC waveguide array is relatively easy. An implementation based on the thermal response (Chapter 9) is sketched in Figure 10.9a–c. An intense and wide Gaussian beam (the pump) is injected in the array, increasing the local temperature and changing in turn the refractive index; the pump beam determines a decrease in extraordinary refractive index via thermo-optic self-defocusing. In our structure (Fig. 10.2a), the thermal NLC response near the isotropic–nematic transition temperature (close to room temperature for PCB) induces a defocusing nonlinearity and made it dominant with respect

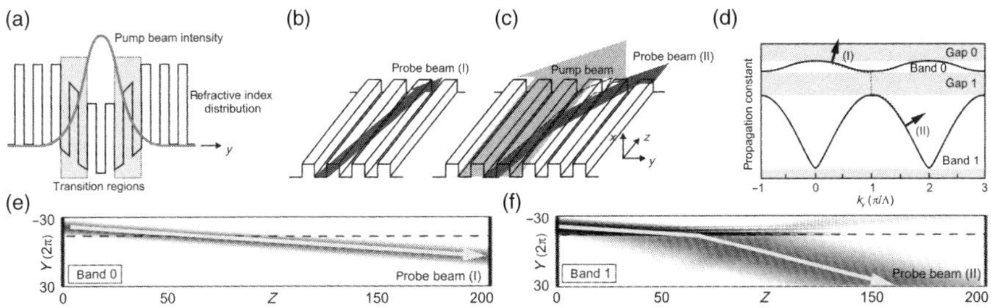

FIGURE 10.9 (a–c) An approach to Landau–Zener tunneling in NLC arrays. The refractive index gradient or acceleration is obtained via the thermal nonlinear response induced by the wide Gaussian beam pump. (d) Schematic drawing of the dispersion (bandgap) diagram. Black arrows, normal to the dispersion curve, indicate a maximum in propagation angle for modes of two different bands (here bands 0 and 1). The LZ transition between these two states is represented by a dashed line. (e) Numerically simulated propagation of FB modes of band 0 at the maximum transverse velocity, and (f) Landau–Zener tunneling between bands 0 and 1. In the transition region (indicated by a dashed line), energy is transferred to band 1 and the beam propagates at a larger angle than in band 0.

to the reorientational one by employing the bias to operate at large orientation angles.

A second low-intensity beam (the probe) is launched into the perturbed photonic structure shown in Figure 10.9a, exciting FB modes able to undergo LZ tunneling to other bands. When a negative index change is produced by the pump in the transition regions (Fig. 10.9a), the probe, which is initially a superposition of FB modes of the upper band (e.g., band 0 in Figure 10.9d), is transferred to the lower band (band 1) and propagates through the lattice at an angle (with z) larger than the initial one. This is due to the larger group velocity, as indicated by the normal to the dispersion curve (black arrow in Figure 10.9d). Thus, the probe beam undergoes all-optical angular steering and spatial switching via LZ tunneling.

Mathematical modeling of the LZ phenomenon requires the solution of three coupled differential equations, including the Euler–Lagrange equation for molecular reorientation due to bias, the wave equation for linear propagation of the probe, and an equation for the pump-driven temperature variation. After some approximations, the model can be recast as a single dimensionless Schrödinger-like equation [52] and, eventually, as the original Zener model [78] predicting tunneling between bands in solid-state physics. Typical numerical simulations are shown in Figure 10.9e,f.

One approach for experimentally verifying LZ tunneling in a lattice consists in measuring the shifts of the intensity profile maxima of the excited FB modes, as they are not the same from band to band [31]. Unfortunately, such an approach cannot be adopted when the lattice period and wavelength are comparable, as it is in the case of micrometric lattices illuminated by near-IR light. Alternatively, as each FB mode in a given band has a maximum propagation angle, LZ tunneling can be detected by observing the change in the probe propagation direction due to the pump-induced acceleration. The higher the band number, the higher the maximum propagation angle (Fig. 10.9d), as shown in Figure 10.9e,f.

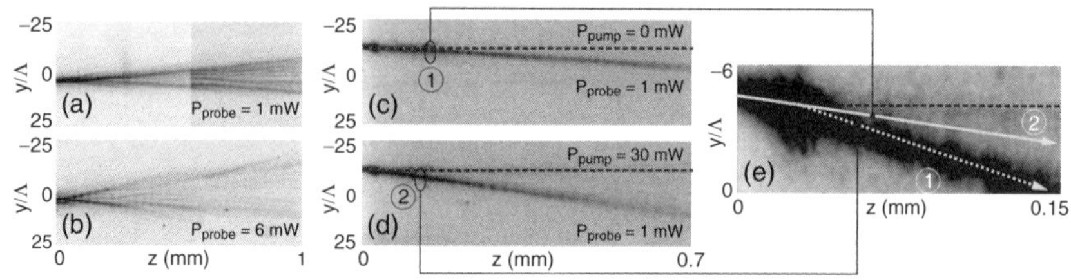

FIGURE 10.10 Experimental results in a cell with $\Lambda = 4$ μm and $d = 6$ μm, biased at 1.9 V. (a) Linear discrete diffraction and (b) self-defocusing of a beam in the array. Increasing the power also increases the beam spreading. (c–e) Landau–Zener tunneling of a 1mW-probe observed when a 30 mW pump was launched along z. (c) The probe, initially in band 0, is injected at the maximum transverse velocity. (d) After tunneling, the probe transfers to band 1 and propagates at a different angle. (e) The zooming-in of cases (c) and (d) underlines the increase in propagation angle after LZ tunneling.

The demonstration of LZ tunneling was performed by using a 1D NLC lattice with $\Lambda = 4$ μm and $d = 6$ μm. A probe at $\lambda = 1064$ nm and power up to 6 mW was used to test the self-defocusing nonlinear response, obtaining a more diffracting beam in yz (Fig. 10.10a,b). Then two mutually incoherent beams from a Nd:YAG laser at $\lambda = 1064$ nm were employed to act as pump and probe. An intense Gaussian pump of waist $w_y = 15$ μm (about four array periods and a corresponding Rayleigh distance of about 1 mm) generated the two transition regions with acceleration, ensuring no overlap with the probe after tunneling. The probe (with waist $w_y = 1.5\Lambda = 6$ μm and power 1 mW) was initially coupled to band 0 at its maximum transverse velocity (for zero pump), discretely diffracting. A pump power above 25 mW was sufficient for all-optical acceleration and band coupling; for instance, at 30 mW (Fig. 10.10d), the probe tunneled from first to second band, propagating at a larger angle in the yz-plane (Fig. 10.10d,e). The direction of the pump in Figure 10.10c–e is indicated by the black dashed line.

10.8 BLOCH OSCILLATIONS

BO were originally discussed for electrons in a periodic potential with an applied electric field [79]. In optics, BO can be obtained in waveguide arrays when a linear gradient is added to the transverse periodicity of the refractive index modulation [26–29]. As an effect of such index distribution, a light beam oscillates periodically in the transverse direction during propagation, retrieving initial position and profile after each full cycle. Linear and nonlinear optical BO were experimentally observed in AlGaAs waveguides [27], polymer arrays [28], and arrays of curved waveguides [64]. Similar effects are accessible in NLC arrays, as well. In order to introduce a linear refractive index gradient, one can apply a voltage linearly varying across the comb-shaped electrode or alter the stripe spacing/width (Fig. 10.11a,b). The simplest approach is to employ an NLC layer with thickness linearly changing with y [i.e., $d = d(y)$], as in Figure 10.11c.

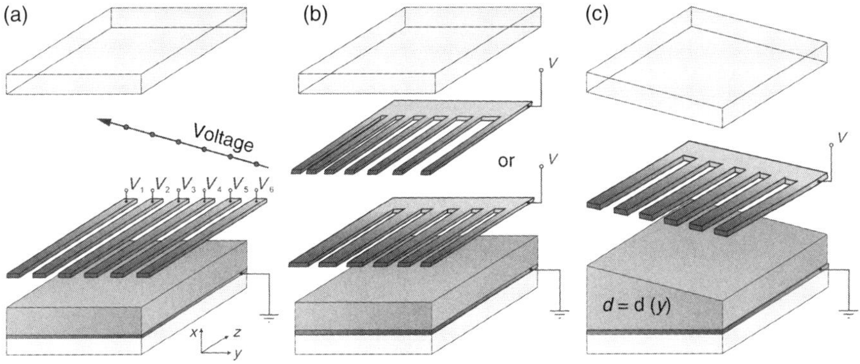

FIGURE 10.11 (a–c) Configurations for observing BO in NLC waveguide arrays. The third option (c) appears the simplest in terms of fabrication.

Figure 10.12a shows the linear index distribution obtained in the NLC midlayer for a thickness varying with a slope of $0°56'$ (thickness change of 1.64 μm for each 100 μm along y) and $d(y = 0) = 3.672$ μm, electrode period of 4 μm, and three applied voltages. Owing to the geometry of the guiding layer, a quasilinear gradient is added to the periodic modulation, resulting in slight differences in the parameters of neighboring channels.

Amplitude and period of the BO can be changed by acting on the array geometry or the bias, thus modifying the effective index distribution. The variation of BO period with voltage is shown in Figure 10.12b–e for a TM-polarized Gaussian beam of waist 10 μm (i.e., a broad beam) launched along z. The injected beam moves

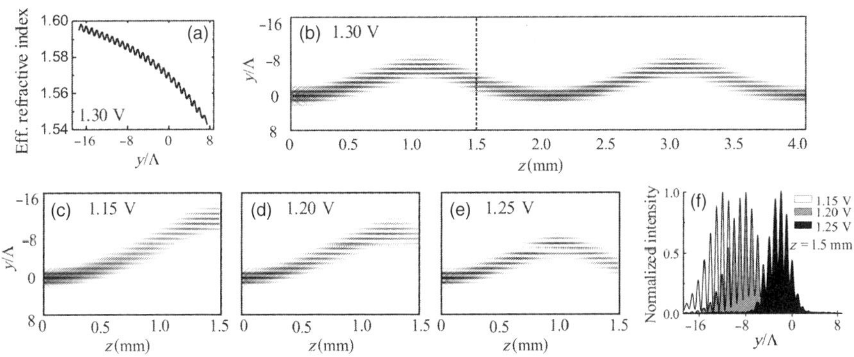

FIGURE 10.12 (a) Refractive index distribution in the middle of an NLC layer with varying thickness for 1.3 V. The periodicity of the ITO electrode is 4 μm and the thickness slope $\alpha = 56'$, with α being the angle between the top glass surface and y-direction. (b–e) BPM simulations of low power BO for a Gaussian beam with input waist of 10 μm and various voltages. In all cases, the input power is low enough to leave the index distribution unaltered by light. (f) The intensity profiles calculated in $z = 1.5$ mm illustrate the bias-controlled change in beam position.

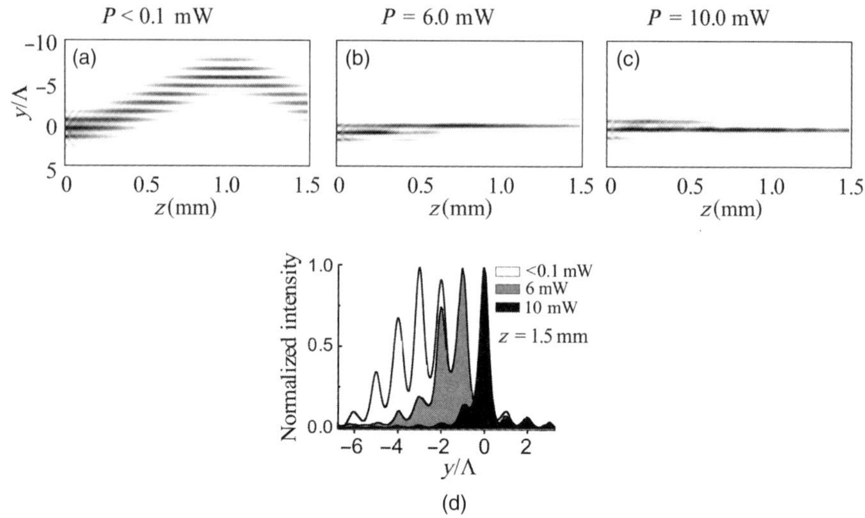

FIGURE 10.13 (a–c) Change of beam propagation with power. A Bloch oscillating beam as in (a) can be trapped in the input channel at high enough power as in (c). (d) Output (at $z = 1.5$ mm) intensity distributions corresponding to the previous cases (a–c).

toward the outermost waveguide (i.e., the channel where the index is higher) until it is Bragg-reflected at the edge of the first Brillouin zone in transformed space; after reflection, the beam moves back toward the innermost guide, undergoing another Bragg reflection. The beam oscillates periodically in the transverse space, with the initial profile reproduced after a complete BO (Fig. 10.12b). Routing can be achieved by acting on the applied voltage to control the output beam position (at $z = 1.5$ mm in Figure 10.12f).

Finally, changes in beam propagation direction can also be induced all-optically, similar to the previously demonstrated nonlinear steering. Figure 10.13a–c shows the calculated discrete propagation of a beam undergoing BO in the linear and nonlinear regime. Figure 10.13d shows the corresponding output intensity profiles (in $z = 1.5$ mm). As nonlinear effects become appreciable, the induced detuning alters the phase matching required for BO, until the input beam remains confined in the launch channel(s).

10.9 CONCLUSIONS

Spatial optical solitons in NLCs have their discrete counterpart in NLC waveguide arrays. One-dimensional arrays can be easily implemented and tuned taking advantage of the NLC electro-optic response, giving rise to a wealth of all-optical discrete phenomena at relatively low powers, from tunable diffraction to self-confinement, angular steering and switching, LZ tunneling, power-controlled BO, and vector breathers. The variety of effects and versatility of the material platform make discrete optical phenomena in NLCs and discrete nematicons an excellent approach toward all-optical signal readdressing and light reconfigurable circuits.

Acknowledgments

We are particularly grateful to Andrea Fratalocchi for his essential contributions to the work reported here.

REFERENCES

1. F. Lederer, S. Darmanyan, and A. Kobyakov. Discrete solitons, in *Spatial Solitons*, ed. S. Trillo and W. Torruellas. Wiley, New York, 2002.
2. F. Lederer, G. I. Stegeman, D. N. Christodoulides, G. Assanto, M. Segev, and Y. Silberberg. Discrete solitons in optics. *Phys. Rep.*, 463:1–126, 2008.
3. A. A. Sukhorukov, Y. S. Kivshar, H. S. Eisenberg, and Y. Silberberg. Spatial optical solitons in waveguide arrays. *IEEE J. Quantum Electron.*, 39:31–50, 2003.
4. R. Morandotti, H. S. Eisenberg, Y. Silberberg, M. Soler, and J. S. Aitchison. Self-focusing and defocusing in waveguide arrays. *Phys. Rev. Lett.*, 86:3296–3299, 2001.
5. H. S. Eisenberg, Y. Silberberg, R. Morandotti, A. R. Boyd, and J. S. Aitchison. Diffraction management. *Phys. Rev. Lett.*, 85:1863–1866, 2000.
6. P. Millar, J. S. Aitchison, J. U. Kang, G. I. Stegeman, A. Villeneuve, G. T. Kennedy, and W. Sibbett. Nonlinear waveguide arrays in AlGaAs. *J. Opt. Soc. Am. B*, 14:3224–3231, 1997.
7. D. Cheskis, S. Bar-Ad, R. Morandotti, J. S. Aitchison, H. S. Eisenberg, Y. Silberberg, and D. Ross. Strong spatiotemporal localization in a silica nonlinear waveguide array. *Phys. Rev. Lett.*, 91:223901, 2003.
8. T. Pertsch, U. Peschel, F. Lederer, J. Burghoff, M. Will, S. Nolte, and A. Tunnermann. Discrete diffraction in two-dimensional arrays of coupled waveguides in silica. *Opt. Lett.*, 29:468–470, 2004.
9. T. Pertsch, T. Zentgraf, U. Peschel, A. Brauer, and F. Lederer. Anomalous refraction and diffraction in discrete optical systems. *Phys. Rev. Lett.*, 88:093901, 2002.
10. N. K. Efremidis, S. Sears, D. N. Christodoulides, J. W. Fleischer, and M. Segev. Discrete solitons in photorefractive optically induced photonic lattices. *Phys. Rev. E*, 66:046602, 2002.
11. D. N. Christodoulides, F. Lederer, and Y. Silberberg. Discretizing light behaviour in linear and nonlinear waveguide lattices. *Nature*, 424:817–823, 2003.
12. R. Iwanow, R. Schiek, G. I. Stegeman, T. Pertsch, F. Lederer, Y. Min, and W. Sohler. Observation of discrete quadratic solitons. *Phys. Rev. Lett.*, 93:113902, 2004.
13. D. N. Christodulides and E. D. Eugenieva. Blocking and routing discrete solitons in two-dimensional networks of nonlinear waveguide arrays. *Phys. Rev. Lett.*, 87:233901, 2001.
14. R. Morandotti, U. Peschel, J. S. Aitchison, H. S. Eisenberg, and Y. Silberberg. Dynamics of discrete solitons in optical waveguide arrays. *Phys. Rev. Lett.*, 83:2726–2729, 1999.
15. W. Królikowski and Y. S. Kivshar. Soliton-based optical switching in waveguide arrays. *J. Opt. Soc. Am. B*, 13:876–887, 1996.

16. O. Bang and P. Miller. Exploiting discreteness for switching in waveguide arrays. *Opt. Lett.*, 21:1105–1107, 1996.
17. D. N. Christodoulides and R. I. Joseph. Discrete self focusing in nonlinear arrays of coupled waveguides. *Opt. Lett.*, 13:794–796, 1988.
18. C. Etrich, F. Lederer, B. Malomed, T. Peschel, and U. Peschel. Optical solitons in media with a quadratic nonlinearity, in *Progress in Optics*. Vol. 41. ed. E. Wolf, Elsevier Science Publishers, Amsterdam, pp. 483–568, 2000.
19. J. Meier, G. I. Stegeman, Y. Silberberg, R. Morandotti, and J. S. Aitchison. Nonlinear optical beam interactions in waveguide arrays. *Phys. Rev. Lett.*, 93:093903, 2004.
20. J. Fleischer, G. Bartal, O. Cohen, T. Schwartz, O. Manela, B. Freedman, M. Segev, H. Buljan, and N. Efremidis. Spatial photonics in nonlinear waveguide arrays. *Opt. Express*, 13:1780–1796, 2005.
21. J. W. Fleisher, M. Segev, N. K. Efremidis, and D. N. Christodoulides. Observation of two-dimensional discrete solitons in optically induced nonlinear photonic lattices. *Nature*, 422:147–150, 2003.
22. A. A. Sukhorukov and Y. S. Kivshar. Generation and stability of discrete gap solitons. *Opt. Lett.*, 28:2345–2347, 2003.
23. D. Mandelik, H. S. Eisenberg, Y. Silberberg, R. Morandotti, and J. S. Aitchison. Observation of mutually trapped multiband optical breathers in waveguide array. *Phys. Rev. Lett.*, 90:253902, 2003.
24. D. Mandelik, H. S. Eisenberg, Y. Silberberg, R. Morandotti, and J. S. Aitchison. Band-gap structure of waveguide arrays and excitation of Floquet-Bloch solitons. *Phys. Rev. Lett.*, 90:053902, 2003.
25. O. Cohen, T. Schwartz, J. W. Fleischer, M. Segev, and D. N. Christodoulides. Multiband vector lattice solitons. *Phys. Rev. Lett.*, 91:113901, 2003.
26. U. Peschel, T. Pertsch, and F. Lederer. Optical Bloch oscillations in waveguide arrays. *Opt. Lett.*, 23:1701–1703, 1998.
27. R. Morandotti, U. Peschel, J. S. Aitchison, H. S. Eisenberg, and Y. Silberberg. Experimental observation of linear and nonlinear optical Bloch oscillations. *Phys. Rev. Lett.*, 83:4756–4759, 1999.
28. T. Pertsch, P. Dannberg, W. Elflein, A. Bräuer, and F. Lederer. Optical Bloch oscillations in temperature tuned waveguide arrays. *Phys. Rev. Lett.*, 83:4752–4755, 1999.
29. G. Lenz, R. Parker, M. C. Wanke, and C. M. de Sterke. Dynamical localization and AC Bloch oscillations in periodic optical waveguide arrays. *Opt. Comm.*, 218:87–92, 2003.
30. H. Trompeter, W. Krolikowski, D. N. Neshev, A. S. Desyatnikov, A. A. Sukhorukov, Y. S. Kivshar, T. Pertsch, U. Peschel, and F. Lederer. Optical Bloch oscillations and Zener tunneling in two-dimensional photonic lattices. *Topical Meeting on Nonlinear Guided Waves and their Applications*, ThD1, OSA, Dresden, Germany, 2005.
31. R. Khomeriki and S. Ruffo. Nonadiabatic Landau-Zener tunneling in waveguide arrays with a step in the refractive index. *Phys. Rev. Lett.*, 94:113904, 2005.
32. B. Wu and Q. Niu. Nonlinear Landau-Zener tunneling. *Phys. Rev. A*, 61:023402, 2000.
33. V. V. Konotop, P. G. Kevrekidis, and M. Salerno. Landau-Zener tunneling of Bose-Einstein condensates in an optical lattice. *Phys. Rev. A*, 72:023611, 2005.
34. T. Pertsch, U. Peschel, and F. Lederer. All-optical switching in quadratically nonlinear waveguide arrays. *Opt. Lett.*, 28:102–104, 2003.

35. R. A. Vicencio, M. I. Molina, and Y. S. Kivshar. Switching of disrete optical solitons in engineered waveguide arrays. *Phys. Rev. E*, 90:026602, 2004.
36. Y. V. Kartashov, L. C. Crasovan, A. S. Zelenina, V. A. Vysloukh, A. Sanpera, M. Lewenstein, and L. Torner. Soliton eigenvalue control in optical lattices. *Phys. Rev. Lett.*, 93:143902, 2004.
37. M. Peccianti, G. Assanto, A. De Luca, C. Umeton, and I. C. Khoo. Electrically assisted self-confinement and waveguiding in Planar Nematic Liquid Crystal cells. *Appl. Phys. Lett.*, 77:7–9, 2000.
38. G. Assanto and M. Peccianti. Spatial solitons in nematic liquid crystals. *IEEE J. Quantum Electron.*, 39:13–21, 2003.
39. G. Assanto, M. Peccianti, and C. Conti. Nematicons: Optical spatial solitons in nematic liquid crystals. *Opt. Photon. News*, 14:44–48, 2003.
40. G. Assanto and M. A. Karpierz. Nematicons: Self-localized beams in nematic liquid crystals. *Liq. Cryst.*, 36:1161–1172, 2009.
41. M. Peccianti, C. Conti, G. Assanto, A. De Luca, and C. Umeton. Routing of highly anisotropic spatial solitons and modulational instability in liquid crystals. *Nature*, 432:733–737, 2004.
42. G. Assanto, M. Peccianti, and A. Fratalocchi. Spatial solitons in nematic liquid crystals: From bulk to discrete. *Opt. Express*, 15:5248–5259, 2007.
43. A. Fratalocchi, G. Assanto, K. A. Brzdakiewicz, and M. A. Karpierz. Discrete propagation and spatial solitons in nematic liquid crystals. *Opt. Lett.*, 29:1530–1532, 2004.
44. A. Fratalocchi, G. Assanto, K. A. Brzdakiewicz, and M. A. Karpierz. Discrete light propagation and self-trapping in liquid crystals. *Opt. Express*, 13:1808–1815, 2005.
45. A. Fratalocchi and G. Assanto. Discrete light localization in one-dimensional nonlinear lattices with arbitrary nonlocality. *Phys. Rev. E*, 72:066608, 2005.
46. K. A. Brzdakiewicz, M. A. Karpierz, A. Fratalocchi, and G. Assanto. Nematic liquid crystal waveguide arrays. *Opto-Electron. Rev.*, 13:107–112, 2005.
47. K. A. Brzdakiewicz, M. A. Karpierz, A. Fratalocchi, and G. Assanto. Discrete light propagation and self-localization in voltage-controlled arrays of channel waveguides in undoped nematic liquid crystals. *Mol. Cryst. Liq. Cryst.*, 453:191–202, 2006.
48. A. Fratalocchi and G. Assanto. Governing soliton splitting in one-dimensional lattices. *Phys. Rev. E*, 73:046603, 2006.
49. A. Fratalocchi, G. Assanto, and M. A. Karpierz. Discrete optics in liquid crystalline lattices. *IEEE LEOS Newsletter*, 2:11–16, 2006.
50. A. Fratalocchi, G. Assanto, K. A. Brzdakiewicz, and M. A. Karpierz. All-optical switching and beam steering in tunable waveguide arrays. *Appl. Phys. Lett.*, 86:051112, 2005.
51. A. Fratalocchi, G. Assanto, K. A. Brzdakiewicz, and M. A. Karpierz. Optical multiband vector breathers in tuneable waveguide arrays. *Opt. Lett.*, 30:174–176, 2005.
52. A. Fratalocchi and G. Assanto. All-optical Landau-Zener tunneling in waveguide arrays. *Opt. Express*, 14:2021–2026, 2006.
53. A. Fratalocchi, G. Assanto, K. A. Brzdakiewicz, and M. A. Karpierz. Optically-induced Zener tunneling in one dimensional lattices. *Opt. Lett.*, 31:790–792, 2006.
54. A. Fratalocchi, K. A. Rutkowska, M. A. Karpierz, and G. Assanto. Light induced angular steering via Floquet-Bloch band-tunnelling in one-dimensional liquid crystalline photonic lattices. *Opto-Electron. Rev.*, 15:210–214, 2007.

55. K. A. Brzdakiewicz, M. A. Karpierz, A. Fratalocchi, and G. Assanto. Bloch oscillations in nematic liquid crystals waveguide arrays. *Proc. SPIE*, 5949:178, 2005.
56. S. Somekh, E. Garmire, A. Yariv, H. L. Garvin, and R. G. Hunsperger. Channel optical waveguide directional couplers. *Appl. Phys. Lett.*, 22:46–48, 1973.
57. S. M. Jensen. The nonlinear coherent coupler. *IEEE J. Quantum Electron.*, 18:1580–1583, 1982.
58. S. Longhi. Discrete diffraction and shape-invariant beams in optical waveguide arrays. *Phys. Rev. A*, 79:033847, 2009.
59. K. Sakoda. *Optical Properties of Photonic Crystals*. Springer, Berlin, 2001.
60. D. Mandelik, R. Morandotti, J. S. Aitchison, and Y. Silberberg. Gap solitons in wave guide arrays. *Phys. Rev. Lett.*, 92:093904, 2004.
61. S. Flach and C. R. Willis. Discrete breathers. *Phys. Rep.*, 295:181–264, 1998.
62. A. A. Sukhorukov and Y. S. Kivshar. Multigap discrete vector solitons. *Phys. Rev. Lett.*, 91:113902, 2003.
63. A. S. Desyatnikov, E. A. Ostrovskaya, Y. S. Kivshar, and C. Denz. Composite band-gap Solitons in nonlinear optically induced lattices. *Phys. Rev. Lett.*, 91:153902, 2003.
64. G. Lenz, I. Talanina, and C. M. de Sterke. Bloch oscillations in an array of curved optical waveguides. *Phys. Rev. Lett.*, 83:963–967, 1999.
65. P. G. de Gennes and J. Prost. *The Physics of Liquid Crystals*, 2nd edn, Oxford University Press, London, 1993.
66. I. C. Khoo. *Liquid Crystals: Physical Properties and Nonlinear Optical Phenomena*. Wiley, New York, 1995.
67. D. A. Dumm, A. Fukuda, and G. R. Luckhurst. *Physical Properties of Liquid Crystals: Nematics*. Inspec, London, 2001.
68. A. Alberucci, A. Piccardi, M. Peccianti, M. Kaczmarek, and G. Assanto. Propagation of spatial optical solitons in a dielectric with adjustable nonlinearity. *Phys. Rev. A*, 82(2):023806, 2010.
69. A. Yariv. *Optical Electronics in Modern Communications*. Oxford Press, New York, 1997.
70. N. W. Ashcroft and N. D. Mermin. *Solid State Physics*. Saunders College Publishing, New York, 1976.
71. M. Mitchell, M. Segev, and D. N. Christodoulides. Observation of multihump multimode solitons. *Phys. Rev. Lett.*, 80:4657–4660, 1998.
72. S. Darmanyan, A. Kobyakov, E. Schmidt, and F. Lederer. Strongly localized vectorial modes in nonlinear waveguide arrays. *Phys. Rev. E*, 57:3520–3530, 1998.
73. J. Hudock, P. G. Kevrekidis, B. A. Malomed, and D. N. Christodoulides. Discrete vector solitons in two-dimensional nonlinear waveguide arrays: Solutions, stability, and dynamics. *Phys. Rev. E*, 67:056618, 2003.
74. J. Meier, J. Hudock, D. N. Christodoulides, G. Stegeman, Y. Silberberg, R. Morandotti, and J. S. Aitchison. Discrete vector solitons in Kerr Nonlinear waveguide arrays. *Phys. Rev. Lett.*, 91:143907, 2003.
75. J. W. Fleisher, T. Carmon, M. Segev, N. K. Efremidis, and D. N. Christodoulides. Observation of discrete solitons in optically induced real time waveguide arrays. *Phys. Rev. Lett.*, 90:023902, 2003.
76. J. Feng. Alternative scheme for studying gap solitons in an infinite periodic Kerr medium. *Opt. Lett.*, 18:1302–1304, 1993.

77. A. B. Aceves, C. De Angelis, T. Peschel, R. Muschall, F. Lederer, S. Trillo, and S. Wabnitz. Discrete self-trapping, soliton interactions, and beam steering in nonlinear waveguide arrays. *Phys. Rev. E*, 53:1172–1189, 1996.

78. C. Zener. Non-adiabatic crossing of energy levels. *Proc. R. Soc. London, Ser. A*, 137:696–702, 1932.

79. F. Bloch. Uber die quantenmechanik der elektronen in kristallgittern. *Z. Physik*, 52:555–600, 1928.

11

Power-Dependent Nematicon Self-Routing

ALESSANDRO ALBERUCCI, ARMANDO PICCARDI, AND GAETANO ASSANTO

Nonlinear Optics and OptoElectronics Lab, University ROMA TRE, Rome, Italy

11.1 INTRODUCTION

Most studies on reorientational nematicons were carried out in the limit of small nonlinear effects, an assumption often adopted in nonlinear optics [1]. Perturbational models in nonlinear optics are based on a nonlinear polarization $\boldsymbol{P}_{NL}$ much smaller than the linear counterpart $\boldsymbol{P}_L$, allowing the use of well-known techniques, for example, coupled-mode theory [2]. In the reorientational nonlinearity of nematic liquid crystals (NLC), perturbative approximations require light-induced variations in the director field to be negligible when compared with the molecular distribution in the absence of illumination. The first theoretical models on nematicons made extensive use of this hypothesis [3–5], making a drastic simplification of the problem. On the other hand, such models are inherently unable to describe a few phenomena expected on physical grounds, such as saturation of the nonlinearity and dependence of nematicon trajectory on power, the latter independent of excitation according to first-order perturbation theory. In particular, nematicon paths change with input power when the optical reorientation becomes comparable with the unperturbed director orientation angle, hence leading to an appreciable variation in soliton walk-off [6].

Nonlinear changes in walk-off are not the only potential cause of light self-steering in NLC. Other mechanisms are based on variations of the beam wave vector. Specifically, when suitable dopants are added to the NLC, the combination of a noncentrosymmetric response and beam nonparaxiality via a nonlinearity enhancement can give rise to a transverse force acting on the beam path [7]. Solitons in NLC can also self-deflect through interactions with the cell boundaries owing to the highly nonlocal response, breaking the translational symmetry, and inducing an index gradient responsible for bouncing off the interface(s) [8, 9].

Nematicons: Spatial Optical Solitons in Nematic Liquid Crystals, First Edition.
Edited by Gaetano Assanto.

Finally, the nonlinear Goos–Hänchen–like shift on total internal reflection (TIR) at graded interfaces makes nematicon trajectory dependent on power [10], and so does their interaction with finite-size linear inhomogeneities in director distribution [11], with a symmetry-breaking analogous to that mentioned earlier.

11.2 NEMATICONS: GOVERNING EQUATIONS

Let us consider an NLC sample with molecular director $\hat{n}$ lying in the plane yz owing to anchoring at the interfaces and to the excitation scheme; θ is the orientation angle of $\hat{n}$ with $\hat{z}$. In this geometry, ordinary and extraordinary (e-) waves are decoupled provided that the spatial spectrum of the beam is not too extended in reciprocal **k**-space. With reference to the extraordinary wave component and indicating with a superscript "(b)" quantities computed in correspondence to the intensity peak, we name H_x the magnetic field along x and set $H_x = A\exp(ik_0 n_e^{(b)} z)$, where A is the slowly varying envelope, k_0 is the vacuum wave number, $n_e^{(b)}$ is the carrier e-refractive index, and z is the direction of propagation. In the paraxial approximation $|\partial_z A| \ll |k_0 n_e^{(b)} A|$ and for a dielectric tensor that varies adiabatically across the transverse profile,[1] light propagation is governed by the Anisotropic NonLinear Schrödinger Equation (ANLSE) [12]

$$2ik_0 n_e^{(b)}\left(\frac{\partial A}{\partial z} + \tan\delta^{(b)}\frac{\partial A}{\partial y}\right) + D_y\frac{\partial^2 A}{\partial y^2} + \frac{\partial^2 A}{\partial x^2} + k_0^2\Delta n_e^2 A = 0, \tag{11.1}$$

where we introduced the dielectric tensor dependent quantities $n_e^2 = \epsilon_{yy} - \epsilon_{yz}^2/\epsilon_{zz}$ (the square of the extraordinary refractive index), $D_y = n_e^2/\epsilon_{zz}$ (the diffraction coefficient along y, in general different from one) and $\delta = \arctan(\epsilon_{yz}/\epsilon_{zz})$ (the walk-off angle); moreover, in Equation 11.1, we defined the perturbation of the extraordinary refractive index $\Delta n_e^2 = n_e^2 - (n_e^{(b)})^2$, containing both linear (elastic effects from boundaries or torque due to quasi-static electric or magnetic fields) and nonlinear (light-induced director reorientation or thermal effects) contributions, the latter responsible for self-focusing (defocusing).

In order to express the electric field corresponding to solutions of Equation 11.1, it is convenient to define the local reference system xts, obtained from xyz through a rotation by an angle $\delta^{(b)}$ around the x-axis; thus, $\hat{t} = \cos\delta^{(b)}\hat{y} - \sin\delta^{(b)}\hat{z}$ and $\hat{s} = \cos\delta^{(b)}\hat{z} + \sin\delta^{(b)}\hat{y}$, being $\hat{t}$ and $\hat{s}$ the directions parallel to the electric field and the Poynting vector, respectively. We stress that the reference system xts can change in propagation: our treatment holds valid only if these variations are adiabatic. Using Maxwell's equations, the nonnegligible electric field components are

$$E_t = \frac{1}{i\omega\epsilon_0 n_\epsilon^2\cos\delta^{(b)}}\left(\frac{\partial H}{\partial z} + \tan\delta^{(b)}\frac{\partial H}{\partial y}\right) - \frac{i\sin\delta^{(b)}}{\omega\epsilon_0\epsilon_{zz}}\frac{\partial H}{\partial y}, \tag{11.2}$$

[1]This implies that walk-off and diffraction can be assumed constant across the beam section; see Chapter 5 for details.

$$E_s = \frac{i \cos \delta^{(b)}}{\omega \epsilon_0 \epsilon_{zz}} \frac{\partial H}{\partial y}; \tag{11.3}$$

hence, for narrow beams, a longitudinal electric field component is present, with an odd (even) profile if H is even (odd), in agreement with the theory of nonparaxial optics developed by Lax, predicting the appearance of a longitudinal field as a first-order correction and longitudinal second-order derivatives at higher orders (neglected in Equation 11.1) [13–15]. It is worth mentioning that, in the limit $|\partial_z A| \ll |k_0 n_e^{(b)} A|$, we get $E_t \approx -Z_0/(n_e^{(b)} \cos \delta^{(b)}) H_x$, with Z_0 the impedance of vacuum.

To complete our model of nonlinear optical propagation in NLC, an equation governing the material response under the action of light is still missing. For a reorientational nonlinearity, the equation stemming from the equilibrium between optical and elastic torques can be found by minimizing the NLC free energy [16, 17], yielding the Euler–Lagrange equation[2]

$$\nabla^2 \theta + \gamma \left\{ (|E_t|^2 - |E_s|^2) \sin[2(\theta - \delta^{(b)})] + 2 \cos[2(\theta - \delta^{(b)})] Re\, (E_t E_s^*) \right\} = 0, \tag{11.4}$$

where $\gamma = \epsilon_0 \epsilon_a / (4K)$ is the interaction strength, with K the scalar Frank's elastic constant [16] and ϵ_a the optical anisotropy. We stress that the reorientational nonlinearity can only be focusing, both in positive and negative uniaxial NLC, because the optical torque acts in order to minimize the dielectric energy stored in the medium [16]. It is also apparent from Equation 11.4 that, for a given profile of the electromagnetic field, the reorientation depends on the normalized parameter γP, P being the beam power; furthermore, for NLC with equal birefringence ϵ_a, reorientational effects scale with the ratio P/K.

The system composed of Equations 11.1–11.4 rules the propagation of reorientational solitons in a rather general way: we can distinguish three different regimes of nematicon evolution, depending on the injected beam power. To this extent, we assume a uniform director distribution $\theta = \theta_0$ in the absence of light (i.e., when $A = 0$) and write $\theta = \theta_0 + \psi$ when $A \neq 0$, ψ representing the all-optical perturbation of the director. In the linear regime, that is, at very low powers, ψ is negligible in each term of Equation 11.1, including self-focusing; thus, the beam spreads according to the diffraction coefficient D_y, with energy flux at an angle $\delta(\theta) = \delta(\theta)|_{\theta_0} = \arctan\left[\epsilon_a \sin 2\theta / \left(\epsilon_a + 2\epsilon_\perp + \epsilon_a \cos 2\theta\right)\right]\big|_{\theta_0}$ with respect to $\hat{z}$ (Fig. 11.1a). As power increases, the all-optical response Δn_e^2 changes, with beam self-trapping but no power-dependent trajectory: this is the nonlinear perturbative regime distinguished by $\psi \ll \theta_0$ (optical reorientation negligible with respect to the unperturbed director distribution), corresponding to first-order nonlinear perturbation (Fig. 11.1b). As power increases further, ψ becomes comparable with θ_0 and the all-optical term affects the trajectory of the self-confined beam: this is the nonperturbative nonlinear regime, where both beam waist and trajectory depend

[2]For thermo-optic effects, we should resort to the heat transmission equation; see Chapter 9.

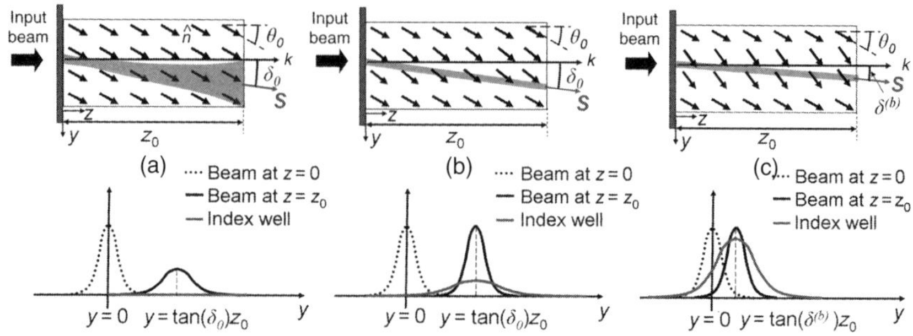

FIGURE 11.1 Light beam propagation in the plane yz. (a) Linear regime. (b) Nonlinear perturbative regime and (c) highly nonlinear regime.

on light power (Fig. 11.1c). We emphasize that the ranges identifying the three regimes depend on θ_0, as discussed in the following sections.

11.2.1 Perturbative Regime

In this section, we address some general properties of nematicons in the perturbative nonlinear regime, identifying two scalar parameters able to measure nonlinearity and nonlinear walk-off as functions of the initial orientation θ_0.

In the perturbative limit, the nonlinear response is not high enough to sustain narrow nematicons where E_s becomes appreciable; as $\psi \ll \theta_0$, we can substitute $\theta \to \theta_0$ in the leftover sinusoidal term of Equation 11.4, obtaining:

$$\nabla^2 \psi + \gamma \kappa^2 \sin\left[2\left(\theta_0 - \delta_0\right)\right] |A|^2 = 0, \tag{11.5}$$

where we set $\kappa = Z_0/(n_e^{(b)} \cos \delta_0)$. Equation 11.5 is a linear Poisson equation, linking the nonlinear perturbation ψ to the beam intensity (proportional to $\left|E_t\right|^2$), which can be solved by using the Green function formalism; incidentally, such feature demonstrates the equivalence between thermal and reorientational nonlinearities in NLC in this regime [8]. After setting $A = \sqrt{P}v$, we get

$$\psi = \gamma \kappa^2 P \sin\left[2\left(\theta_0 - \delta_0\right)\right] \iiint G\left(x, y, z, x', y', z'\right) |v|^2 \mathrm{d}x' \mathrm{d}y' \mathrm{d}z', \tag{11.6}$$

with G the Green function, which takes different shapes according to the examined geometry. We stress that G rules the nonlocality: for an infinitely extended medium, the nonlinear perturbation ψ is infinitely extended because ψ resembles the electrostatic potential of a pointlike charge in empty space [18, 19]. In actual experiments, the sample size is obviously finite, implying a different G with extension defined by the closest boundaries [8, 9].

For small nonlinear perturbations we can linearize Δn_e^2, obtaining $\Delta n_e^2 \approx n_2\left(\theta_0\right) P g\left(x, y, z\right)$ (the perturbation is proportional to power, as expected), with

$g = \iiint G\left(x, y, z, x', y', z'\right) |v|^2 \mathrm{d}x'\mathrm{d}y'\mathrm{d}z'$ and having defined an effective (nonlocal) Kerr coefficient [12]

$$n_2\left(\theta_0\right) = 2\gamma\kappa^2 \sin\left[2\left(\theta_0 - \delta_0\right)\right] n_e^2\left(\theta_0\right) \tan\delta_0. \tag{11.7}$$

Recalling the definition of γ and considering that $\tan\delta_0 \propto \epsilon_a$ for realistic values of the anisotropy, we find that the maximum of n_2 versus θ_0 is proportional to ϵ_a^2 for any given K. Thus, the nonlinear behavior of an NLC mixture depends quadratically on ϵ_a, because of the combined effects of a stronger optical torque and a deeper index well for a fixed perturbation ψ, whereas it is inversely proportional to the elastic constant because larger K imply stronger molecular interactions counteracting the light-induced rotation. Figure 11.2a shows n_2 versus θ_0 for a fixed γ: all curves, each of them corresponding to a different birefringence ϵ_a, show a relative maximum occurring very close to $\theta_0 = 45^\circ$ for small anisotropies, moving toward higher θ_0 for larger ϵ_a. As expected from Equation 11.7, the walk-off $\tan\delta_0$ displays a similar behavior (Fig. 11.2b). Having addressed the nematicon waist with the parameter n_2, we define a scalar figure c_δ to describe the nematicon trajectory in the perturbative limit; it is $c_\delta = d\delta^{(b)}/dP$ calculated in $\theta = \theta_0$. From Equation 11.6 it is straightforward to derive

$$c_\delta = \left.\frac{\mathrm{d}\delta^{(b)}}{\mathrm{d}P}\right|_{\theta_0} = \left.\frac{\mathrm{d}\psi}{\mathrm{d}P}\frac{\mathrm{d}\delta}{\mathrm{d}\theta}\right|_{\theta_0} = \gamma\kappa^2 \max(g) \sin\left[2\left(\theta_0 - \delta_0\right)\right] \left.\frac{\mathrm{d}\delta}{\mathrm{d}\theta}\right|_{\theta_0}, \tag{11.8}$$

where

$$c_\theta = \left.\frac{\mathrm{d}\delta}{\mathrm{d}\theta}\right|_{\theta_0} = \frac{2\epsilon_a\left[\epsilon_a + \left(\epsilon_a + 2\epsilon_\perp\right)\cos 2\theta_0\right]}{\left(\epsilon_a + 2\epsilon_\perp + \epsilon_a \cos 2\theta_0\right)^2 + \epsilon_a^2 \sin^2 2\theta_0}. \tag{11.9}$$

Figure 11.2c shows c_δ and c_θ for the commercial NLC E7: the nonlinear variations in walk-off depend on the trade-off between the sensitivity of δ versus θ (parameter c_θ), with absolute maxima around $\theta_0 = 0^\circ$ or $\theta_0 = 90^\circ$, and the growth rate

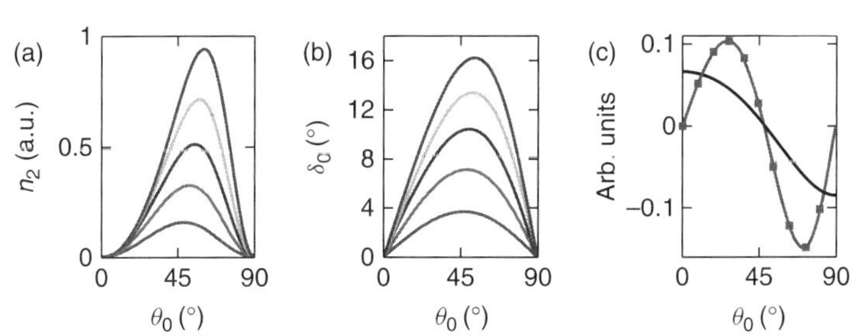

FIGURE 11.2 (a) Nonlinear effective coefficient n_2 versus θ_0 for $n_\perp = 1.5$ and $n_{||}$ from 1.6 (bottom line) to 2 (top line) for a constant γ. (b) Walk-off angle for $n_\perp = 1.5$ and $n_{||}$ from 1.6 (bottom line) to 2 (top line); the NLC mixture E7 corresponds to $n_\perp = 1.5$ and $n_{||} = 1.7$ in the near-infrared. (c) Graphs of c_θ (solid line) and c_δ versus θ_0 (solid line with symbols).

of θ versus power P, maximum for $\theta \approx 45^\circ$ and nearly zero around $\theta_0 = 0^\circ$ or $\theta_0 = 90^\circ$. As a consequence, c_δ is positive for $\theta_0 < 48^\circ$, with a relative maximum for $\theta_0 \approx 26^\circ$; it is negative for $\theta_0 > 48^\circ$ and undergoes a relative minimum for $\theta_0 \approx 71^\circ$.

11.2.2 Highly Nonlinear Regime

Let us consider Equation 11.4 assuming $E_s = 0$ for simplicity; hence, $\nabla^2\theta + \gamma\kappa^2 \sin[2(\theta - \delta)]|A|^2 = 0$. For the sake of generality, we will take θ_0 to be a space-dependent function, as experimentally obtainable, for example, by defining nonuniform anchoring at the boundary (Section 5.3.2) or using patterned electrodes for the application of a low-frequency electric field (Section 5.4) [20]. When ψ becomes comparable with θ_0, it is no longer possible to neglect the all-optical perturbation in the sine term, that is, the integral solution Equation 11.6 can be recast as a nonlinear integral equation

$$\psi = \gamma\kappa^2 P \iiint \sin\left[2\left(\theta_0 + \psi - \delta|_{\theta_0+\psi^{(b)}}\right)\right] G|v|^2 \mathrm{d}x'\mathrm{d}y'\mathrm{d}z'. \tag{11.10}$$

Setting $\epsilon U = |A|^2$ (ϵ is a smallness parameter) and $\theta = \sum_{n=0}^{\infty} \epsilon^n \theta_n$, a standard perturbation approach yields [8]

$$\begin{aligned}
\epsilon^0 &: \nabla^2\theta_0 = 0 \\
\epsilon^1 &: \nabla^2\theta_1 + \gamma\kappa^2 \sin\left[2(\theta_0 - \delta_0)\right] U = 0 \\
\epsilon^2 &: \nabla^2\theta_2 + 2\gamma\kappa^2 \cos\left[2(\theta_0 - \delta_0)\right] \theta_1 U = 0 \\
\epsilon^3 &: \nabla^2\theta_3 + 2\gamma\kappa^2 \left\{\cos\left[2(\theta_0 - \delta_0)\right] \theta_2 U - 2\sin\left[2(\theta_0 - \delta_0)\right] \theta_1^2 U\right\} = 0 \\
&\vdots,
\end{aligned} \tag{11.11}$$

where we neglected all derivatives of δ with respect to θ. Equations 11.11 are an infinite set of Poisson-like equations in agreement with the Born approximation, with a forcing term depending on the perturbation U (equal to $|A|^2$ in the limit $\epsilon \to 1$) and on the lower-order solutions. At order ϵ^0, we get back the equation governing the director distribution in the absence of light. At order ϵ^1, we retrieve Equation 11.5. Recalling that $A = \sqrt{P}v$, in the highly nonlinear regime, the nonlinear effect ψ is no longer directly proportional to power P owing to saturation in reorientation (i.e., a complicate power series with alternating signs).

Let us now turn back to the case of a homogeneous θ_0 and discuss these results with reference to nematicons. In Section 11.2, we defined the transition from perturbative to highly nonlinear regimes referring to the soliton trajectory, that is, comparing nonlinear changes in walk-off with its linear value δ_0. From Equation 11.8 and Figure 11.2c, it is clear that the power P_{tr} corresponding to such transition depends on θ_0 and is inversely proportional to c_δ; hence, P_{tr} is maximum for θ_0 close to 45°. Noteworthy, the higher-order terms in Equation 11.11 also affect the nematicon waist, inducing changes in the effective n_2. The net result consists in a saturation of the nonlinearity, implying a nonmonotonic trend of the nematicon existence curve in the plane *waist-power* [21], at variance with the perturbative theories predicting a soliton waist always decreasing with power [3, 5, 22].

11.2.3 Simplified (1+1)D Model in a Planar Cell

Hereafter, we focus on homogeneous planar cells, with the shortest dimension (thickness L) along x and parallel interfaces in $x = 0$ and $x = L$. The Green function defined in Equation 11.6 reads

$$G(x, x', \rho) = \frac{1}{2\pi} \sum_{n=1}^{\infty} K_0(\rho/L_n) \sin\left(\pi n \frac{x}{L}\right) \sin\left(\pi n \frac{x'}{L}\right) \tag{11.12}$$

where we set $\rho = \sqrt{(y - y')^2 + (z - z')^2}$ and $L_n = L/(n\pi)$; K_0 is the zeroth-order modified Bessel function of the second kind. In the framework of perturbative theories, for the sake of computational savings, in Equation 11.4 the second derivative along the Poynting vector can be neglected (this is rigorous for a soliton in bulk dielectric). This leads to several considerations: first, an a priori knowledge of the propagation direction is needed (an unknown variable in the highly nonlinear regime); second, the assumption of a zero longitudinal second derivative is equivalent to consider the medium local in this direction, an unphysical hypothesis because the Green function is radially symmetric in the plane yz (Eq. 11.12), with appreciable differences between the two models when the breathing period becomes very small; third, losses due to Rayleigh scattering are unavoidable in actual NLC samples, yielding a drop in beam power versus propagation and a weakening of the nonlinear effect, the latter resulting in a curved trajectory (via changes in $\delta^{(b)}$) as well as longitudinal variations of the refractive index Δn_e^2. Therefore, the longitudinal derivative needs to be included in the reorientational equation in order to account for all the properties of the original system (Equations 11.1 and 11.4), that is, nonlocality, self-steering and self-confinement, and their dependence on θ_0 [12].

Numerical simulations of the full 3D model composed by Equations 11.1 and 11.4 can be carried out with standard solvers relying on iterative algorithms; the magnetic field is computed by solving Equation 11.1 (to this extent, the splitting operator can be used by choosing the Crank–Nicolson algorithm for computing diffraction); the electric field components E_t and E_s are calculated from Equations 11.2 and 11.3; finally, the substitution of E_t and E_s into Equation 11.4 allows computing the director distribution (for instance, with an over relaxed Gauss–Seidel scheme), providing the dielectric parameters for the solution of Equation 11.1. In this procedure, the relaxation code in a 3D geometry is the bottleneck on computational time.

A simplified (1 + 1)D model can be obtained by writing the all-optical perturbation in a Fourier series along x

$$\psi(x, y, z) = \sum_{n=1}^{\infty} \psi_n(y, z) \sin\left(\pi n \frac{x}{L}\right). \tag{11.13}$$

Substituting Equation 11.13 into Equation 11.4, neglecting E_s for simplicity (the generalization including the longitudinal field is straightforward) and considering a highly nonlocal response, we obtain

$$\nabla_{yz}^2 \psi_n - \left(\frac{\pi n}{L}\right)^2 \psi_n + \gamma \sin\left[2(\theta_0 + \psi - \delta^{(b)})\right]\Big|_{x_b} I_n(y, z) = 0, \tag{11.14}$$

for every positive integer n, having exploited the orthogonality of the *sine* functions and having defined $I_n = (2/L) \int_0^L |E_t|^2 \sin(\pi n x/L) dx$. Equation 11.14 is an infinite set of Yukawa equations, with the screening length in the nth equation given by $L/(\pi n)$ [12].

By setting $A(x, y, z) = X(x, z)u(y, z)$ (the maximum of X is one), we get $I_n = \kappa^2 I_n^x(z)\, |u|^2$, where $I_n^x = (2/L) \int_0^L |X|^2 \sin(\pi n x/L)\, dx$. If we consider symmetric beams with respect to the cell midplane $x = L/2$ ($|X(a + L/2, z)|^2 = |X(a - L/2, z)|^2$, with $|a| < L/2$), light will propagate in this plane as an equal repulsion from the two parallel boundaries (placed in $x = 0$ and $x = L$) balances

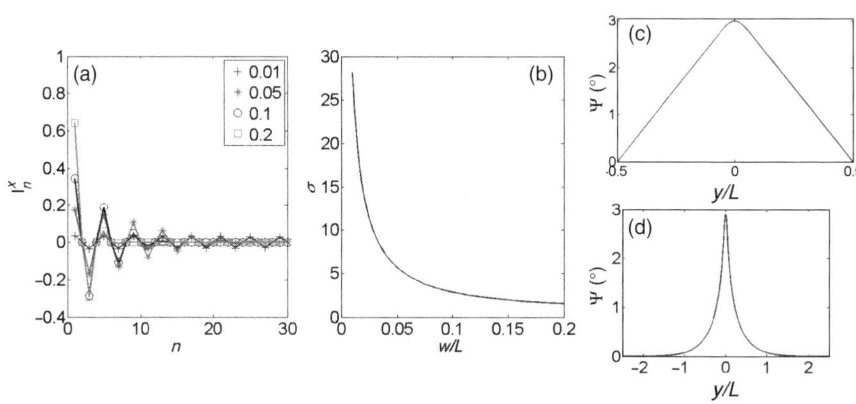

FIGURE 11.3 Fourier series of $X = \exp\left[-(x - L/2)^2/w^2\right]$ and dependence of nonlocality on the implemented model. (a) I_n^x for $w/L = 0.01$ (crosses), 0.05 (stars), 0.1 (circles), and 0.5 (squares), respectively. (b) Fit parameter σ versus normalized beam waist w/L. Nonlinear perturbation ψ versus y/L for the (c) Poisson and the (d) Yukawa equations when $w/L = 0.05$, respectively.

out (Section 11.6); moreover, I_n^x will be nonzero only if n is odd; hence, we can set $n = 2m + 1$ $(m = 0, 1, 2, \ldots)$. As shown in Figure 11.3a, for a bell-shaped beam I_{2m+1}^x is an alternating series and ψ_{2m+1} from Equation 11.14 has the same trend. Equation 11.13 provides $\psi\,(x = L/2, y, z) = \sum_{m=0}^{\infty} \psi_n(y, z) \sin\left[\pi(2m + 1)/2\right]$; therefore, all terms on the right-hand side of the series have the same sign.

Let us now analyze the propagation Equation 11.1. In the highly nonlocal limit, integrating along x yields [12]

$$2ik_0 n_e^{(b)} \left(\frac{\partial u}{\partial z} + \tan\delta^{(b)} \frac{\partial u}{\partial y}\right) + D_y \frac{\partial^2 u}{\partial y^2} + k_0^2\, \Delta n_e^2\big|_{x=L/2}\, u = 0. \tag{11.15}$$

The next step is the introduction of an equivalent two-dimensional all-optical perturbation $\psi_{2D}(y, z)$. Toward this goal, we set $\psi_{2D} = \psi_1$; thus, ψ_{2D} is underestimated with respect to the actual ψ and we have to compensate it by an effective two-dimensional power P_{2D} smaller than the actual power P. Defining the fit parameter σ via $P = \sigma P_{2D}$, it is $\sigma > 1$; assuming equal amplitudes of all ψ_{2m+1}^{2D}, we get a rough estimate for σ:

$$\sigma \approx \sum_{m=0}^{\infty} \left|I_{2m+1}^x\right| / |I_1^x|. \tag{11.16}$$

Figure 11.3b plots σ versus waist for a Gaussian beam. As expected, σ increases for narrower beams due to the spreading of I_n toward larger integers, as demonstrated in Figure 11.3a.

In summary, nonlinear beam propagation in NLC can be effectively modeled in (1 + 1)D by the system of Equation 11.15 and the reorientational equation

$$\nabla_{yz}^2 \psi_{2D} - \left(\frac{\pi}{L}\right)^2 \psi_{2D} + \gamma\kappa^2 \sin\left[2\left(\theta_0 + \psi_{2D} - \delta^{(b)}\right)\right] |u|^2 = 0. \tag{11.17}$$

We point out that Equation 11.17 retains the degree of nonlocality of the full 3D structure. Conversely, removing the term $\partial^2\psi/\partial^2 x$ in Equation 11.4 or 11.5 implies the disappearance of the screening term in Equation 11.17: in such case, the nonlocality range would be the cell size along y [9, 18], that is, a significant misestimate of the nonlinear index well, both in shape and size [8] (Fig. 11.3c).

11.3 SINGLE-HUMP NEMATICON PROFILES

In this section, we discuss light self-confinement in the absence of losses, with specific reference to single-hump nematicons; we solve the nonlinear eigenvalue problem for various powers and initial orientation θ_0, comparing the results of the full 3D model with the simplified 2D model.

11.3.1 (2 + 1)D Complete Model

We consider nematicons with phase fronts normal to $\hat{z}$, that is, solitons that can be excited by beams impinging normally on the input interface of the NLC sample. Shape-preserving nonlinear wave packets will take the form $A = A_s(x, y - \tan\delta^{(b)}z)\exp(ik_0 n_{\mathrm{NL}} z)$, and the corresponding optical perturbation is $\psi = \psi_s(x, y - \tan\delta^{(b)}z)$, with $n_{\mathrm{NL}}k_0$ the variation induced by self-phase modulation on the propagation constant and $\tan\delta^{(b)}$ the nonlinear change in walk-off. The nonlinear eigenvalue problem has the form

$$\left(2n_{\mathrm{NL}}n_e^{(b)} + n_{\mathrm{NL}}^2\right)k_0^2 A_s = \frac{\partial^2 A_s}{\partial x^2} + D_y\frac{\partial^2 A_s}{\partial y^2} + k_0^2\Delta n_e^2 A_s, \tag{11.18}$$

$$\left(1 + \tan^2\delta^{(b)}\right)\frac{\partial^2\psi_s}{\partial y^2} + \frac{\partial^2\psi_s}{\partial x^2} + \kappa^2\gamma\sin\left[2\left(\theta_0 + \psi_s - \delta^{(b)}\right)\right]|A_s|^2 = 0. \tag{11.19}$$

To illustrate the main properties of nonlocal spatial solitons in NLC, the system of Equations 11.18 and 11.19 can be solved in the highly nonlocal limit, that is, when the nonlocal index well can be satisfactorily approximated by a parabola. In fact, setting $\psi_s = \psi^{(b)} - \psi_2\left(\psi_2[(x - L/2)^2 + y^2]\right)$ and neglecting $\tan^2\delta^{(b)}$ with respect to 1, Equation 11.19 provides $\psi_2 = \kappa^2\gamma\sin\left[2\left(\theta_0 + \psi^{(b)} - \delta^{(b)}\right)\right]|A_0|^2/4$, with $A_0 = A_s(x = L/2, y = 0)$. The beam profile is $A_s = A_0\exp\left[-((x - L/2)^2 + y^2)/w_s^2\right]$, where $A_0 = [2Z_0P_s/(\pi n_e^{(b)}\kappa^2 w_s^2)]^{1/2}$ (P_s is the nematicon power), and the soliton existence curve in the plane power-waist takes the expression

$$P_s = \frac{4\pi}{k_0^2 Z_0\gamma\sin\left[2\left(\theta_0 + \psi^{(b)} - \delta^{(b)}\right)\right]\left.\frac{\mathrm{d}n_e}{\mathrm{d}\theta}\right|_{\theta_0+\psi^{(b)}}}\frac{1}{w_s^2}, \tag{11.20}$$

which can be recast as $n_2\left(\theta_0 + \psi^{(b)}\right)P_s w_s^2 = 8\pi n_e^{(b)}/\left(Z_0k_0^2\right)$.

Figure 11.4 shows solutions of system (11.18 and 11.19) for $\theta_0 = 45^\circ$ versus nematicon power. At very low power, the soliton waist is comparable with the sample thickness L; hence, the nonlocality is of the order of unity and the refractive index profile is asymmetric due to the distinct boundary conditions along x (finite side) and y (infinite side). As power increases, the nematicon gets narrower because of a higher nonlinear index well; when the waist becomes much smaller than L, the highly nonlocal regime can be invoked, as confirmed by the conservation of the ψ_s profile versus power. For still higher powers, the soliton waist increases because of the saturation in the reorientational nonlinearity as stated by Equation 11.20, the latter also responsible for changes in soliton walk-off. Finally, consistently with Section 11.2.2, the soliton behavior with power depends on the initial angle θ_0, which determines how soon saturation occurs and the angular span of soliton deflection.

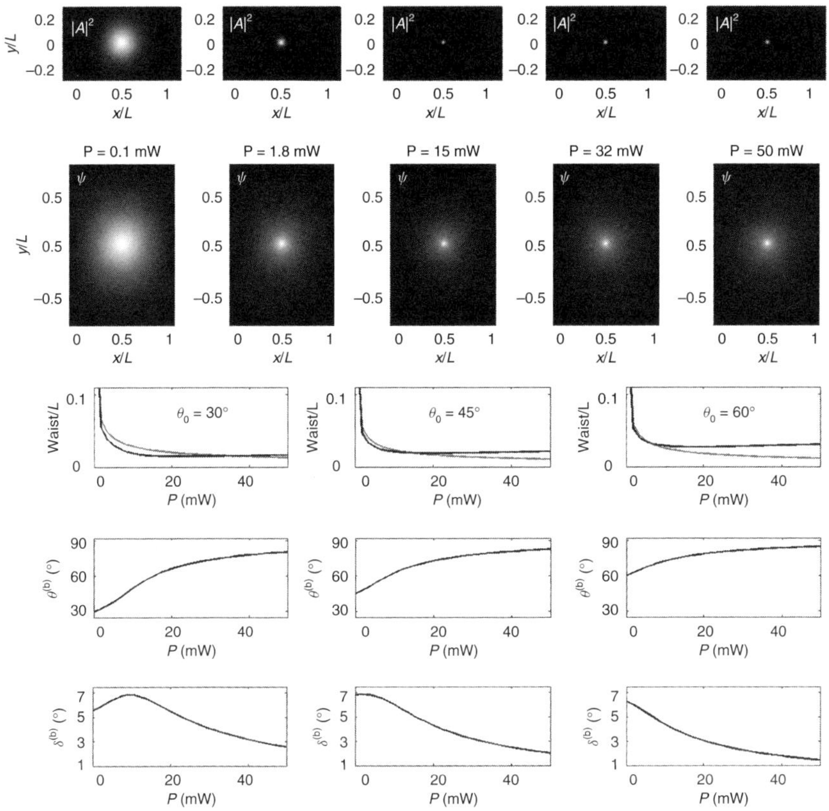

FIGURE 11.4 Numerically calculated nematicon profiles. First (second) row: plots of the intensity $|A|^2$ (nonlinear perturbation ψ) for several input powers and $\theta_0 = 45^\circ$. Third row: soliton existence curve in the plane power-waist, based on the perturbation model (gray lines), and exact solution stemming from Equations 11.18 and 11.19 (black lines) for three values of θ_0. Fourth (fifth) row: behavior of the maximum in director distribution $\theta^{(b)}$ (nematicon walk-off $\delta^{(b)}$) versus input power for three θ_0.

11.3.2 (1 + 1)D Simplified Model

Hereby, we carry out a procedure similar to the one of the former Section 11.3.1 to compute the soliton profile versus power using the simplified (1 + 1)D model described in Section 11.2.3. From Equations 11.15 and 11.17, the nematicon in the midplane $x = L/2$ is governed by the eigenvalue problem

$$\left(2n_{\mathrm{NL}}n_e^{(b)} + n_{\mathrm{NL}}^2\right) k_0^2 u_s = D_y \frac{\partial^2 u_s}{\partial y^2} + k_0^2 \Delta n_e^2 u_s, \tag{11.21}$$

$$\left(1 + \tan^2 \delta^{(b)}\right) \frac{\partial^2 \psi_s}{\partial y^2} - \left(\frac{\pi}{L}\right)^2 \psi_s + \kappa^2 \gamma \sin\left[2\left(\psi_s + \theta_0 - \delta^{(b)}\right)\right] |u_s|^2 = 0. \tag{11.22}$$

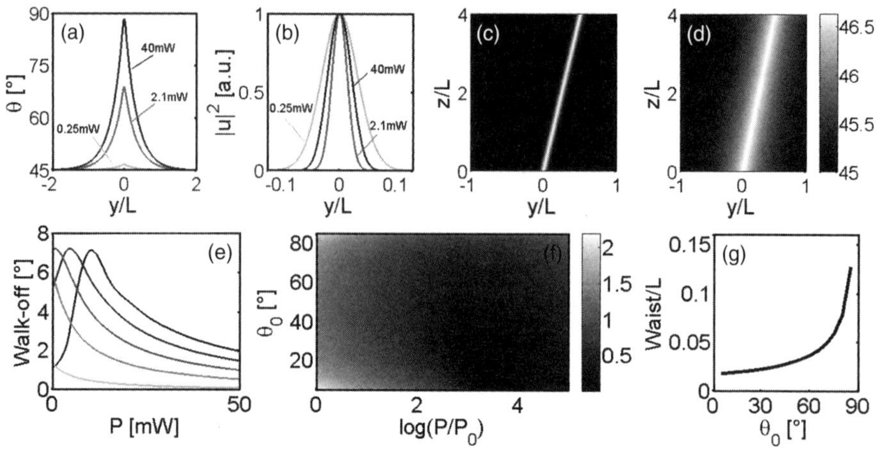

FIGURE 11.5 Two-dimensional simplified model of nematicon propagation. (a) Nonlinear index well and (b) corresponding nematicon profile $|u_s|^2$ computed with Equations 11.21 and 11.22. (c) Intensity profile $|A|^2$ in yz and (d) corresponding ψ_{2D} for $P = 2.2$ mW launched at the input. (e) Walk-off $\delta^{(b)}$ versus power; from the darkest to the brightest color, θ_0 ranges from 5° to 85° in steps of 20°. (f) Plot of $\log\left(w/w_0\right)$ versus power P and initial angle θ_0. (g) Nematicon minimum waist versus θ_0. We took $w_0 = 1$ μm and $P_0 = 0.02$ mW.

Figure 11.5a shows the computed nonlinear index well Δn_e^2 at three excitations: the nonlinear reorientation ψ_s conserves the shape of the complete 3D model (Fig. 11.4) thanks to the presence of the screening term $(\pi/L)^2\,\psi_s$, that is, the nonlocality is conserved even if the dimensionality has been reduced. The corresponding soliton profiles $|u_s|^2$ are graphed in Figure 11.5b: for powers larger than 0.25 mW, the shape is very similar to Gaussian, in agreement with the highly nonlocal approximation. We checked the shape invariance of the calculated profiles by means of a bidimensional anisotropic BPM code (see Section 11.4.1 for details); the results for beam profile and director perturbation are plotted in Figure 11.5c and 11.5d, respectively. Obviously, beam self-steering due to nonlinear changes in walk-off still occurs, but its dynamics are shifted to lower powers, consistently with Section 11.2.3 (Fig. 11.5e). Furthermore, the soliton existence curve in the plane waist-power retains its nonmonotonic behavior because of saturation in the nonlinearity, although the beam dynamics versus initial condition remains the same (Fig. 11.5f). Figure 11.5g plots the minimum soliton waist for a given θ_0, demonstrating that the best confinement takes place for small θ_0 as it stems from the highly perturbative regime discussed in Section 11.2.2.

11.4 ACTUAL EXPERIMENTS: ROLE OF LOSSES

In Section 11.3, we computed nematicon shape and trajectory versus power and initial director angle θ_0, assuming no losses in the sample. In typical experiments, the input can be controlled in profile, power, and polarization, whereas the beam evolution in yz is monitored by collecting the out-scattered light. The latter is a substantial cause of attenuation even in undoped NLC; hence, it affects actual solitons and their properties, even though many of the properties discussed in Section 11.3 hold valid (e.g., dependence of self-focusing and self-steering on θ_0, nonlocality dependence on the geometry, etc.). In this section, we account for scattering losses in Equations 11.15 and 11.17, comparing the results with experimental measurements.

11.4.1 BPM (1 + 1)D Simulations

Figure 11.6 displays the numerical simulations of a Gaussian beam input with waist $w_0 = 5$ μm propagating in NLC with attenuation $\alpha = 5$ cm^{-1}, the latter directly related to experimental measurements: for a given θ_0, the beam spreads at low power (diffraction dominates over self-focusing); for increasing power, self-effects become strong enough to counteract diffraction, going from self-focusing (i.e., the waist increases more gradually than in the linear regime) to self-confinement (i.e., the waist diminishes along z). Such dynamics take place at lower powers for initial orientation θ_0 close to 45°, as predicted by Figure 11.2a. At variance with the exact soliton solutions studied in Section 11.3, the beam trajectory is no longer a straight line but bends because of power attenuation along z: in fact, the perturbation ψ decays exponentially along z, thereby reducing the nonlinear walk-off. Hence, for $z \gg 1/\alpha$, the beam propagates at δ_0 with respect to z, that is, the evolution goes from highly nonlinear to perturbative nonlinear regimes along z. Figure 11.7 graphs ψ corresponding to the beam in Figure 11.6. Such curves underline the differences arising in the highly nonlinear regime between the case

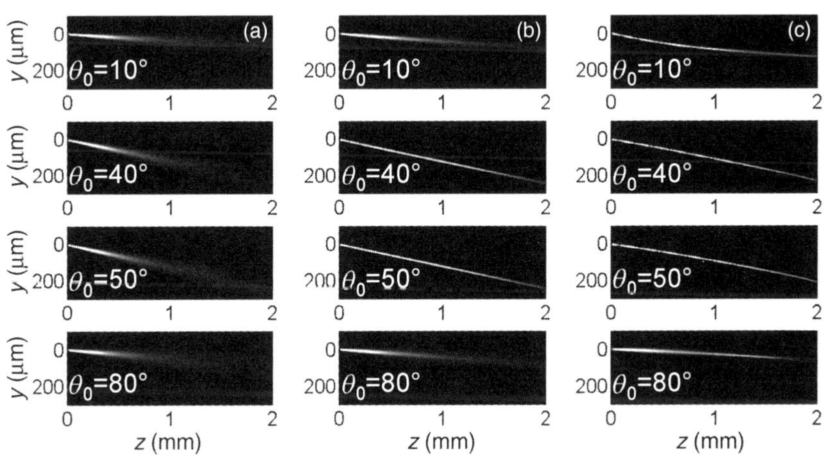

FIGURE 11.6 Evolution of $|u|^2$ in the plane yz, computed from Equations 11.15 and 11.17 in E7 for $P = 1$ μW (a), 1 mW (b), and 5 mW (c). The input waist is 5 μm, the thickness $L = 100$ μm, and the wavelength $\lambda = 1064$ nm.

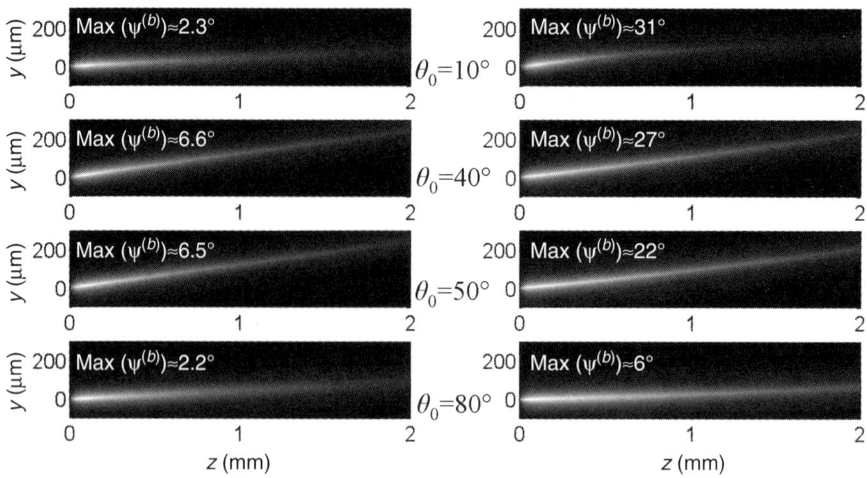

FIGURE 11.7 Distribution of ψ in the plane yz for $P = 1$ mW (perturbative regime) and $P = 5$ mW (highly nonlinear regime). Maxima are computed along the propagation coordinate z. Parameters are as in Figure 11.6.

$\theta_0 = \pi/4 - \Delta\theta$ and $\theta_0 = \pi/4 + \Delta\theta$ $(0 < \Delta\theta < \pi/4)$: in the perturbative regime, self-focusing would be of equal strength in the two cases (Fig. 11.2a), whereas at high powers, the nonlinear effects are much stronger for small θ_0 than for large θ_0 because of saturation when all-optical reorientation becomes comparable with the initial angle θ_0. This is visible in Figure 11.8, plotting waist versus z for several θ_0. As regards the trajectory, differences versus θ_0 are apparent even in the nonlinear perturbative regime, as the nonlinear walk-off variations are positive (negative) when θ_0 is smaller (larger) than 45° (Fig. 11.9).

11.4.2 Experiments

We verified our theoretical and numerical results by injecting a 5-μm waist TEM_{00} beam at wavelength $\lambda = 1064$ nm in an NLC cell and observing its evolution by acquiring the image of out-scattered light (see Fig. 11.10). Two comb electrodes realized in Indium Tin Oxide are deposited on each glass–NLC interface, in order to set and tune by voltage V the angle θ_0 in the midplane $x = L/2$; each finger of the combs is much smaller than L, ensuring a constant director angle inside the cell (Chapter 5). To quantify the dependence between θ_0 and V, we measured the walk-off in the linear regime versus bias; the results are shown in Figure 11.11 (points), whereas the solid lines stem from the reorientational equation, neglecting electric field components along x and assuming an exponential decay in the applied field toward the center $x = L/2$ [12]. Figure 11.10 plots the acquired evolution profiles in the plane yz. As expected, at low power, the beam diffracts according to D_y, forming an angle δ_0 with respect to the z-axis. As the power increases, the beam undergoes self-focusing and eventually self-confinement (top row in Fig. 11.8);

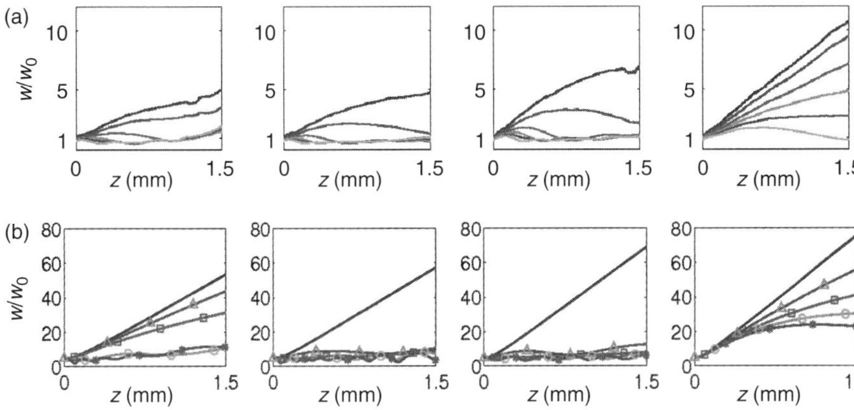

FIGURE 11.8 Comparison between (top) experimentally measured and (bottom) numerically calculated waist versus z. Initial θ_0 are $10^\circ, 30^\circ, 60^\circ$, and 80°, from left to right column, respectively. The plotted values are normalized to the input waist w_0. In the experimental lines, the input powers from top to bottom are 1, 2, 4, 6, 8, and 10 mW, respectively; in the simulations, they are 0.001, 0.5, 1, 3, and 5 mW, respectively.

self-focusing depends on θ_0 and is maximum for $\theta_0 \approx 45^\circ$. Solitons have bent trajectories because of nonlinear changes in walk-off, with negative and positive curvatures for $\theta_0 < 45^\circ$ and $\theta_0 > 45^\circ$, respectively, in agreement with the theoretical predictions (Fig. 11.9).

11.5 NEMATICON SELF-STEERING IN DYE-DOPED NLC

In this section, we investigate nematicons and their self-steering in dye-doped NLC. Dyes consist of molecules absorbing in a specific spectral range, altering the optical nonlinearity via the so-called Janossy effect [23, 24]. Such effect is modeled by inserting a material-dependent enhancement factor η in the expression of the optical torque; in the reorientation equation, for $\eta > 1$, the all-optical response is amplified; thus, the power can be scaled down by a factor $1/\eta$ in order to achieve the same nonlinear effects[3]. In this case, we consider an E7 mixture doped with 1% in weight of 1-amino-anthraquinone (1-AAQ) [23, 25]. The input interface (placed in $z = 0$) is such to induce planar director orientation: at rest, the director angle θ versus z goes from $\theta_{in} = 90^\circ$ to $\theta_0 = 45^\circ$, the latter determined by anchoring at the two interfaces parallel to the plane yz; we also recall that the extension of the transition region along z is equal to the cell thickness L (in our case $L \approx 100$ μm). Thus, we can write $\tilde{\theta}(z) = \theta_0 + (\theta_{\text{in}} - \theta_0)\exp(-z/L)$, being $\tilde{\theta}$ the director distribution in the absence of illumination.

Figure 11.12a–c shows the expected behavior on the basis of the standard Janossy effect: linear diffraction at low power, perturbative and highly nonlinear

[3]Given the resonant nature of the dye, the increase in losses has to be accounted for.

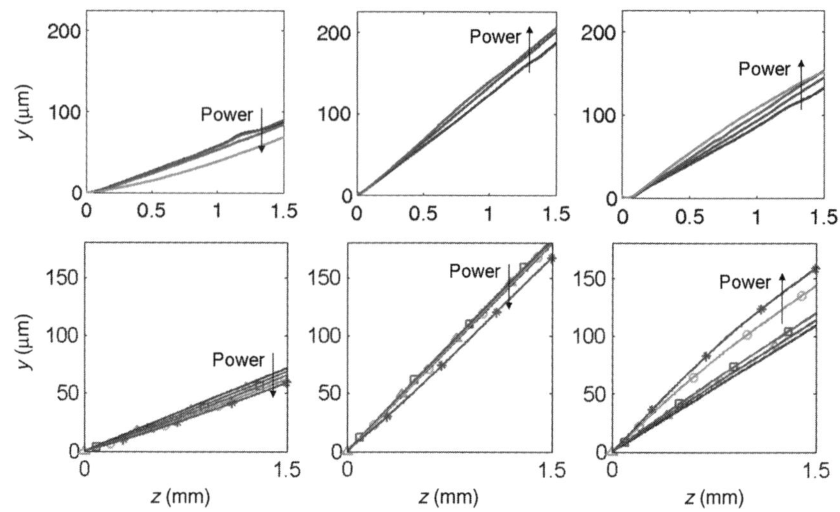

FIGURE 11.9 Comparison between experimentally acquired and numerically calculated trajectories. Measured (top row) and numerical (bottom row) waist versus z. Angles θ_0 are 80°, 40°, and 20°, from left to right column, respectively. In the measurements, the input powers are 1, 5, and 10 mW from top to bottom on the leftmost column and from bottom to top in the other two columns, respectively. The bottom (top) line in the leftmost (rightmost) column corresponds to a power of 60 mW (30 mW). The simulation powers are 0.001 mW (lines without symbols), 0.5 mW (triangles), 1 mW (squares), 3 mW (circles), and 5 mW (asterisks), respectively. Small discrepancies in the middle column are ascribable to experimental uncertainty on θ_0 (most relevant in proximity to the maximum walk-off angle, see Figure 11.2b) and a different power range.

effects at higher powers, with the corresponding changes in beam waist and trajectory in analogy to the case of undoped NLC. The main difference between the doped and undoped NLC should be the lower power needed to observe nonlinear effects and the propagation losses due to dye absorption. We launched along z in the sample described above an extraordinarily polarized fundamental Gaussian beam at $\lambda = 442.5$ nm with a waist of about 5 μm. We name ϕ the angle of the Poynting vector with $\hat{z}$. The linear walk-off is nearly 9° because of the higher NLC birefringence in the blue spectrum. Losses are measured to be $\alpha_{dye} \approx 50\ \text{cm}^{-1}$ because of the dye. Photos of the beam evolution are displayed in Figure 11.12d: clearly, the path is not as anticipated based on standard effects. Even though self-confinement takes place above a low input power of about $P_0 = 80$ μW (i.e., $\eta \approx 50$, consistently with values reported in literature [23]), the output slope of the self-confined waves ϕ_{output} (i.e., ϕ measured for z much larger than the absorption length) is not δ_0, as it should in the linear limit. In particular, as power is increased, the beam gets closer to the z-axis; when P_0 approaches 240 μW, we observe a threshold-like effect with the sudden appearance of an ordinary component diffracting along z; at the same time, the extraordinary component remains self-confined but takes negative ϕ_{output}, the latter reaching its maximum $\phi_{output} = -30°$ for $P_0 = 400$ μW (Fig. 11.13a and c). Figure 11.12e shows photos of the red light (acquired through

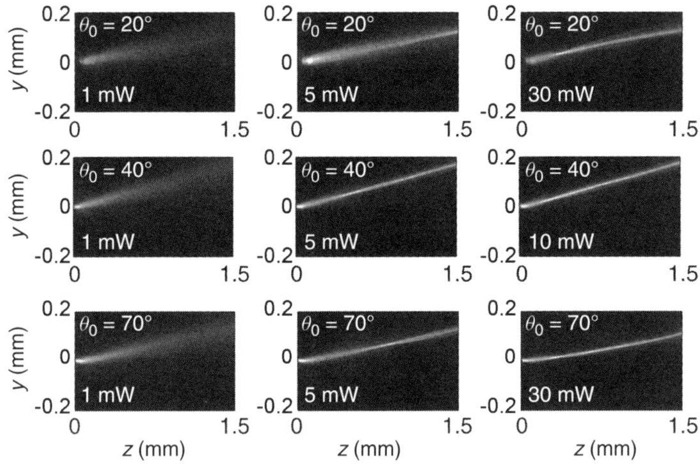

FIGURE 11.10 Acquired intensity profiles in the sample for various input powers and initial director angle θ_0. The initial waist is 5 μm, $L = 100$ μm, and $\lambda = 1064$ nm.

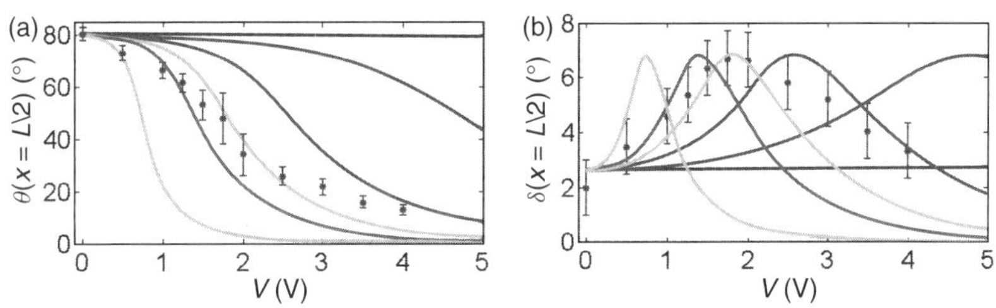

FIGURE 11.11 Reorientation angle in the cell midplane (a) and walk-off angle versus applied bias; points and solid lines are experimental and theoretical data, respectively. In (a) and (b), theoretical curves from left to right correspond to $\kappa = 5 \times 10^4$, 1×10^5, 1.3×10^5, 2×10^5, 4×10^5, and 2×10^6 m^{-1}, respectively.

a red filter at 632.8 nm) photoemitted by the dye and eventually guided by the blue soliton; the photoluminescence is stronger corresponding to the extraordinary blue component forming a nematicon.

The above-described phenomena are in contrast with the hypothesis that self-bending is only caused by nonlinear changes in walk off. In fact, they can be explained only if an effective force directed along negative y acts on the beam wave vector, thereby inducing the observed deflections even for large z. In other words, the light–matter coupling has to produce a symmetry breaking between $\hat{y}$ and $-\hat{y}$, consistently with the power dependence of the deflection angle.

Thermal effects cannot explain the observations, given the lack of self-focusing on the ordinary component. A source of asymmetry could be the inherent anisotropy in reorientational dynamics that, for angles $\theta \neq 0$ and in conjunction with the boundary at the input, could misalign the beam intensity peak and the maximum of

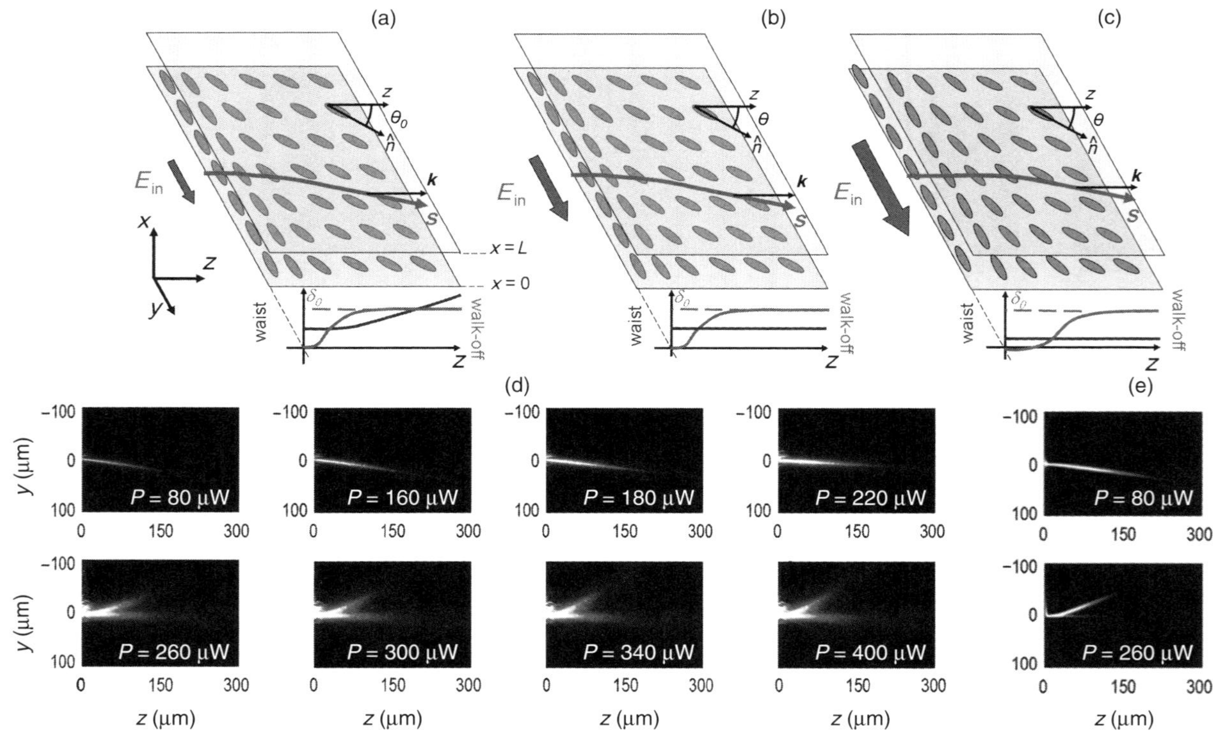

FIGURE 11.12 Expected nematicon behavior in a dye-doped NLC: for increasing power, the beam goes from (a) diffraction to (b–c) self-confinement, either in the (b) perturbative or (c) in the highly nonlinear regimes, respectively, with nonlinear response decaying exponentially in propagation owing to attenuation. (d) In actual experiments at 442.5 nm, the nematicon path and output slope depend on power, with an abrupt change in trajectory for $P_0 = 260$ μW and the appearance of an ordinary component. (e) Images of the light emitted (and scattered) in the red portion of the spectrum, due to photoluminescence of 1-AAQ.

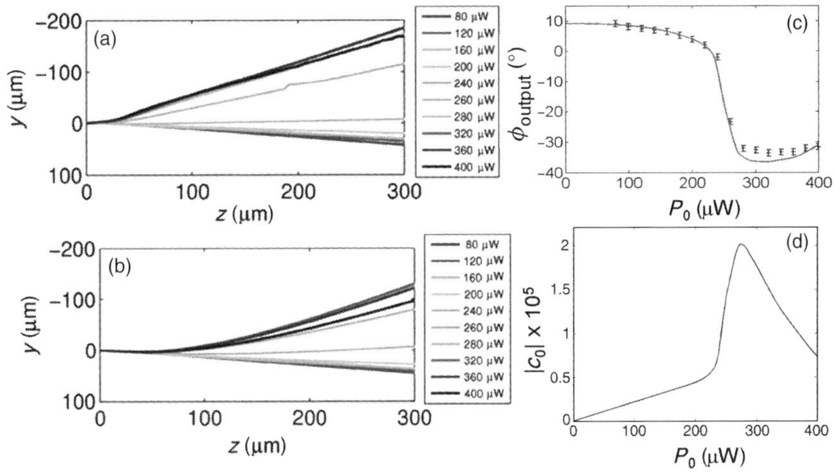

FIGURE 11.13 Comparison between measured trajectories and numerical results stemming from Equations 11.24 and 11.25. (a) Acquired and (b) computed nematicon trajectories; power increases from bottom to top. (c) Comparison between measured (points) and predicted (solid line) trend of the output angle ϕ_{output} versus power. (d) Fit coefficient c_0 versus P_0.

the nonlinear index well Δn_e^2. Numerical simulations, however, confirm that, even if such effects took place, they could account for deflections up to 1°, in contrast to the experimental data. A source of asymmetry is the optical field itself: in fact, as described by Equation 11.3, in the nonparaxial regime the longitudinal field E_s becomes relevant and exhibits an odd parity profile if the transverse field E_t is even, as in our case. Noteworthy, solitons in doped NLC are narrow owing to the enhancement η. As the presence of E_s alone is not sufficient to explain the observations, the material itself has to be sensitive to the sign of E_s, that is, needs to be noncentrosymmetric. We model such noncentrosymmetry by writing

$$\eta = \frac{\eta_0}{1 + I/I_{\text{sat}}} + \Delta\eta, \tag{11.23}$$

with I the intensity and I_{sat} its saturation value. The first term on the right-hand side of Equation 11.23 accounts for saturation in the Janossy response at high peak intensities. The second term models the noncentrosymmetric response and we write $\Delta\eta = c_0 \text{Im}(E_s e^{-i\beta z})$, where β is the propagation constant of the self-confined wave and c_0 is found by fitting the data. In the highly nonlocal limit, the soliton profile can be assumed Gaussian, that is, $E_t \propto \exp[-(y - y_b)^2/w_b^2]$, where w_b and y_b are soliton waist and position, respectively, both of them varying with z. Under such approximations and defining the power in each section $z = \text{const}$ as $P = P_0 \exp(-\alpha_{\text{dye}} z)$, the beam evolution can be calculated by solving the ODE system [7, 26]

$$\frac{d^2 w_b}{dz^2} = -\left(\sigma \frac{\sqrt{P}}{w_b}\right)^2 \left[w_b - w_0(P)\right], \tag{11.24}$$

$$\frac{d^2 y_b}{dz^2} = \frac{d \tan \delta^{(b)}}{dz} + \chi \frac{P_0^{3/2}}{w_b^2} e^{-\alpha_{\text{dye}} z}. \tag{11.25}$$

where $w_0(P)$ is the soliton waist in the highly nonlocal and lossless ($\alpha_{\text{dye}} = 0$) case for a power P, whereas σ and χ are two constants depending on the dielectric properties of the medium [7].

The second term on the right-hand side of Equation 11.25 represents the transverse force responsible for the wave vector deviation [7]. Figure 11.13b shows the calculated trajectories: owing to losses, the wave packet self-steers in the proximity of the input interface, with paths in good agreement with the model and consistent with the change in c_0 graphed in Figure 11.13d. c_0 undergoes a Freedericksz-like threshold, suggesting that a torque tilts the molecular director out of the plane yz, thus explaining the appearance of the ordinary component when the sudden deviation occurs in the beam. Note that, according to the Mauguin limit, coupling between extraordinary and ordinary components can only occur for fast variations in the director distribution as compared with the wavelength: nonadiabatic variations are permitted just near the input interface, at distances smaller than the thickness L.

11.6 BOUNDARY EFFECTS

From Section 11.2.3–11.5, we limited ourselves to the study of nematicons propagating in the cell midplane $x = L/2$. Here, we discuss, both experimentally and theoretically, nematicons launched away from this plane, addressing the role of a finite cell (along x) and the consequent break in translational symmetry.

Let us consider a planar cell with director homogeneously distributed in the absence of external excitations, lying on the plane yz and forming an angle θ_0 with $\hat{z}$. Assuming that the nonlinear index well conserves parity with respect to y (i.e., for an infinite sample along y), for small wave vector deviations, the application of the Ehrenfest's theorem to Equation 11.1 provides [27, 28]

$$\frac{d^2 \langle \boldsymbol{r} \rangle}{dz^2} = \frac{d \tan \delta^{(b)}}{dz} \hat{y} - p \sum_{m=0}^{\infty} V_m \left(\langle x \rangle\right) \left\langle (x - \langle x \rangle)^m \right\rangle \hat{x}, \tag{11.26}$$

where we set $\boldsymbol{r} = x\hat{x} + y\hat{y}$, $\langle f \rangle = \iint f |A|^2 dx dy / \iint |A|^2 dx dy$, and $V_m = (1/m!) \partial^{m+1} V_{\text{eff}} / \partial x^{m+1} |_{\langle \boldsymbol{r} \rangle}$, having defined the effective potential $V_{\text{eff}} = -\Delta n_e^2$ and $p = k_0 / (2 n_e^{(b)})$. In the following discussion, we neglect nonlinear changes in walk-off in order to focus on the role played by the boundaries. In the highly nonlocal limit, in the series on the RHS of Equation 11.26, only the term for $m = 0$ survives, that is, the nematicon trajectory obeys geometric optics [29] (Chapter 5). To decouple the beam motion along x and in the plane yz, we exploit the separation of variables by setting $|A|^2/\int |A|^2 \mathrm{d}x\mathrm{d}y = \varphi_x(x,z)\varphi_y(y,z)$; thus, it is straightforward to get a simplified two-dimensional model for nematicon trajectory in the plane xz

$$\frac{\mathrm{d}^2 \langle x\rangle}{\mathrm{d}z^2} = -pW_0\left(\langle x\rangle\right), \tag{11.27}$$

where $W_0 = \partial V_{\mathrm{eq}}/\partial x|_{\langle x\rangle}$, being $V_{\mathrm{eq}}(x,z) = \int \varphi_y(y) V_{\mathrm{eff}}(x,y,z)\,\mathrm{d}y$ the equivalent 1D potential. In deriving Equation 11.27, we neglected the dependence of φ_y on z and dropped the dependence on $\langle y\rangle$, the latter determined only by linear walk-off δ_0 due to the y-invariance of the nonlinear index well.

To complete the model, we also need an expression for the nonlinear force W_0. To this extent, we use the perturbative expansion Equation 11.11 in conjunction with the Green formalism in a planar cell of finite thickness along x, as stated in Equation 11.12 for unbound geometries in yz. After the transformation $y \to y - z\tan\delta_0$, neglecting the partial derivatives along the propagation coordinate in the walk-off reference system of Equation 11.4, the Green function reads:

$$G\left(x, x', |y-y'|\right) = \frac{1}{\pi n}\sum_{n=1}^{\infty} \sin\left(\pi n\frac{x}{L}\right)\sin\left(\pi n\frac{x'}{L}\right)\mathrm{e}^{-\pi n\frac{|y-y'|}{L}}. \tag{11.28}$$

To obtain a closed form, for the nematicon we take a Gaussian ansatz, consistent with the high nonlocality [22]

$$|A(x,y,z)| = \left[\frac{2Z_0P}{\left(\pi\cos\delta_0 n_e^{(b)}\kappa^2 w^2\right)}\right]^{1/2} \mathrm{e}^{-[(x-\langle x\rangle(z))^2+(y-\tan\delta_0)^2]/w^2(z)}. \tag{11.29}$$

Computing the convolution integral (Eq. 11.6) employing the Green function (Eq. 11.28) provides for θ_1:

$$\theta_1(x,y) = \gamma\kappa^2 P\sin[2(\theta_0-\delta_0)]\sum_{n=1}^{\infty}\frac{1}{\pi n}R_n^y(y)R_n^x(\langle x\rangle)\sin\left(\pi n\frac{x}{L}\right), \tag{11.30}$$

with

$$R_n^x(\langle x\rangle) \approx \sqrt{\pi}\frac{w}{L}\sin\left(\frac{\pi n\langle x\rangle}{L}\right)\mathrm{e}^{-\left(\frac{\pi n w}{2L}\right)^2}, \tag{11.31}$$

$$R_n^y(y) = \frac{\sqrt{\pi}}{2}\frac{w}{L}\mathrm{e}^{\left(\frac{\pi m w}{2L}\right)^2}$$

$$\left[\mathrm{erfc}\left(\frac{y}{w}+\frac{\pi m}{2}\frac{w}{L}\right)\mathrm{e}^{\pi m\frac{y}{L}} + \mathrm{erfc}\left(-\frac{y}{w}+\frac{\pi m}{2}\frac{w}{L}\right)\mathrm{e}^{-\pi m\frac{y}{L}}\right]. \tag{11.32}$$

Equation 11.30 with Equations 11.31 and 11.32 represent a closed-form solution for invariant nematicons along z propagating in planar cells, in the limit of high nonlocality and the perturbative regime, for beams launched in $x = L/2$ with appropriate waist (Eq. 11.20). In the more general case, these equations model a self-induced index well depending on beam position $\langle x \rangle$ via the coefficients 11.31, that is, indicating that the nonlinear potential loses its symmetry with respect to $x = \langle x \rangle$ when beams are not injected in the midplane $x = L/2$. As a consequence, the gradient of the nonlinear index center is nonzero on the beam axis, yielding a nonzero equivalent force $W_0 \neq 0$ on the wave packet. In the perturbative regime, we can set $\Delta n_e^2 \approx 2n_e^{(b)} \Delta n_e$; then, from Equation 11.30, we can compute V_{eq}:

$$V_{\mathrm{eq}}(x) = -2n_e^{(b)} \left.\frac{\partial n_e}{\partial \theta}\right|_{\theta_0} \int_{-\infty}^{\infty} \theta_1(x, y) \mathrm{e}^{-2y^2/w^2} \mathrm{d}y$$

$$= -n_2(\theta_0) P \sum_{n=1}^{\infty} \frac{1}{\pi n} S_n R_n^x(\langle x \rangle) \sin\left(\pi n \frac{x}{L}\right), \tag{11.33}$$

where we introduced $S_n = (\pi)^{-1/2} \mathrm{e}^{\left(\frac{\pi n w}{\sqrt{2}L}\right)^2} F(\pi n w/L)$, having defined the function $F(x) = \int_{-\infty}^{\infty} \mathrm{erfc}(x') \mathrm{e}^{-(x'-x)^2} \mathrm{d}x'$. Finally, W_0 reads

$$W_0(\langle x \rangle) = -n_2(\theta_0) \frac{P}{L} \sum_{n=1}^{\infty} S_n R_n^x(\langle x \rangle) \cos\left(\frac{\pi n \langle x \rangle}{L}\right). \tag{11.34}$$

We stress once again that expression (11.34) was derived in the perturbative regime only for the sake of simplicity. The same derivation can be carried out considering the square terms in P stemming from θ_1^2 and θ_2 [8]; it can be found that the shape of the self-induced well (i.e., the effective force W_0) is nearly unchanged owing to the high nonlocality, with the peak of the optical perturbation ψ changing by less than 10% for $\theta_0 = 45^\circ$ and $P = 2$ mW.

Force $-pW_0$ versus beam position $\langle x \rangle$ is plotted in Figure 11.14: the effective potential acting on the nematicon is anharmonic and even with respect to $\langle x \rangle = L/2$, with the force increasing as the beam gets closer to one of the interfaces. Noteworthy, the sign of the force (negative for $\langle x \rangle > L/2$, positive or zero otherwise) is such as to repel out the beam from the boundaries, thus favoring light trapping within the NLC layer. The solutions of Equation 11.27 together with Equation 11.34 predict periodic oscillations in the nematicon trajectory; some examples are graphed in Figure 11.14b for a zero initial velocity (i.e., input beam with a zero wave vector component along x) and various input positions $\langle x \rangle_0$. The trajectory tends to a sinusoidal form in the case of small displacements from the midplane. Conversely, the oscillation period Λ decreases as the nematicon is launched further and further away from the equilibrium position $x = L/2$; furthermore, Λ depends on excitation because of its nonlinear origin, decreasing

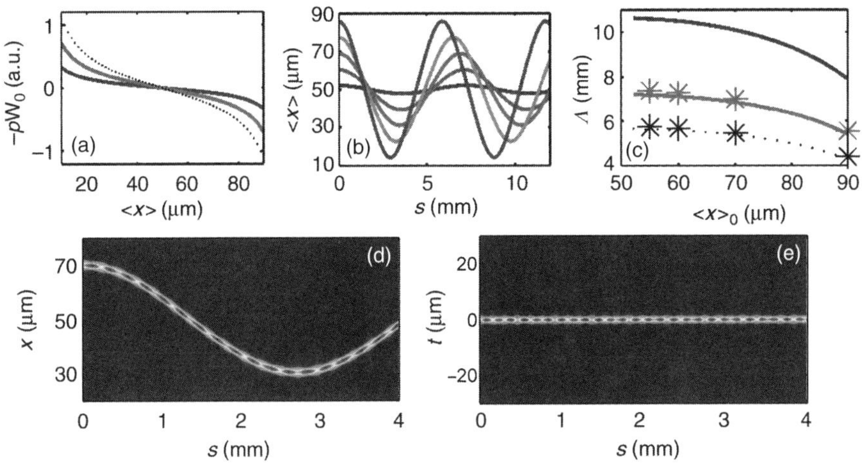

FIGURE 11.14 (a) Force $-pW_0$ acting on the nematicon versus its position for excitations $P = 1, 2,$ and 3 mW from lowest to highest (dots) absolute values, respectively. (b) Soliton trajectories in the plane xs for $P = 2$ mW and various launch positions. (c) Oscillation period Λ (solid lines and points are theoretical predictions and numerical results, respectively) versus input position $\langle x \rangle_0$ with zero initial momentum, the powers are as in (a): the largest (lowest) power corresponds to the shortest (longest) period. (d) and (e) Evolution a Gaussian beam with input waist of 2.8 μm launched in $\langle x \rangle_0 = 70$ μm with power 3 mW and wavelength 632.8 nm in the plane (d) xs and (e) ts. Here, $L = 100$ μm.

approximately by $P^{-1/2}$ (Fig. 11.14c) [8]. We numerically simulated the PDE system of the reorientation equation and the NLSE in a 3D geometry, neglecting the second-order derivatives along the Poynting vector. We rotated the reference system xyz by δ_0 around the x-axis to obtain xts. Figure 11.14d and 11.14e shows that nematicons undergo oscillations in good agreement with the theoretical predictions based on Equations 11.27 and 11.34.

A series of experiments were carried out to verify these findings, using an $L_z =$ 4-mm-long NLC cell of thickness $L = 100$ μm and width > 1 cm along y, with the commercial mixture E7 (Fig. 11.15a and 11.15b). The glass interfaces were treated to ensure $\theta_0 = 30^\circ$. The soliton evolution in yz as well as the output profiles in xy ($z = L_z$) were imaged via CCD cameras. A small offset with respect to $x = L/2$ and an angular tilt ξ (in our experiments, $\langle x \rangle_0 \approx 58$ μm and $\xi \approx 0.6^\circ$, respectively) were impressed on the input wave vector to maximize the nematicon x-displacement versus power; the beam at 1.064 μm was linearly polarized parallel to $\hat{y}$ to maximize coupling to the extraordinary wave. Clearly, the soliton is expected to interact with the boundary-driven potential and oscillate for a fraction of the period Λ, shifting along x in $z = L_z$ as power changes. The photographs of the output profiles in yz are superimposed in Figure 11.15c for several input powers (from 0.5 to 6 mW), demonstrating the predicted power-dependent repulsion from the boundary. Such nonlinear transverse dynamics along x are in excellent agreement with the integration of Equation 11.27 (using Equation 11.34 for the force) considering

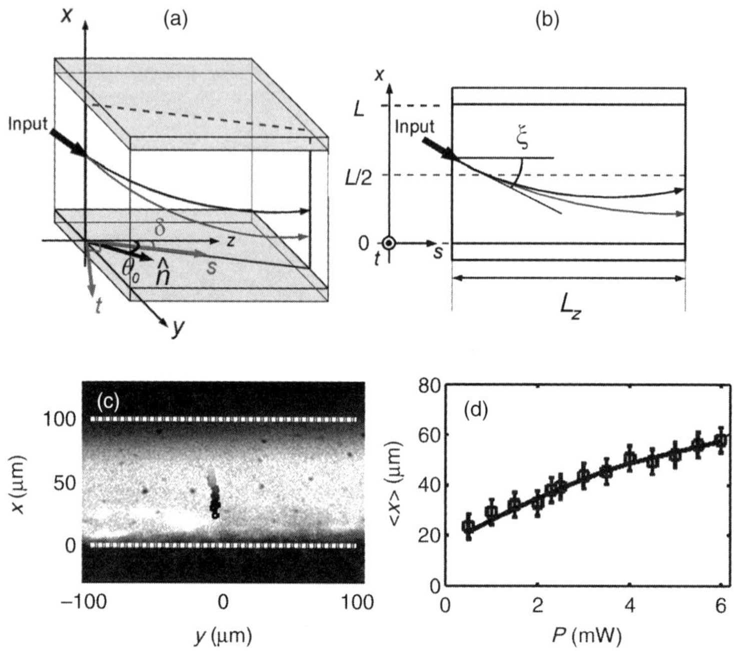

FIGURE 11.15 (a) 3D sketch and (b) side view in plane xs of the planar cell employed in the experiments. Nematicons are launched with an initial tilt with respect to $\hat{x}$: the self-trapped beams are repelled from the interfaces, with a repulsive force increasing with power (black line corresponds to a power larger than for gray line). (c) Collected and superimposed photographs of spatial soliton profiles at the cell output plane xy for various powers, ranging from 0.5 mW (darkest symbol) to 6 mW (brightest symbol). (d) Experimental (squares) and calculated (solid line) output positions versus excitation. To fit the experimental data (in the presence of scattering losses), we assumed a power coupling of 50%.

the actual sample parameters and a nonzero initial velocity $\left.\frac{\mathrm{d}\langle x\rangle}{\mathrm{d}z}\right|_{z=0} = \tan\xi$, as displayed in Figure 11.15d [27].

11.7 NEMATICON SELF-STEERING THROUGH INTERACTION WITH LINEAR INHOMOGENEITIES

Till now, we addressed power self-steering of nematicons when they propagate in a homogeneous molecular dielectric. In this section, we discuss how dielectric inhomogeneities affect the dependence of the beam trajectory on power. We address the nonlinear Goos–Hänchen shift on TIR at an interface, that is, a power-dependent lateral beam shift at a graded interface between two nonlinear NLC regions. We also illustrate power-dependent nematicon self-ejection from a finite index well. In both cases, perturbations in director distribution are impressed via the application of suitable biases.

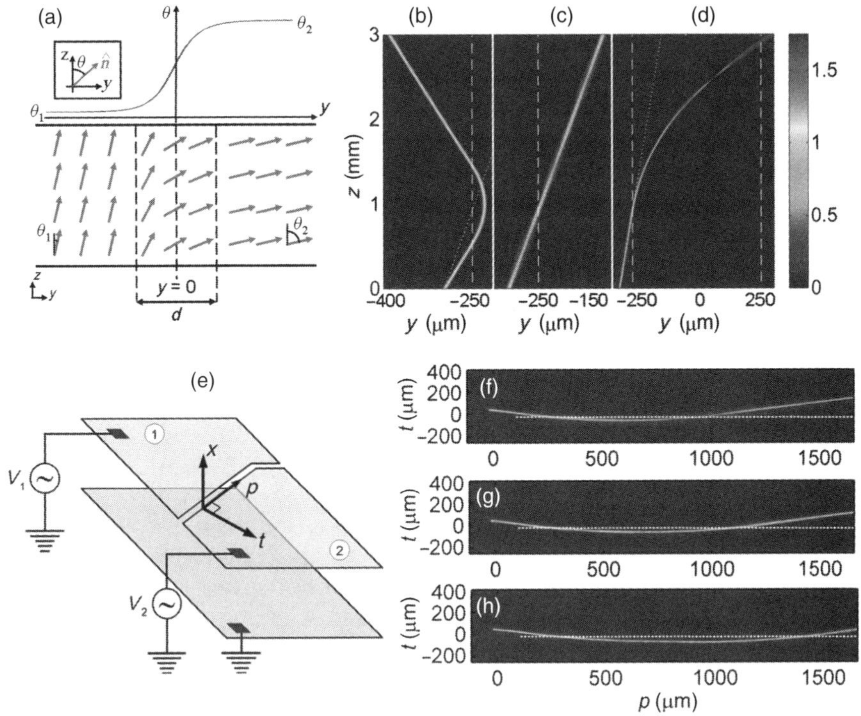

FIGURE 11.16 Nonlinear Goos–Hänchen shift. (a) Graph of a dielectric interface in a planar geometry. Nematicon trajectories for $P = 3$ mW and (b) $\theta_1 = 60^\circ$ and $\theta_2 = 10^\circ$, (c) $\theta_1 = \theta_2 = 10^\circ$, and (d) $\theta_1 = 10^\circ$ and $\theta_2 = 70^\circ$. Dashed lines mark the graded interface boundaries, whereas dotted lines are the nematicon trajectory for $\theta_1 = \theta_2$ in the perturbative nonlinear regime. (e) Sketch of the double-electrode configuration exploited in the experiments. Acquired intensity profiles for (f) $P = 1.6$ mW, (g) 4.4 mW, and (h) 9.3 mW. The applied biases are $V_1 = 1.5$ V and $V_2 = 0$ V, and the wavelength is 1064 nm.

11.7.1 Interfaces: Goos–Hänchen Shift

Let us consider a nematicon propagating in a planar cell and encountering a dielectric interface in its path, as sketched in Figure 11.16a. We assume that the director orientation changes linearly from θ_1 for $y < -d/2$ to θ_2 for $y > d/2$. The transition width d has to be comparable with the shortest cell dimension L, in agreement with Equation 11.12. Figure 11.16b–d shows the results of BPM simulations, with nematicons surviving the interaction thanks to the adiabatic character of the dielectric perturbation as compared to the soliton waist. Self-confinement changes within the refractive barrier because the effective Kerr coefficient given by Equation 11.7 follows the variations in θ.

If the index contrast across the interface is high enough to provide TIR (for wave vectors along $\hat{z}$, the condition $\theta_1 > \theta_2$ is sufficient to ensure TIR), in the highly nonlinear regime, we expect the reflected beam path to depend on power because of nonlinear walk-off. In particular, for $\theta_2 < \theta_1 < 45^\circ$, the changes in walk-off are positive (if optical powers are low enough to ensure a director angle belonging to the domain where walk-off angle is monotonically growing with θ, see Fig. 11.2c); therefore, nematicons undergoing TIR move toward positive y, giving rise to a nonlinear Goos–Hänchen shift [10].

To verify the predictions, we performed experiments in a cell with two (top) electrodes separated by a 50-μm gap and a common ground (bottom) electrode. Voltages V_1 and V_2 applied to these electrodes could lift the molecular director out of the plane yz by different amounts in the two NLC regions, defining a graded interface (Fig. 11.16e). The gap was realized in order to provide an incidence angle of 79.5° with respect to the interface normal $\hat{p}$ for beams launched with wave vector along z. Although this geometry is intrinsically 3D as walk-off acts also in xz, we can use our results to qualitatively explain the nematicon behavior: when the two biases are $V_1 = 1.5$V and $V_2 = 0$V, the director is lifted out of the plane yz only in the first region; hence, when the soliton propagates within the graded potential, the optical torque increases the walk-off and the self-confined beam penetrates deeper inside the second region, as shown in Figure 11.16f–h, with a larger lateral shift on TIR. We stress that, for a given position and transverse velocity, the force due to the index gradient depend slightly on power as we are in the highly nonlocal case (i.e., Equation 11.27 holds valid) [12]. Finally, the beam waist is much narrower than d; thus, the asymmetry in the nonlinear index well Δn_e^2 provoked by spatial variations of the effective Kerr effect can be neglected [30, 31].

11.7.2 Finite-Size Defects: Nematicon Self-Escape

After the interaction of a nematicon with an interface, we direct our attention to a localized defect in the dielectric tensor. We consider a perturbing region invariant along z and centered in $y = 0$, a nematicon with wave vector parallel to the symmetry axis $\hat{z}$ and initial position in $y = 0$. If the coefficients c_δ (see Equation 11.8) corresponding to the unperturbed (i.e., in the absence of light) director distribution $\theta_0(y)$ are positive, we expect the force stemming from nonlinear walk-off to be able to "push" the nematicon out of the linear trapping potential, where no restoring forces are present; thereby, above a certain excitation, the nematicon should self-escape the trap [11, 32].

For the experiments carried out by Peccianti et al. [11], we employed the cell sketched in Figure 11.17a, similar to the one employed in Section 11.7.1 but with the two applied voltages V_1 and V_2 of opposite signs, in order to have the low-frequency electric field mainly in the plane yz and in the gap between the top electrodes [11]. This geometry allows us to consider molecular reorientation only in the plane yz. Figure 11.17b–d plots the experimental results: for excitations up to 4 mW, the propagation in the perturbation regime allows the beam to remain trapped in the linear potential. As power is increased, the nematicon begins to bend toward positive y, until it eventually escapes the confining index well for powers larger than 13 mW, as shown in Figure 11.17d.

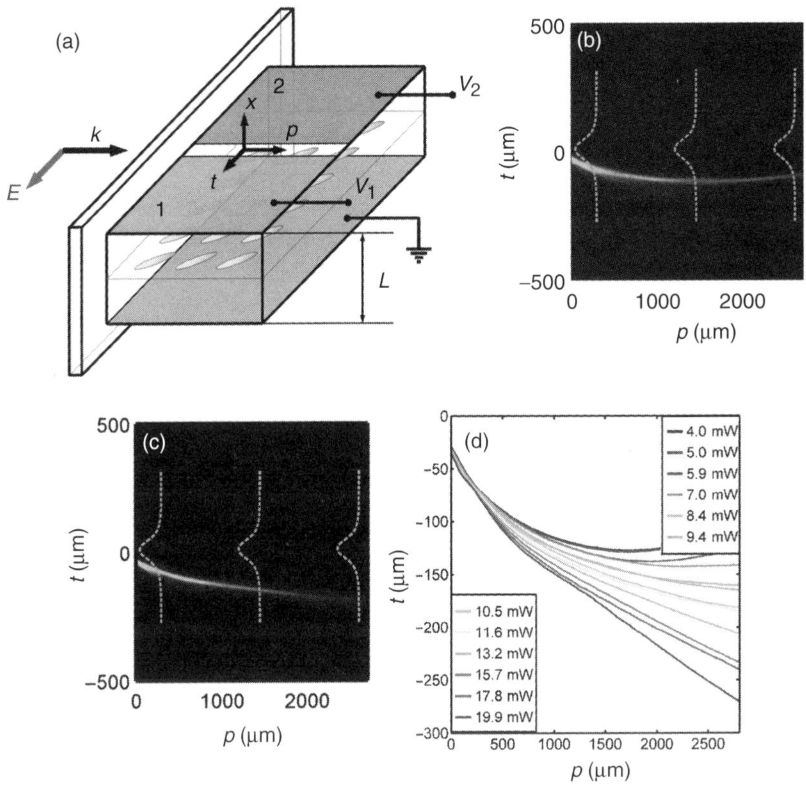

FIGURE 11.17 Nematicon self-escape from a linear potential. (a) Sketch of the cell. Measured nematicon evolution in the plane *tp* for (b) $P = 4$ mW and (c) $P = 13.2$ mW; the dashed lines represent the (attractive) linear index well with n_e peaked in $y = 0$. (d) Nematicon trajectories in the observation plane *tp* for various excitations. Here, the wavelength is 1064 nm, the beam is launched in $y = 0$, and its initial waist is about 4 µm.

11.8 CONCLUSIONS

In this chapter, we discussed nematicon properties in the highly nonlinear regime, the latter easily accessible in NLC because of their large reorientational response. When no losses are present, nematicon evolution strongly depends on initial conditions, as we discussed with reference to saturation effects and self-steering via nonlinear walk-off. We generalized our findings to include the role of scattering losses, unavoidable in the nematic phase, and compared the theoretical predictions with experimental results. We showed how nematicons can undergo nonlinear changes in trajectory because of interaction with the boundaries, nonparaxiality and asymmetry in light–matter interaction in guest–host dye-doped NLC, interaction with inhomogeneities in the dielectric tensor. These results point out several routes toward the realization of all-optical guided-wave signal processors, allowing the design of communication networks where light itself can define and modify the topology of the implemented circuits.

Acknowledgments

Special thanks to Marco Peccianti and Malgosia Kaczmarek for their important role in the work here discussed. A.A. thanks Regione Lazio for financial support.

REFERENCES

1. R. W. Boyd. *Nonlinear Optics*. Academic Press, Boston, MA, 1992.
2. S. Trillo and S. Wabnitz. Nonlinear nonreciprocity in a coherent mismatched directional coupler. *Appl. Phys. Lett*., 49:752, 1986.
3. C. Conti, M. Peccianti, and G. Assanto. Route to nonlocality and observation of accessible solitons. *Phys. Rev. Lett*., 91:073901, 2003.
4. M. Peccianti, C. Conti, G. Assanto, A. De Luca, and C. Umeton. Routing of anisotropic spatial solitons and modulational instability in nematic liquid crystals. *Nature*, 432:733, 2004.
5. C. Conti, M. Peccianti, and G. Assanto. Spatial solitons and modulational instability in the presence of large birefringence: the case of highly non-local liquid crystals. *Phys. Rev. E*, 72:066614, 2005.
6. M. Peccianti and G. Assanto. Observation of power-dependent walk-off via modulational instability in nematic liquid crystals. *Opt. Lett*., 30:2290–2292, 2005.
7. A. Piccardi, A. Alberucci, and G. Assanto. Self-turning self-confined light beams in guest-host media. *Phys. Rev. Lett*., 104:213904, 2010.
8. A. Alberucci and G. Assanto. Propagation of optical spatial solitons in finite-size media: interplay between nonlocality and boundary conditions. *J. Opt. Soc. Am. B*, 24(9):2314–2320, 2007.
9. I. Kaminer, C. Rotschild, O. Manela, and M. Segev. Periodic solitons in nonlocal nonlinear media. *Opt. Lett*., 32(21):3209–3211, 2007.
10. M. Peccianti, G. Assanto, A. Dyadyusha, and M. Kaczmarek. Nonlinear shift of spatial solitons at a graded dielectric interface. *Opt. Lett*., 32:271–273, 2007.
11. M. Peccianti, A. Dyadyusha, M. Kaczmarek, and G. Assanto. Escaping solitons from a trapping potential. *Phys. Rev. Lett*., 101(15):153902, 2008.
12. A. Alberucci, A. Piccardi, M. Peccianti, M. Kaczmarek, and G. Assanto. Propagation of spatial optical solitons in a dielectric with adjustable nonlinearity. *Phys. Rev. A*, 82(2):023806, 2010.
13. M. Lax, W. H. Louisell, and W. B. McKnight. From Maxwell to paraxial wave optics. *Phys. Rev. A*, 11(4):1365–1370, 1975.
14. A. Alberucci and G. Assanto. Nonparaxial (1+1)D spatial solitons in uniaxial media. *Opt. Lett*., 36(2):193–195, 2011.
15. A. Alberucci and G. Assanto. On beam propagation in anisotropic media: one-dimensional analysis. *Opt. Lett*., 36(3):334–336, 2011.

16. P. G. DeGennes and J. Prost. *The Physics of Liquid Crystals*. Oxford Science, New York, 1993.
17. I. C. Khoo. *Liquid Crystals: Physical Properties and Nonlinear Optical Phenomena*. Wiley, New York, 1995.
18. C. Rotschild, O. Cohen, O. Manela, M. Segev, and T. Carmon. Solitons in nonlinear media with an infinite range of nonlocality: first observation of coherent elliptic solitons and of vortex-ring solitons. *Phys. Rev. Lett.*, 95:213904, 2005.
19. C. Rothschild, B. Alfassi, O. Cohen, and M. Segev. Long-range interactions between optical solitons. *Nat. Phys.*, 2:769, 2006.
20. A. Piccardi, M. Peccianti, G. Assanto, A. Dyadyusha, and M. Kaczmarek. Voltage-driven in-plane steering of nematicons. *Appl. Phys. Lett.*, 94(9):091106, 2009.
21. A. Alberucci and G. Assanto. Nematicons beyond the perturbative regime. *Opt. Lett.*, 35(15):2520–2522, 2010.
22. A. W. Snyder and D. J. Mitchell. Accessible solitons. *Science*, 276:1538, 1997.
23. I. Jánossy and T. Kósa. Influence of anthraquinone dyes on optical reorientation of nematic liquid crystals. *Opt. Lett.*, 17(17):1183–1185, 1992.
24. L. Marrucci and D. Paparo. Photoinduced molecular reorientation of absorbing liquid crystals. *Phys. Rev. E*, 56:1765–1772, 1997.
25. H. Inoue, T. Hoshi, J. Yoshino, and Y. Tanizaki. The polarized absorption spectra of some α-aminoanthraquinones. *Bull. Chem. Soc. Jpn.*, 45:1018–1021, 1972.
26. C. Conti, M. Peccianti, and G. Assanto. Observation of optical spatial solitons in a highly nonlocal medium. *Phys. Rev. Lett.*, 92:113902, 2004.
27. A. Alberucci, M. Peccianti, and G. Assanto. Nonlinear bouncing of nonlocal spatial solitons at the boundaries. *Opt. Lett.*, 32(19):2795–2797, 2007.
28. C. P. Jisha, A. Alberucci, R.-K. Lee, and G. Assanto. Optical solitons and wave-particle duality. *Opt. Lett.*, 36(10):1848–1850, 2011.
29. A. Alberucci, A. Piccardi, U. Bortolozzo, S. Residori, and G. Assanto. Nematicon all-optical control in liquid crystal light valves. *Opt. Lett.*, 35(3):390–392, 2010.
30. A. B. Aceves, J. V. Moloney, and A. C. Newell. Theory of light-beam propagation at nonlinear interfaces. i. equivalent-particle theory for a single interface. *Phys. Rev. A*, 39(4):1809–1827, 1989.
31. R. Barboza, A. Alberucci, and G. Assanto. Large electro-optic beam steering with nematicons. *Opt. Lett.*, 36(14):2725–2727, 2011.
32. G. Assanto, A. A. Minzoni, M. Peccianti, and N. F. Smyth. Optical solitary waves escaping a wide trapping potential in nematic liquid crystals: Modulation theory. *Phys. Rev. A*, 79:033837, 2009.

12

Twisted and Chiral Nematicons

URSZULA A. LAUDYN AND MIROSLAW A. KARPIERZ
Warsaw University of Technology, Warsaw, Poland

12.1 INTRODUCTION

The uniqueness of nematicons is connected with the reorientational nonlinearity, by which light can change the initial orientation of liquid crystals. Such orientation, being the result of long-range interactions between molecules, is introduced at the boundaries and can be modified by electric or magnetic fields [1–3]. Solitary waves in nematic liquid crystals (NLCs) were investigated in capillaries [4–6], but the most often used configurations are planar cells with two parallel glass plates. In such cells, NLC can be homogeneously oriented, for example, in the homeotropic texture (where molecules are perpendicular to the glass plates) [7–9] or in the homogenic (planar) texture (where molecules are parallel to the glass plates) [10, 11]. Configurations in which the mean direction of the molecules (molecular director $\vec{n}$) varies across the cell thickness, as in planar cells biased by a low-frequency external electric field [12–14], seem to be promising. A nonhomogeneous initial orientation can also be introduced by the boundary conditions, as in hybridly aligned nematics (molecules parallel to one plate and perpendicular to the second plate) or in twisted nematic liquid crystals (TNLC, in which the molecules are parallel to both plates but twisted within the layer thickness). This twisted orientation also characterizes chiral nematic liquid crystals (ChNLC, cholesteric liquid crystals).

Nematicons in TNLC and ChNLC have properties similar to standard nematicons in other configurations (Chapter 1): they are observed over distances of a few millimeters, they exist in a wide range of optical powers above a few milliWatts, they can be redirected by external electric fields, their excitation depends on light polarization, they can guide other signals, and their mutual interactions are attractive. However, at variance with other orientations, a cell

Nematicons: Spatial Optical Solitons in Nematic Liquid Crystals, First Edition.
Edited by Gaetano Assanto.

filled by ChNLC offers the possibility of propagating nematicons (without external fields) in several independent layers. By choosing the birefringence and the period of orientation, either the independent propagation of nematicons or interaction characteristic for systems with discrete diffraction can be observed.

12.2 CHIRAL AND TWISTED NEMATICS

ChNLC consist of molecules with twisted alignment across the cell [1–3, 15, 16]. This results in a helical structure of finite pitch p, the latter defined as *the distance along the helical axis after which the molecular orientation has turned by an angle* 2π (Fig. 12.1). The ChNLC periodicity is $p/2$ as directors $\vec{n}$ and $-\vec{n}$ are equivalent (for nonpolar molecules). Chirality can be induced by mixing a nematic liquid crystal with a chiral dopant. The pitch p of such a mixture can be easily modified because it depends on the concentration η of the chiral dopant: $p = 1/(H\eta)$, where H is the helical twisting power determined by the chemical structure of the chiral material [2, 16]. In the investigated ChNLC cells, the molecular director is parallel to the surrounding glass plates (interfaces) and twisted within the layer thickness. A similar orientation can be obtained for TNLCs, where the NLC molecules are anchored parallel to the glass plates, and these plates are mutually rotated by some angle. However, the maximum rotation in TNLCs is $\pi/2$, that is, corresponding to $p/4$.

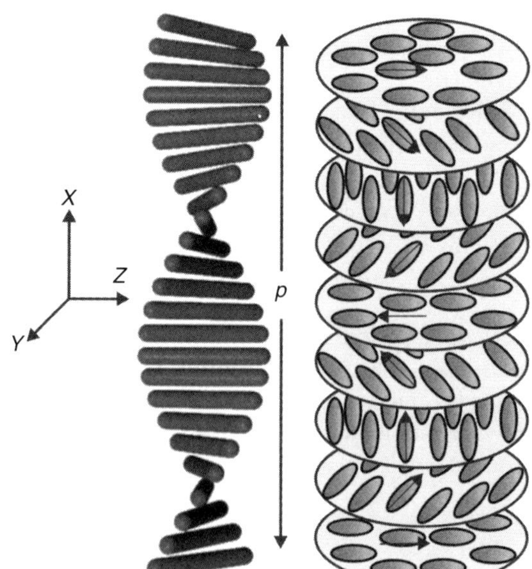

FIGURE 12.1 Orientational order in chiral nematic liquid crystals.

The expression for the free energy of deformed ChNLCs is quite similar to that in NLC, and its density takes the form [1–3, 15, 16]

$$f_F = \frac{1}{2}K_{11}(\nabla \cdot \vec{n})^2 + \frac{1}{2}K_{22}(\vec{n} \cdot \nabla \times \vec{n} + G)^2 + \frac{1}{2}K_{33}(\vec{n} \times \nabla \times \vec{n})^2, \qquad (12.1)$$

where K_{ii} are the elastic (Frank) constants for the three basic spatial distortions of the molecules. In the term related to a twist deformation, the parameter $G = 2\pi/p$ describes chirality (in nematics $G = 0$). Energy minimization in unperturbed ChNLCs leads to $\vec{n} \cdot \nabla \times \vec{n} = -G$, which can be fulfilled for example, for a director $\vec{n} = (0, \cos\theta, \sin\theta)$ with the orientation angle $\theta = \theta_0 + Gx$. Such an orientation corresponds to a helical structure with an axis along the x direction (as in Figure 12.1).

External fields and surface interactions can easily deform the initial helical configuration. The electrical permittivity of ChNLCs for fields parallel to the helical axis is different from the average permittivity in perpendicular directions. Therefore, an external electric field also causes the reorientation of the whole helical axes [17–19]. It should be noted that the deformation occurs in two stages: first, the molecules are tilted toward the electric field direction and then, untwisting of the helical structure occurs.

Optical spatial solitons or nematicons are investigated in configurations in which a light beam propagates parallel to the glass plates. We assume a beam propagating in the z-direction with a dominating E_y component of the electric field (as in Figure 12.2). For light polarized along y, the refractive index varies across the cell from the ordinary value n_o in those planes in which the molecules are parallel to the z-axis (i.e., where $\theta = \pi/2$) to the extraordinary value n_e in those where the molecules are parallel y ($\theta = 0$). This can be compared to light propagation in a graded index waveguide. Therefore, the light is guided in a thin layer where

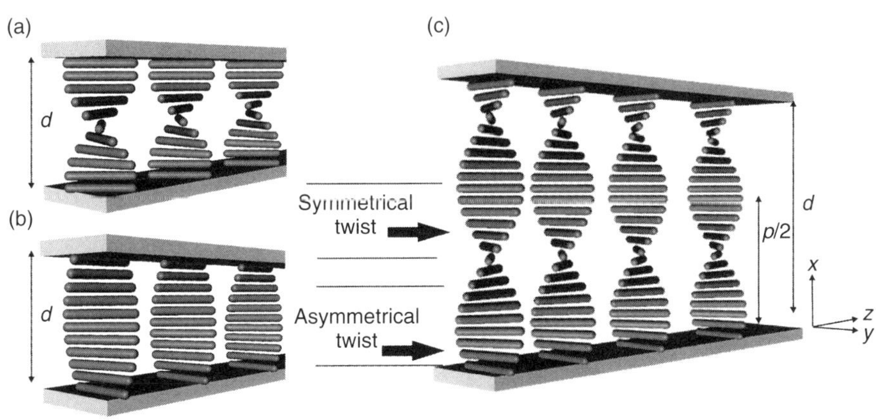

FIGURE 12.2 Configuration of (a) symmetrically and (b) asymmetrically twisted TNLCs; (c) configuration of ChNLC layers with the indication of potential input beams.

$\theta = 0$. By acting on the anchoring conditions, guiding index layers along x can be formed within the cell thickness, with location depending on θ at the glass/NLC interface and on the pitch p. For anchoring conditions such that $\theta_0 = \pi/2$, the largest refractive index is at the edges of the cell: if the refractive index of the glass is lower than the extraordinary index n_e, the beam is guided at the NLC/glass boundary, partially extending into the isotropic medium (glass) and partially into the twisted liquid crystals. The asymmetry in molecular twist causes the appearance of beam walk-off.

12.3 THEORETICAL MODEL

The problem of light propagation in TNLCs and ChNLCs is quite complex, and its quantitative description requires an exact treatment of electromagnetic fields in anisotropic media [20]. However, for a qualitative description, the behavior of a light beam propagating in ChNLCs can be simply described by using the following approximated model [21–23]. This model assumes that the beam profile in the x-direction is invariant during propagation, that is, it assumes that the beam is guided by the linear waveguide created by the pitch. The results are in good agreement with experiments, and the description is adequate for both chiral and twisted structures.

The optic axis changes across the ChNLC cell, generally implying that all components of the electromagnetic fields are nonzero, even for linearly polarized input beams. However, only E_y and E_z cause molecular reorientation in the yz-plane. Along x, when the beam width ensures a good overlap with the guided mode (i.e., for a proper ratio of pitch and beam waist), the beam does not diffract appreciably in that direction. In this case, it can be assumed that the reorientation of the molecules is only in the yz-plane and is mainly connected with the twist deformation. Consequently, the electric permittivity tensor has the components $\varepsilon_{xy} = \varepsilon_{yx} = \varepsilon_{xz} = \varepsilon_{zx} = 0$ and $\varepsilon_{xx} = n_0^2$. In this case, Maxwell's equations for monochromatic electromagnetic waves lead to

$$\left[k_0^2\varepsilon_{yy} + \frac{\partial^2}{\partial x^2} + \frac{\partial^2}{\partial y^2}\frac{\varepsilon_{yy}}{\varepsilon_{xx}} + \frac{\partial^2}{\partial z^2} + \frac{\partial^2}{\partial y \partial z}\frac{\varepsilon_{yz}}{\varepsilon_{xx}}\right]E_y$$
$$= \left[-k_0^2\varepsilon_{yz} - \frac{\partial^2}{\partial y^2}\frac{\varepsilon_{yz}}{\varepsilon_{xx}} - \frac{\partial^2}{\partial y \partial z}\left(\frac{\varepsilon_{zz}}{\varepsilon_{xx}} - 1\right)\right]E_z, \tag{12.2}$$

$$\left[k_0^2\varepsilon_{zz} + \frac{\partial^2}{\partial x^2} + \frac{\partial^2}{\partial y^2} + \frac{\partial^2}{\partial z^2}\frac{\varepsilon_{zz}}{\varepsilon_{xx}} + \frac{\partial^2}{\partial y \partial z}\frac{\varepsilon_{yz}}{\varepsilon_{xx}}\right]E_z$$
$$= \left[-k_0^2\varepsilon_{yz} - \frac{\partial^2}{\partial z^2}\frac{\varepsilon_{yz}}{\varepsilon_{xx}} - \frac{\partial^2}{\partial y \partial z}\left(\frac{\varepsilon_{yy}}{\varepsilon_{xx}} - 1\right)\right]E_y, \tag{12.3}$$

where $k_0 = \omega/c$ is the vacuum wave number.

For the beam propagating in the z-direction, it is convenient to introduce the ansatz

$$E_y = A(y, z)\psi(x)\exp(\mathrm{i}\omega t - \mathrm{i}k_0 N z), \tag{12.4}$$

$$E_z = A(y, z)\phi(x)\exp(\mathrm{i}\omega t - \mathrm{i}k_0 N z), \tag{12.5}$$

where $\phi(x)\exp(\mathrm{i}\omega t - \mathrm{i}k_0 N z)$ and $\psi(x)\exp(\mathrm{i}\omega t - \mathrm{i}k_0 N z)$ are components of the planar waveguide mode with an effective refractive index N. The modal envelopes $\phi(x)$ and $\psi(x)$ fulfill the equations

$$\left[\frac{\partial^2}{\partial x^2} + k_0^2\left(\varepsilon_{yy}^{(0)} - N^2\right)\right]\psi = -k_0^2\varepsilon_{yz}^{(0)}\phi, \tag{12.6}$$

$$\left[\frac{\partial^2}{\partial x^2} + k_0^2\varepsilon_{zz}^{(0)}\left(\frac{N^2}{\varepsilon_{xx}} - 1\right)\right]\phi = k_0^2\varepsilon_{yz}^{(0)}\left(\frac{N^2}{\varepsilon_{xx}} - 1\right)\psi, \tag{12.7}$$

where $\varepsilon^{(0)}$ is the electric permittivity corresponding to the initial orientation of the nematics. In the presence of such fields, the Euler–Lagrange equation for energy minimization has the form

$$\frac{\partial^2\theta}{\partial x^2} + \frac{\varepsilon_0\Delta\varepsilon}{4K_{22}}\left[\left(\phi^2 - \psi^2\right)\sin(2\theta) + 2\psi\phi\cos(2\theta)\right] = 0, \tag{12.8}$$

where $\Delta\varepsilon = n_e^2 - n_o^2$ is the optical anisotropy of the electric permittivity. Reorientation changes the local value of the electric permittivity tensor, which depends on the orientation angle θ as $\varepsilon_{yy} = n_o^2 + \Delta\varepsilon\cos^2\theta$, $\varepsilon_{zz} = n_o^2 + \Delta\varepsilon\sin^2\theta$, $\varepsilon_{yz} = \varepsilon_{zy} = \Delta\varepsilon\sin\theta\cos\theta$.

For the paraxial approximation electric field components of the light beam $E_z \ll E_y$, the slowly varying complex amplitude A fulfills the equation obtained from integration of Equation 12.2 over the cross section

$$-2\mathrm{i}k_0 N\frac{\partial}{\partial z}A + \frac{\partial^2}{\partial y^2}\gamma_1 A - 2\mathrm{i}k_0 N\frac{\partial}{\partial y}\gamma_2 A + k_0^2\varepsilon_{xx}\left(\gamma_1 - \gamma_1^{(0)}\right)A = 0, \tag{12.9}$$

where the coefficients γ_1 and γ_2 depend on the orientation angle θ as

$$\gamma_1 = 1 + \frac{\Delta\varepsilon}{\varepsilon_{xx}}\frac{\int\psi^2\cos^2\theta\,\mathrm{d}x}{\int\psi^2\mathrm{d}x}, \tag{12.10}$$

$$\gamma_2 = \frac{\Delta\varepsilon}{2\varepsilon_{xx}}\frac{\int\psi^2\sin\theta\cos\theta\,\mathrm{d}x}{\int\psi^2\mathrm{d}x}. \tag{12.11}$$

At low optical power (the linear case), the coefficients γ have initial values equal to $\gamma^{(0)}$. As light intensity increases (the nonlinear case), the dominant E_y component of a linearly polarized beam induces reorientation in the yz-plane. The liquid crystal molecules are forced to reorient parallel to the electric field, which increases the

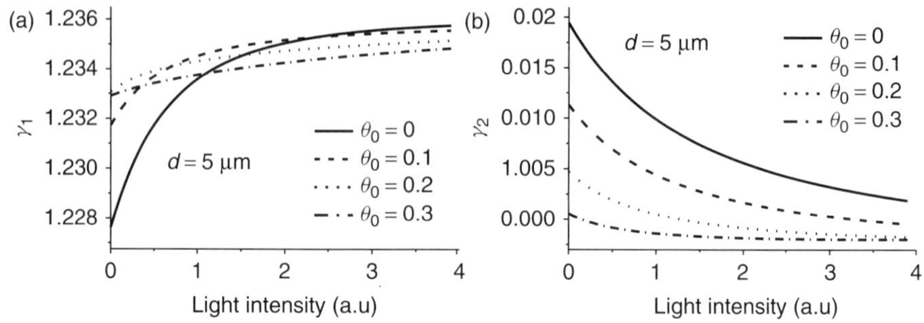

FIGURE 12.3 Dependence of nonlinear coefficients (a) γ_1 and (b) γ_2 on light intensity. The calculations were performed for TNLC films with orientation $\theta = \theta_0 + \pi x/2d$ and thickness $d = 5$ μm for different orientations at the boundary θ_0. Increasing θ_0 causes the beam to propagate farther from the cell boundary, that is, the configuration becomes more symmetric.

γ_1 coefficient. It modifies the diffractive broadening of the beam in the y-direction, whereas $(\gamma_1 - \gamma_1^{(0)})$ is responsible for self-focusing of the beam and the creation of a spatial soliton. It should be noted that the coefficient γ_1 has a saturable form and can be calculated for a given liquid crystal layer, as shown in Figure 12.3a.

The second nonlinear coefficient γ_2 is responsible for beam walk-off. This coefficient is zero in the symmetric case, that is, when the light beam profile ψ is symmetrically located around the layer in which $\theta = 0$. This occurs with beams propagating inside the cell. If the beam is guided at the cell boundary (partially in the glass plate), the coefficient $\gamma_2 \neq 0$ defines the walk-off angle. In such a case, at low intensities, γ_2 is the largest, and an increase in intensity causes a decrease in γ_2 (as shown in Figure 12.3b). It means that for higher intensities, the beam walk-off is smaller, that is, the beam direction changes with light power (Chapter 11).

12.4 EXPERIMENTAL RESULTS

12.4.1 Nematicons in a Single Layer

Self-focusing has been observed in a ChNLC cell filled by capillary effect with 6CHBT (4-trans-4′-n-hexyl-cyclohexyl-isothiocyanatobenzene) doped with a chiral material. 6CHBT possesses low absorption and high nonlinear response [24, 25], with refractive indices $n_o = 1.51$ and $n_e = 1.67$ in the near infrared. The sample consisted of two glass plates glued together, with a gap controlled by a spacer. An alignment layer was deposited on the top plate to control the alignment of the liquid crystal molecules. The beam propagation in the cell was observed with a microscope objective and a CCD camera. The thickness of the cell was 50 μm, and the pitch of the ChNLC was about 25 μm. The long molecular axes were approximately parallel to the plane of the layers constituting the helical structure. The input beam waist was estimated to be about 2 μm by measuring the divergence during linear propagation in the NLC. Typical experimental results obtained with

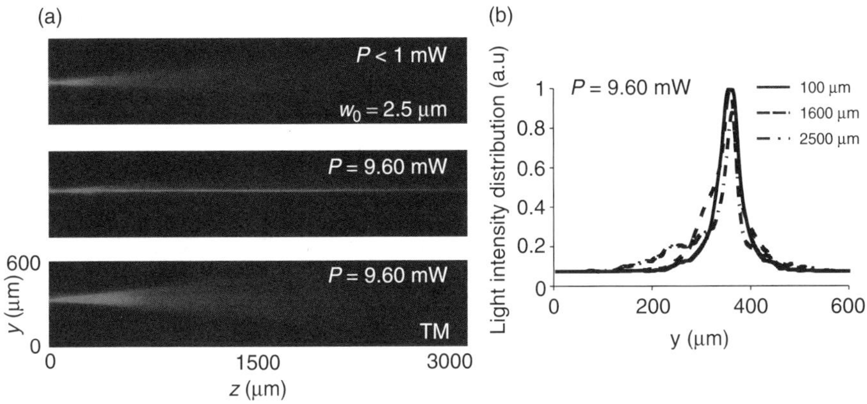

FIGURE 12.4 Experimental results of spatial soliton creation in ChNLC layer for Ti:Sapphire laser: (a) light beam propagation for different inputs of light power (marked on photos) and centrally launching light beam, the last picture was taken for TM polarization; (b) light intensity profiles for different values of propagation distance.

a Ti:Sapphire laser ($\lambda = 793$ nm) are presented in Figure 12.4 [26, 27]. Increasing the input power led to self-focusing and, finally, for an average power $P = 9.6$ mW, a spatial soliton was formed. The solitary beam had a transverse intensity distribution that remained unchanged after a propagation distance of about 3 mm (i.e., over 80 times the Rayleigh length).

Owing to the finite thickness of each layer in the chiral structure, self-focusing can balance diffraction and give rise to self-trapped solitons only in a limited waist range [22]. Owing to structural anisotropy connected with chiral orientation, the beam walks-off from the initial direction of propagation while changing its polarization. Indeed, as shown in Figure 12.5, crossing from the TE-like to TM-like polarization causes the nematicon to propagate in different directions. However, for the TM-like polarization (for which the Freedericksz threshold prevents nematicon formation), the light beam propagates again along the z-axis.

12.4.2 Asymmetric Configuration

It should be noted that beam walk-off was also observed in asymmetrically twisted liquid crystals [28–30]. Typical pictures are presented in Figure 12.6a. First, light propagation in a homogeneous medium was investigated to verify that the beam was launched parallel to the y-axis. This was achieved by applying an external field, as an applied field with E_y polarization renders the medium homogeneous. Next, without an external electric field, the direction of beam propagation was changed, showing walk-off. In the linear case, when the light power was too low to induce reorientation, beam diffraction was observed. Increasing the power modified the twisting angle and increased the effective refractive index. Light started to change its direction of propagation and eventually created a solitary wave via self-focusing. The beam was guided close to the TNLC/glass boundary. As a consequence, the

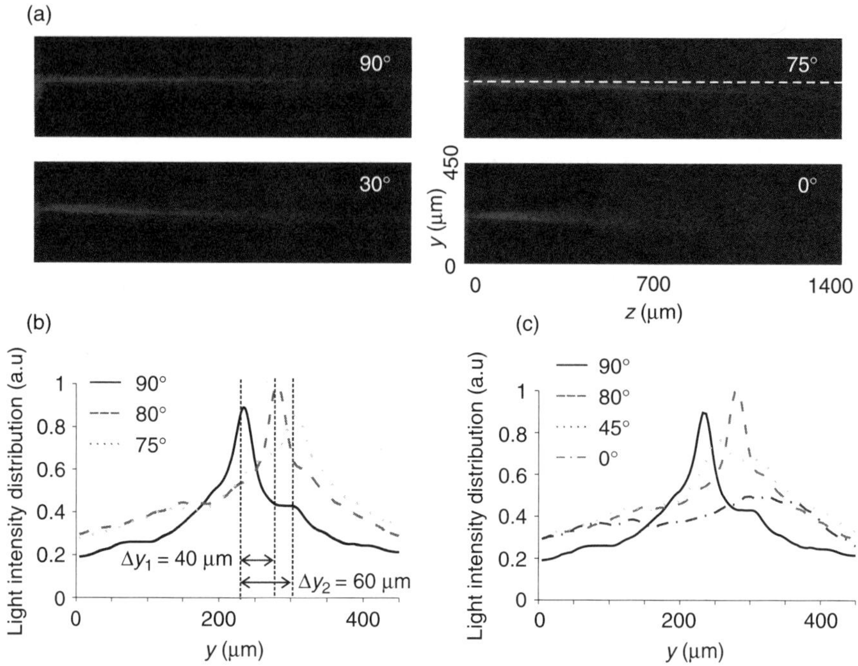

FIGURE 12.5 (a) Experimental results showing beam walk-off by changing the polarization of the input beam. (b) Normalized light intensity profiles for three different polarizations for which the soliton changed its direction. (c) Normalized light intensity profiles for different polarizations switching from TE to TM-like.

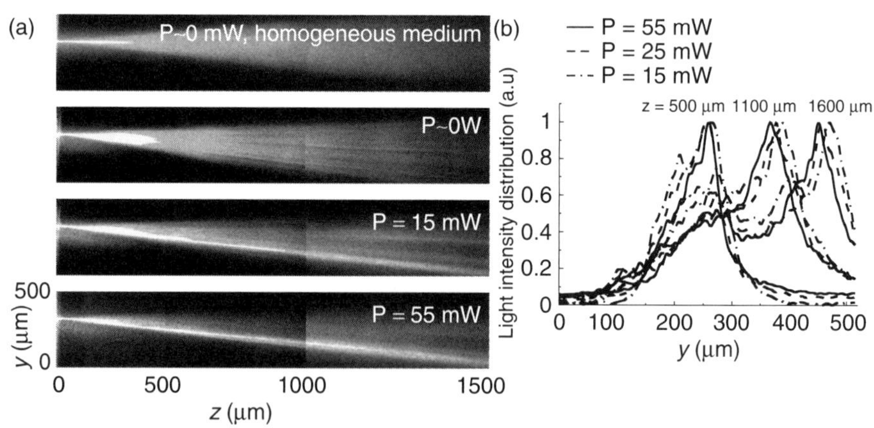

FIGURE 12.6 Experimental results on nematicon creation with an argon laser ($\lambda = 514$ nm) beam in asymmetrically configured TNLCs. (a) Beam propagation for various input powers (marked on photos) and (b) normalized intensity profiles for various propagation distances and powers.

effective direction of the birefringence axis was not perpendicular to the initial beam direction, showing walk-off.

A comparison of size and shape of such a nematicon for various optical powers and at different distances are presented in Figure 12.6b. The intensity cross section of the scattered light was normalized to its maximum value. The size and shape of the obtained nematicons do not change significantly with increasing input power. However, an increase in power can modify the walk-off. The changes in beam direction versus power are weak but measurable. Effectively, at a distance of 1 mm, the beam can change position by approximately 20 μm as the power goes from 15 to 55 mW.

12.4.3 Multilayer Propagation

A light beam propagates in the region where the ChNLC molecules are parallel to the electric field. The twist is periodic versus the cell thickness, which implies that there are a few high-index regions where the beam can propagate. The number of such layers is determined by the cell thickness and the chiral pitch: it corresponds to $2d/p$.

In this configuration, the width of the cell ($d = 50$ μm) and the ChNLC pitch ($p = 25$ μm) are such that there are four layers with molecules oriented in the same way (Fig. 12.7a). As a result, there are four layers with a thickness of about 12 μm in which nematicons can be excited independently [26, 27]. By changing the vertical position along the cell, it is possible to independently launch as many solitons as the number of layers in the structure. The vertical position in a ChNLC cell is controlled by means of a microscope slide with micrometric patterns fixed to the (x, y, z) stage and a second CCD camera mounted at the butt of the cell.

The vertical position of the ChNLC cell was changed for a defined TE-like polarization and input power high enough (about 10 mW) to form nematicons. Typical results are shown in Figure 12.7b for the input positions marked on the photos. The position $\Delta x = 0$ corresponds to a light beam propagating at the boundary between a glass plate and an NLC. A nematicon is formed in the first layer marked $\Delta x = 10$ μm and can be generated in each of the subsequent layers. Indeed, four nematicons were formed in distinct layers, about 10–12 μm away from each other.

12.4.4 Influence of an External Electric Field

In a separate set of experiments, the influence of an external electric field on nematicon propagation was also investigated [31, 32]. An electric field (voltage) along the x-axis, thus parallel to the helix axis, leads to a bend deformation and director reorientation in the xy-plane. When a nematicon was generated in one of the middle guiding layers (for an input power of 15.4 mW and $\lambda = 514$ nm), it induced a channel waveguide with confinement along y. Applying an electric field along x changed the director orientation, and consequently, the soliton started to propagate at some angle to the z-axis. Figure 12.8a shows such a case corresponding to $P = 15.4$ mW and measured over a distance of $z = 600$ μm. A further increase

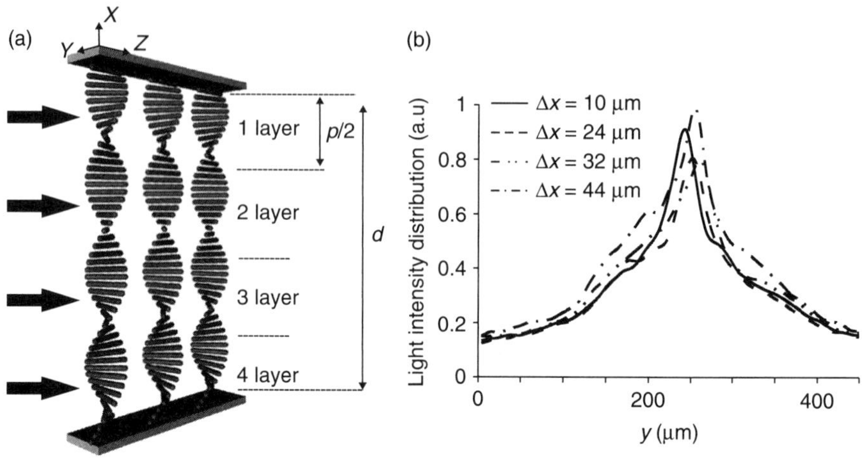

FIGURE 12.7 (a) Schematic configuration of the chiral nematic liquid crystal cell with marked layers and (b) normalized intensity profiles for each position and after a propagation distance $\Delta z = 1$ mm.

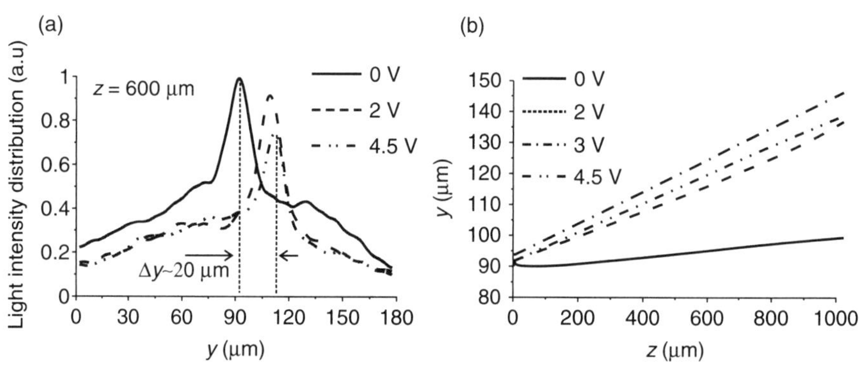

FIGURE 12.8 Experimental results showing (a) the normalized light intensity distribution for three voltages for which the soliton changed direction and (b) nematicon trajectory for various applied voltages.

in the electric field at fixed optical power increased the relative contribution of the electrically induced reorientation and destroyed the helical structure. As a result, the beam did not follow the molecule orientation, resulting in linear diffraction. Figure 12.8b plots the soliton trajectory as the voltage varies from 0 to 4.5 V. The displacement changed by 20 μm at a distance $z = 600$ μm. The maximum steering was estimated to be about 3°, which is comparable with the birefringent walk-off in this material (the maximum walk-off was calculated to be 3.5°).

Furthermore, when the soliton propagated in the guiding layers nearest to the boundaries, the voltage required for steering shifted to higher values. This is caused by the fact that the anchoring conditions strongly inhibit molecular reorientation. In the middle layers, the director orientation changes are much less affected by the boundary conditions.

The polarization dependence of an input beam on steering was also investigated [31]; it was found that a change in input polarization alters the soliton direction for given beam, layer, and electric field. Changing the polarization causes the nematicon to propagate in a different direction relative to the z-axis. For an input polarization along y and a voltage of approximately 3.5 V, the beam propagates at an angle to the z-axis as a consequence of reorientation of the molecules. Changing the input polarization but keeping the voltage constant makes the beam propagate at an angle smaller than the z-axis.

Usually, in these kinds of experiments, one can see a diffractive background. This occurs because the field-induced reorientation changes the direction of the extraordinary axis in the xy-plane by the tilt angle, that is, a y-polarized input beam contains a small component along the new ordinary axis. With no voltage and sufficient power to create a nematicon, the wave has only an E_y component. With an external field, the nematicon propagates at a tilt angle relative to the yz-plane, with fields polarized along the new director. Furthermore, the small ordinary component diffracts in propagation.

12.4.5 Guiding Light by Light

Similar to other nematicons, it was verified in TNLCs and ChNLCs that a solitary wave can confine a second low power probe, for example, from a He–Ne laser [23, 30, 32]. Nematicon propagation in a ChNLC or TNLC cell introduces changes in the refractive index distribution and leads to the formation of an optical waveguide (Chapter 1). This can be verified by injecting a second copolarized low-power probe, generally at a different wavelength. When the pump beam diffracts, the probe beam diffracts as well (Fig. 12.9a). When a high-power TE-polarized beam (from an argon laser, in this case) forms a spatial soliton, the copropagating low-power probe is also confined (Fig. 12.9b). Moreover, the probe follows exactly the direction of the soliton. The pump beam causes a focusing "lensing" effect, and the probe becomes increasingly confined, too.

12.4.6 Nematicon Interaction

The interaction between two identical nematicons was also verified in ChNLCs [27]. If two nematicons are launched close to each other in one layer, they can attract and merge into one single self-trapped beam. This happens because of the spatially nonlocal response of liquid crystals, analogous to former experiments in planarly aligned NLC cells, for example, References 33, 34 and Chapters 1–2.

Two beams with $w_0 = 2$ µm and separated by 25 µm were launched in one layer of ChNLC to form two identical nematicons. The refractive perturbation created by one beam diffuses and affects the other beam, causing two initially parallel solitons to attract one another. Depending on the initial geometry and the strength of nonlinearity, the two-beam interaction can have different outputs, that is, the solitons can either drag each other or pass through each other. The distance at which two solitons collide decreases with the input power. Solitons can not only

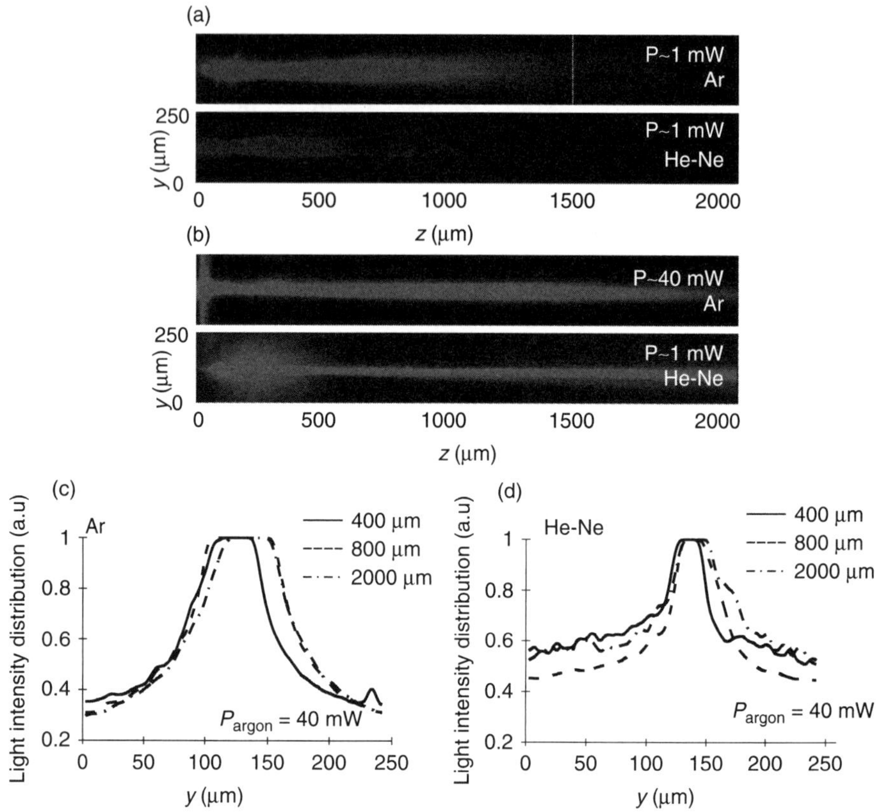

FIGURE 12.9 Spatial soliton and optically induced waveguide in a ChNLC cell. (a) A linearly diffracting pump beam from an Ar laser and the corresponding propagation of a probe He–Ne beam, (b) soliton propagation for a high power Ar beam and the corresponding nondiffractive propagation of a probe beam, (c,d) normalized light intensity profiles at various propagation distances for (c) high power Ar beam and (d) colaunched low power He–Ne probe beam. Flat tops of the intensity profiles are caused by saturation of the CCD camera.

interact with each other but also merge in a single self-confined beam. An example of time evolution of soliton–soliton interaction is also reported. Two beams of 10 mW propagate in the *yz*-plane with initial separation of 25 μm. Owing to the slow ChNLC time response, after illumination, no nonlinear behavior is visible and diffraction is observed with an overlap of the two waves as they propagate. After 1 s, two solitons are formed and propagate, maintaining their initial separation (Fig. 12.10a). In a few seconds, the solitons start to attract each other. Eventually, owing to the nonlocality of the NLC, after 35 s, they collapse into one beam (Fig. 12.10b) giving rise to a stable solution.

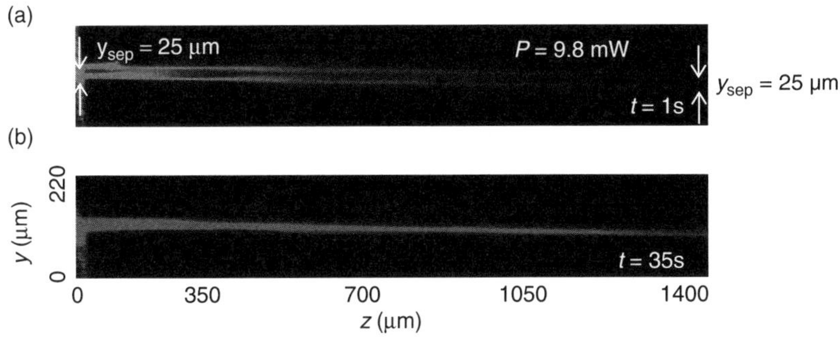

FIGURE 12.10 Time evolution of two interacting solitons of power 9.8 mW each. (a) Soliton formation after 1 s and separation equal to the initial one and (b) soliton fusion into one beam after 35s.

12.5 DISCRETE DIFFRACTION

The results presented in the previous section were obtained in a single layer of the ChNLC cell. However, the interaction between solitons propagating in different layers is also possible [35]. Owing to the fact that the refractive index distribution changes periodically along the x-direction, that is, perpendicular to the glass plates, the analyzed cell can be treated as an array of planar waveguides. In such structure, it is possible to obtain the conditions for which the standard continuous diffraction is substituted by a discrete one by way of a coupling between waveguides side by side. Discrete diffraction has numerous interesting features (Chapter 10). In ChNLCs, a beam can diffract in two ways: continuously in one direction (in the yz-plane) and discretely in the other (in the xz-plane). The reorientational nonlinearity can suppress both types of diffraction and form nematicons.

To enable the experimental observation of discrete diffraction, the coupling length L_c defining the distance at which light switches from one waveguide to the adjacent one should be shorter than the sample length. The coupling length can be modified by changing the amplitude and period of the refractive index modulation, that is, by changing pitch and birefringence in ChNLCs. Low birefringence ChNLCs with a pitch $p < 10$ μm should be used, as presented in Figure 12.11a. Numerical simulations of beam diffraction in such configuration are shown in Figure 12.11c. Note that discrete diffraction does not depend on beam waist but on the refractive index modulation. To obtain comparable amounts of discrete (along x) and continuous (along y) diffractions, the initial beam size along y needs to be properly engineered (in the simulations, the input beam is much wider in y than in x).

The creation of spatial solitons requires an increase of refractive index, which must be higher for shorter coupling lengths. In the presence of reorientational nonlinearity, the maximum changes in refractive index are limited by the birefringence. The ratio between the nonlinear index changes necessary to create solitons

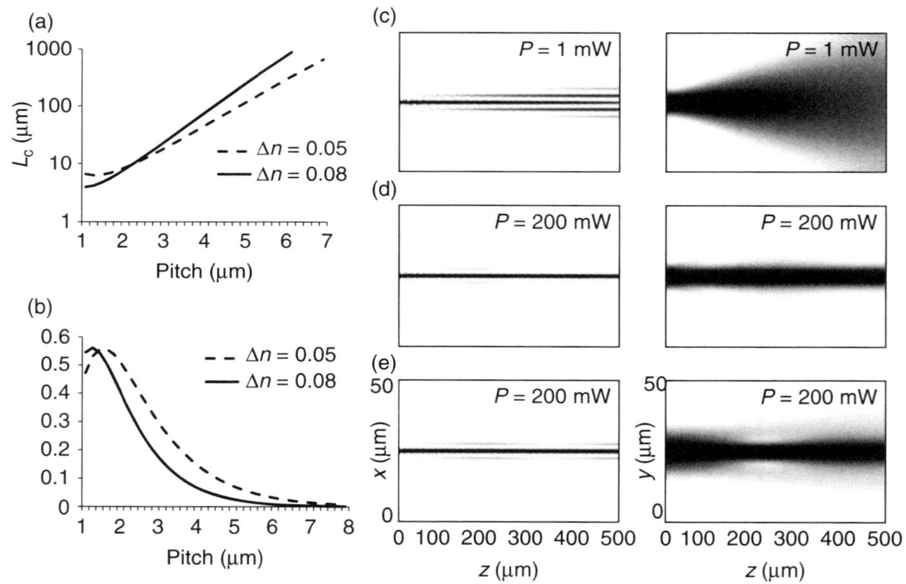

FIGURE 12.11 (a) Dependence of coupling length L_c on pitch in ChNLCs with two values of birefringence Δn; (b) dependence of the ratio between nonlinear refractive changes necessary to observe solitons and maximum possible changes in pitch for two ChNLCs with different birefringence; (c–e) numerical simulations of beam propagation in ChNLC with pitch $p = 5$ μm and birefringence $\Delta n = 0.08$: diffraction in the xz-plane (left) and in the yz-plane (right) for (c) the linear case (power $P =$ 1mW); (d) nonlinear case ($P = 200$mW) for initial beam size 1 μm ×4 μm (FWHM); (e) nonlinear case for beam size 1 μm × 8 μm.

and the maximum possible changes versus pitch is graphed in Figure 12.11b. A lower ratio requires lower reorientation; in turn a lower input power is necessary to form nematicons. As a consequence, pitch and birefringence cannot be too small. In addition, for low birefringence ChNLCs, the elastic constants are larger than in 6CHBT. Keeping in mind the previous discussion, the powers necessary to excite discrete nematicons are much larger than in the configurations discussed earlier.

The nematicon shown in Figure 12.11d was generated by a beam diffracting with comparable divergences in both dimensions, that is, its size in y was chosen so that Rayleigh and coupling lengths were equal. However, if these two divergences were different, the creation of spatial solitons would be more difficult (as in the example in Figure 12.11e) and sometimes even impossible.

In summary, discrete nematicons seem to be very peculiar of ChNLCs but require larger powers and more accurately controlled excitation conditions than standard nematicons.

12.6 CONCLUSIONS

TNLC and ChNLC form convenient geometry for creation of spatial solitons. They do not need an external electric field, and optimization of nematicon parameters can be done by choosing the pitch and the birefringence of NLCs. The properties of observed solitons are typical for nematicons. They can also be easily steered by external fields, by other beams, as well as by changing the polarization or light power. The prospective properties of ChNLCs are connected with the existence of multiple-layered structures created by periodical orientation of molecules. For high birefringent ChNLCs, it allows to propagate independent nematicons in different positions across the cell. On the other hand, for low birefringent ChNLCs it can be an interesting structure for application of discrete diffraction. This, among others, can be the basis of three-dimensional elements for light beam switching and routing.

Acknowledgments

We are grateful to Edward Nowinowski-Kruszelnicki, Marek Sierakowski, Katarzyna Jaworowicz, Michal Kwasny, Katarzyna Rutkowska, and Filip Sala for extensive contributions to this work. This work was supported financially by the National Science Centre.

REFERENCES

1. P. G. de Gennes and J. Prost. *The Physics of Liquid Crystals*, 2nd edn. Oxford University Press, London, 1995.
2. D. K. Yang and S. T. Wu. *Fundamentals of Liquid Crystal Devices*. Wiley, New York, 2006.
3. I. C. Khoo. *Liquid Crystals*, 2nd edn. Wiley, New York, 2007.

4. E. Braun, L. P. Faucheux, and A. Libchaber. Strong self-focusing in nematic liquid crystals. *Phys. Rev. A*, 48:611–622, 1993.
5. D. W. McLaughlin, D. J. Muraki, M. J. Shelley, and X. Wang. A paraxial model for optical self-focussing in a nematic liquid crystal. *Physica D*, 88:55–81, 1995.
6. M. Warenghem, J. F. Henninot, and G. Abbate. Non linearly induced self waveguiding structure in dye doped nematic liquid crystals confined in capillaries. *Opt. Express*, 2:483–490, 1998.
7. M. A. Karpierz, M. Sierakowski, M. Swillo, and T. R. Wolinski. Self focusing in liquid crystalline waveguides. *Mol. Cryst. Liq. Cryst.*, 320:157–163, 1998.
8. M. A. Karpierz, in *Soliton Driven Photonics*, eds. A. D. Boardman and A. P. Sukhorukov, Kluwer, Dordrecht, Netherlands, p. 41, 2001.
9. M. A. Karpierz. Solitary waves in liquid crystalline waveguides. *Phys Rev. E*, 66:036603, 2002.
10. A. Piccardi, A. Alberucci, and G. Assanto. Soliton self-deflection via power-dependent walk-off. *Appl. Phys. Lett.*, 96:061105, 2010.
11. A. Piccardi, A. Alberucci, and G. Assanto. Self-turning Self-confined light beams in Guest-Host media. *Phys. Rev. Lett.*, 104:213904, 2010.
12. M. Peccianti, G. Assanto, A. De Luca, C. Umeton, and I. C. Khoo. Electrically assisted self-confinement and waveguiding in planar nematic liquid crystal cells. *Appl. Phys. Lett.*, 77:7–9, 2000.
13. G. Assanto and M. Peccianti. Spatial solitons in nematic liquid crystals. *IEEE J. Quantum Electron.*, 39:13–21, 2003.
14. M. Peccianti, C. Conti, G. Assanto, A. De Luca, and C. Umeton. Routing of highly anisotropic spatial solitons and modulational instability in liquid crystals. *Nature*, 432:733–737, 2004.
15. P. Oswald and P. Pieranski. *Nematic and Cholesteric Liquid Crystals: Concepts and Physical Properties Illustrated by Experiments*. Taylor & Francis, London, 2005.
16. H. S. Kitzerrow and C. Bahr. *Chirality in Liquid Crystals*. Springer, New York, 2001.
17. R. B. Meyer. Effects of electric and magnetic fields on the structure of cholesteric liquid crystals. *Appl. Phys. Lett.*, 12:281–283, 1968.
18. A. Chanishvili, G. Chilaya, and D. Sikharulidze. Electro-optic effect in an optically active nematic chiral liquid crystal structure. *Appl. Opt.*, 33:3482–3485, 1994.
19. L. M. Blinov. Electro-optical effects in liquid crystals. *Sov. Phys. Usp.*, 17:658, 1975.
20. F. A. Sala and M. A. Karpierz. Numerical simulation of beam propagation in a layer filled with chiral nematic liquid crystals. *Photon. Lett. Pol.*, 1:163–165, 2009.
21. G. Assanto and M. A. Karpierz. Nematicons: self localised beams in nematic liquid crystals. *Liq. Cryst.*, 36:1161–1172, 2009.
22. U. A. Laudyn, M. Kwasny, K. Jaworowicz, K. Rutkowska, and M. A. Karpierz. Self focusing and nematicons in chiral nematic liquid crystals. *Proc. SPIE*, 7141:71410F, 2008.
23. U. A. Laudyn, K. Jaworowicz, and M. A. Karpierz. Spatial solitons in chiral nematics. *Mol. Cryst. Liq. Cryst.*, 489:214–221, 2008.
24. W. Baran, Z. Raszewski, R. Dabrowski, and J. Kedzierski. Some physical properties of mesogenic 4-(trans-4′-n-Alkylcyclohexyl) isothiocyanatobenzenes. *Mol. Cryst. Liq. Cryst.*, 123:237–245, 1985.

25. R. Dabrowski, J. Dziaduszek, and T. Szczucinski. Mesomorphic characteristics of some new homologous series with the isothiocyanato terminal group. *Mol. Cryst. Liq. Cryst.*, 124:241–257, 1985.

26. U. A. Laudyn, M. Kwasny, and M. A. Karpierz. Nematicons in chiral nematic liquid crystal cell. *Appl. Phys. Lett.*, 94:091110, 2009.

27. U. A. Laudyn, M. Kwasny, and M. A. Karpierz. Nematicons interaction in chiral nematic liquid crystals. *Mol. Cryst. Liq. Cryst.*, 527:92–97, 2010.

28. M. A. Karpierz, M. Sierakowski, and T. R. Woliñski. Light beam propagation in twisted nematics nonlinear waveguides. *Mol. Cryst. Liq. Cryst.*, 375:313–320, 2002.

29. K. Jaworowicz, K. A. Brzdakiewicz, M. A. Karpierz, and M. Sierakowski. Spatial solitons in twisted nematic layer. *Mol. Cyst. Liq. Cryst.*, 453:301–307, 2006.

30. U. A. Laudyn, M. Kwasny, K. Jaworowicz, K. A. Rutkowska, M. A. Karpierz, and G. Assanto. Nematicons in twisted liquid crystals. *Photon. Lett. Poland*, 1:7–9, 2009.

31. U. A. Laudyn, M. Kwasny, and M. A. Karpierz. Electric-field steering in soliton direction of propagation in chiral nematic liquid crystals. *Opt. Commun.*, 283:1463–1466, 2010.

32. U. A. Laudyn, M. Kwasny, and M. A. Karpierz. Properties of spatial solitons in chiral nematic liquid crystal cells. *Photon. Lett. Poland*, 1:157–159, 2009.

33. M. Peccianti, K. A. Brzdakiewicz, and G. Assanto. Nonlocal spatial soliton interactions in bulk nematic liquid crystals. *Opt. Lett.*, 27:1460–1462, 2002.

34. M. Peccianti, C. Conti, G. Assanto, A. De Luca, and C. Umeton. All optical switching and logic gating with spatial solitons in liquid crystals. *Appl. Phys. Lett.*, 81:3335–3337, 2002.

35. M. Kwasny, U. A. Laudyn, P. Jung, and M. A. Karpierz. Possibility of discrete beam propagation in chiral nematic liquid crystal. *Photon. Lett. Poland*, 1:160–162, 2009.

13

Time Dependence of Spatial Solitons in Nematic Liquid Crystals

JEROEN BEECKMAN AND KRISTIAAN NEYTS

Department of Electronics and Information Systems, Ghent University, Ghent, Belgium

13.1 INTRODUCTION

Although quite a number of applications of spatial solitons can be envisaged, the main application is their use as a dynamic optical interconnection. Depending on the speed at which the optical interconnection can be redirected from one output to another, different possibilities arise. If the switching time is in the order of a second, then typically only reconfigurable interconnects are possible, for example, protective switching (when one optical path fails, the optical signal is switched to a backup optical path). More interesting is the use as high-speed optical modulators, but then typically switching times in the order of a nanosecond are necessary. It is clear that the switching speed will determine for which application solitons can be of practical use, but in general one can state: the faster the switching speed, the better.

In this chapter, when we speak about switching time, we refer to the time it takes for the soliton to form when the optical beam is switched on. For applications as reconfigurable interconnects, the switching time is actually the time it takes to switch the optical signal from one output to another, but as this is much harder to describe theoretically, we stick to the simpler problem of switching on (or off) the soliton beam. Obviously, the temporal behavior of the soliton will be determined by the optical nonlinearity used for the soliton formation and different nonlinearities are important in liquid crystals with completely different typical timescales. Therefore, in the first part of this chapter, we give an overview of the behavior of the two most important optical nonlinearities in liquid crystals, namely, the reorientational and the thermal nonlinearity. The other nonlinearities are summarized briefly. In the second part, results for the soliton formation time are presented for the reorientational nonlinearity.

Nematicons: Spatial Optical Solitons in Nematic Liquid Crystals, First Edition.
Edited by Gaetano Assanto.

13.2 TEMPORAL BEHAVIOR OF DIFFERENT NONLINEARITIES AND GOVERNING EQUATIONS

13.2.1 Reorientational Nonlinearity

The reorientational nonlinearity occurs because the liquid crystal director orients in order to decrease the total free energy of the system. Let us suppose that the system is in equilibrium, with no static or optical fields present. When switching on an electric field, the system will try to minimize its energy according to this novel situation, leading to director reorientation. This process is governed on one hand by the driving force (the electric field) and on the other hand by a counteracting force originating from the viscosity of the material. In general, the free energy of the liquid crystal consists of different terms:

1. The elastic energy arises from spatial nonuniformity in the director orientation. This energy is often expressed in terms of the Frank–Oseen equation. The elastic forces between molecules are responsible for it and any deformation of the uniform orientation of the director $\overline{n}$ results in a nonzero elastic energy f_d per unit volume.

$$f_d = \frac{1}{2}K_1\left(\nabla.\overline{n}\right)^2 + \frac{1}{2}K_2\left[\left(\overline{n}.\nabla\times\overline{n}\right)+q\right]^2 + \frac{1}{2}K_3\left(\overline{n}\times\nabla\times\overline{n}\right)^2. \quad (13.1)$$

   In this equation, K_1, K_2, and K_3 are the elastic constants for splay, twist, and bend, respectively. q is a parameter that is zero for liquid crystals that align uniformly, whereas it is nonzero for chiral nematics that orient spontaneously in a twisted way.
2. The thermotropic energy is related to order parameter variations of the liquid crystal. The order parameter of the liquid crystal is a material property of the specific liquid crystal material at a certain temperature. In certain situations, the local order parameter can deviate from this value and these variations are usually only significant close to the nematic–isotropic transition temperature. The variations can occur because of strong electric fields or strong deformations of the director orientation resulting in liquid crystal defects. The order parameter variations can usually be ignored for reorientational solitons. Thermal nonlinear effects on the contrary are entirely caused by order parameter variations due to local heating (Chapter 9).
3. The electric and magnetic energy is due to electric and magnetic fields. As the anisotropy of liquid crystals for magnetic field is very small, we only focus on the electric fields originating from applied voltages $\overline{E}^{\,\mathrm{s}}$ or electromagnetic fields $\overline{E}^{\,\mathrm{o}}$. The energy related to (quasi-)static (s) and optical (o) electric fields is given, respectively, by (with $\overline{D}$ the dielectric displacement vector):

$$f_{\mathrm{s}} = \frac{1}{2}\overline{D}^{\,\mathrm{s}}\cdot\overline{E}^{\,\mathrm{s}}, \quad (13.2)$$

$$f_o = \frac{1}{2}\overline{D}^{o} \cdot \overline{E}^{o}. \tag{13.3}$$

4. The surface energy is related to the anchoring of the liquid crystal molecules along a preferential direction that is imposed by the surface boundary. Usually, the energy term related to deviations from this preferential direction is so important that this energy term can be neglected by fixing the orientation of the liquid crystal near the boundary.

Any change in time of the energy terms listed earlier will result in a reaction of the liquid crystal (LC) in order to minimize the energy in the new situation. This reaction will take a certain time. In the case that a reorientation of the director is necessary to reach the new equilibrium orientation, the time evolution is almost entirely due to the viscous forces in the anisotropic fluid and not to inertia of the molecules. The calculation of the reorientation in time is governed by the time-dependent version of the Euler–Lagrange equation:

$$\frac{\partial}{\partial x_i}\left(\frac{\partial F}{\partial \frac{\partial \theta}{\partial x_i}}\right) - \frac{\partial F}{\partial \theta} = \frac{\partial \mathcal{D}}{\partial \left(\frac{\partial \theta}{\partial t}\right)}. \tag{13.4}$$

In this equation, x_i denotes the three spatial coordinates x, y, and z and summation is necessary over repeated indices. The angle θ is the inclination of the director. In the general case, one can also define a similar equation for the azimuthal angle ϕ. To shorten the notation, we will further denote a derivative $\partial/\partial x_i$ as a subscript, i. This equation is derived from the theory of Ericksen and Leslie [1, 2] and contains the Rayleigh dissipation function $\mathcal{D}$ for an anisotropic fluid, which basically describes the rate of viscous dissipation per unit volume. The Rayleigh dissipation function is given by Ericksen and Leslie [1]:

$$\begin{aligned}
\mathcal{D} = {} & \alpha_1 n_j n_i n_k n_m A_{km} u_{i,j} + \alpha_4 A_{ij} u_{i,j} \\
& + \frac{1}{2}\left(\alpha_3 - \alpha_2\right)\left(n_i N_j - n_j N_i\right) u_{i,j} \\
& + \frac{1}{2}\left(\alpha_5 + \alpha_6\right)\left(n_j A_{ik} n_k + n_i A_{jk} n_k\right) u_{i,j} \\
& + \frac{1}{2}\left(\alpha_2 + \alpha_3\right)\left[n_i (N_j + A_{jk} n_k) + n_j (N_i - A_{ik} n_k)\right] u_{i,j} \\
& + \left(\alpha_3 - \alpha_2\right) N_i \frac{\partial n_i}{\partial t} + \left(\alpha_2 + \alpha_3\right) A_{ij} n_j \frac{\partial n_i}{\partial t}.
\end{aligned} \tag{13.5}$$

In the latter equation, the standard summation convention over repeating indices has to be enforced. n_i indicate the components of the director, u_i are the fluid velocities in different directions, and α_i are the Leslie viscosity coefficients. The Leslie coefficients are not independent as $\alpha_2 + \alpha_3 = \alpha_6 - \alpha_5$. The Rayleigh dissipation function contains the rate of strain tensor

$$A_{ij} = \frac{1}{2}\left(u_{i,j} + u_{j,i}\right), \tag{13.6}$$

and the angular velocity of the director relative to the fluid

$$N_i = \frac{\partial n_i}{\partial t} - \frac{1}{2}\left(u_{i,j} - u_{j,i}\right) n_j. \tag{13.7}$$

Different coefficients for the viscosity exist and the Leslie coefficients are related to the Miesowicz viscosities in the following way [3]: $2\eta_1 = -\alpha_2 + \alpha_4 + \alpha_5, 2\eta_2 = \alpha_2 + 2\alpha_3 + \alpha_4 - \alpha_5, \eta_3 = \alpha_4/2, \eta_{12} = \alpha_1$, and $\gamma_1 = \alpha_3 - \alpha_2$. These Miesowicz coefficients are important because their physical meaning can be easily understood. Moreover, they can be measured experimentally. They express the viscosity of the liquid crystal during shear flow when the orientation of the director is kept constant [4, 5]. The meaning of the first three coefficients is explained in Figure 13.1. The physical meaning of coefficient η_{12} is more difficult to visualize. The first three Miesowicz coefficients are related to the force that is necessary to move the upper plate at a certain speed. This force will be different for the different orientations of the liquid crystal. In practice, it is difficult to keep the alignment of the liquid crystal fixed during the shear flow, so typically strong magnetic fields are used to maintain the alignment of the liquid crystal constant [6, 7]. Different methods exist to measure the different viscosity terms, and an overview can be found in Reference 8. The most important coefficient, however, is γ_1, which is referred to as the *rotational viscosity* because it is related to the viscosity of rotation of the molecules around their axes. It characterizes the viscous torque associated with the angular velocity of the director. The switching times of liquid crystals in displays or other applications such as soliton generation are mainly determined by this rotational viscosity. For this reason, a large amount of publications have dealt with the measurement of this coefficient. A straightforward way of measuring this coefficient is by measuring the torque exerted on a sample by $\overline{n}$ rotating with a constant angular velocity [9]. A simpler method is to derive γ_1 from the switching

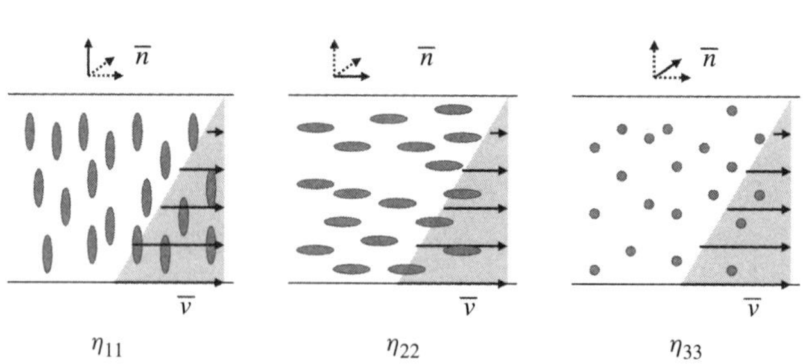

FIGURE 13.1 Director orientation according to which the coefficients η_1, η_2, and η_3 are defined.

and/or relaxation time in a planar cell when applying electric or magnetic fields. The switching time is approximately proportional to $\gamma_1 d^2$ with d the thickness of the cell (Eq. 13.20). The behavior of the liquid crystal can then be easily observed under an optical polarization microscope [10, 11].

In most cases, when switching of the liquid crystal does not occur very fast, the flow of the liquid crystal material can be neglected (i.e., velocities u_i are zero). When the flow cannot be neglected, an additional set of equations is necessary to calculate the temporal behavior of the liquid crystal, namely, the Navier–Stokes equations for an anisotropic fluid, given by:

$$\rho \frac{\partial u_i}{\partial t} = F_i + \sigma_{ji,j}. \tag{13.8}$$

The first term includes the density of the fluid ρ and describes the inertia of the system. This term can often be neglected, and this approximation is known as the *Berreman/van Doorn simplification*. F_i refers to external body forces, and also this term can be neglected as mentioned before. The only term remaining contains the derivative of the dynamic stress tensor σ_{ji}, which can be found from the Rayleigh dissipation function:

$$\sigma_{ij} = \frac{\partial \mathcal{D}}{\partial u_{i,j}}. \tag{13.9}$$

In order to study the temporal dynamics of the system, one thus has to solve Equation 13.4 and $\sigma_{ji,j} = 0$ simultaneously.

Let us now consider the case of flow not occurring; then the dissipation function reduces to:

$$\mathcal{D} = \frac{1}{2}\left(\alpha_3 - \alpha_2\right) \sum_{k=x,y,z} \left(\frac{\partial}{\partial t} n_k\right)^2. \tag{13.10}$$

13.2.2 Thermal Nonlinearity

As explained in the previous section, the thermal nonlinearity in liquid crystals mainly arises from order parameter changes that are related to the local temperature changes of the liquid crystal (Chapter 9). Let us suppose that a beam with certain intensity profile $I(x, y)$ is launched into the liquid crystal and that diffraction can be neglected. Owing to absorption, the intensity will decrease exponentially along the propagation direction: $I(x, y, z) = I(x, y)\exp(-\alpha z)$. The heat dissipated in the material per unit volume $P(x, y, z)$ because of absorption of the light is then simply given by $\alpha I(x, y, z)$.

The dynamic temperature distribution T due to heat dissipation in a material is governed by two constants: the specific heat capacity C_V and the thermal conductivity k. The heat capacity describes the amount of heat per unit mass that is necessary to increase the temperature by one degree, whereas the heat conductivity expresses the material ability to conduct heat. The equation that describes the temperature variation when a certain power P is dissipated is given by:

$$k\nabla^2 T + P = \rho C_V \frac{\partial T}{\partial t}, \tag{13.11}$$

Let us first consider a special case, assuming that the intensity profile has rotational symmetry in the xy plane and that the heat dissipation at $z = 0$ is given by $\alpha I(r)$, with r the radial distance from the origin. If the beam profile extends over a certain distance x_0, then the heat conduction, governed by the first term in Equation 13.11, can be neglected if we consider timescales shorter than τ, for which

$$\tau \ll \frac{\rho C_V x_0^2}{k}. \tag{13.12}$$

The heat equation then reduces to:

$$\frac{\partial T}{\partial t} \approx \frac{P}{\rho C_V} = \frac{\alpha I(r)}{\rho C_V}. \tag{13.13}$$

The temperature has the same profile as the heat dissipation as it cannot spread through conduction and the temperature increases linearly in time. Considering typical values for liquid crystals: $\rho \approx 10^3$ kg/m^3, $C_V \approx 2.5 \times 10^3$ J/(kgK)^3 [12], $k = 0.14$ W/(mK) [13], and a beam waist of 3 μm, this regime is only valid for a value $\tau \ll 160$ μs. For a Gaussian beam with a total optical power P_{opt}, the temperature variation at the beam center is then:

$$T(t, r = 0) \approx t\frac{\alpha I(0)}{\rho C_V} = t\frac{\alpha}{\rho C_V}\frac{2P_{\text{opt}}}{x_0^2 \pi}. \tag{13.14}$$

In order to calculate the temperature evolution for longer timescales, the heat equation reformulated in cylindrical coordinates is given by:

$$k\frac{1}{r}\frac{\partial}{\partial r}\left(r\frac{\partial T}{\partial r}\right) + P(r) = \rho C_V \frac{\partial T}{\partial t}. \tag{13.15}$$

We can assume that the optical properties react almost instantaneously to temperature variations, which means that the temporal evolution of thermal solitons is governed by Equation 13.15. The mathematical solution of this equation is handled in different books and articles for different inputs and boundary conditions [14–16], so we will not further elaborate it. It is interesting to note that, for laser pulses with a duration smaller than τ, the heat profile and thus the refractive index profile are local during the duration of the pulse, after which the profiles spread and attenuate. Longer pulses will give rise to a large nonlocality, the width of which is mainly determined by the boundary conditions. These boundary conditions determine how the heat flows out of the geometry, typically either convectively or conductively.

13.2.3 Other Nonlinearities

Liquid crystals exhibit a number of optical nonlinearities that are not present in other material systems, such as the collective molecular reorientation in the nematic phase and the high thermal nonlinearity due to the changes in order parameter. Other nonlinearities such as the pure electronic nonlinearities (Chapter 14) can also be found in other materials. Up to now, soliton generation has been demonstrated with reorientational nonlinearities in the nematic phase ([17], see also Chapter 1), with thermal nonlinearities ([18], see also Chapter 9), and with photosensitive azobenzene liquid crystals [19]. Light bullets have been demonstrated by including electronic nonlinearities also [20]. All these nonlinearities have their own characteristic response times. For most of the effects, the switching-on time is dependent on the driving force, that is, the magnitude of the optical electric field or the optical power. This means that one can utilize a so-called overdriving scheme that is used also in liquid crystal displays: initially, a strong electric is applied to achieve a rapid initial response, after which the (lower) electric field is applied with a value that corresponds to the necessary condition to excite, for example, a stable soliton. Consequently, it is not the switching-on time that determines the speed of the system but the switching-off time. Table 13.1 shows the different response times in terms of the switching-off times for different optical nonlinearities in liquid crystals. Obviously, the electronic nonlinearities are the fastest as they are related directly to the behavior of the electron wave functions.

13.3 FORMATION OF REORIENTATIONAL SOLITONS

In the rest of this chapter, we limit the description to the time dependence of the reorientational nonlinearity. In order to investigate the time dependence of the reorientation of the director, we consider only the two-dimensional situation in the plane $z = 0$ where the beam enters the liquid crystal layer, as shown in Figure 13.2. If the director can only rotate in the xz plane, then its orientation can be described by the tilt angle θ: $\overline{n} = \sin\theta \overline{1}_x + \cos\theta \overline{1}_z$. In this case, the different terms of the

TABLE 13.1 Typical Response Times for Different Optical Nonlinearities in Liquid Crystals [21]

Nonlinear Effect	Response Time
Electronic polarizability	Less than picoseconds
Molecular reorientation (isotropic phase)	100 ns
Molecular reorientation (nematic phase)	Milliseconds to seconds
Thermal	100 ms
Dye enhanced and trans–cis isomerization	Seconds

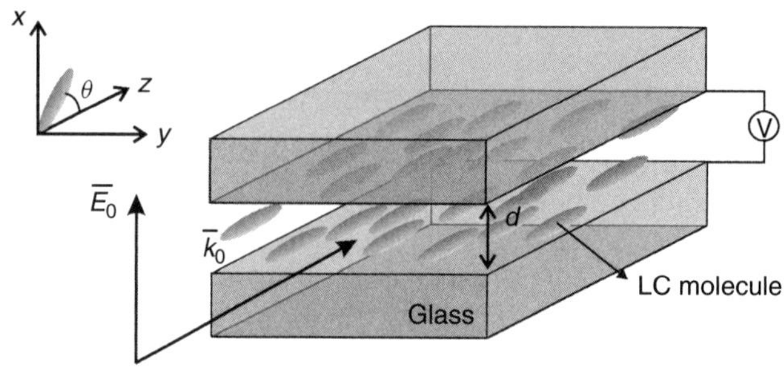

FIGURE 13.2 Configuration for the investigation of the time dependence of liquid crystal reorientation under the influence of static and optical electric fields.

free energy reduce to:

$$f_{\mathrm{d}} = \frac{1}{2}\left(K_1 \cos^2\theta + K_3 \sin^2\theta\right)\left(\frac{\partial\theta}{\partial x}\right)^2 + \frac{1}{2}K_2\left(\frac{\partial\theta}{\partial y}\right)^2, \tag{13.16}$$

$$f_{\mathrm{s}} = \frac{1}{2}\left(\varepsilon_{\perp}^{\mathrm{s}} + \Delta\varepsilon^{\mathrm{s}}\sin^2\theta\right)|E^{\mathrm{s}}|^2, \tag{13.17}$$

$$f_{\mathrm{o}} = \frac{1}{2}\left(\varepsilon_{\perp}^{\mathrm{o}} + \Delta\varepsilon^{\mathrm{o}}\sin^2\theta\right)|E^{\mathrm{o}}|^2. \tag{13.18}$$

Using the above equations and Equation 13.10 in the time-dependent version of the Euler–Lagrange equation (Eq. 13.4), one finally obtains:

$$\begin{aligned}\gamma_1\frac{\partial\theta}{\partial t} &= \left(K_1\cos^2\theta + K_3\sin^2\theta\right)\frac{\partial^2\theta}{\partial x^2} + \frac{1}{2}\left(K_3 - K_1\right)\sin^2\theta\left(\frac{\partial\theta}{\partial x}\right)^2\\ &\quad + K_2\frac{\partial^2\theta}{\partial y^2} + \frac{1}{2}\varepsilon_0\sin 2\theta\left(\Delta\varepsilon^{\mathrm{s}}|E^{\mathrm{s}}|^2 + \Delta\varepsilon^{\mathrm{o}}|E^{\mathrm{o}}|^2\right).\end{aligned} \tag{13.19}$$

Equation 13.19 can be solved numerically by using a time-stepping algorithm. At every time step, the static electric field has to obey Maxwell's equation $\nabla\cdot(\overline{\overline{\varepsilon}}^{\,s}\cdot\overline{E}^{\,s}) = 0$.

13.3.1 Bias Voltage Switching Time

Many publications on nematicons deal with configurations in which a bias voltage is necessary. Nematic LCs are widely used in display applications, and, obviously, in this type of application, the switching speed of the individual pixels is an important specification for displaying video contents fluently. In essence, the LC cell for the generation of spatial solitons is very similar to a pixel in a display, so one can use the extended literature on liquid crystal displays to find estimates for the switching-on τ_{on} and switching-off times τ_{off} of the LC cell. The switching speed in display applications is determined in terms of rise and fall times based on the initial slope of the switching. This initial switching often has an exponential evolution $\exp(-t/\tau)$. The following estimates can be found for switching a planarly oriented cell without twist [22]:

$$\tau_{off} \approx \frac{\gamma_1 d^2}{\pi^2 K}, \tag{13.20}$$

$$\tau_{on} \approx \frac{\gamma_1 d^2}{\Delta\varepsilon V^2 - \pi^2 K}. \tag{13.21}$$

In Equations 13.20 and 13.21 it is assumed that $K_{11} \approx K_{33} = K$ and V is the applied voltage. The switching time of a cell depends on the material parameters γ and K. But, more importantly, the switching times depend quadratically on the thickness of the LC layer. In displays, the LC layer thickness is typically in the order of 3–5 μm, which results in switching times in the order of tens of milliseconds. In τ_{on}, the applied voltage is also present: increasing the applied voltage leads in principle to shorter switching times. However, to achieve certain gray levels, a certain voltage is necessary. Therefore, a common trick in displays is to apply a high voltage for a short time, until the pixel reaches the required gray level, after which the correct voltage is applied. Obviously, the switching-on speeds are not the major problem in displays. The switching-off time is critical as it cannot be reduced by applying a voltage and is governed by surface actions and elastic forces within the liquid crystal.

Equations 13.20 and 13.21 for the switching times provide a good insight in the temporal dynamics of a liquid crystal cell but are not applicable here because the bias voltage for the generation of nematicons is rather small and close to the threshold voltage for switching, for which the earlier equations are not valid. Equations 13.20 and 13.21 apply only to the transition between a high voltage and zero. Figure 13.3 shows results obtained from numerical simulations based on Equation 13.19. Figure 13.3a shows the temporal evolution of θ in the middle of a 75-μm-thick cell when a voltage around the threshold is turned on. Note that no laser-induced director reorientation is considered, so that θ is invariant with respect to y. Figure 13.3b shows the evolution of the rise time versus the applied voltage. The rise time is based on the evolution of the midtilt, that is, the tilt distribution in the middle of the cell (i.e., in the plane $x = 0$), and is arbitrarily defined as the *time* for the midtilt to reach 99% of its steady-state value. The switching time is long—on the order of 100 s for 1 V—for two reasons. First, the cell is very thick compared to LC displays; second, the voltage is around threshold. This results in a static electric field of 0.013 V/μm, which gives rise to a very small driving force to reorient the molecules and consequently to a slow process.

In this one-dimensional case, a scaling of the cell thickness by a factor α results in a scaling of the time by a factor α^2, so the results in Figure 13.3 are valid for any cell thickness provided that the time is scaled appropriately.

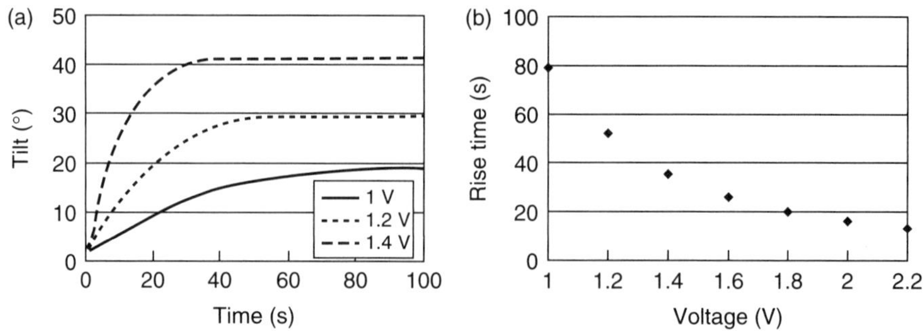

FIGURE 13.3 (a) Evolution of θ in the middle of a 75 μm-thick liquid crystal layer when switching-on the voltage at $t = 0$ s. (b) Corresponding rise time, that is, the time for the midtilt to reach 99% of its steady-state value versus voltage. *Source:* Taken from Reference 23.

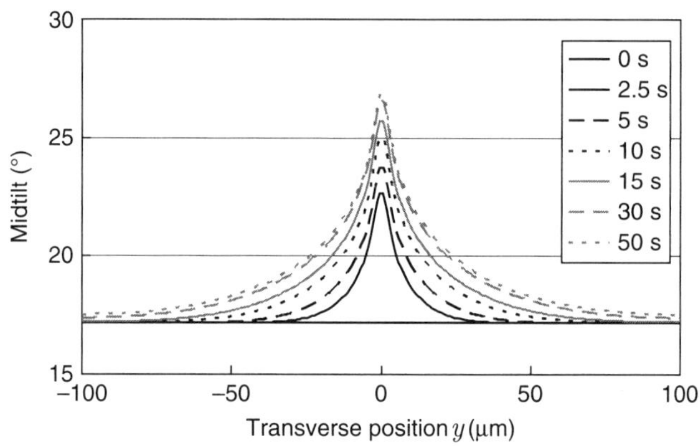

FIGURE 13.4 Evolution of the midtilt when turning on a 4.41 mW laser beam at $t = 0$ s. The voltage across the cell is 1 V. *Source:* Taken from Reference 23.

13.3.2 Soliton Formation Time

Figure 13.4 shows the evolution of the midtilt after turning on the laser beam. The optical beam is a 3-μm-waist TEM_{00} beam with polarization along the x-axis. The tilt increase in the middle gives rise to a nonhomogeneous increase of the refractive index for light polarized along the x-axis. This self-induced mechanism is responsible for the collimation of the beam through self-focusing.

The initial reorientation is relatively fast and takes place on a timescale of a few seconds. Indeed, the reorientation in the middle reaches more than 50% of the final reorientation after 2.5 s. More importantly, it can be seen that the angular profile in the vicinity of the beam center stays almost unchanged after 2.5 s. After this, the reorientation of the molecules further away from the center occurs until an overall steady state is reached with the width of the induced tilt profile much larger than that after a few seconds and, a fortiori, than the solitonlike beam itself. This

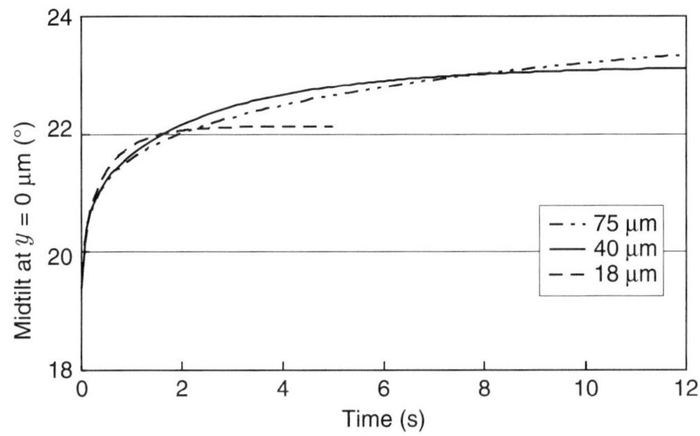

FIGURE 13.5 Simulation of the evolution of the midtilt at $y = 0$ μm when a 2.25-mW laser beam is turned on at $t = 0$ s and for different cell thicknesses. *Source:* Taken from Reference 23.

reorientation occurs on a much larger timescale, and the steady state is reached after about 50 s. This means that the reorientation is initially fast and the nonlocality of the nonlinear effect is small. Then, the nonlocality increases in time until a maximum is achieved. This effect is plausible as only a low energy is required to reorient the liquid crystal in the immediate vicinity of the optical beam. The complete reorientation, whose spatial extent is much wider, requires a much larger energy and, consequently, a longer time.

The long time to reach the steady state in thick cells (about 50 s) originates mainly from the large scale of the nonlocal reorientation. Hence, it can be expected that the reorientation time will be much smaller in thinner cells, owing to a less wide molecular reorientation. This is illustrated in Figure 13.5, which shows the evolution of the maximum tilt—that is, the tilt at $(x, y) = (0, 0)$, where the light intensity is maximum. For the 75-μm-thick cell the reorientation takes about 50 s, as illustrated before, but for the thinnest cell (18 μm), the reorientation already reaches its steady-state value after 2 s. For even thinner cells, the time for the overall molecular reorientation to settle is even smaller. However, the initial dynamics are quite similar for thick and thin cells. Indeed, reorientation in the vicinity of the light beam completes in less than 2 s.

13.3.3 Experimental Observation of Soliton Formation

In order to experimentally examine the temporal behavior of soliton formation, the setup in Figure 13.6 was used. A He–Ne laser beam (633 nm) is injected into a liquid crystal cell by use of a 40× microscope objective. The beam propagation in the cell can be observed with the combination of a lens and a CCD camera by collecting the scattered light. Besides light propagation, it is also possible to observe the laser-induced molecular reorientation. This is achieved by using the polarized light from a Xenon lamp. The polarized light traveling through the LC layer undergoes a polarization change depending on the molecular orientation. In

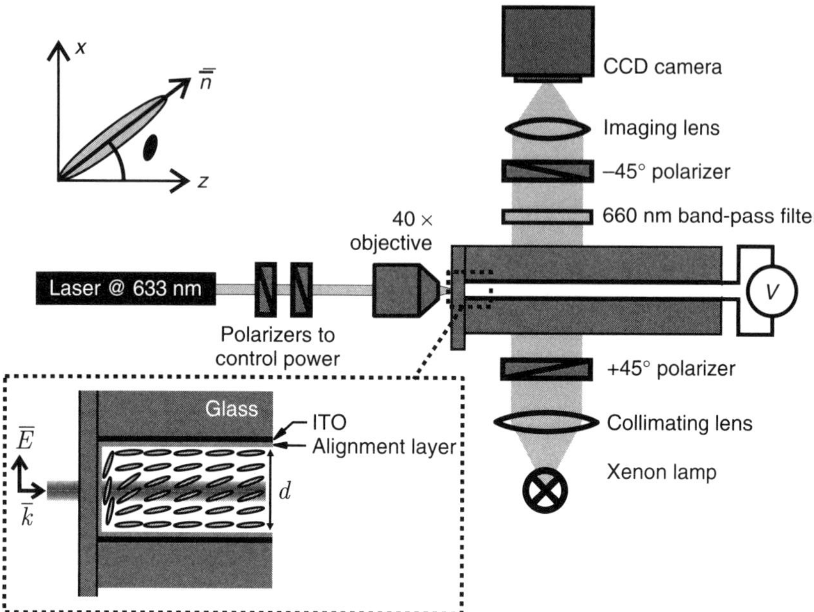

FIGURE 13.6 Experimental setup and indication of axes. *Source:* Taken from Reference 23.

this way, the self-induced waveguide can be observed. For such purpose, two crossed polarizers and a band-pass filter (660 nm) are employed, the directions of polarizer and analyzer being $\pm 45^\circ$ with respect to the z-axis.

When the optical power is turned on, the beam evolves from a diffracting regime (Figure 13.7a when the molecules are not reoriented yet) to a solitonlike propagation regime (Figure 13.7b). The evolution of the intensity profile of light scattered from the beam after a propagation distance of 1.6 mm is shown in Figure 13.7c. At 0 ms, the width of the beam is large and the intensity is low. After 1 s, the width of the beam has become narrower, the peak intensity higher and the intensity profile reaches a steady state.

Figure 13.8 shows the polarized transmission image of the cell for different powers of the laser beam. The latter enters from the left side, where the input of the cell is visible. The dark region is caused by glue on top of the interface between the two glass plates. Owing to anisotropy, there is a phase retardation of about $25\times2\pi$ between ordinary and extraordinary polarizations after propagation through the 75-μm-thick cell.[1] The laser-induced molecular reorientation reduces the retardation and consequently changes the transmission through the crossed polarizers in the region where the soliton-like beam propagates. In Figure 13.8a, the effect of the reorientation is small, but it increases with increasing optical power (Figure 13.8b and c). With increasing molecular reorientation, the image becomes

[1]This is because of the large thickness of the cell (75 μm). Nematic LC displays typically have a thickness that is 10–20 times smaller.

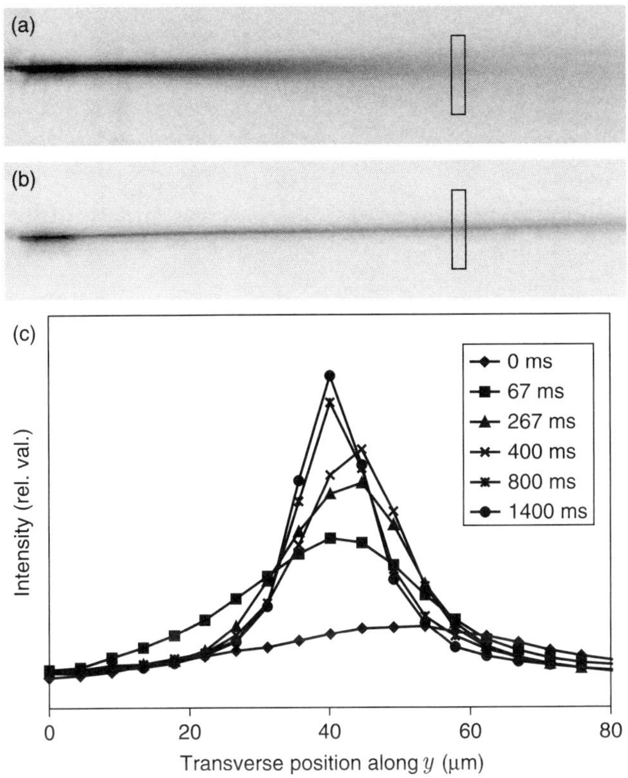

FIGURE 13.7 (a and b) Steady-state light propagation in the cell for a voltage of 1 V: (a) the diffraction regime, that is, for low optical power, and (b) the soliton regime at relatively high optical power (here 2.6 mW). The rectangles show the location where the light intensity is measured. (c) Temporal evolution of the beam intensity for a 2.6-mW optical beam. Light is turned on after the voltage-induced molecular orientation (orientation at rest) reached steady state. *Source:* Taken from Reference 23.

black, white, and black again as can be seen in Figure 13.8c. This is because the retardation is larger than π. This method of visualizing the self-induced waveguide is clearly not sensitive enough for the small reorientation occurring at 2.6 mW. For this purpose, other approaches are more sensitive, based on interferometry [24] or Raman scattering [25]; to demonstrate the time effects, the described method is useful as the molecular reorientation can be immediately estimated from the transmission profiles.

The time evolution of the transmission close to the entrance of the 75-μm-thick cell is illustrated in Figure 13.9 for two different optical powers. Owing to the glue at the entrance, it is not possible to define the distance exactly, but it is estimated to be a few hundred micrometers. The two pictures show that the time to completely reorient the liquid crystal is on the order of 50 s. This is in contrast to Figure 13.7c where the formation of the solitonlike beam takes on the order of 1 s. However, it clearly coincides with the findings from numerical simulations. It takes less than

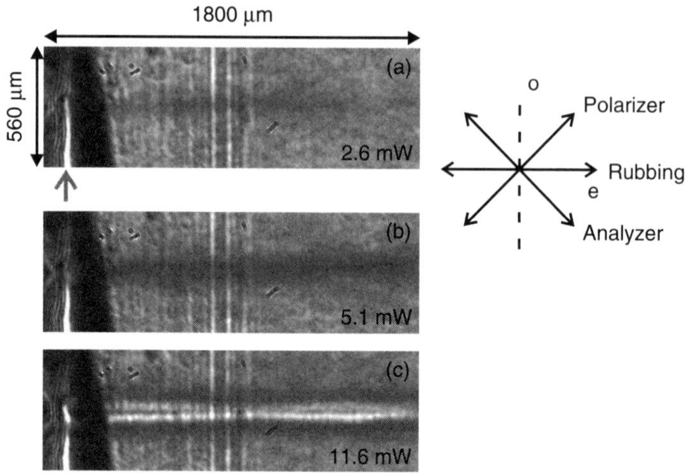

FIGURE 13.8 Steady-state transmission of Xenon light through the cell between crossed polarizers, when a laser beam with increasing power from (a) to (c) is launched in the cell. The gray arrow indicates the estimated entrance of the cell. *Source:* Taken from Reference 23.

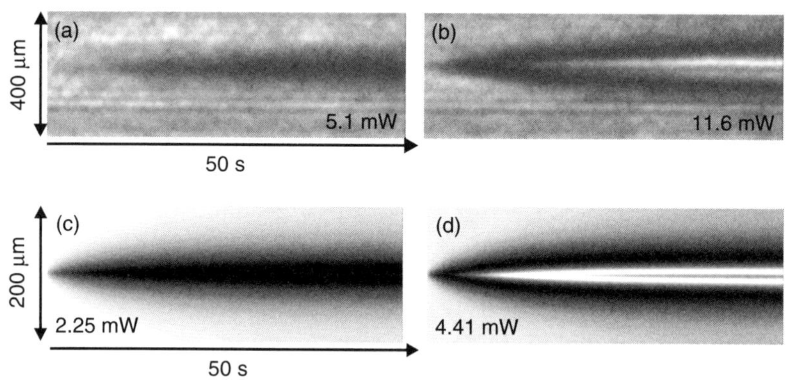

FIGURE 13.9 Experimental (a and b) and numerical (c and d) evolution of the transmission through the cell after switching on the laser beam. *Source:* Taken from Reference 23.

a second to reorient the liquid crystal in the vicinity of the beam (although not visible in the transmission experiments) and a few tens of seconds to reorient the liquid crystal further away, related to the large nonlocality of the nonlinearity.

As it is clear from the results presented before, the total reorientation of the liquid crystal can be roughly subdivided into two parts. The beam first induces a more or less local reorientation that constitutes a self-induced waveguide. After that, the reorientation relaxes across the whole cell and a slow process of increasing the nonlocal reorientation occurs. The first process occurs in less than 2 s, whereas the second process takes about 50 s to end (for a 50-μm-thick cell). This observation

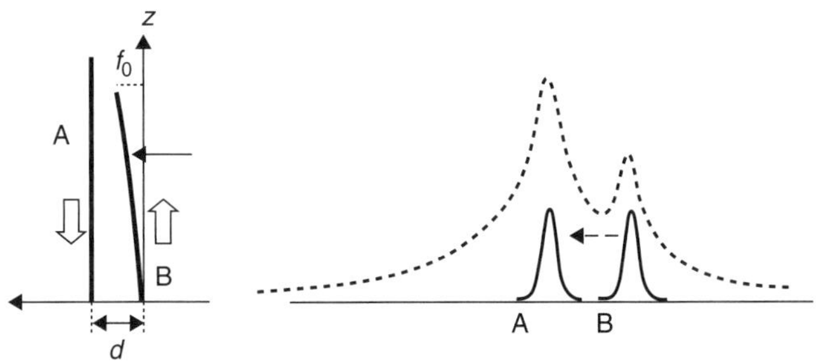

FIGURE 13.10 Interaction of counterpropagating solitons. A: CW beam, B: pulsed beam. Bold curves represent the solitons, the dashed curve represents the refractive index profile, the white arrows indicate the directions of propagation, and the black arrow indicates the direction of deflection. *Source:* Taken from Reference 26.

provides a clever way to control the nonlocality of the nonlinearity. When using a continuous beam, after 50 s, the nonlocality is large. However, when chopping the laser beam into pulses of about 2 s and having sufficient time between different pulses, one obtains a local reorientation profile. In this way, the soliton is created, but the director does not get the time to fully relax. By controlling duty cycle of the gated beam, the nonlocality of the nonlinearity can be controlled.

In the work of Henninot et al. [26], this technique was used to investigate the interaction between two counterpropagating solitons as shown in Figure 13.10. One beam (A) is launched from an optical fiber into the LC layer from the top, whereas the other beam (B) is launched from the bottom. Beam A is a CW beam, which means that its self-induced waveguide profile is wide and highly nonlocal. Beam (B) originates from the same laser with equal power, but the beam is chopped with a frequency of 0.25 Hz. Hence, the reorientation profile is more local and this beam can be used as a *weak* soliton to probe. The experiment revealed that, indeed, the trajectory of the *heavy* soliton B was not influenced by the presence of the *weak* soliton A. The lateral deviation from the original trajectory of beam A was used to determine the nonlocal refractive index variation induced by the CW beam B (Chapter 9).

13.3.4 Influence of Flow Effects

In the calculation examples mentioned earlier, the influence of flow in the liquid crystal has been ignored. However, it is known that in certain configurations the flow has an important contribution to the time dependence of the LC switching. In pi cells, for example, the flow can improve the switching [27] by a small factor. Much more dramatic are the effects in vertically aligned liquid crystals with negative $\Delta\varepsilon$. Typically, increasing the voltage leads to a faster switching speed. However, voltages higher than a certain threshold lead to a reverse flow phenomenon and a dramatic increase in switching time [28, 29]. This reverse flow is also known as *backflow*. Owing to backflow, an unwanted twisting of the director occurs and the time

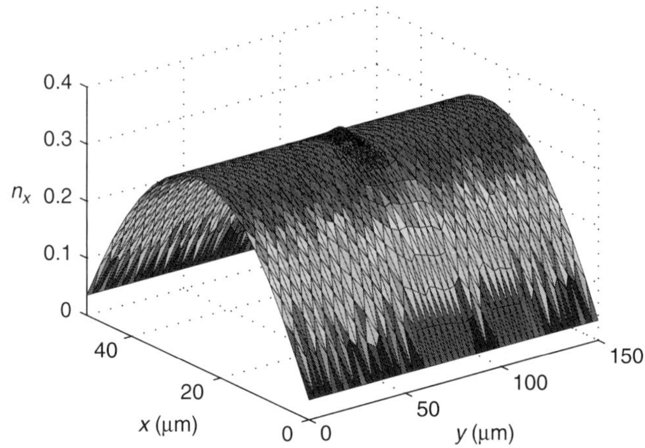

FIGURE 13.11 n_z component of the director in a 50-μm-thick cell with 1 V bias voltage, 17 ms after turning on the optical electric field.

to reach the steady state may increase by a factor 100 or higher. Also, in twisted nematic cells, the occurrence of backflow leads to a noticeable difference in switching speed compared to what one would expect in the absence of flow [30–32].

Considering the configuration of the previous sections (Figure 13.4), it is expected that turning on the laser beam will not generate a large amount of flow in the liquid crystal. In order to test the influence of flow on the switching behavior, a 2D modeling program was used that incorporates the flow equations [33, 34]. Figure 13.11 shows the director orientation in terms of the n_z component of the director 17 ms after the light is turned on (with 5 mW power). The reorientation of the director is almost a local effect as can be seen from Figure 13.11. The time of 17 ms corresponds more or less to the maximum fluid velocity observed in liquid crystal of about 2×10^{-7} m/s.

Figure 13.12 shows the different velocity components. The x component of the velocity is the easiest to understand. This speed component shows a strong outward flow above and below the region where the light beam is launched. Because the molecules tilt more, there is a shear flow that repels the molecules to the side of the cell. A simplistic explanation for this effect is that the molecules occupy more space in the x direction when they tilt. An increase in tilt will thus result in a flow outward. The z component of the flow can be explained in a similar manner. Important to note is that v_z is antisymmetric with respect to x. The average velocity in the z direction is zero; otherwise, there would be a net flow of material. This is because the pressure of the optical fields is neglected in the simulations. The v_y component mainly shows an inward flow from the side, which means that the molecules are mainly flowing away from the region of the light beam along the x-direction and flowing in from the y-direction. The question that now arises is whether these values for the flow affect the switching speed or not.

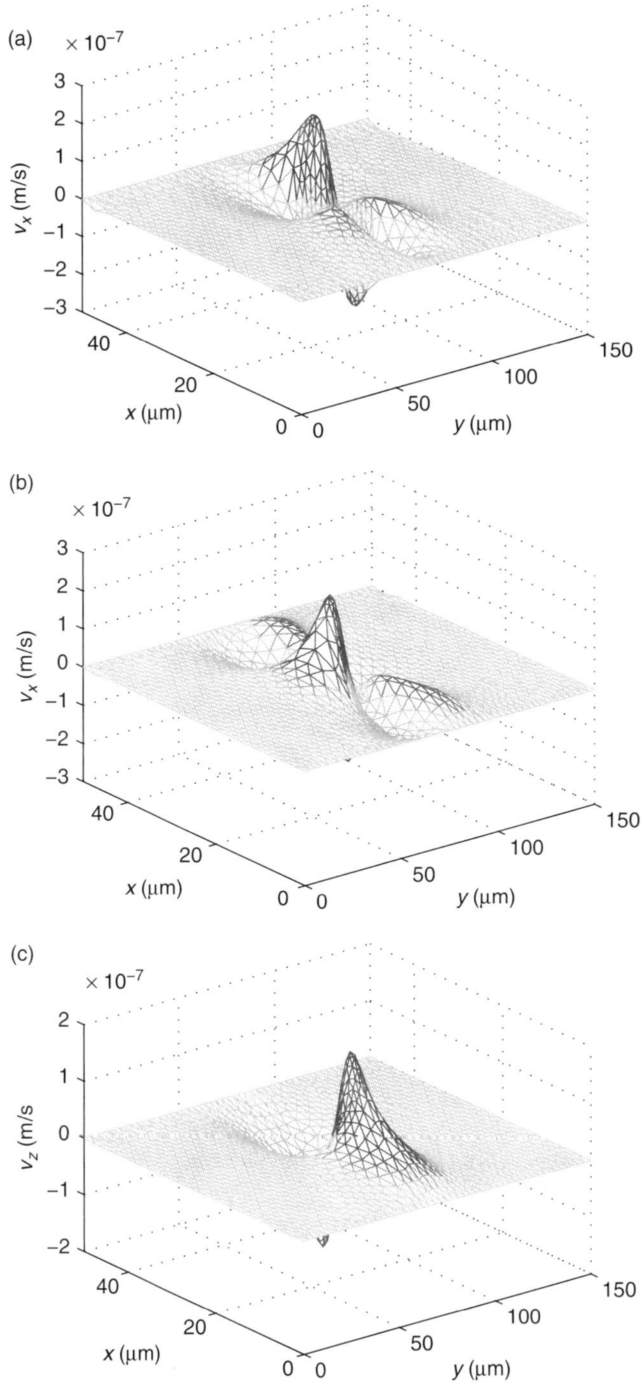

FIGURE 13.12 (a–c) Velocity components of the liquid crystal flow, 17 ms after turning on the optical electric field.

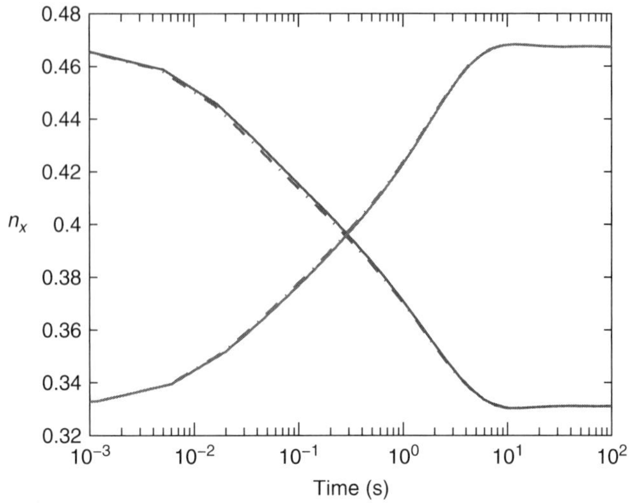

FIGURE 13.13 Time evolution of the n_x component of the director when switching the beam on and off. For the solid curve, the flow is ignored, while for the dashed curve, the flow is taken into account.

Figure 13.13 shows the evolution of the n_x component of the director versus time for the same parameters. When the light beam is turned on, the n_x component increases until it reaches steady state after about 10 s. The solid line shows the result when the flow is neglected, and the dashed curve shows the result including flow. It is clear that for these values the influence of flow is negligible. The same conclusion can be drawn when studying the curve for switching off. The flow remains small enough in both cases to have only little influence on the switching behavior.

13.4 CONCLUSIONS

Liquid crystals exhibit different optical nonlinear effects that can be used to generate spatial optical solitons. It is clear that these nonlinearities have different timescales that affect not only the soliton formation time but also the time it takes to steer or switch soliton beams. In this chapter, we have mainly considered the reorientational nonlinearity as most of the nematicon work has been devoted to this nonlinearity. The formation of the nematicon occurs in less than a few seconds, whereas the actual full reorientation takes a few tens of seconds. During this process, the refractive index change evolves from being nearly local to highly nonlocal, which can be used as a way to control nonlocality.

REFERENCES

1. J. L. Ericksen. Equation of motion for liquid-crystals. Q. J. Mech. Appl. Math., 29:203–208, 1976.

2. F. M. Leslie. Some constitutive equations for liquid crystals. *Arch. Ration. Mech. An.*, 28(4):265, 1968.
3. H. G. Walton and M. J. Towler. On the response speed of pi-cells. *Liq. Cryst.*, 27(10): 1329–1335, 2000.
4. M. Miesowicz. The three coefficients of viscosity of anisotropic liquids. *Nature*, 158:27, 1946.
5. A. Buka and L. Kramer Eds. Pattern Formation in Liquid Crystals. Springer-Verlag, New York, 1996.
6. W. W. Beens and W. H. Dejeu. Flow-measurements of the viscosity coefficients of 2 nematic liquid-crystalline azoxybenzenes. *J. Phys.-Paris*, 44(2):129–136, 1983.
7. F. Hennel, J. Janik, J. K. Moscicki, and R. Dabrowski. Improved miesowicz viscometer. *Mol. Cryst. Liq. Cryst.*, 191:401–405, 1990.
8. D. Dunmur, A. Fukuda, and G. Luckhurst Eds. Physical Properties of Liquid Crystals: Nematics. INSPEC, London, 2001.
9. H. Kneppe, F. Schneider, and N. K. Sharma. Rotational viscosity -gamma-1 of nematic liquid-crystals. *J. Chem. Phys.*, 77(6):3203–3208, 1982.
10. P. R. Gerber. Measurement of the rotational viscosity of nematic liquid-crystals. *Appl. Phys. A*, 26(3):139–142, 1981.
11. H. Kneppe and F. Schneider. Determination of the rotational viscosity coefficient-gamma-1 of nematic liquid-crystals. *J. Phys. E Sci. Instrum.*, 16(6):512–515, 1983.
12. G. Cordoyiannis, D. Apreutesei, G. H. Mehl, C. Glorieux, and J. Thoen. High-resolution calorimetric study of a liquid crystalline organo-siloxane tetrapode with a biaxial nematic phase. *Phys. Rev. E*, 78(1):011708, 2008.
13. G. Ahlers, D. S. Cannell, L. I. Berge, and S. Sakurai. Thermal-conductivity of the nematic liquid-crystal 4-n-pentyl-4′-cyanobiphenyl. *Phys. Rev. E*, 49(1):545–553, 1994.
14. K. Y. Kung and H. M. Srivastava. Analytic transient solutions of a cylindrical heat equation with oscillating heat flux. *Appl. Math. Comput.*, 195(2):745–753, 2008.
15. K. T. Chiang, K. Y. Kung, and H. M. Srivastava. Analytic transient solutions of a cylindrical heat equation with a heat source. *Appl. Math. Comput.*, 215(8):2877–2885, 2009.
16. J. Holman. *Heat Transfer*, *McGraw -Hill Series in Mechanical Engineering*, 10th edn, 2006.
17. M. Peccianti, A. De Rossi, G. Assanto, A. De Luca, C. Umeton, and I. C. Khoo. Electrically Assisted Self-confinement and Waveguiding in Planar Nematic Liquid Crystal Cells. *Appl. Phys. Lett.*, 77:7–9, 2000.
18. F. Derrien, J. Henninot, M. Warenghem, and G. Abbate. A Thermal (2D+1) Spatial Optical Soliton in a Dye Doped Liquid Crystal. *J. Opt. A-Pure Appl. Opt.*, 2:332–337, 2000.
19. S. V. Serak, N. V. Tabiryan, M. Peccianti, and G. Assanto. Spatial soliton all-optical logic gates. *IEEE Photon. Techol. Lett.*, 18(9–12):1287–1289, 2006.
20. I. B. Burgess, M. Peccianti, G. Assanto, and R. Morandotti. Accessible light bullets via Synergetic Nonlinearities. *Phys. Rev. Lett.*, 102:203903, 2009.
21. I. C. Khoo. *Liquid Crystals*, *Wiley Series in Pure and Applied Optics*. 2nd edn, Wiley, Hoboken, New Jersey, 2007.
22. E. Lueder. *Liquid Crystal Displays: Addressing Schemes and Electro-Optical Effects*, *Wiley Series in Display Technology*. John Wiley & Sons, Chichester, 2001.

23. J. Beeckman, K. Neyts, X. Hutsebaut, C. Cambournac, and M. Haelterman. Time dependence of soliton formation in planar cells of nematic liquid crystals. *IEEE J. Quantum Electron.*, 41:735–740, 2005.
24. X. Hutsebaut, C. Cambournac, M. Haelterman, J. Beeckman, and K. Neyts. Measurement of the self-induced waveguide of a solitonlike optical beam in a nematic liquid crystal. *J. Opt. Soc. Am. B*, 22:1424–1431, 2005.
25. M. Warenghem, J. F. Blach, and J. F. Henninot. Measuring and monitoring optically induced thermal or orientational non-locality in nematic liquid crystal. *Mol. Cryst. Liq. Cryst.*, 454:297–314, 2006.
26. J. F. Henninot, J. F. Blach, and M. Warenghem. Experimental study of the nonlocality of spatial optical solitons excited in nematic liquid crystal. *J. Opt. A: Pure Appl. Opt.*, 9:20–25, 2007.
27. P. D. Brimicombe, and E. P. Raynes. The influence of flow on symmetric and asymmetric splay state relaxations. *Liq. Cryst.*, 32(10):1273–1283, 2005.
28. L. Y. Chen and S. H. Chen. Crucial influence of the azimuthal alignment on the dynamics of pure homeotropic liquid crystal cells. *Jpn. J. Appl. Phys.*, 39(4):368–370, 2000.
29. P. J. M. Vanbrabant, N. Dessaud, and J. F. Stromer. Temperature influence on the dynamics of vertically aligned liquid crystal displays. *Appl. Phys. Lett.*, 92(9):091101, 2008.
30. C. Z. van Doorn. Dynamic behavior of twisted nematic liquid-crystal layers in switched fields. *J. Appl. Phys.*, 46(9):3738–3745, 1975.
31. D. W. Berreman. Liquid-crystal twist cell dynamics with backflow. *J. Appl. Phys.*, 46(9):3746–3751, 1975.
32. F. Z. Yang, Y. M. bong, L. Z. Ruan, and J. R. Sambles. Dynamical process of switch-off in a supertwisted nematic cell. *J. Appl. Phys.*, 96(1):310–315, 2004.
33. R. James, E. Willman, F. A. Fernández, and S. E. Day. Finite-element modeling of liquid-crystal hydrodynamics with a variable degree of order. *IEEE Trans. Electron Dev.*, 53:1575–1582, 2006.
34. R. James, E. Willman, F. A. Fernandez, and S. E. Day. Computer modeling of liquid crystal hydrodynamics. *IEEE Trans. Magn.*, 44(6):814–817, 2008.

14

Spatiotemporal Dynamics and Light Bullets in Nematic Liquid Crystals

MARCO PECCIANTI
Institute for Complex Systems, ISC-CNR, Rome, Italy

We see only what we know.

Johann Wolfgang von Goethe (1749–1832)

14.1 INTRODUCTION

Electromagnetic wave packets tend to spread out as they evolve, whatever dimension (time or space) is involved. The fundamental cause of this phenomenon is the propagation at different velocities or directions of the wave packet frequency components. Hence, similar to the beam diffracting in space, pulses spread in time because of the so-called group-velocity-dispersion (GVD), as the energy associated to the various Fourier components of the field tends to temporally shift in propagation. One of the most challenging goals in the study of multidimensional nonlinear field–matter interactions is the generation of wave packets confined in both time and space or *light bullets*. This term was coined in 1990 by Silberberg [1] to stress the particle nature of such nonlinear objects. We refer to these self-confined waves, or spatiotemporal solitons (STSs), as (2+1+1)D solitary waves, where the "2" refers to the confinement along the two transverse spatial dimensions, the former "1" relates to the confinement in the time coordinate and the latter "1" refers to the spatial propagation direction.

The formation of STSs requires the simultaneous nonlinear compensation of GVD and diffraction, respectively, occurring in the (1+1)D temporal and (2+1)D spatial counterparts. Indeed, the number of dimensions in which the nonlinear dynamics evolve is a central issue in the soliton stability as it determines the strength of the nonlinear focusing in the competitive action with the linear

Nematicons: Spatial Optical Solitons in Nematic Liquid Crystals, First Edition.
Edited by Gaetano Assanto.

diffraction and dispersion [2]. The general nematicon phenomenology is an interesting example of the general stability problem arising as dimensionality of the nonlinear propagation increases.

In a simple picture (as explained in the previous chapters), mechanisms such as saturations [3, 4], absorption [5–7], and the nonlocality [8, 9] can introduce higher-order terms in the nonlinear response, stabilizing the otherwise unstable (2+1)D Kerr solitary wave [10, 11].

The current investigation on (2+1+1)D optical bullets is then facing two major challenges: (i) the formalization of relevant models predicting the stable propagation of STS and (ii) the identification of suitable optical media or systems where such models apply. A number of specific nonlinear regimes are known to theoretically allow STS. For example, conditions for stable STS propagation have been theoretically derived in quadratic media, originating by the energy exchange between a fundamental frequency (FF) and a second harmonic (SH). However, the proposed regime implies a number of experimental constraints such as the requirements of a large enough anomalous GVD for both the FF and the SH [12–14], difficult to find in known optical systems. Similarly, it has been calculated that nonparaxiality, higher-order dispersion [15], or self-generation in nonlocal nonlinearities mediated by the rectification [16–20] could provide a suitable collapse arrest mechanism for (2+1+1)D soliton propagation. The feasibility of such regimes is still unclear and the formalization of consistent ways to realize an experimental verification is still a subject of intense investigation.

14.1.1 (2+1+1)D Nonlinear Wave Propagation in Kerr Media

We can formalize the spatiotemporal nonlinear dynamics considering a scalar field propagating along Z, centered at frequency ω_0 and having field complex envelope $\psi(X, Y, Z, t)$. We refer to a Kerr-like field–matter interaction, which is described by a nonlinear perturbation of the refractive index. Naming $\tilde{\psi}(X, Y, Z, \omega) = F_t\{\psi\}$ where F_t is the Fourier's operator acting along t, we can write the generalized nonlinear Schrödinger equation (NLS) ruling the propagation:

$$2\mathrm{i}k_0\frac{\partial\tilde{\psi}}{\partial Z} + \nabla^2_{XY}\tilde{\psi} + (k^2 - k_0^2)\tilde{\psi} = 0, \tag{14.1}$$

where $k_0 = 2\pi n(\omega_0)/\lambda$ is the field wave number, where $n(\omega_0)$ is the refractive index at frequency ω_0. $k = 2\pi[n(\omega) + \Delta n_{\mathrm{NL}}]/\lambda$ represents the general dependence of the optical wavenumber from the frequency and the nonlinear index perturbation Δn_{NL}. By using the common approach to Taylor expanding the frequency dependence of the wave number [21] around ω_0, we can recast Equation 14.1 in the time domain form:

$$\mathrm{i}\frac{\partial\psi}{\partial Z} + \frac{1}{2k_0}\nabla^2_{XY}\psi + \mathrm{i}\beta_1\frac{\partial\psi}{\partial t} - \frac{\beta_2}{2}\frac{\partial^2\psi}{\partial t^2} + \frac{\Delta n_{\mathrm{NL}}k_0}{n(\omega_0)}\psi = 0, \tag{14.2}$$

where $\beta_1 = \partial k/\partial\omega$ and $\beta_2 = \partial^2 k/\partial\omega^2$ are the group velocity and the GVD parameters, respectively. By considering a reference frame moving at the group velocity $v_g = 1/\beta_1$, that is redefining the temporal coordinates as $T = t - z\beta_1$, we get the (3+1)D evolution model:

$$\mathrm{i}\frac{\partial\psi}{\partial Z} + \frac{1}{2k_0}\nabla^2_{XY}\psi - \beta_2\frac{\partial^2\psi}{\partial T^2} + \frac{\Delta n_{\mathrm{NL}} k_0}{n(\omega_0)}\psi = 0. \tag{14.3}$$

Bright (2+1+1)D solitons can be stable general solution of the Equation 14.3 for specific nonlinear response Δn_{NL}. For example, Mihalache and coworkers [22] predicted that nonlocal media can support optical bullet with temporal duration within the scale of the nonlocal response time. The latter constraint makes those wave packets hard to be experimentally investigated, because of the high response time of the nonlocal nonlinearity, usually ranging from milliseconds to seconds scale in optical systems such as thermal [23, 24], photorefractive [25, 26], and reorientational [27–27].

As detailed in the next sections, a scenario of feasible verification conditions for optical bullets propagating in nonlocal media has been recently identified when multiple nonlinear media responses we taken into account.

14.2 OPTICAL PROPAGATION UNDER MULTIPLE NONLINEAR CONTRIBUTIONS

14.2.1 Multiple Nonlinearities and Space–Time Decoupling of the Nonlinear Dynamics

The relatively large magnitude of several nonlocal nonlinearities (e.g., molecular, photovoltaic, or thermal) is usually connected to an inherent storage and accumulation of the energy of the propagating field. This is also the main cause of their characteristic large response time. Whatever the inner nature of such nonlinearity is, it usually coexists with other nonlinear mechanisms operating at different temporal scales. Considering, for example, the nematic liquid crystal (NLC), we can express the nonlinear index perturbation induced by an optical field as the superposition of contributions originating from the optical-induced reorientation $\Delta n_\theta\left(\langle|\psi|^2\rangle_\tau\right)$ and the electronic instantaneous Kerr response $n_2|\psi|^2$, that is, $\Delta n_{\mathrm{NL}} = \Delta n_\theta\left(\langle|\psi|^2\rangle_\tau\right) + n_2|\psi|^2$, where $\langle\rangle_\tau$ is a temporal average operator of window τ to account for the slow response of molecular reorientation.

If we consider the propagation of a pulse of duration $T_0 \ll \tau$, its temporal profile is only affected by the instantaneous nonlinear response. If such response has no significant effect on the field spatial profile, a complete decoupling between the spatial and the temporal evolutions of the propagation is realized, the former governed by the slow nonlinear response, the latter ruled by the fast nonlinear process.

In this condition, the description of the dynamics of the field decouples in time and space, with a remarkable simplification of the field–matter model, which becomes *accessible* [30].

14.2.2 Suitable Excitation Conditions

The scenario described earlier can be easily reached by considering the propagation of a train of short pulses, like the one generated by mode-locked lasers. Hence, we assume a Gaussian light beam propagating along Z with wave vector $\boldsymbol{\beta} = \beta\hat{z}$ consisting of a train of pulses of duration T_0 and pulse-to-pulse separation σ:

$$\psi(\boldsymbol{r}, T) = A(X, Y, Z) \sum_n \xi(Z, T - n\sigma) \, \mathrm{Exp}\left[\mathrm{i}(\beta Z - \omega_0 t)\right]. \tag{14.4}$$

$A(X, Y, Z)$ is the normalized field spatial profile with $\mathbf{r} = (X, Y, Z)$, the reference in physical coordinates. In the regime considered, both T_0 and σ are much smaller than the characteristic nonlocal response time τ. Hence, by neglecting the spatial perturbation induced by the fast nonlinear response in Equation 14.3, the train profile evolution ξ is completely governed by the model [21]

$$\mathrm{i}\frac{\partial \xi}{\partial Z} - \frac{\beta_2}{2}\frac{\partial^2 \xi}{\partial T^2} + \gamma|\xi|^2\xi = 0, \tag{14.5}$$

where $\gamma = n_2\omega_0/c\pi W_0^2$ is the effective temporal nonlinearity that depends on the electronic Kerr coefficient n_2, the optical angular frequency ω_0, and the beam transverse effective area πW_0^2, where W_0 is the optical beam waist. Being $\gamma > 0$ for the electronic Kerr nonlinearity, the Equation 14.5 supports *stable* bright solitons for $\beta_2 < 0$ and dark solitons $\beta_2 > 0$ [21].

In both cases, if the pulses are also confined in a stable (2+1)D nonlocal soliton, the total field distribution is a (2+1+1)D soliton solution of the propagation (in the commonly accepted extension of the term to nonintegrable models) and consists in a train of *stable* light bullets. Their stability stems from decoupling the (2+1+1)D problem into a (2+1)D spatial (nonlocal) and a (1+1)D temporal (Kerr) cases, respectively, providing stable self-trapping. The condition of negligible spatial phenomena induced by the instantaneous Kerr response can be formalized from the simple assumption that in a spatial soliton the diffraction is compensated by the self-focusing. The diffraction angle for a paraxial Gaussian beam is $\Theta_d = 2\lambda_0/\left(\pi n_0 W_0\right)$, whereas the Kerr self-focusing angle can be expressed as $\Theta_f = \sqrt{2n_2 P_{\mathrm{peak}}/\left(\pi n_0 W_0^2\right)}$, with P_{peak} the field peak power and n_0 and n_2 the linear index and the Kerr coefficient, respectively. If P_{peak} is lower than the Kerr soliton critical power [10], the overall diffraction can be approximated as $\Theta_d^* = 2\sqrt{\Theta_d - \Theta_f}$. For a typical nematicon case where $W_0 < 10$ μm and temporal soliton dispersion length in the millimeter scales (as shown in the following section, this is a very conservative assumption for the temporal dynamics), it is straightforward to estimate that $\Theta_d - \Theta_d^* \ll \Theta_d$, that is, Kerr self-focusing, can be neglected.

14.3 ACCESSIBLE LIGHT BULLETS

14.3.1 From Nematicons to Spatiotemporal Solitons

We now address in physical quantities the specific case of a nonlocal soliton sustained by a pulsed excitation and propagating in NLCs. We consider a planarly aligned NLC cell with thickness along X and director aligned in the plane YZ. The optical field $A(X, Y, Z)$, linearly polarized in the plane YZ, propagates in the medium at an angle θ with respect to the molecular director. From Equation 14.3, the evolution of A is then governed by the model:

$$\mathrm{i}\frac{\partial A}{\partial Z} + \frac{1}{2k_0}\nabla^2_{XY}A + \frac{\Delta n_{\mathrm{NL}}(\theta)k_0}{n(\omega_0)}A = 0. \tag{14.6}$$

The angular distribution of the director with respect to the field wave vector can be directly derived from NLC ruling model [9, 27, 31]

$$K\frac{\partial^2\theta}{\partial Z^2} + K\nabla^2_{\perp}\theta + \sin(2\theta)\frac{\varepsilon_0(n^2_{||} - n^2_{\perp})|A|^2}{4}\left\langle\sum_n |\xi\,(Z, T - n\sigma)\,|^2\right\rangle_T = 0. \tag{14.7}$$

Naming η the medium viscosity, if the excitation is removed, the relaxation of a perturbed distribution is ruled by the diffusion model:

$$\eta\frac{\partial\theta}{\partial T} = K\nabla^2_{\perp}\theta. \tag{14.8}$$

Approximating the transverse angle perturbation with the Gaussian distribution $\theta = \hat{\theta}(T)\mathrm{Exp}\left[-(r/W_N)^2\right]$, Equation 14.8 can be recasted as:

$$\frac{\mathrm{d}\hat{\theta}}{\mathrm{d}T} = K\frac{4}{\eta}\left(\frac{r^2}{W_N^4} - \frac{1}{W_N^2}\right)\hat{\theta}. \tag{14.9}$$

In the highly nonlocal regime, the soliton waist W_0 is significantly smaller than W_N, hence $r^2/W_N^4 << 1/W_N^2$ for the relevant range of r. In this case, Equation 14.9 has the closed-form solution $\hat{\theta} = \hat{\theta}_0\exp(-t/\tau)$, where $\tau = \eta W_N^2/4K$ is the perturbation relaxation time. Realistic values for τ in NLC cells of thickness around 100 μm lie between the milliseconds and seconds scale [29, 30, 32]. Hence, trains of short pulses having repetition rate in the kiloHertz range or above are suitable to establish the regime mentioned earlier (Figure 14.1).

In order to estimate the temporal properties of a pulse propagating confined in a nematicon, we formalize the wavelength dependence of the NLC indexes using a single resonance model:

$$n_{\perp} = 1 + \Gamma_{\perp}\lambda^2/(\lambda^2 - \lambda^2_{\perp}), n_{||} = 1 + \Gamma_{||}\lambda^2/(\lambda^2 - \lambda^2_{||}), \tag{14.10}$$

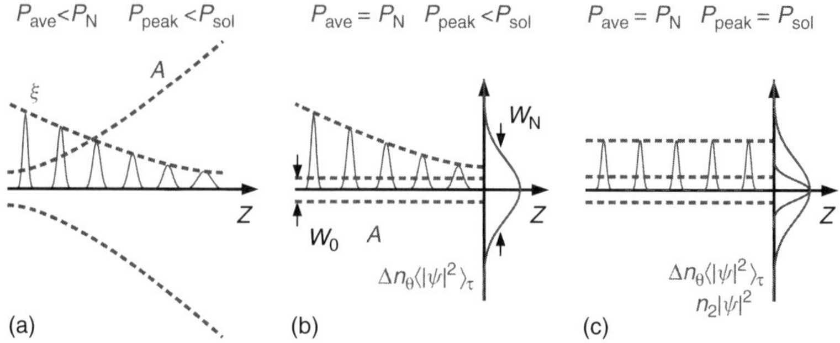

FIGURE 14.1 Sketch of the accessible bullet formation in NLC. (a) The pulse train disperses and diffracts in the linear regime. (b) For an average train power P_{ave} equal to the nonlocal soliton characteristic power P_N the train disperses within the soliton waveguide. The nonlocal feature is highlighted by the diffusion of the index perturbation $\Delta n_\theta\left(\langle|\psi|^2\rangle_\tau\right)$ from the illuminated region (i.e., $W_0 \ll W_N$). (c) If the pulse peak power P_{peak} matches the temporal soliton characteristic power P_{sol}, the temporal confinement occurring within the nonlocal soliton results in the formation of a train of optical bullets. The instantaneous index perturbation $n_2|\psi|^2$ matches the temporal intensity profile, and it is spatially local.

where $\Gamma_\perp$ and $\Gamma_{||}$ are the strengths and $\lambda_\perp$ and $\lambda_{||}$ are the wavelengths of the resonances for $n_\perp$ and $n_{||}$, respectively. For example, for the E7, a best-fit approach determines $\lambda_\perp = 129$ nm, $\lambda_{||} = 182$ nm, and $\Gamma_\perp = 2.983 \times 10^{-5}$ and $\Gamma_{||} = 2.083 \times 10^{-5}$ [30, 32]. Then, by looking at the fundamental nonlinear modes of the system (Eqs 14.6 and 14.7), we can derive the properties of the soliton waveguides, effective index, area, GVD, and equivalent nonlinear coefficient γ as the power changes. In Figure 14.2, we considered a standard cell thickness $L = 100$ μm with anchoring conditions $\theta(X = L/2) = \theta(X = -L/2) = \theta_0 = \pi/6$ and NLC elastic constants $K_{11} = K_{22} = K_{33} = 10^{-11} N$ [27, 28].

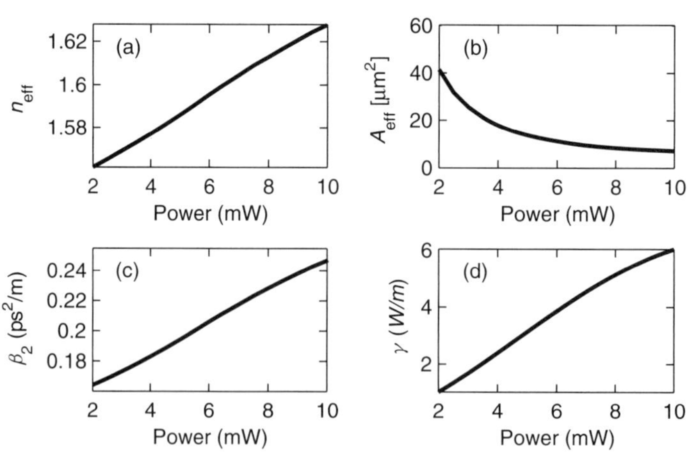

FIGURE 14.2 Calculated soliton waveguides (a) effective index, (b) effective area, (c) GVD, and (d) nonlinear Kerr coefficient for an average excitation power $P_{ave} = 2.3$ mW at wavelength $\lambda = 850$ nm.

It is worth noticing that for many nematics an absorption resonance is usually present in the UV range. Hence, visible and near-IR dispersion are usually normal, forbidding the formation of temporal bright soliton. However, the introduction of a small amount of a resonant dye can easily change the medium dispersion magnitude and sign around specific wavelengths.

14.3.2 Experimental Conditions for Accessible Bullets Observation

In Figure 14.3a, the effect of a diluted dye with $\Gamma_{res} = 10^{-9}$ and resonance at $\lambda_{res} = 930$ nm on the E7 GVD can be appreciated. By assuming an excitation centered at $\lambda = 850$ nm, the dye absorption reverses the sign of the GVD and significantly affects its magnitude even for low concentrations, with negligible contributions to the medium losses in our case.

By calculating the nonlinear eigenmodes of the system (Eqs. 14.6 and 14.7), for the case of weak doping, we determine the dispersion of the soliton waveguide (Figure 14.3b) versus the soliton average power.

Figure 14.4a and 14.4b shows the temporal evolution of a $T_0 = 100$ fs hyperbolic secant pulse $\xi = \sqrt{P_{peak}}\text{sech}\left(T/T_0\right)$ centered at $\lambda = 850$ nm obtained from the Z-integration of the system (Eqs 14.6 and 14.7). The average power is $P_{ave} =$ 2.3 mW, which results in $\beta_2 = -1.54$ ps^2/m, that is, a soliton peak power of $P_{sol} = |\beta_2|/T_0^2\gamma = 126.3$ W and a dispersion length of $L_d = T_0^2/|\beta_2| = 6$ mm. In particular, as expected, for $P_{peak} = P_{sol}/10$, the pulse disperses, whereas for $P_{peak} = P_{sol}$, a temporal soliton is formed. The latter case corresponds to a pulse energy of 25 pJ that imposes an excitation repetition rate of 92 MHz. As the spatial confinement is guaranteed by the average illumination properties, a train of *light bullets* is then formed.

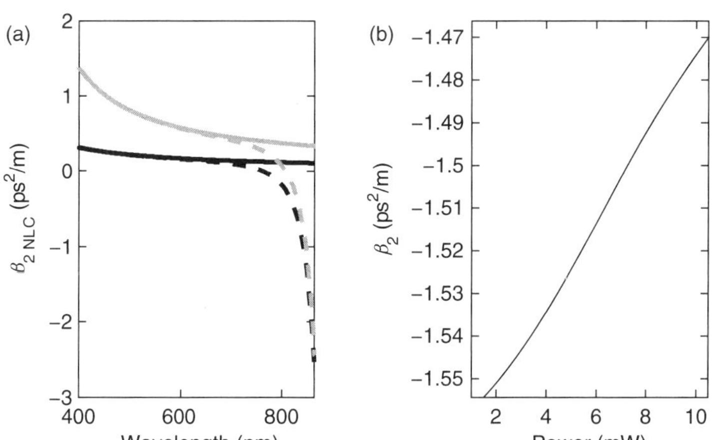

FIGURE 14.3 (a) $\beta_{2\parallel}$ and $\beta_{2\perp}$ in the undoped E7 NLC (solid line) and for the case of diluted doping ($\Gamma_{res} = 10^{-9}$) with a dye resonating at $\lambda_{res} = 930$ nm (dashed line). (b) Calculated power dependence of the soliton waveguide GVD in the presence of the dye doping at $\lambda = 850$ nm.

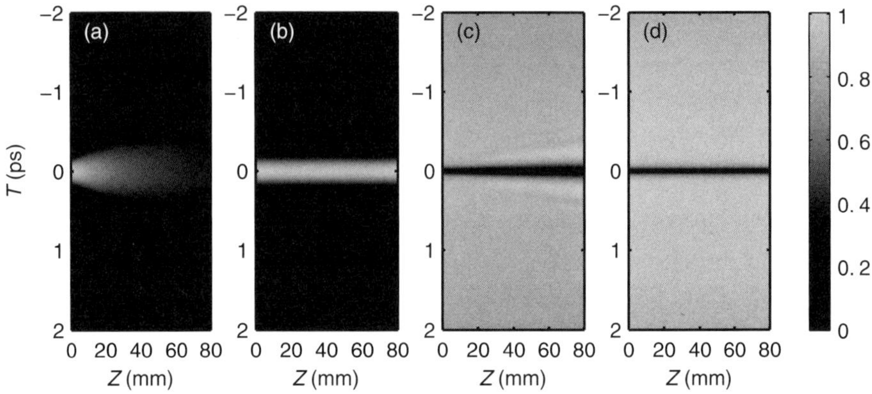

FIGURE 14.4 Temporal evolution of the intensity (normalized units) of the bright pulse for $T_0 = 100$ fs and $\beta_2 = -1.5\ \text{ps}^2/\text{m}$: (a) at low peak power $P_0 = 12.63$ W, the pulse disperses in propagation, whereas (b) at high peak power $P_0 = 126.3$ W, a bright soliton forms. The average power of the train is 2.3 mW in both cases. For comparison, the case (c) and (d) shows the evolution of a dark notch for $T_0 = 70$ fs at the same average power 2.3 mW without resonant dye, that is, $\beta_2 = 0.166\ \text{ps}^2/\text{m}$, with (c) $P_0 = 2.78$ W and (d) $P_0 = 27.8$ W.

In Figure 14.4c and 14.4d, an equivalent case for a dark temporal soliton propagating confined in a spatial soliton is presented. The soliton is formed in an undoped E7-NLC and sustained by a 2.3 mW powered CW excitation, that is, $\beta_2 = 0.166\text{ps}^2/\text{m}$. The input field, $\xi = \sqrt{P_{\text{peak}}}\tanh\left(T/T_0\right)\text{Exp}\left[-(T/T_p)^2\right]$, is a realistic excitation commonly adopted to investigate temporal dark solitons propagation. In the example, $T_0 = 70$ fs, $T_p = 5$ ps (which provides a suitable pedestal), whereas the linear and nonlinear propagation examples are obtained for $P_{\text{peak}} = P_{\text{sol}}/10$ (Fig. 14.4c) and $P_{\text{peak}} = P_{\text{sol}}$ (Fig. 14.4d), respectively. In this case, the train repetition rate is not a fixed parameter and depends on the chosen pedestal duration T_p. We refer to this case as *antibullet* [30] as it results from the combination of a bright spatial soliton solution with a dark temporal one. The examples presented earlier represent cases of stable (3+1)D solitary waves as they are complete solution of the general nonlinear ruling model and occurring in feasible experimental conditions. It is important to stress here that although we addressed the required balance between average power and peak power (to sustain the described regimes) in terms of source repetition rate, such balance can be reached by other means. In particular, the large difference in terms of response speed between the two involved nonlinearities in NLCs allows for a more convenient and practical approach. Once a dense pulse train is obtained from the source with the required pulse energy, a submillisecond on–off modulation (shorter than the nonlocal nonlinearity response time) is applied to the excitation: its duty cycle determines the desired average power for an arbitrary source repetition rate.

14.4 TEMPORAL MODULATION INSTABILITY IN NEMATICONS

In the demonstrated decoupled regime, we can translate all the phenomenology peculiar to the 1D nonlinear Schrödinger equation in the domain of localized temporal effect in spatial nonlocal solitons. In particular, it is possible to explore the possibility of creating dense train of optical bullets by exploiting the temporal modulational instability mechanism. At such purpose, we consider the propagation of a flat-top pulse within a nonlocal soliton waveguide (Fig. 14.5a). To stabilize the random occurrence of pulses typical of the MI phenomenon, we assume the presence of a weak CW background having wavelength slightly shifted with respect to the central flat-top wavelength.

In order to get a significant amount of anomalous dispersion, we assume a dye resonance with $\lambda_{res} = 930$ nm and a pulse central wavelength $\lambda = 850$ nm (Fig. 14.3b). For an average soliton power of $P_{ave} = 2.3$ mW, it turns out $\beta_2 \approx -1.54$ ps^2/m. P_{ave} fixes the combined averaged power of pulsed pump and CW background. The CW background frequency shift depends on the peak power of the pulsed pump as the latter fixes the frequency shift of the maximum of the modulational instability gain $f_p = \sqrt{2\gamma P_{peak}/|\beta_2|}/(2\pi)$ [21]. Figure 14.5 shows the resulting propagation of a 200 W 7 ps-wide (super-Gaussian) flat top, on a 2.3 mW background (note that in this example the pump does not carry any significant average contribution) centered at wavelength $\lambda_{bg} = 856.8$ nm. In this case, a dense train of optical bullets with 2.8 THz repetition rate emerges in propagation with a maximum contrast at the $Z = 39$ mm. In this case, we assumed a white optical noise contribution of the same magnitude of the CW background.

This generation scheme requires the output section to be specifically determined (further propagation results in the mutual interaction of the formed temporal solitons). However, the position of this specific section depends on the power of the CW background; hence, it can be tuned by changing the ratio between the average power of the pulsed excitation (i.e., its repetition rate) and the CW background [32].

14.5 SOLITON-ENHANCED FREQUENCY CONVERSION

The decoupled propagation regime described in the previous paragraphs has been recently demonstrated. In particular, third harmonic (TH) frequency conversion enhanced by a nematicon localization has been demonstrated, by exciting the third-order nonlinearity of an NLC using femtosecond pulses obtained by an optical parametric amplifier [33]. Because NLCs exhibit local inversion symmetry, considering the interaction between an FF at wavelength λ_{FF} with its TH, the source terms for the extraordinary and the ordinary component of the TH reduce, respectively, to:

$$P_{THe} = \chi_{2222}\xi_{FFe}^3 + 3\chi_{1122}\xi_{FFo}^2\xi_{FFe},$$
$$P_{THo} = \chi_{1111}\xi_{FFo}^3 + 3\chi_{2211}\xi_{FFe}^2\xi_{FFo}. \tag{14.11}$$

The confinement within a nonlocal soliton implies input fields preferentially e-polarized; hence, we can consider only the first source term of $P_{\mathrm{TH}e}$ ($eee \rightarrow e$ type I) and the second of $P_{\mathrm{TH}o}$ ($eoe \rightarrow o$ type II) relevant. The corresponding wave vector mismatches are

$$\begin{aligned}\Delta k_{eee\rightarrow e} &= 3k_{\mathrm{FF}e} - k_{\mathrm{TH}e} = \\ &= \left[n_e(\lambda_{\mathrm{FF}}, \theta) - n_e(\lambda_{\mathrm{FF}}/3, \theta)\right] \frac{6\pi}{\lambda_{\mathrm{FF}}},\end{aligned} \tag{14.12}$$

$$\begin{aligned}\Delta k_{eoe\rightarrow o} &= 2k_{\mathrm{FF}e} + k_{\mathrm{FF}o} - k_{\mathrm{TH}o} = \\ &= \left[2n_e(\lambda_{\mathrm{FF}}, \theta) + n_o(\lambda_{\mathrm{FF}}) - 3n_o(\lambda_{\mathrm{FF}}/3)\right] \frac{2\pi}{\lambda_{\mathrm{FF}}},\end{aligned} \tag{14.13}$$

where the momenta of the extraordinary components depend on the NLC orientation. For the E7 NLC, in Figure 14.6a, the resulting phase mismatch versus the angle between o- and e- wave vectors and the director is shown.

The experimental demonstration has been performed by coupling in the NLC cell a 40 nJ, 150 fs pulsed pump beam at $\lambda = 1500$ nm of peak power $P_{\mathrm{peak}} \approx$ 250 kW. In the experimental Figure 14.6b, the propagation occurs at an angle of $\theta \approx 31^\circ$. Type II generation is nearly in phase matching and a diffracting o-polarized THG—that is, without appreciable walk-off, emerges in propagation. Type I generation is not observed because it is not in phase matching. Conversely, if a collinear and copolarized 6 mW CW pump at $\lambda = 1550$ nm is superimposed to the pulsed excitation, a relatively strong e-TH resulting from a type I process

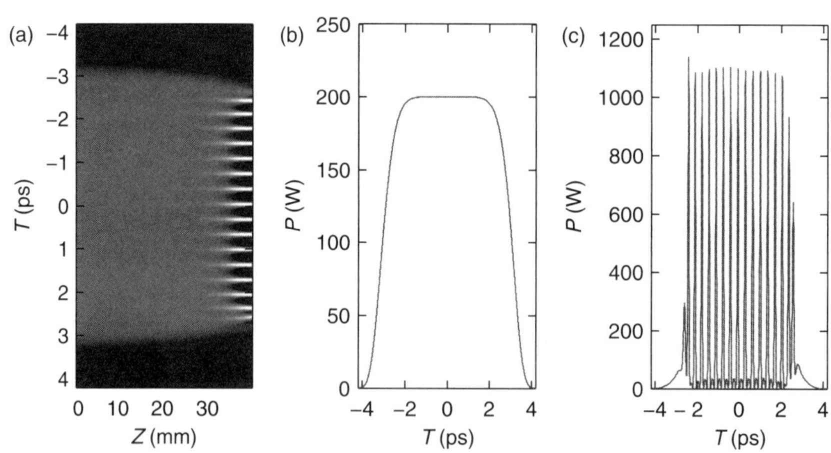

FIGURE 14.5 (a) Propagation of 200 W flat top of duration $\tau_{bg} = 3.3$ ps with a superimposed 2.3 mW background, 2.8 THz shifted in frequency. (b) Input pulse temporal profile. (c) Pulse profile after $Z = 39$ mm of propagation.

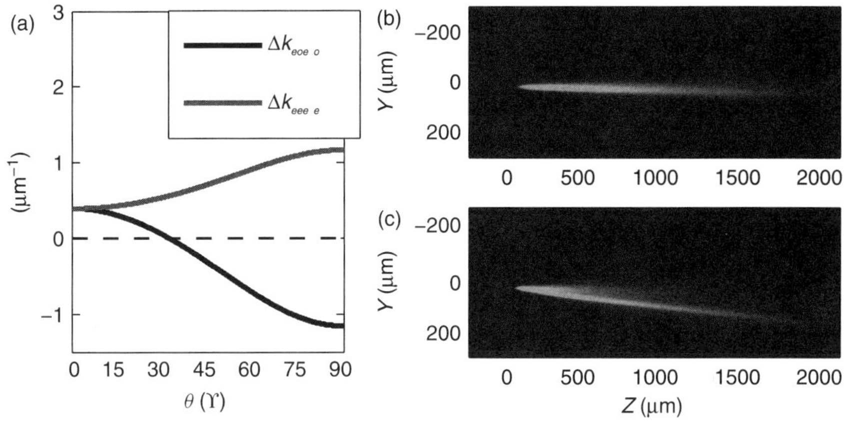

FIGURE 14.6 (a) Dependence of the phase mismatch $\Delta k_{eee\rightarrow e}$ and $\Delta k_{eoe\rightarrow o}$ from the reorientation angle θ. (b) Type II generation in bulk NLC. (c) Nematicon enhanced Type I generation.

is generated within the excited nonlocal soliton, the conversion being strongly enhanced by the tight spatial confinement. The confinement of the TH indicates its e- polarization. As the spatial profile of the TH perfectly matches the soliton waveguide, a decoupled nonlinear regime (spatial nonlocal, temporal local) is realized.

14.6 CONCLUSIONS

In conclusion, the synergic interplay between slow-nonlocal and fast-local nonlinearities enables the access to a range of completely new experimental observable phenomena thanks to the stability arising in properly tailored illumination conditions. These results carry a large potential impact not only limited to optical systems possessing nonlinear and nonlocal optical responses (e.g., thermo-optic and photorefractive dielectrics) but also extended to physical systems where multiple nonlinear mechanisms coexist (e.g., matter waves in fluids, plasmas, and Bose–Einstein condensates).

Acknowledgments

I would like to thank Dr Alessia Pasquazi for the enlightening discussions regarding phase-matching considerations in nonlinear processes and Prof. Roberto Morandotti who made some of the measurements presented in this chapter possible. The contribution of Mr Ian Bruce Burgess to Sections 14.3 is also gratefully acknowledged. Finally, I thank Prof. Gaetano Assanto for bringing significant expertise to part of the work presented in this chapter.

REFERENCES

1. Y. Silberberg. Collapse of optical pulses. *Opt. Lett.*, 15(22):1282–1284, 1990.
2. E. A. Kuznetsov and V. E. Rubenchik. Soliton stability in plasmas and hydrodynamics. *Phys. Rep.*, 142(3):103–165, 1986.
3. J. E. Bjorkholm and A. Ashkin. Cw self-focusing and self-trapping of light in sodium vapor. *Phys. Rev. Lett.*, 32(4):129–132, 1974.
4. M. Segev, B. Crosignani, A. Yariv, and B. Fischer. Spatial solitons in photorefractive media. *Phys. Rev. Lett.*, 68:923–926, 1992.
5. A. Dubietis, E. Gaizauskas, G. Tamosauskas, and P. Di Trapani. Light filaments without self-channeling. *Phys. Rev. Lett.*, 92(25):253903, 2004.
6. A. Pasquazi, S. Stivala, G. Assanto, J. Gonzalo, J. Solis, and C. N. Afonso. Near-infrared spatial solitons in heavy metal oxide glasses. *Opt. Lett.*, 32(15):2103–2105, 2007.
7. A. Pasquazi, S. Stivala, G. Assanto, J. Gonzalo, and J. Solis. Transverse nonlinear optics in heavy-metal-oxide glass. *Phys. Rev. A*, 77(4):043808, 2008.
8. C. Conti, M. Peccianti, and G. Assanto. Route to nonlocality and observation of accessible solitons. *Phys. Rev. Lett.*, 91(7):073901, 2003.
9. M. Peccianti, C. Conti, G. Assanto, A. De Luca, and C. Umeton. Routing of anisotropic spatial solitons and modulational instability in liquid crystals. *Nature*, 432(7018):733–737, 2004.
10. Y. S. Kivshar and G. P. Agrawal. *Optical Solitons: From Fibers to Photonic Crystals*. Academic Press, San Diego, CA, 2003.
11. S. Trillo and W. E. Torruellas. *Spatial Solitons*. Springer-Verlag, Berlin, 1998.
12. A. A. Kanashov and A. M. Rubenchik. On diffraction and dispersion effect on three wave interaction. *Physica D*, 4(1):122–134, 1981.
13. B. A. Malomed, P. Drummond, H. He, A. Berntson, D. Anderson, and M. Lisak. Spatiotemporal solitons in multidimensional optical media with a quadratic nonlinearity. *Phys. Rev. E*, 56(4):4725–4735, 1997.
14. B. A. Malomed, D. Mihalache, F. Wise, and L. Torner. Spatiotemporal optical solitons. *J. Opt. B: Quantum Semiclass. Opt.*, 7(5):R53–R72, 2005.
15. G. Fibich, B. Ilan, and G. Papanicolaou. Self-focusing with fourth-order dispersion. *SIAM J. Appl. Math.*, 62(4):1437–1462, 2002.
16. L. C. Crasovan, J. P. Torres, D. Mihalache, and L. Torner. Arresting wave collapse by wave self-rectification. *Phys. Rev. Lett.*, 91(6):63904, 2003.
17. H. Leblond. Bidimensional optical solitons in a quadratic medium. *J. Phys. A: Math. Gen.*, 31:5129–5143, 1998.
18. H. Leblond. Propagation of optical localized pulses in chi2 crystals: A (3+1)-dimensional model and its reduction to the NLS equation. *J. Phys. A: Math. Gen.*, 31:3041–3066, 1998.
19. J. P. Torres, S. L. Palacios, L. Torner, L.-C. Crasovan, D. Mihalache, and I. Biaggio. Method for generating solitons sustained by competing nonlinearities by use of optical rectification. *Opt. Lett.*, 27(18):1631–3, 2002.
20. J. P. Torres, L. Torner, I. Biaggio, and M. Segev. Tunable self-action of light in optical rectification. *Opt. Commun.*, 213(4–6):351–356, 2002.
21. G. P. Agrawal. *Nonlinear Fiber Optics*, 3rd edn. Academic Press, New York, 2001.

22. D. Mihalache, D. Mazilu, F. Lederer, B. A. Malomed, Y. Kartashov, L. C. Crasovan, and L. Torner. Three-dimensional spatiotemporal optical solitons in nonlocal nonlinear media. *Phys. Rev. E*, 73(2):4–7, 2006.
23. F. W. Dabby and J. R. Whinnery. Thermal self-focusing of laser beams in lead glasses. *Appl. Phys. Lett.*, 13(8):284–286, 1968.
24. C. Rotschild, B. Alfassi, O. Cohen, and M. Segev. Long-range interactions between optical solitons. *Nat. Phys.*, 2(11):769–774, 2006.
25. G. Duree, J. Shultz, G. Salamo, M. Segev, A. Yariv, B. Crosignani, P. Di Porto, E. Sharp, and R. Neurgaonkar. Observation of self-trapping of an optical beam due to the photorefractive effect. *Phys. Rev. Lett.*, 71(4):533–536, 1993.
26. M. Mitchell, Z. G. Chen, M. F. Shih, and M. Segev. Self-trapping of partially spatially incoherent light. *Phys. Rev. Lett.*, 77(3):490–493, 1996.
27. P. G. de Gennes and J. Prost. *The Physics of Liquid Crystals*, 2nd edn. Oxford University Press, London, 1995.
28. I. C. Khoo. Nonlinear optics of liquid crystalline materials. *Phys. Rep.*, 471(5–6):221–267, 2009.
29. M. Peccianti and G. Assanto. Nematic liquid crystals: A suitable medium for self-confinement of coherent and incoherent light. *Phys. Rev. E*, 65(3):35603, 2002.
30. I. B. Burgess, M. Peccianti, G. Assanto, and R. Morandotti. Accessible light bullets via synergetic nonlinearities. *Phys. Rev. Lett.*, 102(20):203903, 2009.
31. M. Peccianti, C. Conti, and G. Assanto. Optical modulational instability in a nonlocal medium. *Phys. Rev. E*, 68:25602, 2003.
32. M. Peccianti, I. B. Burgess, G. Assanto, and R. Morandotti. Space-time bullet trains via modulation instability and nonlocal solitons. *Opt. Express*, 18(6):5934–41, 2010.
33. M. Peccianti, A. Pasquazi, G. Assanto, and R. Morandotti. Enhancement of third-harmonic generation in nonlocal spatial solitons. *Opt. Lett.*, 35:3342–4, 2010.

15

Vortices in Nematic Liquid Crystals

ANTONMARIA A. MINZONI

Fenomenos Nonlineales y Mecánica, Department of Mathematics and Mechanics, Universidad Nacional Autónoma de México, Mexico D.F., Mexico

LUKE W. SCIBERRAS

School of Mathematics and Applied Statistics, University of Wollongong, Wollongong, New South Wales, Australia

NOEL F. SMYTH

School of Mathematics and Maxwell Institute for Mathematical Sciences, University of Edinburgh, Edinburgh, Scotland, United Kingdom

ANNETTE L. WORTHY

School of Mathematics and Applied Statistics, University of Wollongong, Wollongong, New South Wales, Australia

15.1 INTRODUCTION

The study of optical vortices has a long history in both linear and nonlinear media. The book by Kivshar and Agrawal [1] provides a complete review of this extensive work for vortex propagation in local media. Although this extensive body of knowledge for vortex propagation in local media exists, the vortex evolution in nonlocal media, such as thermal nonlinear media and nematic liquid crystals, has been studied only recently [1]. Another recent topic of study is vortex-type structures in coupled waveguides, which provide an underlying lattice for the vortex propagation and lead to a discrete analog of a vortex [1]. It is the purpose of this chapter to describe the main analytical tools for the asymptotic analysis of the stability and propagation of optical vortices in nonlocal, nonlinear media, with an emphasis on nematic liquid crystals. To this end, we begin by recalling the basic ideas about optical vortices in the paraxial approximation, which is the limit useful for the parameter ranges for experimental studies of optical vortices in nematic liquid crystals.

Nematicons: Spatial Optical Solitons in Nematic Liquid Crystals, First Edition.
Edited by Gaetano Assanto.

In the linear limit in the paraxial approximation, the electric field E of the light of an optical vortex is described by Schrödinger's equation

$$\mathrm{i}\frac{\partial E}{\partial z} = \frac{1}{2}\nabla^2 E. \tag{15.1}$$

Here, z is the direction of propagation and the Laplacian gives the diffractive effects in the (x, y) plane orthogonal to the direction of propagation. The initial condition at $z = 0$ is an input vortex experimentally produced by a mask whose thickness increases around its center by a multiple of 2π. Through the use of this mask, the rays will have a phase difference around the center of the mask, forcing the amplitude to be zero at the center and forcing circulation in the angular variable.

Elementary solutions of Schrödinger's equation (Eq. 15.1) are obtained in the form

$$E(r, \varphi, z) = \mathrm{e}^{\mathrm{i}\sigma z}\mathrm{e}^{\mathrm{i}m\varphi} f_m(r, \sigma), \tag{15.2}$$

where r and φ are polar coordinates whose origin is at the center of the mask and the function f_m satisfies Bessel's equation of integer order m. The function $f_m(r, \sigma)$ is zero at the origin, has a maximum away from the origin, and decays to 0 at a rate $r^{-1/2}$ in an oscillatory fashion. As Equation 15.1 is linear, the evolution of an initial condition

$$E(r, \varphi, 0) = ar^m \mathrm{e}^{-r/w}\mathrm{e}^{\mathrm{i}m\varphi} \tag{15.3}$$

can be obtained as a superposition in σ of the solution (Eq. 15.2). Here, m is a positive integer, termed the charge of the vortex. This representation shows that the vortex diffracts as z increases, decaying in amplitude at the rate $z^{-1/2}$. In this way, the vortex beam broadens as the total power is conserved. Moreover, as happens for all solutions of the linear Schrödinger's equation, the vortex beam develops a phase that is r dependent. We thus see that vortices in linear media diffract away, as expected.

The next step is to consider the nonlinear counterpart of Equation 15.1 with a cubic nonlinearity. However, such media solutions that correspond to localized initial conditions, such as vortices, become singular in finite time [1]. This collapse behavior is avoided in saturable media [1]. For these saturable media, approximate vortex-type solutions of the form 15.2 can be determined variationally, resulting in families of vortex solutions whose width w and angular speed σ are functions of the amplitude a. It has been found that when Equation 15.2 with $m = 1$ is used as an initial condition for the saturable nonlinear Schrödinger (NLS) equation, it evolves into a vortex for a short distance in z and then narrows at two diagonally opposite points on the vortex. It subsequently breaks up into two lump-type beams. The spin angular momentum of the vortex is transferred to the orbital angular momentum of the beams [1]. The vortex for the saturable NLS equation is then unstable. A similar result is obtained for larger values of m. This unstable behavior is further established by solving the linearized stability problem numerically. The governing equations are linearized about the numerically found steady vortex, resulting in a linear stability problem whose eigenvalues are found numerically [2]. In this manner, it was found that there is an azimuthal instability with the fastest growing mode having wave number $l = 2$ for $m = 1$ [2]. This result explains the splitting instability of the vortex. However, this numerical analysis gives no hint of the detailed instability mechanism involved.

A first, crude argument to show the azimuthal instability in the limit of a large vortex for an NLS equation is as follows. Consider the saturable NLS equation

$$\mathrm{i}\frac{\partial u}{\partial z} + \frac{1}{2}\nabla^2 u + f(|u|^2)u = 0, \tag{15.4}$$

where $f(|u|^2)$ approaches a constant as $|u| \to \infty$. This equation has the Lagrangian

$$L = \int_0^z \int_{-\infty}^{\infty} \int_{-\infty}^{\infty} \left[\mathrm{i}\left(u^* u_z - u u_z^*\right) - |\nabla u|^2 + F(|u|^2)\right] \mathrm{d}x\mathrm{d}y\mathrm{d}z. \tag{15.5}$$

Here, $F'(x) = f(x)$ and the * superscript denotes the complex conjugate. An approximate vortex solution takes the form

$$u(r, \varphi, z) = v(\varphi, z) g(r - R), \tag{15.6}$$

where R is the radius of the vortex and the function $g(\xi)$ is peaked around $\xi = 0$. Substitution of this vortex into the Lagrangian (Eq. 15.5) and integration in the radial variable gives a new Lagrangian for $v(\varphi, z)$ of the form

$$\overline{L} = \int_0^z \int_0^{2\pi} \left[\mathrm{i}A\left(v^* v_z - v v_z^*\right) - B|v|^2 - \frac{A}{R^2}|v_\varphi|^2 + h(|v|^2)\right] \mathrm{d}\varphi\mathrm{d}z. \tag{15.7}$$

Here,

$$A = \int_0^{\infty} g^2(r - R)\, r\mathrm{d}r, \quad B = \int_0^{\infty} g'^2(r - R)\, r\mathrm{d}r,$$

$$h(|v|^2) = \int_0^{\infty} F[g^2(r - R)|v|^2]\, r\mathrm{d}r. \tag{15.8}$$

Variations of the Lagrangian (Eq. 15.7) with respect to v^* show that v satisfies the one-dimensional NLS equation in the azimuthal direction

$$\mathrm{i}\frac{\partial v}{\partial z} + \frac{A}{R^2}\frac{\partial^2 v}{\partial \varphi^2} - Bv + h'(|v|^2)v = 0. \tag{15.9}$$

The angular part of the vortex solution then has the form

$$v = a e^{i(m\varphi + \sigma z)}, \tag{15.10}$$

where σ satisfies the nonlinear dispersion relation

$$\sigma = -\frac{m^2 A}{R^2} - B + h'(a^2). \tag{15.11}$$

The expression 15.10 is a uniform wave train. It is modulationally unstable for sufficiently long-wave-sideband-type perturbations [3]. In this case, wave numbers with $m = O(1)$ are equivalent to long wave sidebands, as the dispersion relation for the linearized equation (Eq. 15.9) shows that the effective wave number is $\sqrt{A}m/R$, which is small for large R.

The question of the behavior of the cluster of solitary waves that results from vortex destabilization has also been widely studied in local media [1]. However, in nonlocal media, the behavior of such clusters has been shown to be very different [2, 4]. In fact, in such media, the solitary wave clusters merge, forming a stable vortex that transforms orbital angular momentum into vortex spin. One question that remains to be addressed is that of the linking of the azimuthal solitary waves, which result from the numerically observed saturation of the azimuthal instabilities by the nonlocality, to suitably bifurcated branches from the basic $m = 1$ vortex.

Finally, it is possible to reformulate the classical problems of vortex dynamics from fluid mechanics to the context of nematic liquid crystals and then explore the results of the stabilizing effect of the nonlocality of the liquid crystal.

15.2 STABILIZATION OF VORTICES IN NONLOCAL, NONLINEAR MEDIA

The results of the previous section show that the only way to stabilize a large vortex is to decrease the nonlinear interaction between the wave propagating around the vortex and the long-wave instability. It is clear that local media cannot decrease the size of the nonlinearity unless the medium becomes linear. However, it has been found that optical vortices can be stabilized in nonlocal, nonlinear media, as the balance required for stability is possible in such media. This stabilization in nonlocal, nonlinear media was first studied numerically by Yakimenko et al. [2] for optical vortices in nematic liquid crystals. The equations for the envelope E of the electric field of the light beam and the optical axis orientation θ of the nematic molecules are [5, 6]

$$i\frac{\partial E}{\partial z} + \frac{1}{2}\nabla^2 E + \theta E = 0, \tag{15.12}$$

$$\nu\nabla^2\theta - \theta = -|E|^2. \tag{15.13}$$

The parameter ν is called the nonlocality parameter and is a measure of the elastic response of the nematic. As ν increases, the nematic becomes more nonlocal, in the sense that the response of the nematic extends (far) beyond the waist of the beam.

In the study by Yakimenko et al. [2], a steady solution of the nematic Equations 15.12 and 15.13 was sought in the form

$$E = f(r)\mathrm{e}^{\mathrm{i}m\varphi}, \tag{15.14}$$

where $f(0) = 0$ and $f(r) \to 0$ as $r \to \infty$. The coordinates (r, φ) are the standard plane polar coordinates in the (x, y) plane. This steady solution was found by numerically solving the boundary value problem for $f(r)$ and then finding the corresponding θ. The nematic Equations 15.12 and 15.13 were then linearized about this numerically determined steady vortex for $m = 1, 2, 3$, and the corresponding eigenvalue problem was solved numerically to determine the linearized stability. It was found for $m = 1$ that an azimuthal instability with angular dependence $\exp(2i\varphi)$ is present. This instability leads to the breakup of the vortex into two beams that rotate around a center and move away from each other, as illustrated in Figure 15.1. It was also shown that this instability is eliminated for large enough ν, $\nu \sim 100$, leading to a stable vortex. However, for $m = 2$ and $m = 3$, the corresponding instabilities with azimuthal dependencies $\exp(3i\varphi)$ and $\exp(4i\varphi)$ are not suppressed by increasing nonlocality ν. Although giving the stability picture for the nonlocal vortex, this numerical study does not give any indication of the mechanisms involved.

To study this vortex stability in a nonlocal, nonlinear nematic liquid crystal theoretically, let us use a Green's function to solve the director Equation 15.13 and then rewrite the electric field Equation 15.12 in the form

$$\mathrm{i}\frac{\partial E}{\partial z} + \frac{1}{2}\nabla^2 E + E\int_{-\infty}^{\infty}\int_{-\infty}^{\infty} K_0\left(\left[\frac{(x-x')^2+(y-y')^2}{\nu}\right]^{1/2}\right) \times |E(x', y')|^2\,\mathrm{d}x'\mathrm{d}y' = 0. \tag{15.15}$$

Here, K_0 is the modified Bessel function of the second kind of order 0, which is the Green's function for the director Equation 15.13.

To assess the effect of the nonlocality measured by ν, let us replace the kernel K_0 in Equation 15.15 by a Gaussian

$$\mathrm{e}^{-\left[(x-x')^2+(y-y')^2\right]/\nu}. \tag{15.16}$$

This is a commonly used kernel in studies of beam and vortex propagation in nonlocal, nonlinear media [7]. However, caution should be exercised with this Gaussian kernel, as it does not arise for any real medium [8, 9]. In particular, it does not contain the logarithmic singularity of the real kernel for a nematic liquid crystal. For this Gaussian kernel, the nonlocal term in the modified electric field

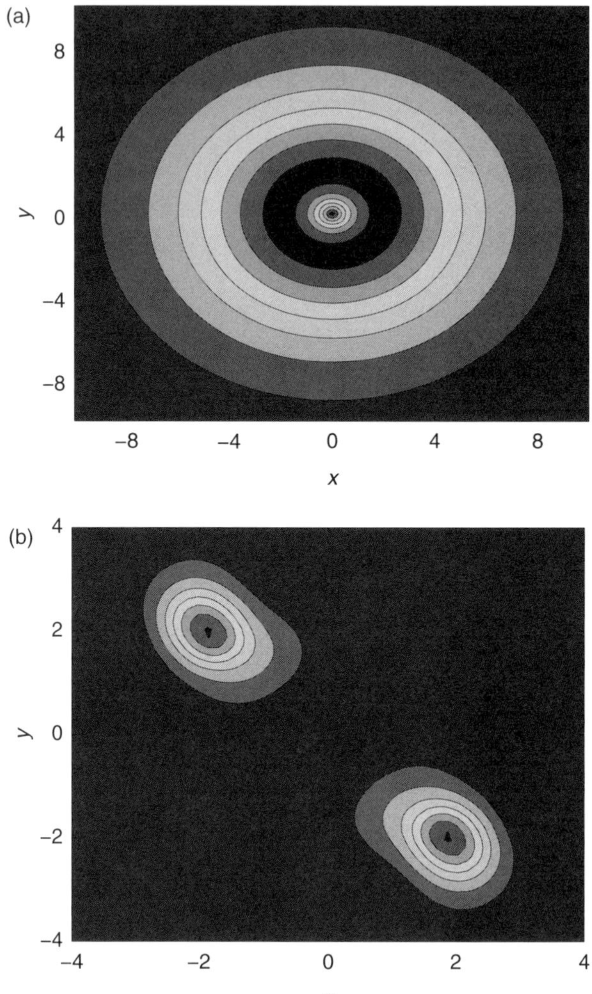

FIGURE 15.1 Numerical solution of nematicon equations (Eqs. 15.12 and 15.13) for $|E|$ for the initial condition $E = 2r\exp(-r/2 + \mathrm{i}\varphi)$, with $\nu = 0.5$. (a) $z = 0$ and (b) $z = 10$. *Source*: Reproduced from Fig. 1 in Reference 4, with permission.

Equation 15.15 can be approximated for large ν as

$$\int_{-\infty}^{\infty}\int_{-\infty}^{\infty} \mathrm{e}^{-|\mathbf{x}-\mathbf{x}'|^2/\nu}|E(x', y')|^2\,\mathrm{d}x'\mathrm{d}y'$$
$$= \int_{-\infty}^{\infty}\int_{-\infty}^{\infty}\left[1 - \frac{1}{\nu}\left(|\mathbf{x}|^2 - 2\mathbf{x}\cdot\mathbf{x}' + |\mathbf{x}'|^2\right)\right]|E(x', y')|^2\,\mathrm{d}x'\mathrm{d}y'$$
$$+\, O(1/\nu^2). \tag{15.17}$$

For simplicity, the coordinates (x, y) have been written in the vector form $\mathbf{x}$. The NLS Equation 15.15 for the electric field is then approximated by

$$\mathrm{i}\frac{\partial E}{\partial z} + \frac{1}{2}\nabla^2 E + PE - \frac{P}{\nu}|\mathbf{x}|^2 E$$
$$- \frac{E}{\nu}\int_{-\infty}^{\infty}\int_{-\infty}^{\infty}\left[-2\mathbf{x}\cdot\mathbf{x}' + |\mathbf{x}'|^2\right]|E(x', y')|^2\,\mathrm{d}x'\mathrm{d}y' = 0. \qquad (15.18)$$

The power

$$P = \int_{-\infty}^{\infty}\int_{-\infty}^{\infty} |E|^2\,\mathrm{d}x\mathrm{d}y \qquad (15.19)$$

is a conserved quantity for Equations 15.12 and 15.13 or Equation 15.15 or 15.18. We further note that the approximate Equation 15.18 for large ν is a linear Schrödinger's equation for E with harmonic oscillator and linear potentials. The approximation 15.18 is the Snyder–Mitchell approximation [10] modified by a linear potential. This linear potential can be reabsorbed into the quadratic term by a phase transformation of E. Therefore, to leading order, we can consider vortex solutions of the Snyder–Mitchell approximation [10]

$$\mathrm{i}\frac{\partial u}{\partial z} + \frac{1}{2}\nabla^2 u - \frac{P}{\nu}|\mathbf{x}|^2 u = 0, \qquad (15.20)$$

where

$$P = \int_{-\infty}^{\infty}\int_{-\infty}^{\infty} |u|^2\,\mathrm{d}x\mathrm{d}y. \qquad (15.21)$$

Let us seek vortex-type solutions in the Snyder–Mitchell approximation 15.20

$$u_0 = v(r)\mathrm{e}^{\mathrm{i}(m\varphi + \sigma z)}. \qquad (15.22)$$

The profile $v(r)$ then satisfies the linear eigenvalue problem

$$\frac{1}{2r}\frac{\mathrm{d}}{\mathrm{d}r}\left(r\frac{\mathrm{d}v}{\mathrm{d}r}\right) - \sigma v - \frac{m^2}{2r^2}v - \frac{P}{\nu}r^2 v = 0. \qquad (15.23)$$

Equation 15.23 has vortex-type solutions. The solutions with no nodes are analogs of the nonlocal vortices discussed in Reference 9. We now perturb this steady vortex by a perturbation g, so that $u = u_0 + g$. If this perturbation does not change the power P, we have to leading order that

$$\int_{-\infty}^{\infty}\int_{-\infty}^{\infty}\left(u^* g + u g^*\right)\,\mathrm{d}x\mathrm{d}y = 0. \qquad (15.24)$$

Then, the linearization of the Snyder–Mitchell approximation 15.20 with $u = u_0 + g$ gives the perturbation equation

$$\mathrm{i}\frac{\partial g}{\partial z} + \frac{1}{2}\nabla^2 g - \frac{P}{\nu}|\mathbf{x}|^2 g = 0. \tag{15.25}$$

This linear equation implies that if g is initially small, it will remain small as the solutions of this equation are bounded. Therefore, in the Snyder–Mitchell approximation, vortices are always stable. The Snyder–Mitchell approximation is then not appropriate for the study of the stability change of vortices with ν, and the full Equation 15.15 must be reexamined in detail.

To this end, let us rewrite Equation 15.15 using the additional property of the Bessel functions [12]

$$\begin{aligned} G(r,\varphi,r',\varphi') &= K_0\left(\left[\sqrt{r^2 + r'^2 - 2rr'\cos(\varphi-\varphi')}\right]/\sqrt{\nu}\right) \\ &= \sum_{l=-\infty}^{\infty} \mathrm{e}^{il(\varphi-\varphi')} G_l(r,r'), \end{aligned} \tag{15.26}$$

where

$$G_l(r,r') = \begin{cases} I_l(r/\sqrt{\nu})K_l(r'/\sqrt{\nu}), & 0 \le r \le r', \\ K_l(r/\sqrt{\nu})I_l(r'/\sqrt{\nu}), & r' < r < \infty. \end{cases} \tag{15.27}$$

I_l and K_l are the modified Bessel functions of the first and second kinds of order l. We now show how the azimuthal dependence of Green's function for the nematic Equation 15.15 can resonantly couple with the vortex with an additional azimuthal perturbation, resulting in instability. It is then necessary to study the behavior of these resonances as a function of the nonlocality. This resonant coupling can be studied using a suitable modulation theory, as detailed later.

The nonlocal Equation 15.15 has the Lagrangian density

$$\begin{aligned} L = \mathrm{i}r\left(E^*E_z - EE_z^*\right) - r|E_r|^2 - \frac{1}{r}|E_\varphi|^2 \\ + r|E|^2 \int_0^{2\pi}\int_0^{\infty} G(r,\varphi,r',\varphi')|E(r',\varphi')|^2\, r'\mathrm{d}r'\mathrm{d}\varphi'. \end{aligned} \tag{15.28}$$

Let us consider the $m = 1$ vortex, so that the trial function to obtain an approximate vortex solution is chosen to be [4]

$$E = are^{-r/w(z,\varphi)}\mathrm{e}^{\mathrm{i}\sigma+\mathrm{i}\varphi} + \mathrm{i}g(z,\varphi)\mathrm{e}^{\mathrm{i}\sigma+\mathrm{i}\varphi}. \tag{15.29}$$

The amplitude a and phase σ are functions of φ and z, respectively. The conjugate variables w and g are also dependent on φ and z. When the width $w \to 0$ at a particular value of φ, the vortex is pinched at this point, which indicates that the vortex is unstable. Therefore, to examine possible instabilities, we expand w as

$$w(z, \varphi) = w_0(z) + w_1(z)\cos\varphi + w_2(z)\cos 2\varphi + \cdots \tag{15.30}$$

Thus, as w_i increases, $i \geq 1$, w can become zero, leading to the breakup of the vortex. The growth of w_2 will then lead to the splitting of the vortex into two beams. The shelf of radiation under the vortex given by g is also dependent on the azimuthal variable φ to take account of the distortion of the vortex due to the change of its width. The shelf of radiation g is assumed to be nonzero in the region $w - R/2 \leq r \leq w + R/2$ centered at the vortex, where the width R will be determined later.

Substituting the trial function (Eq. 15.29) into the Lagrangian (Eq. 15.28) and integrating into r from 0 to ∞ gives the averaged Lagrangian

$$\mathcal{L} = \int_0^\infty L\, r\mathrm{d}r = \mathcal{L}_0 + \mathcal{L}_\nu. \tag{15.31}$$

Here, $\mathcal{L}_0$ is the local contribution and $\mathcal{L}_\nu$ is the nonlocal contribution. In detail, [4]

$$\begin{aligned}\mathcal{L}_0 = &-\left(\frac{3}{4}a^2w^4 + 2\Lambda_1 g^2\right)\frac{\partial\sigma}{\partial z} - 4aw^3\frac{\partial g}{\partial z} + 4gw^3\frac{\partial a}{\partial z} + 12agw^2\frac{\partial w}{\partial z}\\ &- \frac{3}{8}w^2a^2 - \frac{1}{4}w^2\left(\frac{\partial a}{\partial\varphi}\right)^2 - \frac{1}{2}aw\frac{\partial a}{\partial\varphi}\frac{\partial w}{\partial\varphi} - \frac{3}{8}a^2\left(\frac{\partial w}{\partial\varphi}\right)^2\\ &+ 2wg\frac{\partial a}{\partial\varphi} + 2ag\frac{\partial w}{\partial\varphi} - 2aw\frac{\partial g}{\partial\varphi} - \Lambda_2\left[\left(\frac{\partial g}{\partial\varphi}\right)^2 + g^2\right]\\ &- \frac{1}{2}a^2w^2\frac{\partial\sigma}{\partial\varphi} - 2aw\frac{\partial\sigma}{\partial\varphi}\frac{\partial g}{\partial\varphi} + 2wg\frac{\partial\sigma}{\partial\varphi}\frac{\partial a}{\partial\varphi} + 2ag\frac{\partial\sigma}{\partial\varphi}\frac{\partial w}{\partial\varphi}\\ &- \Lambda_2 g^2\left[\left(\frac{\partial\sigma}{\partial\varphi}\right)^2 + 2\frac{\partial\sigma}{\partial\varphi}\right] - \frac{1}{4}a^2w^2\left(\frac{\partial\sigma}{\partial\varphi}\right)^2\end{aligned} \tag{15.32}$$

and

$$\mathcal{L}_\nu = \int_0^\infty |E(z, r, \varphi)|^2 \int_0^{2\pi}\int_0^\infty G(r, \varphi, r', \varphi')|E(z, r', \varphi')|^2\, r'\mathrm{d}r'\mathrm{d}\varphi' r\mathrm{d}r. \tag{15.33}$$

The first term $\mathcal{L}_0$ comes from a positive Lagrangian (Eq. 15.28), and therefore, it can only produce stable behavior under perturbations. Hence, the only destabilizing term is contained in $\mathcal{L}_\nu$. The quantities Λ_1 and Λ_2 are related to the radius of the shelf of radiation g and are given by Minzoni et al. [4]

$$\Lambda_1 = wR \quad \text{and} \quad \Lambda_2 = \ln\left(\frac{w + R/2}{w - R/2}\right). \tag{15.34}$$

As remarked in the discussion after the expansion 15.27, we need to study the resonance between the vortex and the shelf g. We thus calculate $\mathcal{L}_\nu$ to leading order, keeping only quadratic terms in g, to obtain

$$\begin{aligned}
\mathcal{L}_\nu = & \int_0^\infty |E_0(z,r,\varphi)|^2 \int_0^{2\pi}\int_0^\infty G(r,\varphi,r',\varphi')|E_0(z,r',\varphi')|^2\, r'\mathrm{d}r'\mathrm{d}\varphi' r\mathrm{d}r \\
& + \int_0^\infty g^* E_0(z,r,\varphi) \int_0^{2\pi}\int_0^\infty G(r,\varphi,r',\varphi')E_0^*(z,r',\varphi')g\, r'\mathrm{d}r'\mathrm{d}\varphi' r\mathrm{d}r \\
& + \int_0^\infty g E_0^*(z,r,\varphi) \int_0^{2\pi}\int_0^\infty G(r,\varphi,r',\varphi')E_0(z,r',\varphi')g^*\, r'\mathrm{d}r'\mathrm{d}\varphi' r\mathrm{d}r \\
= & \mathcal{L}_\nu^0 + \mathcal{L}_\nu^g,
\end{aligned} \tag{15.35}$$

where E_0 is the vortex component and g is the shelf term of the trial function (Eq. 15.29). To calculate the first term, we note that the integral of the Green's function is just the solution of the radial equation for the director forced by $a^2 r'^2 \exp(-2r'/w)$. In the limit of large nonlocality ν, this forcing acts as a point source as $w \ll \sqrt{\nu}$. This gives a solution that is flat where the vortex has a node, giving no contribution to the integral. The only contribution to the integral then comes from the tail and is small for ν large. We then have that

$$\mathcal{L}_\nu^0 = \frac{a^4 w^6}{32\sqrt{\nu}}. \tag{15.36}$$

Using this result, the variational equations from the averaged Lagrangian (Eq. 15.31) give the steady-state amplitude–width relation for the vortex in the form

$$a_0^2 w_0^4 = 48\sqrt{\nu}. \tag{15.37}$$

To examine the interaction terms, one can, in principle, substitute the Fourier expansion for g

$$g = \sum_{l=-\infty}^{\infty} g_l \mathrm{e}^{\mathrm{i}l\varphi} \tag{15.38}$$

into the interaction averaged Lagrangian (Eq. 15.35) and calculate the resulting integrals approximately. However, this procedure is easier to implement on noting that the integral

$$I(r,\varphi) = \int_0^{2\pi} \int_0^{\infty} G(r,\varphi,r',\varphi') E_0^*(r',\varphi') g\, r' \mathrm{d}r' \mathrm{d}\varphi' \tag{15.39}$$

is the solution of

$$\nu \nabla^2 I - qI = E_0^* g. \tag{15.40}$$

This solution is regular at the origin and decays as $r \to \infty$. Expanding I in a Fourier series gives

$$I(r,\varphi) = \sum_{l=-\infty}^{\infty} Q_l(r,z) \mathrm{e}^{\mathrm{i}\sigma} \mathrm{e}^{\mathrm{i}l\varphi}. \tag{15.41}$$

The Q_l satisfy

$$\frac{\mathrm{d}^2 Q_l}{\mathrm{d}r^2} + \frac{1}{r}\frac{\mathrm{d}Q_l}{\mathrm{d}r} - \frac{(l-1)^2}{r^2} Q_l - \frac{1}{\nu} Q_l = \frac{ar}{\nu} \mathrm{e}^{-r/w} g_l(r,z). \tag{15.42}$$

It follows from this equation that when $l \neq 1$, the solution for Q_l is $O(1/\nu)$. This result could also be obtained from the exact solution for Q_l in terms of Bessel functions. However, it is easier to obtain this order estimate for Q_l directly from the governing Equation 15.42 by expanding the solution in powers of ν^{-1}. The term Q_l/ν then introduces an $O(\nu^{-2})$ correction. On the other hand, the $l = 1$ term can be approximated as for Equation 15.36. As the contribution of g_1 is $O(\Lambda_1)$, the contribution is again $O(\nu^{-1})$ and not $O(\nu^{-1/2})$. It should be noted that the solution for Q_l is not symmetric in l. However, the resulting averaged Lagrangian is symmetric in l due to the contribution of $E_0 g^*$ in expression 15.35 for $\mathcal{L}_\nu$. The contribution for $l > 0$ of the term $E_0 g^*$ adds with the contribution of $E_0^* g$ for $l < 0$, giving a symmetric dependence on l in the final averaged Lagrangian. The averaged Lagrangian (Eq. 15.35) then becomes

$$\mathcal{L}_\nu = \frac{a^2 w^6}{\nu} g^*(\varphi) \int_0^{2\pi} \sum_{|l|\geq 2}^{\infty} \frac{202\mathrm{e}^{-2}}{256(l^2-1)} \mathrm{e}^{\mathrm{i}l(\varphi-\varphi')} g(\varphi')\, \mathrm{d}\varphi' + \text{c.c.}$$

$$+ \frac{\Lambda_1^2}{\nu} g^*(\varphi) \int_0^{2\pi} \mathrm{e}^{\mathrm{i}(\varphi-\varphi')} g(\varphi')\, \mathrm{d}\varphi' + \text{c.c.}, \tag{15.43}$$

where c.c. denotes the complex conjugate of the preceding term. With this final term in the averaged Lagrangian calculated, we can now determine the stability of the vortex.

This dependence of the solution on the modes is not captured by the Snyder–Mitchell approximation. In this approximation, all the modes behave in the same manner as the nonlocal term is replaced by a radially symmetric effective potential. This difference accounts for the stable behavior of vortices in the Snyder–Mitchell approximation, whereas the full equations show that the vortices can be unstable, depending on the value of ν.

The Lagrangians $\mathcal{L}_0$ and $\mathcal{L}_\nu$ are then averaged with respect to the angle and z variables to obtain modulation equations for w and g. These are then linearized about the vortex fixed point given by Equation 15.37. The linearized perturbation in amplitude a is related to that in width w using the mass conservation equation. Eliminating the amplitude perturbation in this manner, we finally obtain

$$8a_0 w_0^2 \frac{\partial w_1}{\partial z} = \left(2\Lambda_1 w_0^{-2} + \Lambda_2\right) g_1 - 2\Lambda_2 \frac{\partial^2 g_1}{\partial \varphi^2} - \frac{2a_0^2 w_0^6}{\nu} \sum_{|l| \geq 2}^{\infty} \frac{202\mathrm{e}^{-2}}{256(l^2-1)} \int_0^{2\pi} \mathrm{e}^{\mathrm{i}l(\varphi-\varphi')} g_1(\varphi')\, d\varphi' - \frac{\Lambda_1^2}{\nu} \int_0^{2\pi} \mathrm{e}^{\mathrm{i}l(\varphi-\varphi')} g_1(\varphi')\, \mathrm{d}\varphi', \tag{15.44}$$

$$16 w_0^2 \frac{\partial g_1}{\partial z} = 3a_0 \frac{\partial^2 w_1}{\partial \varphi^2}. \tag{15.45}$$

Here, the 1 subscript denotes the linear perturbation in the variable. The stability of the vortex is then reduced to the linear eigenvalue problem for the system in Equations 15.44 and 15.45. The modes are of the form

$$w_1 = \mathrm{e}^{in\varphi} W(z) \quad \text{and} \quad g_1 = \mathrm{e}^{in\varphi} G(z). \tag{15.46}$$

Substitution of these modes into the system in Equations 15.44 and 15.45 results in an n-dependent system of ordinary differential equations for W and G. These ordinary differential equations have oscillatory solutions for $|n| \geq 2$, provided the determinant

$$\Delta_\nu(n) = \frac{405\mathrm{e}^{-2}}{64\nu(n^2-1)} a_0^2 w_0^6 - 1 - 3\Lambda_1 w_0^2 - 3\Lambda_2 \left(n^2+1\right) \leq 0, \tag{15.47}$$

and a similar determinant relation for $|n| = 1$. From the fixed point relation 15.37, it follows that for constant steady amplitude a_0, the numerator of the destabilizing term in Equation 15.47 scales as $\nu^{3/4}$, whereas the denominator scales as ν. It is therefore apparent that as ν becomes large, the destabilization term decays as $\nu^{-1/4}$, so that the vortex stabilizes for sufficiently large ν.

15.3 VORTEX IN A BOUNDED CELL

A novel experimental scenario is that in which nonlinear beams propagate in a cell for which no pretilting field is applied [13]. In these cases, the cell walls are "rubbed" to obtain a fixed orientation angle of the nematic molecules at the walls. Elastic forces then transmit this angle into the bulk of the nematic. The cell walls act as a repulsive force on a beam. This interaction with the walls can produce complicated beam trajectories, as studied using modulation theory for a nematicon (solitary wave) in References 14 and 15. The same modulation theory approach was further used to study the motion of a vortex in a finite cell [16]. In this section, we describe the main approximations used to study optical vortices in finite cells and use modulation theory to qualitatively predict the motion of vortices under different conditions.

In the absence of a pretilting static electric field, the equations governing a vortex in a nematic cell are

$$\mathrm{i}\frac{\partial E}{\partial z} + \frac{1}{2}\nabla^2 E + \theta E = 0, \tag{15.48}$$

$$\nu\nabla^2\theta = -|E|^2. \tag{15.49}$$

Here, E is the electric field of the optical beam and θ is the deviation of the optical director from its initial pretilt angle due to rubbing. These equations hold in some region Ω in the transverse (x, y) variables with $\theta = 0$ on the boundary of Ω. Poisson's equation (Eq. 15.49) can be solved for θ using a Green's function G. On substituting this solution for θ into Equation 15.48, we obtain a single equation for the electric field E, which has the Lagrangian

$$\int_0^z \int\int_\Omega L\,\mathrm{d}\mathbf{x}\mathrm{d}z, \tag{15.50}$$

where

$$L = \mathrm{i}\left(E^* E_z - E E_z^*\right) - |\nabla E|^2 - |E|^2 \int\int_\Omega G(\mathbf{x}, \mathbf{x}')|E(\mathbf{x}', z)|^2\,\mathrm{d}\mathbf{x}'. \tag{15.51}$$

The kernel $G(\mathbf{x}, \mathbf{x}')$ is the Green function for the Poisson's equation (Eq. 15.49) in Ω, with source $\mathbf{x}'$ and $G(\mathbf{x}, \mathbf{x}') = 0$ for $\mathbf{x}$ on the boundary of Ω.

Let us assume that the nonlocality ν is large so that a vortex is stable [2]. We shall assume a trial function for E to be of the form

$$E(\mathbf{x}, z) = a|\mathbf{x} - \mathbf{X}(z)|\mathrm{e}^{-|\mathbf{x}-\mathbf{X}(z)|/w}\mathrm{e}^{\mathrm{i}\arg((\mathbf{x}-\mathbf{X}(z))/|\mathbf{x}-\mathbf{X}(z)|)}\mathrm{e}^{\mathrm{i}\mathbf{V}\cdot(\mathbf{x}-\mathbf{X}(z))+\mathrm{i}\sigma}. \tag{15.52}$$

¡?xmltex ?>The amplitude a and width w are taken to satisfy the amplitude–width relation for a steady vortex, which will be determined. The position $\mathbf{X}(z) = (X(z), Y(z))$ and velocity $\mathbf{V}(z)$ of the vortex are given by the Euler–Lagrange equations obtained by substituting the trial function (Eq. 15.52) into the Lagrangian (Eq. 15.50) and then averaging.

The determination of the Green's function G is, in principle, standard. However, there are a few closed-form expressions for G that can be used to understand, in simple terms, the motion of the vortex. The simplest instance for which the Green's function can be explicitly determined is Ω being the half space $y > 0$. In this case, the method of images provides the desired Green's function with only one image point. In fact, the fundamental solution for the whole space $-\infty < x, y < \infty$ is

$$S(\mathbf{x}, \mathbf{x}') = \frac{1}{2\pi\nu} \ln |\mathbf{x} - \mathbf{x}'|, \tag{15.53}$$

where $\mathbf{x}' = (\xi, \eta)$ being the source. For the half space $y > 0$, the image is located at $\hat{\mathbf{x}}' = (\xi, -\eta)$. The Green's function for the half space is then

$$G(\mathbf{x}, \mathbf{x}') = S(\mathbf{x}, \mathbf{x}') + G_R(\mathbf{x}, \hat{\mathbf{x}}'), \tag{15.54}$$

where

$$G_R(\mathbf{x}, \hat{\mathbf{x}}') = -\frac{1}{2\pi\nu} \ln |\mathbf{x} - \hat{\mathbf{x}}'|. \tag{15.55}$$

This Green's function is then used to calculate the averaged Lagrangian. This cannot be done explicitly. However, the integrals can be approximated if we assume that the scale of variation of the vortex is small compared with that of the director field. In this case, we can use the approximation

$$\begin{aligned} \int\int_\Omega G(\mathbf{x}, \mathbf{x}')|E(\mathbf{x}', z)|^2 \, \mathrm{d}\mathbf{x}' &\approx G(\mathbf{x}, \mathbf{X}(z)) \int\int_\Omega |E(\mathbf{x}', z)|^2 \, \mathrm{d}\mathbf{x}' \\ &= G(\mathbf{x}, \mathbf{X}(z)) a^2 w^4 M_0 \end{aligned} \tag{15.56}$$

on using the trial function (Eq. 15.52). The integral M_0 is

$$M_0 = 2\pi \int_0^\infty r^3 \mathrm{e}^{-2r} \, \mathrm{d}r = \frac{3\pi}{4}. \tag{15.57}$$

The assumption that the size of the region Ω is much larger than the width w of the vortex has been made to extend the limit of integration in M_0 to ∞.

In the case of the half space $y > 0$, with the Green's function being given by Equation 15.54, the nonlinear term, that is, Equation 15.56, is

$$\int\int_\Omega G(\mathbf{x}, \mathbf{x}')|E(\mathbf{x}', z)|^2 \, \mathrm{d}\mathbf{x}' = \frac{a^2 w^4 M_0}{2\pi\nu} \left[\ln |\mathbf{x} - \mathbf{X}(z)| - \ln |\mathbf{x} - \hat{\mathbf{X}}(z)| \right], \tag{15.58}$$

where $\hat{\mathbf{X}}(z) = (X(z), -Y(z))$. The corresponding term in the averaged Lagrangian is then

$$\begin{aligned}\mathcal{L}_\nu &= -\int\int_\Omega |E(\mathbf{x}, z)|^2 \int\int_\Omega G(\mathbf{x}, \mathbf{x}')|E(\mathbf{x}', z)|^2 \,\mathrm{d}\mathbf{x}'\mathrm{d}\mathbf{x} \\ &= -\frac{a^2 w^4 M_0}{2\pi\nu}\int\int_\Omega |E(\mathbf{x}-\mathbf{X}(z), z)|^2 \left[\ln|\mathbf{x}-\mathbf{X}(z)| - \ln|\mathbf{x}-\hat{\mathbf{X}}(z)|\right] \mathrm{d}\mathbf{x}.\end{aligned} \tag{15.59}$$

Changing variables using $\mathbf{x} = \mathbf{X}(z) + \tilde{\mathbf{x}}$ and then assuming that the region Ω is large compared with the size of the vortex results in, after changing to polar coordinates,

$$\mathcal{L}_\nu = -\frac{a^4 w^8}{\nu} M_0 \left[M_0 \ln w + M_1\right] - \frac{a^4 w^8}{2\pi\nu} M_0^2 \ln|\mathbf{X}(z) - \hat{\mathbf{X}}(z)|. \tag{15.60}$$

The second term in $\mathcal{L}_\nu$ arises from approximating the Green's function by its value at the vortex center. The integral M_1 is

$$M_1 = \int_0^\infty \mathrm{e}^{-2r} r^3 \ln r \,\mathrm{d}r. \tag{15.61}$$

The remaining terms in the averaged Lagrangian are calculated as in Section 15.2. The final averaged Lagrangian for the half space $y > 0$ is then

$$\mathcal{L} = \mathcal{L}_0 + \mathcal{L}_\nu, \tag{15.62}$$

where $\mathcal{L}_0$ given by

$$\mathcal{L}_0 = M_0\left[-\frac{3}{4}a^2 w^4 \dot{\sigma} - \frac{3}{8}a^2 w^2 + \frac{1}{4}a^2 w^3 \left(\mathbf{V}\cdot\dot{\mathbf{X}} - \frac{1}{2}\mathbf{V}\cdot\mathbf{V}\right)\right]. \tag{15.63}$$

and $\mathcal{L}_\nu$ is given by Equation 15.60.

We have assumed that the vortex is small compared with the size of the cell and that its motion is slow. Let us further assume that it is far from the cell walls. In this case, the variational equations for a and w decouple from those for the position and velocity of the vortex. Variations with respect to a and w give the amplitude–width relation for a steady vortex as

$$a^2 w^6 = \frac{3}{4}\frac{\nu}{M_0}. \tag{15.64}$$

This steady vortex relation is different from that, expression 15.37, which applies when there is an external biasing electric field. To simplify the analysis and resulting equations, let us assume that the vortex profile is in its equilibrium shape given by the amplitude–width relation (Eq. 15.64). In effect, we are assuming that the time evolution of the amplitude and width of the vortex are slow compared to its motion.

A similar assumption was made in Chapter 7. Under this assumption, variations with respect to **V** and **X** give the position of the vortex being determined by

$$\dot{\mathbf{X}} = \mathbf{V},$$

$$\dot{\mathbf{V}} = -\nabla_{\mathbf{X}} \frac{a^2 w^4}{2\pi\nu} M_0 G_R(\mathbf{X}(z), \hat{\mathbf{X}}(z)) = -\nabla_{\mathbf{X}} \frac{a^2 w^4}{2\pi\nu} M_0 \ln|2y(z)|. \qquad (15.65)$$

The potential in the last equation of Equation 15.65 is then infinitely repulsive at the cell wall. Therefore, if a vortex travels toward the wall, it is bounced back away from it. The component of the motion parallel to the wall is not altered by the image. This scenario is different from that of a vortex in an ideal fluid, which can move parallel to a wall with a velocity that depends on its strength [17].

We then expect that a vortex, or a beam, in a closed cell will be repelled by the walls and will reach an equilibrium position due to momentum being shed in diffractive radiation as it evolves [16, 18]. We now show that this is indeed the case.

Mathematically, the simplest possible closed cell is a circular cylinder of radius R. In this case, the regular part of the Green's function is [19]

$$G_R(\mathbf{X}(z), \hat{\mathbf{X}}(z)) = -\frac{1}{2\pi\nu} \ln\left[\frac{R}{|\hat{\mathbf{X}}|\left|\mathbf{X} - R^2\hat{\mathbf{X}}/|\hat{\mathbf{X}}|^2\right|}\right]. \qquad (15.66)$$

Using this representation and the fact that the vortex is assumed to be much smaller than the circular cell, $w \ll R$, we can approximate the integral for $\mathcal{L}_\nu$ as before to obtain

$$\mathcal{L}_\nu = \frac{\pi}{4\nu} M_0^2 a^4 w^8 \ln\left|\frac{R^2 - |\mathbf{X}(z)|^2}{Rw}\right|. \qquad (15.67)$$

The final averaged Lagrangian is then $\mathcal{L}_0 + \mathcal{L}_\nu$. The repulsive nature of the cell boundary $|\mathbf{X}(z)| = R$ is clear.

In the simple case with no angular velocity $\mathbf{X} = (x(z), 0)$, $\mathbf{V} = (p(z), 0)$, the equation of motion obtained from the variational equations for the averaged Lagrangian is the oscillator equation

$$\ddot{x} + \frac{M_0}{\nu} a^2 w^5 \frac{x}{R^2 - x^2} = 0. \qquad (15.68)$$

This equation shows the oscillatory motion of the vortex about the center of the cell due to the repulsion of the cell walls. The vortex will eventually settle to the center of the cell due to the effect of radiation loss.

The case when the vortex has angular momentum is analyzed using polar coordinates $r = |\mathbf{X}|$ and polar angle φ. It is then found that the equations for the trajectory of the vortex can be expressed in the Kepler form

$$\frac{\mathrm{d}}{\mathrm{d}z}\left(r^2\dot{\varphi}\right) = 0, \tag{15.69}$$

$$\frac{\mathrm{d}^2 r}{\mathrm{d}z^2} - \frac{L^2}{r^3} + \frac{a^2 w^5}{\nu}\frac{r}{R^2 - r^2} = 0. \tag{15.70}$$

The angular momentum $L = r^2\dot{\varphi}$ is then a conserved quantity. Equation 15.70 shows that the origin is repulsive, as is the cell wall $r = R$. Equation 15.70 has bounded solutions that oscillate in the radial variable about the fixed point r_f, with

$$r_f = \frac{\nu^{1/4}\sqrt{RL}}{\sqrt{a}w^{5/4}} \tag{15.71}$$

for $R \gg w$, and that advance in the angular direction with angular velocity

$$\dot{\varphi} \sim \frac{L}{r_f^2}. \tag{15.72}$$

Again, loss to shed diffractive radiation will decrease the angular momentum and the vortex will then slowly spiral into the center of the circular cell.

Arbitrary closed, smooth regions, Ω, can be mapped onto a circle via conformal mapping [20]. Even if this mapping is not explicit, some general qualitative statements can be made about the motion of a vortex. The Green's function H for a region Ω takes the form

$$H(\mathbf{X}, \hat{\mathbf{X}}) = G(f(\mathbf{X}), f(\hat{\mathbf{X}})), \tag{15.73}$$

where $f(z)$, $z = x + \mathrm{i}y$, and maps the region Ω into the interior of the circle, with G being the Green's function for the circle. Therefore, the potential for the motion of the vortex, which is the regular part R of the Green's function, has as level lines in Ω the images of circles under the inverse mapping of $f(z)$. The center of the circle is mapped to a point z_0 in Ω. The vortex is then bounced away from the cell walls, the boundary of Ω, until it eventually settles in equilibrium at z_0.

A rectangular or polygonal region Ω can be explicitly mapped onto the half plane using the Schwarz–Christoffel map and then onto the circle [20]. Although the conformal map is not explicit in general, the qualitative behavior of the vortex is as before—it settles to the center of the cell. For the special case of a rectangular cell, the motion of the vortex can be approached in a different manner. Green's function can be found using a Fourier series, which is then summed using Poisson's formula, resulting in a solution in terms of infinitely many images [15]. These images are placed on a lattice obtained by periodically extending the region Ω and reflecting the Green's function source in each wall [15]. To ensure the zero-boundary condition, the images have alternating signs. This image solution explains qualitatively why the motion of an optical vortex near a corner in the domain Ω is more complicated due to the influence of an infinite series of images, which has a small effect on the motion away from the corner.

In principle, one can formulate nematic vortex problems as for the classical vortices of perfect fluid flow [17]. However, the corresponding solutions will not be the same as those for fluid flow, as already seen earlier for a single nematic vortex close to an infinite straight wall.

15.4 STABILIZATION OF VORTICES BY VORTEX–BEAM INTERACTION

In saturable media, the interaction between a dark vortex and a beam has been studied [1, 21]. Using two wavelengths (colors) of coherent light, a beam or vortex in one color can interact with a beam or vortex in another color [22]. The interaction between these two colors can have a significant effect on the evolution of the beams and/or vortices. For instance, it has been shown that a dark vortex in one color can form a waveguide for a beam in another color [1, 21]. In this work, the interaction of a bright vortex with a bright beam in another color was not considered, as bright vortices are unstable in local media. However, bright vortices are stable in nonlocal media, such as a liquid crystal, for sufficiently large nonlocality [2]. The interaction of an $m = 1$ vortex with a beam was studied by Yakimenko et al. [2]. Subsequently, an asymptotic theory was developed, which showed that a beam can stabilize a bright vortex in a nematic liquid crystal, with the stability region greatly enlarged to $\nu \geq O(10)$ [23, 24]. Let us now consider the main ideas of this analysis.

As in the previous section, the equations governing a beam in one color (wavelength) u and a vortex in another color (wavelength) v are [22–24]

$$\mathrm{i}\frac{\partial u}{\partial z} + \frac{1}{2}D_u \nabla^2 u + 2A_u u\theta = 0, \tag{15.74}$$

$$\mathrm{i}\frac{\partial v}{\partial z} + \frac{1}{2}D_v \nabla^2 v + 2A_v v\theta = 0, \tag{15.75}$$

$$\nu\nabla^2\theta - 2q\theta = -2A_u|u|^2 - 2A_v|v|^2. \tag{15.76}$$

A slightly different scaling of these equations has been used from that in the previous section. Also, the diffraction coefficients D_u and D_v and the coupling coefficients between the light and the nematic A_u and A_v have been left general, rather than both being scaled to 1. In experiments, these coefficients are very close. For instance, in the experiments of Alberucci et al. [22], the diffraction coefficients are 0.805 for red light and 0.823 for near-infrared light. The Lagrangian for the

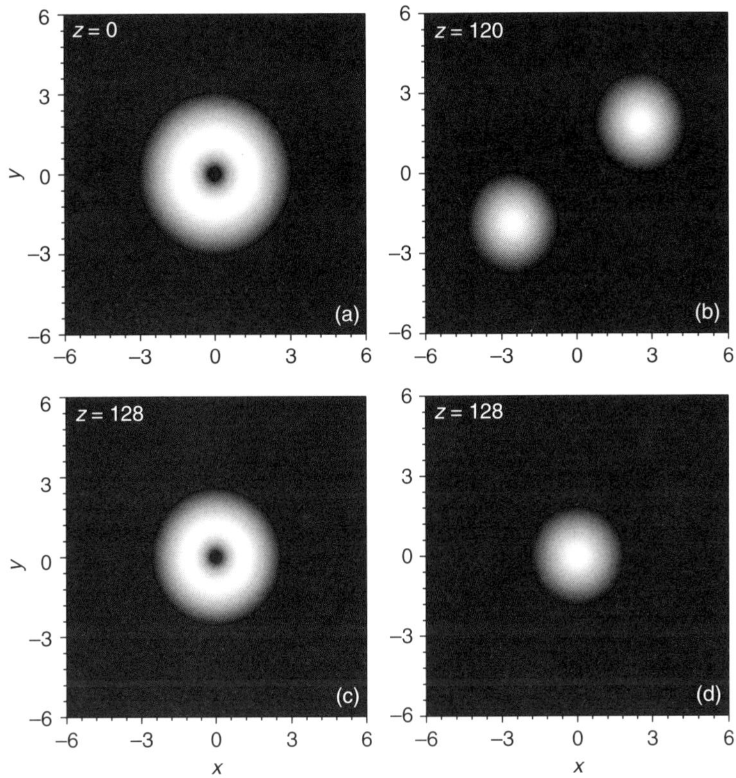

FIGURE 15.2 Numerical solution of two-color equations (Eqs. 15.74–15.76) for the initial condition (15.79 and 15.80) with $g = 0$. (a,b) Input and output intensities for an unstable vortex in color v with no beam in color u. (c,d) Vortex in color v (c) stabilized by nematicon in color u (d). Here, $\nu = 4$. *Source*: Reproduced from Fig. 1 in Reference 24, with permission.

two color equations (Eqs. 15.74–15.76) is

$$L = \mathrm{i}\left(u^* u_z - u u_z^*\right) - D_u|\nabla u|^2 + 4A_u|u|^2\theta + \mathrm{i}\left(v^* v_z - v v_z^*\right) - D_v|\nabla v|^2 + 4A_v|v|^2\theta - \nu|\nabla\theta|^2 - 2q\theta^2. \tag{15.77}$$

As the diffraction and coupling coefficients differ by only a few percent, the simplifying assumption $D_u = D_v = 1$ and $A_u = A_v = 1$ will be made.

The numerical results displayed in Figure 15.2 show that the beam in the second color stabilizes the vortex even for the low value $\nu = 4$ of nonlocality. It is clear from the figure that the width of the stabilized vortex is larger than that of the supporting beam (nematicon). We recall from the previous section that a vortex is stabilized for large ν by the relatively large deformation of the optical axis in its center. It is then plausible to expect stabilization of a vortex if the optical axis in its core is deformed by a beam in the other color. Let us now verify this qualitative picture with a modulation analysis.

We see from the stability result (Eq. 15.47) that the destabilizing term for a single-color vortex is

$$\frac{405}{64}\frac{\mathrm{e}^{-2}}{\nu(n^2-1)}a_0^2 w_0^6. \tag{15.78}$$

This destabilizing term decreases as ν increases and as w_0 decreases. Numerical results show that for a two-color vortex–beam pair, as the vortex stabilizes, the width of the vortex decreases because of the effect of the beam in the other color. To determine the stability of the vortex analytically, we then need to calculate the effect of the beam on the width of the vortex and show that its width decreases as the beam amplitude increases.

Suitable trial functions for the beam and an $m = 1$ vortex are

$$u = a_u \operatorname{sech}\frac{r}{w_u}\,\mathrm{e}^{\mathrm{i}\sigma_u}, \tag{15.79}$$

$$v = a_v r \mathrm{e}^{-r/w_v}\mathrm{e}^{\mathrm{i}\varphi+\mathrm{i}\sigma_v} + \mathrm{i}g\mathrm{e}^{\mathrm{i}\varphi+\mathrm{i}\sigma_v}, \tag{15.80}$$

with the director being given by

$$\theta = \alpha \operatorname{sech}^2\frac{r}{\beta}. \tag{15.81}$$

Here, (r, φ) are the polar coordinates based on the center of the vortex. Substituting these trial functions into the Lagrangian (Eq. 15.77) and averaging by integrating in r from 0 to ∞ and φ from 0 to 2π gives the averaged Lagrangian

$$\mathcal{L} = \mathcal{L}_u + \mathcal{L}_v + \mathcal{L}_{uv}. \tag{15.82}$$

Here, $\mathcal{L}_u$ is the contribution due to the beam, $\mathcal{L}_v$ is the contribution due to the vortex, and $\mathcal{L}_{uv}$ is the interaction-averaged Lagrangian. From the work of the previous sections, $\mathcal{L}_u$ is given by Equation 15.32 and $\mathcal{L}_v$ is given by Equation 15.35. It is then only necessary to calculate the interaction term $\mathcal{L}_{uv}$. This term is responsible for the change in the steady amplitude–width relation for the vortex from that given by Equation 15.37.

The numerical solution of Figure 15.2 shows that the vortex is stabilized by the beam in the second color, and this stabilization is accompanied by a decrease in the vortex's width. We therefore need to estimate the decrease in the width of the vortex as a result of its interaction with the beam in the second color. To this end, we consider the Lagrangian (Eq. 15.77) and average the interaction term in this Lagrangian to obtain

$$\mathcal{L}_{uv} = 4\int_{-\infty}^{\infty}\int_{-\infty}^{\infty}\left[\theta_u|v|^2 + \theta_v|u|^2\right]\,\mathrm{d}x\mathrm{d}y. \tag{15.83}$$

This interaction term can be rearranged by integrating by parts and using the symmetry of Green's function for the director Equation 15.76 to obtain

$$\mathcal{L}_{uv} = 16 \int_{-\infty}^{\infty} \int_{-\infty}^{\infty} \theta_v |u|^2 \, \mathrm{d}x \mathrm{d}y. \tag{15.84}$$

To calculate θ_v, we use the same approximations as used to derive the averaged Lagrangians 15.35 and 15.36 in Section 15.2. These approximations assume that the v color vortex is narrow in comparison with $\sqrt{\nu}$, which is the size of the width of the director field. The equation for θ_v in this limit of $w \ll \sqrt{\nu}$ can be approximated as

$$\nu \frac{\mathrm{d}^2\theta_v}{\mathrm{d}r^2} - 2q\theta_v = -2a_v^2 r^2 \mathrm{e}^{-2r/w_v}, \tag{15.85}$$

$$\text{with} \quad \frac{\mathrm{d}\theta_v}{\mathrm{d}r} = 0 \quad \text{at} \quad r = 0, \theta_v \to 0 \quad \text{as} \quad r \to \infty. \tag{15.86}$$

This equation has the solution

$$\theta_v = \begin{cases} \dfrac{a_v^2 w_v^3}{2\sqrt{2q\nu}}, & r \leq w_v \\ A_w \mathrm{e}^{-\sqrt{2q/\nu}(r - w_v)}, & r > w_v. \end{cases} \tag{15.87}$$

For θ_v to be continuous, we have

$$A_w = \frac{a_v^2 w_v^3}{2\sqrt{2q\nu}}. \tag{15.88}$$

The interaction averaged Lagrangian is then

$$\mathcal{L}_{uv} = 16 A_w a_u^2 w_u^2. \tag{15.89}$$

The interaction of the vortex with the beam is weaker than the self-interaction of the beam. We can then use the steady-state amplitude–width relation for an isolated beam [25]

$$a_u^2 = \frac{2\nu I_{22} I_{42}}{A^4 B^6 \beta^2 w_u^6}. \tag{15.90}$$

Here,

$$I_{22} = \frac{1}{3}\ln 2 + \frac{1}{6}, \quad I_{42} = \frac{2}{15}\ln 2 + \frac{1}{60},$$

$$A = \frac{I_2\sqrt{2}}{\sqrt{I_{x32}}}, \qquad B = \sqrt{2I_2}, \tag{15.91}$$

$$I_2 = \ln 2, \; I_{x32} = 1.352314016\ldots.$$

Taking variations of the resulting averaged Lagrangian with respect to a_v and w_v then gives the steady-state amplitude–width relation

$$\frac{e^2 H_v^2}{8\sqrt{\nu}} w_v^2 + (a_u w_u)^2 w_v - \frac{3}{2} = 0, \tag{15.92}$$

where

$$H_v = e^{-1} a_v w_v. \tag{15.93}$$

Clearly, as the beam power $a_u^2 w_u^2 \to 0$, we recover the amplitude–width relation (Eq. 15.37) for a free vortex, once the different scalings of the single-color Equations 15.12 and 15.13 and the two-color Equations 15.74–15.76 are accounted for. However, as the power $a_u^2 w_u^2$ of the beam increases, we have

$$w_v = \frac{3}{2} (a_u w_u)^{-1}, \tag{15.94}$$

which gives the narrowing of the vortex by the beam, as observed in numerical solutions. This beam interaction and narrowing causes the vortex to be stable for smaller ν than for a free vortex, as the destabilizing term in Equation 15.78 becomes smaller. Minzoni et al. [24] showed that this modulation analysis gave a stability threshold in excellent agreement with the numerical threshold of $\nu = 5$, which is much lower than the threshold $\nu \sim 100$ for an isolated vortex [2, 4].

It is expected that this mechanism of stabilization by a beam in another color will be able to stabilize other types of coherent structures that are unstable at relatively low values of nonlocality ν. In the following section, the stabilization mechanism for azimuthally dependent vortices is explored numerically. This study again shows the effectiveness of a beam in another color in stabilizing more complicated structures.

15.5 AZIMUTHALLY DEPENDENT VORTICES

Let us now consider the evolution of vortex-like initial conditions of the form

$$E = ar^m e^{-r/w} e^{im\varphi}. \tag{15.95}$$

The appropriate governing equations are now Equations 15.12 and 15.13, as the nematic is pretitled by a static external electric field. These vortices were considered by Yakimenko et al. [2] and were found to be unstable for ν values of up to 200. The instability for $\nu < 200$ manifests itself by the vortex breaking up into $m + 1$ lumps, which rotate in a circular motion. The radius of the circle becomes larger as z increases, with the motion of the individual lumps becoming slower. A typical unstable example is shown in Figure 15.3. A modulation theory analysis of this vortex evolution predicts that for sufficiently large ν values, the vortex is stable.

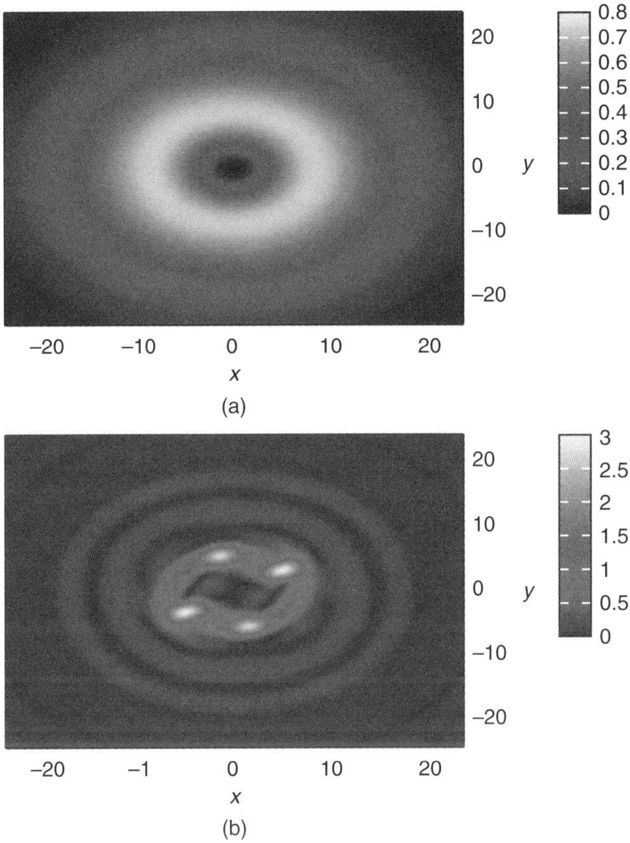

FIGURE 15.3 Solution of Equations 15.12 and 15.13 for $|E|$ for the initial condition 15.95, with $a = 0.02$ and $w = 3.0$ for $m = 3$ and $\nu = 5$. (a) $z = 0$ and (b) $z = 50$.

To test this prediction, the nematicon Equations 15.12 and 15.13 were solved for $\nu = 400$ and a vortex with $m = 3$ was found to be stable up to $z \sim 400$, which is a typical nondimensional length of a nematic cell [11]. This stable vortex evolution is shown in Figure 15.4.

It has also been shown that a collection of beams rotating in a circle merges into a vortex when the nonlocality ν is large due to the overlap of the radiation shelves generated by each beam as it evolves [26]. Skupin et al. [7] showed that azimuthally dependent vortices exist for suitably chosen parameter values in the infinite nonlocality limit put forth by Snyder and Mitchell [10]. These vortex solutions have rectangular shapes that rotate around a dark center and were termed azimuthons.

We now describe the role of these azimuthons in the transition from a vortex to a cluster of beams rotating around a common center. Let us take the specific case $m = 3$. The azimuthal instability with mode $n = 3$ for the $m = 3$ vortex (Eq. 15.95) is responsible for shrinking the vortex width in the form $w = w_0 + w_3 \cos 3\varphi$, with

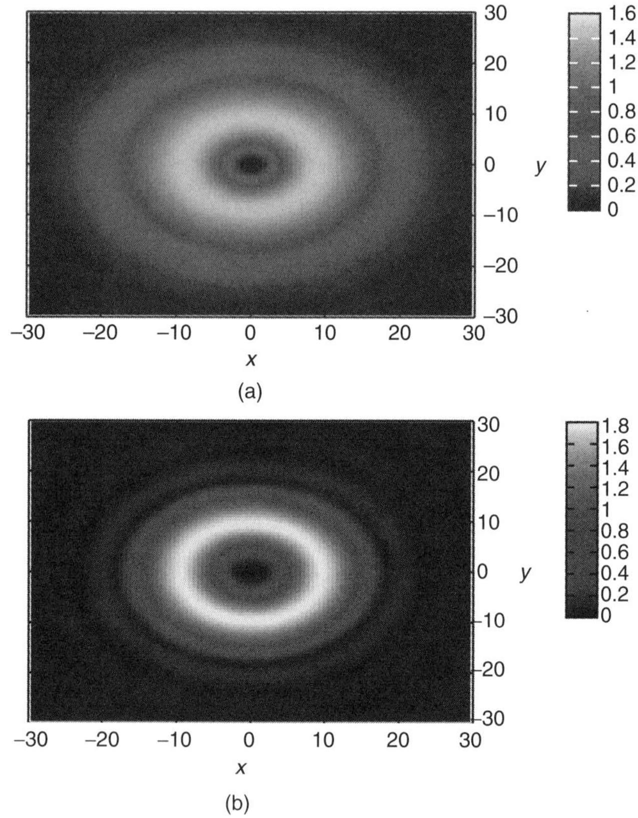

FIGURE 15.4 Solution of Equations 15.12 and 15.13 for $|E|$ for the initial condition 15.95, with $a = 0.04$ and $w = 3.0$ for $m = 3$ and $\nu = 400$. (a) $z = 0$ and (b) $z = 400$.

instability occurring as w_3 increases. However, in principle, it is possible for the instability to saturate. If such a saturation occurs, the maximum of the vortex amplitude in the radial variable will be at

$$r = R(\varphi) = 3\left(w_0 + w_3 \cos 3\varphi\right). \tag{15.96}$$

This is the polar equation of the corresponding level line, which is triangular. To test this possibility, an initial condition with triangular level lines and a vortex angular dependence was used, this being

$$v = a_v r^3 \mathrm{e}^{-(r-R(\varphi))^2/w_v^2} \mathrm{e}^{3\mathrm{i}\varphi}, \tag{15.97}$$

where $R(\varphi) = r_0 + b\cos 3\varphi$. The nematicon equations (Eqs 15.12 and 15.13), which apply when there is a pretilting external electric field, and Equations 15.48 and 15.49, which apply when there is no pretilting electric field, were then solved numerically for this initial condition. Under both operating conditions, it was found

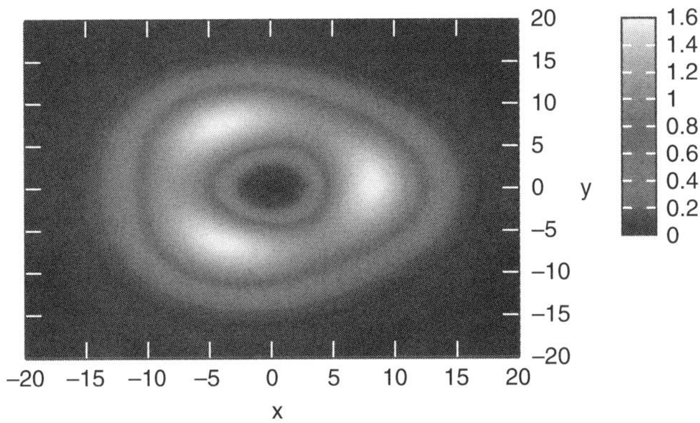

(a)

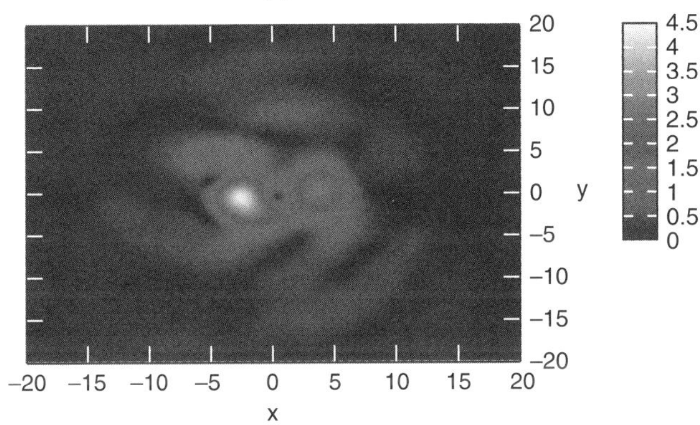

(b)

FIGURE 15.5 Solution of two-color equations (Eqs. 15.98–15.100) for $|v|$ for the initial conditions 15.97 and 15.101, with $a_v = 0.01$, $w_v = 6.0$, $a_u = 1.5$, $w_u = 5.0$. $r_0 = 1.0$, $b = 0.5$, and $\nu = 100$. (a) $|v|$ at $z = 0$ and (b) $|v|$ at $z = 400$.

that azimuthons are unstable and broke up due to a parametric instability. An example of an unstable one-color azimuthon is shown in Figure 15.8, where it forms part of the following discussion of two-color azimuthon and beam vector solitary waves.

Finally, thc instability mechanism was also tested using a stabilizing beam in a second color. In this case, the equations governing a beam in one color (wavelength) u and a vortex in another color (wavelength) v are [22–24]

$$\mathrm{i}\frac{\partial u}{\partial z} + \frac{1}{2}\nabla^2 u + u\theta = 0, \tag{15.98}$$

$$\mathrm{i}\frac{\partial v}{\partial z} + \frac{1}{2}\nabla^2 v + v\theta = 0, \tag{15.99}$$

$$\nu\nabla^2\theta - \theta = -|u|^2 - |v|^2. \tag{15.100}$$

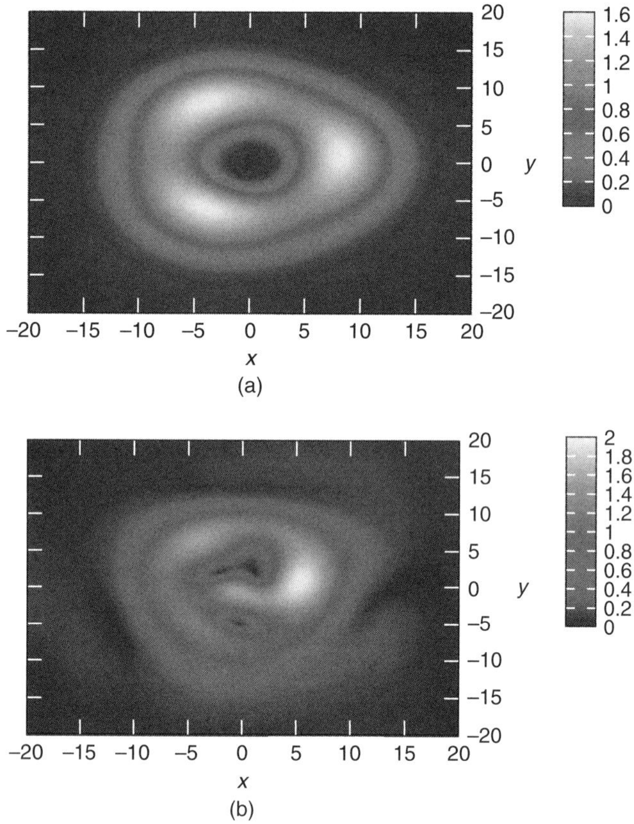

FIGURE 15.6 Solution of two-color equations (Eqs. 15.98–15.100) for $|v|$ for the initial conditions 15.97 and 15.101, with $a_v = 0.01$, $w_v = 6.0$, $a_u = 1.5$, $w_u = 5.0$. $r_0 = 1.0$, $b = 0.5$, and $\nu = 200$. (a) $|v|$ at $z = 0$ and (b) $|v|$ at $z = 400$.

In these governing equations, the simplifying assumption that the diffraction and coupling coefficients for each color are equal has been made. As discussed in Section 15.4, these coefficients only differ by a few percent points in experiments. The initial condition for v is Equation 15.97, and the initial condition for u is

$$u = a_u \operatorname{sech} r/w_u. \tag{15.101}$$

Figures 15.5–15.7 show the change of stability of the azimuthon as the nonlocality ν increases. For low values of ν, the azimuthon shows the typical vortex instability and it breaks up into beams (Fig. 15.5). As the nonlocality increases, the azimuthon becomes stable for longer distances and shows more complicated breakup behavior (Fig. 15.6). Finally, for very large values of the nonlocality, the azimuthon is stable for typical cell lengths (Fig. 15.7). It should be noted, however, that the values of the nonlocality ν for which the azimuthon is stable over a cell length are much

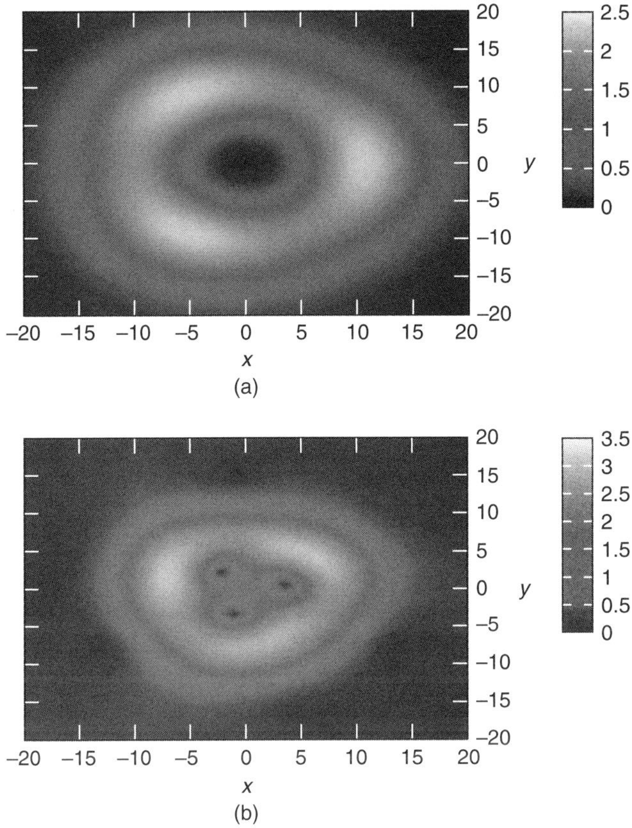

FIGURE 15.7 Solution of two-color equations (Eqs. 15.98–15.100) for $|v|$ for the initial conditions 15.97 and 15.101, with $a_v = 0.007$, $w_v = 8.0$, $a_u = 1.5$, $w_u = 5.0$. $r_0 = 1.0$, $b = 0.5$, and $\nu = 1000$. (a) $|v|$ at $z = 0$ and (b) $|v|$ at $z = 400$.

larger than typical experimental values [11]. The vital role of the second color beam in stabilizing the azimuthon is illustrated in Figure 15.8. Comparing Figures 15.7 and 15.8 shows that the azimuthon is stable for $\nu = 1000$ in the presence of a second color beam, but is unstable for this large nonlocality in the absence of the beam.

Finally, in a recent work, it was shown that guided azimuthally, dependent modes can become azimuthons in a nonlinear, nonlocal medium [27]. In this context, it is of interest to consider the influence of an external voltage bias, such as considered by Assanto et al. [11], on the behavior of azimuthons in liquid crystals.

15.6 CONCLUSIONS

As well as solitary wave beams, nematicons, nematic liquid crystals can support the propagation of optical vortex beams. In local media, such optical vortices

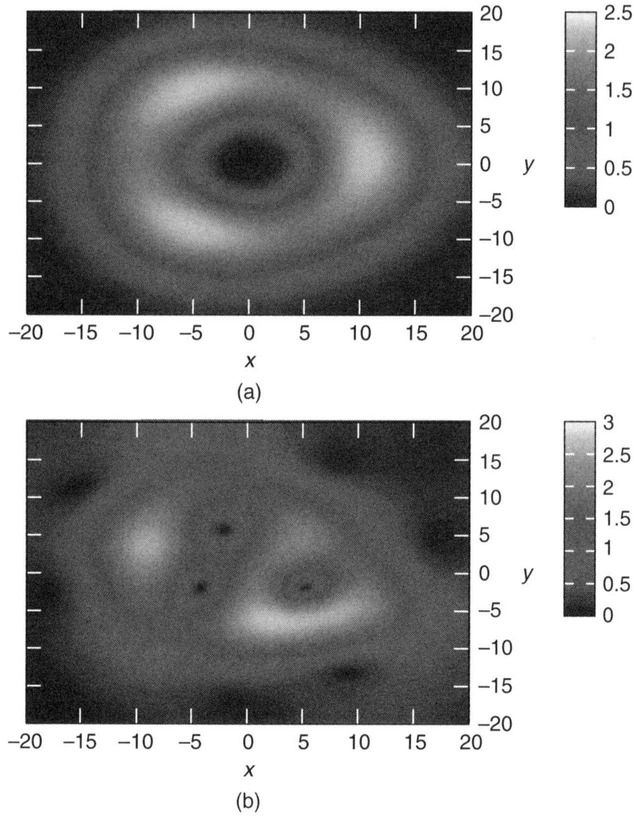

FIGURE 15.8 Solution of two-color equations (Eqs. 15.98–15.100) for $|v|$ for the initial conditions 15.97 and 15.101, with $a_v = 0.007$, $w_v = 8.0$, $a_u = 0.0$, $w_u = 5.0$. $r_0 = 1.0$, $b = 0.5$, and $\nu = 1000$. (a) $|v|$ at $z = 0$ and (b) $|v|$ at $z = 400$.

are unstable due to a dominant azimuthal $n = 2$ instability that splits the vortex into two solitary wave beams. However, a nematic liquid crystal is a nonlocal, nonlinear medium, and the nonlocal nature of the interaction of the optical field with the nematic stabilizes the vortex, as long as the nonlocality ν is sufficiently large. This stabilization effect of the nematic, or optical axis, on a vortex is most dramatically revealed through the interaction of a nematicon with a copropagating vortex. The nematicon deforms the optical axis, that is, makes the director rotation nonzero, at the center of the vortex, thus stabilizing it. This stabilization effect of the copropagating nematicon reduces the stability threshold for the vortex in the nonlocality ν by a factor of 10. Copropagating nematicons can be used to stabilize other vortex structures. For example, azimuthons, vortices with azimuthal structure, are stabilized by a copropagating nematicon. Indeed, a copropagating nematicon allows these azimuthally dependent structures to take complicated forms.

The analysis of the propagation of vortices in a finite nematic cell is more complicated than that in an infinite cell because of the necessity of the director satisfying anchoring conditions. The same general conclusion that nonlocality stabilizes a vortex also holds. A neglected technique that has been shown to be useful for finite cells is the method of images [15]. The advantage of this technique is that it allows the anchoring conditions for the nematic to be incorporated in a mathematically simple manner. The method of images has been exploited extensively in the fluid mechanics literature, and many of the solutions hence derived could be transferred over to the nonlinear optics of nematicons and vortices.

Azimuthally dependent vortices can be obtained, in principle, as bifurcating structures due to the development of azimuthal instabilities. An analytical description of this bifurcation requires a trial function that incorporates both the evolution of the instability and the modulation of the shape of the vortex. This is an unresolved problem.

Acknowledgments

This research was supported by the Royal Society of London under Grant No. JP090179.

REFERENCES

1. Y. S. Kivshar and G. Agrawal. *Optical Solitons: From Fibers to Photonic Crystals*. Academic Press, San Diego, CA, 2003.
2. A. I. Yakimenko, Y. A. Zaliznyak, and Y. S. Kivshar. Vortex solitons in nonlocal self-focusing nonlinear media. *Phys. Rev. E*, 71:065603, 2005.
3. G. B. Whitham. *Linear and Nonlinear Waves*. John Wiley & Sons, New York, 1974.
4. A. A. Minzoni, N. F. Smyth, A. L. Worthy, and Y. S. Kivshar. Stabilization of vortex solitons in nonlocal nonlinear media. *Phys. Rev. A*, 76:063803, 2007.
5. C. Conti, M. Peccianti, and G. Assanto. Route to nonlocality and observation of accessible solitons. *Phys. Rev. Lett.*, 91:073901, 2003.
6. C. Conti, M. Peccianti, and G. Assanto. Observation of optical spatial solitons in a highly nonlocal medium. *Phys. Rev. Lett.*, 92:113902, 2004.
7. S. Skupin, M. Grech, and W. Krolikowski. Rotating soliton solution in nonlocal nonlinear media. *Opt. Express*, 16:9118–9131, 2008.
8. Z. Xu, Y. V. Kartashov, and L. Torner. Upper threshold for stability of multipole-mode solitons in nonlocal nonlinear media. *Opt. Lett.*, 30:3171–3173, 2005.
9. S. Skupin, O. Bang, D. Edmundson, and W. Krolikowski. Stability of two-dimensional spatial solitons in nonlocal nonlinear media. *Phys. Rev. E*, 73:066603, 2006.
10. A. W. Snyder and M. J. Mitchell. Accessible solitons. *Science*, 276:1538–1541, 1997.
11. G. Assanto, A. A. Minzoni, M. Peccianti, and N. F. Smyth. Optical solitary waves escaping a wide trapping potential in nematic liquid crystals: modulation theory. *Phys. Rev. A*, 79:033837, 2009.

12. M. Abramowitz and I. A. Stegun. *Handbook of Mathematical Functions with Formulas, Graphs and Mathematical Tables*. Dover Publications, New York, 1972.

13. Q. Shou, Y. Liang, Q. Jiang, Y. Zheng, S. Lan, W. Hu, and Q. Guo. Boundary force exerted on spatial solitons in cylindrical strongly nonlocal media. *Opt. Lett*., 34:3523–3525, 2009.

14. A. Alberucci, G. Assanto, D. Buccoliero, A. Desyatnikov, T. R. Marchant, and N. F. Smyth. Modulation analysis of boundary induced motion of nematicons. *Phys. Rev. A*, 79:043816, 2009.

15. A. A. Minzoni, L. W. Sciberras, N. F. Smyth, and A. L. Worthy. Propagation of optical spatial solitary waves in bias-free nematic liquid crystal cells. *Phys. Rev. A*, 84:043823, 2011.

16. A. A. Minzoni, N. F. Smyth, and Z. Xu, Stability of an optical vortex in a circular nematic cell. *Phys. Rev. A*, 81:033816, 2010.

17. H. Lamb. *Hydrodynamics*. Dover Publications, New York, 1945.

18. G. Assanto, C. García-Reimbert, A. A. Minzoni, N. F. Smyth, and A. L. Worthy. Lagrange solution for three wavelength solitary wave clusters in nematic liquid crystals. *Physica D*, 240:1213–1219, 2011.

19. R. Courant and D. Hilbert. *Methods of Mathematical Physics*, Vol. 1. Interscience Publishers, New York, 1965.

20. M. J. Ablowitz and A. S. Fokas. *Complex Variables. Introduction and Applications*. Cambridge University Press, Cambridge, 1997.

21. A. S. Desyatnikov, Y. S. Kivshar, and L. Torner. *Prog. Opt*., 47:291–391, 2005.

22. A. Alberucci, M. Peccianti, G. Assanto, A. Dyadyusha, and M. Kaczmarek. Two-color vector solitons in nonlocal media. *Phys. Rev. Lett*., 97:153903, 2006.

23. Z. Xu, N. F. Smyth, A. A. Minzoni, and Y. S. Kivshar. Vector vortex solitons in nematic liquid crystals. *Opt. Lett*., 34:1414–1416, 2009.

24. A. A. Minzoni, N. F. Smyth, Z. Xu, and Y. S. Kivshar. Stabilization of vortex-soliton beams in nematic liquid crystals. *Phys. Rev. A*, 79:063808, 2009.

25. A. A. Minzoni, N. F. Smyth, and A. L. Worthy. Modulation solutions for nematicon propagation in non-local liquid crystals. *J. Opt. Soc. Am. B*, 24:1549–1556, 2007.

26. G. Assanto, A. A. Minzoni, and N. F. Smyth. Light self-localization in nematic liquid crystals: modelling solitons in nonlocal reorientational media. *J. Nonlin. Opt. Phys. Mater*., 18:657–691, 2009.

27. Y. Zhang, S. Skupin, F. Maucher, A. G. Pour, K. Lu, and W. Krolikowski. Azimuthons in weakly nonlinear waveguides of different symmetries. *Opt. Express*, 18:27846–27857, 2010.

16

Dispersive Shock Waves in Reorientational and Other Optical Media

TIM R. MARCHANT
School of Mathematics and Applied Statistics, University of Wollongong, Wollongong, New South Wales, Australia

16.1 INTRODUCTION

Nematic liquid crystals are an example of a nonlinear nonlocal optical medium in which rod-like molecules orientate in response to a light beam, or some other applied electric fields. This is because the electric field induces dipoles in the organic molecules. Nematic liquid crystals have received much attention from researchers, as nonlinear effects can be observed over small length scales. A series of elegant experiments have shown that stable optical spatial solitary waves, so-called nematicons, can propagate in nematic liquid crystals [1, 2]. Such solitary waves have generated much interest because of their possible applications as reconfigurable "circuits" for all-optical information processing [3–5].

Nematic liquid crystals are just one example [6] in a variety of optical dielectrics that support spatial solitary waves (Chapter 1). Other optical media include photorefractives [7], quadratic media [8], glasses [9], thermal media [10], and colloids [11]. In all these media, solitary waves occur because of a balance between diffraction (in the spatial domain) or dispersion (in the temporal domain) and nonlinear (self-) focusing. In all cases, a nonlinear interaction between the medium and the light alters the refractive index of the medium.

The nonlinear Schrödinger (NLS) equation is central to our theoretical understanding of nonlinear optics and the field of fiber-based telecommunications. In this context, the NLS equation was first derived in Reference 12 to describe the evolution of the slowly varying envelope of an optical pulse. Derived asymptotically from Maxwell's equations, it assumes slow variation in the carrier envelope and that the refractive index is linearly dependent on the optical intensity (the Kerr

Nematicons: Spatial Optical Solitons in Nematic Liquid Crystals, First Edition.
Edited by Gaetano Assanto.

dependence). The NLS soliton solution occurs because of a balance between dispersion and nonlinear effects. As the NLS equation is integrable, the NLS solitons undergo elastic collisions and a vast array of analytical solution techniques are available for its analysis, such as the inverse scattering transform. Nonlocal effects are important for some of these optical applications, such as nematic liquid crystals, and the governing equations then become a coupled Schrödinger–Poisson system.

For nonlinear dispersive (or diffractive) wave equations, initial discontinuities are resolved through dispersion into a dispersive shock wave (DSW). Also termed as an *undular bore* or a *collisionless shock*, a DSW is a modulated wave train consisting of solitary waves at its leading edge and linear waves at its trailing edge. The DSW then gives a smooth transition across the discontinuity. Mathematically DSW solutions are derived using Whitham's modulation theory [13], who first applied the technique to the Korteweg–de Vries (KdV) equation. Modulation theory develops partial differential equations, which describe the evolution of a slowly varying wave train. In general, if these modulation equations are hyperbolic, the wave train is modulationally stable and the DSW is found as a simple wave solution. If the modulation equations are elliptic, the wave train suffers modulational instability (MI).

The Schrödinger–Poisson system can be of defocusing (stable DSW) or focusing (unstable DSW) type, depending on the properties of the physical medium. The type depends on whether the refractive index of the medium increases or decreases with increasing light intensity. Reorientational nematic liquid crystals tend to be focusing (Chapter 1) whereas photorefractives, Bose–Einstein condensates, and thermal media can be of either type. Hence, a vast array of optical media with different properties are available for experimentation and investigation by researchers. DSW for focusing NLS-type equations are subject to MI and do not persist at long length or timescales. However, recent experiments show that the DSW do occur for focusing media. In Reference 14, it has been shown that a DSW develops for a focusing photorefractive medium; the MI can be suppressed by using partially incoherent light. Moreover, the development of DSW is also observable in nematic liquid crystals; in References 15–17, MI was studied in nematic liquid crystals over typical experimental length scales.

This chapter reviews the existing experimental, numerical, and theoretical results for the development of DSW in nematic liquid crystals and a range of other optical media and also analyze in detail a method to obtain approximate analytical results for DSW governed by focusing-type equations.

16.2 GOVERNING EQUATIONS AND MODULATIONAL INSTABILITY

A coupled Schrödinger–Poisson system

$$\mathrm{i}\frac{\partial E}{\partial z} + \frac{1}{2}\nabla^2 E + 2q\theta E = 0, \quad \nu\nabla^2\theta - 2\theta = -2|E|^2, \tag{16.1}$$

is an appropriate model for optical waves in a nonlinear nonlocal medium. The scalings in Equation 16.1 are chosen to be consistent with Reference 18. E is the electric field. In nematic liquid crystals, the angle θ is the deviation of the molecular director from its pretilt. Even though in other optical applications, θ can have other physical meanings, we term θ the director orientation. ν is a measure of the nonlocality and q is used to classify the equation type; $q = 1$ is the focusing case and $q = -1$ is the defocusing case. The Laplacian ∇^2 is in the (x, y) plane and the governing equations support (2 + 1)D spatial solitary waves with the z coordinate a time-like variable. The physical meaning of θ in thermal media is temperature, whereas in colloidal media it is particle density. In photorefractive materials, nonlocal effects are due to charge transport, whereas, for Bose–Einstein condensates, they are caused by longrange particle interactions.

In the local limit, where $\nu = 0$, $\theta = |E|^2$, and the NLS equation

$$\mathrm{i}\frac{\partial E}{\partial z} + \frac{1}{2}\nabla^2 E + 2q|E|^2 E = 0 \tag{16.2}$$

is obtained. Numerical and analytical results for photorefractive crystals and Bose–Einstein condensates are typically considered in the local limit.

A model related to Equation 16.1 is the nonlocal NLS equation

$$i\frac{\partial E}{\partial z} + \frac{1}{2}\nabla^2 E + E\int_{-\infty}^{\infty}\int_{-\infty}^{\infty} R(x - x', y - y')|E(x', y', z)|^2\,dx'dy' = 0, \tag{16.3}$$

for various response functions R [19–21]. Solutions of Equation 16.1 in (2 + 1)D correspond to a response function R, given in terms of modified Bessel functions, which results in solutions of the nonlocal NLS equation (Eq. 16.3) being difficult to find. As a consequence, most of the analysis of this nonlocal NLS equation has been done for simplified response functions, such as the Gaussian and exponential types.

In the case of (1 + 1)D, the governing equations (Eqs 16.1 and 16.3) are equivalent. In this case, the director equation, the second of Equation 16.1, can be solved using Green's functions and the result substituted into the first of Equation 16.1, resulting in the nonlocal NLS equation

$$iE_z + \frac{1}{2}E_{xx} + \frac{\gamma E}{2}\int_{-\infty}^{\infty} e^{-\gamma|x-x'|}|E(x', z)|^2\,dx' = 0, \tag{16.4}$$

where $\gamma = (2/\nu)^{1/2}$. It is possible to obtain approximate solitary wave solutions to Equation 16.4 in terms of Bessel functions, in the limit of large ν (or θ small) [22].

The continuous wave (cw) solution for Equation 16.1 is

$$E = a_m \mathrm{e}^{\mathrm{i}(kx-\omega z)}, \quad \theta = a_m^2, \quad \omega = \frac{k^2}{2} - 2qa_m^2. \tag{16.5}$$

The cw for the defocusing case is stable, so we consider the focusing case with $q = 1$. The electric field and director angle are perturbed by small quantities

$$E = a\mathrm{e}^{\mathrm{i}2a_m^2 z}(1 - \epsilon), \quad \theta = a_m^2(1 + \theta_1), \quad \text{where} \quad |\epsilon|, |\theta_1| \ll 1. \tag{16.6}$$

A modal expansion for the perturbation of the form

$$\epsilon = A\mathrm{e}^{\lambda z}\mathrm{e}^{\mathrm{i}\mu x} + B\mathrm{e}^{\lambda z}\mathrm{e}^{-\mathrm{i}\mu x}, \tag{16.7}$$

is used, which is then substituted into Equation 16.1, giving

$$\lambda = \pm\frac{|\mu|}{2}\left[\frac{16a_m^2}{(2 + \nu\mu^2)} - \mu^2\right]^{1/2}. \tag{16.8}$$

Hence, the cw is unstable if

$$0 < \mu^2 < \frac{-1 + (1 + 16\nu a_m^2)^{1/2}}{\nu}. \tag{16.9}$$

This stability result is the same, suitably rescaled, as Equation 39 in Reference 23. It implies that the cw is subject to MI developing from perturbations of small wave number or long wavelengths. In the local limit, $\nu \to 0$ the NLS instability band $\mu \in (0, 8a_m^2)$ is obtained [4, 13]. In the nonlocal limit, $\nu \to \infty$ the instability band (Eq. 16.9) becomes $\mu \in (0, 4a_m \nu^{-\frac{1}{2}})$ with a maximum rate of instability of $\lambda_{\max} = (2/\nu)^{\frac{1}{2}} a_m$. This result shows that increased nonlocality suppresses MI as the instability band and the maximum growth rate are both reduced.

An appropriate initial condition for the development of a DSW in (1 + 1)D is

$$E = \begin{cases} a_m \mathrm{e}^{\mathrm{i}(kx - \omega z)}, & x < 0, \\ b_m \mathrm{e}^{\mathrm{i}(kx - \omega z)}, & x > 0, \end{cases} \qquad \theta = \begin{cases} a_m^2, & x < 0, \\ b_m^2, & x > 0, \end{cases} \tag{16.10}$$

at $z = 0$. To the left and right of the initial shock, two background cw with different amplitudes occur. This initial condition is experimentally feasible [14], who created an initial step condition by blocking half the initial light beam using a razor blade, which implies an initial condition of the form (Eq. 16.10) with $b_m = 0$.

16.3 EXISTING EXPERIMENTAL AND NUMERICAL RESULTS

In Reference [23], the development of MI in a (1 + 1)D nonlocal NLS equation (Eq. 16.3) has been considered. A stability analysis showed that MI always occurs for the focusing case, whereas in the defocusing case, the occurrence of MI is dependent on the particular form of the response function. Three main types of response kernel were considered: the exponential function, which corresponds to the (1 + 1)D version of Equation 16.1, and the Gaussian and rectangular response kernels. The MI gain profiles showed that the maximum gain decreases as the degree of nonlocality increases, which confirms that nonlocality can suppress the development of MI but cannot eliminate it. A modulational transverse stability analysis of long, one-dimensional beams for the (focusing) nematicon equations was undertaken in Reference 24. Again, nonlocality was found to suppress MI.

In Reference 15, the development of MI for nematic liquid crystals has been considered. The experimental results showed that a wide input beam develops a periodic pattern, as the propagation distance increases, which is progressively amplified. Analysis of the FFT spectrum of intensity showed that the maximum gain corresponds well to the predictions of a stability analysis. From a practical viewpoint, the z location at which MI occurs in this and other experimental scenarios is of great interest, as it determines whether or not DSW can be observed in focusing media. Figure 3 in Reference 15 showed that MI develops at $z \approx 300$ μm for an input beam with power $P = 352$ mW. [This power level is very high for experiments involving nematicons, with a power level of $P = O(10)$ mW much more typical; see References 1, 25, and 26. The reason that such a high power level was used was that MI was not observable, over typical experimental length scales, using lower power levels.] In Reference 15 for the MI gain λ, the relation $\lambda \propto P^{\frac{1}{2}}$ is demonstrated, which implies that the onset of MI will occur at millimeters length scales for typical power levels.

In Reference 27, the development of DSW in nonlocal media has been considered, both in the focusing and defocusing cases. In the focusing case, it was shown that nonlocality suppresses the development of MI, so that the development of a DSW could be seen for a certain length scale, before MI destroys the bore structure. The defocusing case is modulationally stable, so that the shock develops into a stable DSW consisting of dark solitary waves. Experimental results were also presented for the development of a shock in a defocusing thermal medium, with good qualitative comparisons obtained with numerical results. In Reference 28, the generation of a DSW in a defocusing thermal medium has been investigated, finding a good qualitative comparison between their experimental and numerical results. In Reference 29, the transverse stability of dark solitary waves and DSW in a nonlocal defocusing medium has been theoretically examined and it has been shown that the nonlocal response has a strong stabilizing effect.

In Reference 30, experimental results for a defocusing photorefractive crystal have been carried out, where a bright hump was imposed on a uniform low-intensity background. Results were presented for the development of DSWs with both line and circular geometries. The DSWs are qualitatively similar to the development of undular bores for the KdV equation and other nonlinear wave equations. In Reference 10, a defocusing thermal medium was considered, which has significant nonlocal effects. An initial condition of a Gaussian hump was used, and a two-dimensional DSW with circular geometry was obtained. The experimental results were qualitatively similar to those for the photorefractive medium. Numerical results for a nonlocal version of the NLS equation, with a Gaussian response function, were also presented. It was found that the effect of nonlocality is to shorten the length of the DSW. In References 14 and 31, DSW development in a focusing photorefractive media was considered. Partially spatially incoherent light is used to suppress MI and the development of a 1D DSW is captured. The shock width was found to be a linear function of nonlinearity and coherence length.

In Reference 32, a generalized focusing NLS equation with three additional terms representing third-order dispersion, self-steepening, and intrapulse stimulated Raman scattering (SRS) has been considered. Authors found an analytical kink solution of this equation that resolves an optical shock in a smooth, monotonic manner with no oscillatory waves. This kink solution is a viscous bore solution, with the SRS giving the required loss, as opposed to the oscillatory, and frictionless, DSW. Authors further considered the numerical evolution of pulses composed of kink–antikink pairs and a super-Gaussian profile and found that there is a redshift in the pulse spectra, which is typical of SRS. In Reference 33, a NLS equation with an SRS term has been considered as well, finding out a kink (viscous bore) solution, which generalized one found in Reference 32. This kink solution could be either monotone or oscillatory, depending on the parameter choices. Lastly, the kink was found to be unstable in both the focusing and defocusing cases.

16.4 ANALYTICAL SOLUTIONS FOR DEFOCUSING EQUATIONS

Few analytical solutions are available for DSW in nonlocal defocusing media. In the local limit, the Schrödinger–Poisson equation becomes the NLS equation and exact solutions for DSW are then available. The analytical results for the defocusing NLS equation, and related local NLS-type equations, give a useful qualitative indication of the likely nature of DSW in nonlocal media.

In Reference 34 the evolution of a shock for the defocusing NLS equation has been considered. In general, the evolution of an initial discontinuity leads to the development of two waves separated by a plateau region. The two waves can be either a DSW or a rarefaction wave, which is nonoscillatory. The occurrence of a vacuum point, at which the light intensity is zero at a given location in the DSW, was also analyzed. In Reference 35, a method to obtain analytical results for dispersive wave equations has been developed, which are defocusing in nature but not integrable. The method allows many analytical features to be obtained from the hyperbolic modulation equations, such as the amplitude of the lead solitary wave and the width of the DSW region.

In Reference 36, the development of optical shocks governed by a defocusing NLS equation with saturable nonlinearity has been considered. Analytical approximations were found for the rarefaction wave and also the leading and trailing edges of the DSW. Figure 16.1 shows the numerical evolution of a DSW for the defocusing NLS equation with saturable nonlinearity, illustrating the complexity of DSW in defocusing media. A plateau region of constant amplitude has formed, and two DSW are developing on either side of the plateau region. One DSW matches the plateau region to the mean level on the left of the Figure, $\rho = 10$, whereas the other DSW matches the plateau region to the mean level on the right, $\rho = 1$. Moreover, a vacuum point, where $\rho = 0$, has formed in the DSW on the right.

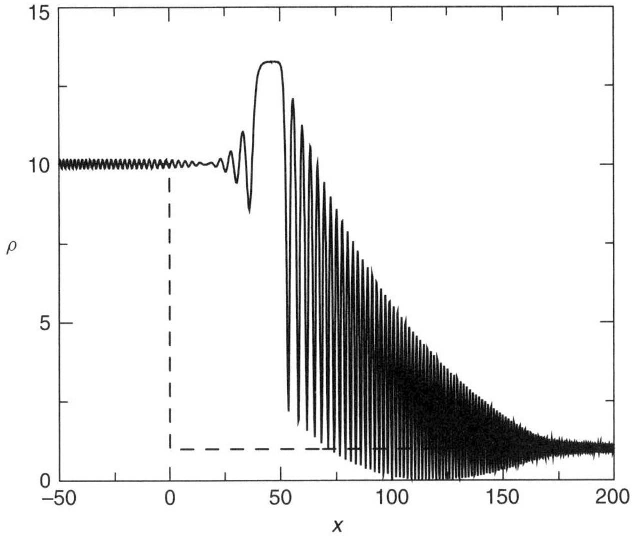

FIGURE 16.1 The light intensity ρ versus x. Shown are numerical solutions from the defocusing NLS equation with saturable nonlinearity. *Source:* The figure is from Reference 36.

In Reference 37, the diffraction of light propagating past a reflecting wire in a photorefractive media has been considered. The governing equation was a defocusing NLS equation with saturable nonlinearity and a potential term representing the obstacle. Analytical solutions were developed to describe the optical "ship waves" generated at the obstacle.

For particular choices of initial conditions, a single DSW occurs as a solution of the defocusing NLS equation, and in Reference 34 a simple wave solution to describe this case has been found. The hydrodynamic representation is used where

$$E = \sqrt{\rho}e^{i\phi}, \quad v = \phi_x, \tag{16.11}$$

and ρ and v are analogs of the hydrodynamic density and velocity, respectively. The initial conditions considered were

$$\begin{aligned} &\rho - 1, \quad v = 0, \quad x < 0, \\ &u = \rho_0, \quad v = v_0 = 2(\sqrt{\rho_0} - 1), \quad x > 0, \end{aligned} \tag{16.12}$$

and $\rho_0 < 1$, so a step down in amplitude is considered. The defocusing NLS equation has a traveling wave solution defined in terms of cnoidal functions (see Equation 4 in Reference 34), which is a function of the modulus m. The two limits of the cnoidal wave solution are $m = 1$, which is the soliton solution, and $m = 0$, which is a linear wave. The simple wave solution is an expansive fan of characteristics with the other Riemann invariants remaining constant throughout the bore.

The bore solution is

$$\overline{\rho} = a_1^2 - (a_1^2 - a_3^2)\frac{E(m)}{K(m)}, \quad \overline{v} = U + \frac{a_2 a_3 \Pi(\nu_1, m)}{a_1 K(m)},$$
$$a_1 = \frac{1}{4}(4 + r_2 - c), \quad a_2 = \frac{1}{4}(4 - r_2 + c),$$
$$a_3 = \frac{1}{4}(4 - c), \quad r_2 = \frac{4c - 2}{2 - c + m(c + 2)},$$
$$c = 2(2\sqrt{\rho_0} - 1), \quad \nu_1 = \frac{a_1^2 - a_3^2}{a_1^2}, \quad 0 \leq m \leq 1. \tag{16.13}$$

E, K, and Π are complete elliptic functions of the first, second, and third kinds, respectively, $\overline{\rho}$ is the mean light intensity, $\overline{v}$ is the mean velocity whereas the peak and trough envelopes to the DSW are given by a_2^2 and a_3^2, respectively. The solution is completed by the formula for the characteristics in the fan, given by

$$\frac{x}{t} = \frac{c}{2} - \frac{(2 - r_2)(r_2 - c)K(m)}{2((r_2 - c)K(m) + E(m)(c - 2))}, \tag{16.14}$$
$$\text{where } \frac{c}{2} \leq \frac{x}{t} \leq \frac{(c+2)}{4} + \frac{4(2 - c)}{6 - c}.$$

Equations 16.13 and 16.14 represent the analytical DSW for the defocusing NLS equation. For $\rho_0 < 0.25$, a vacuum point occurs in the DSW, which means that the amplitude $\rho = 0$ at a particular location in the bore.

16.5 ANALYTICAL SOLUTIONS FOR FOCUSING EQUATIONS

The usual method to find the DSW of nonlinear wave equations is to construct the associated modulation equations, and if these equations are hyperbolic, find the DSW as a simple wave solution [38]. The modulation equations for the focusing version of Equation 16.1 are elliptic, hence subject to MI, and so do not possess a simple wave solution. However, a number of experimental results show that a DSW forms in focusing media, for a certain length scale, z, before MI dominates and the bore structure is destroyed. These emerging experimental results are a strong motivation for the development of analytical results to describe DSW in focusing media.

In Reference 39, the modulation equations for the focusing NLS equation have been considered and an analog of the simple wave solution has been found. This self-similar solution describes the decay of a uniform initial condition, which is perturbed at a given point. An oscillatory structure develops with solitons in the central region (at the location of the initial perturbation) and linear waves at the edge of the disturbance. Analytical results for the soliton amplitude and the width of the oscillatory region are found and discussed, as is the characteristic time for the development of instabilities. In Reference 40, modulation equations for the Landau–Lifshitz equation have been developed, which describes ferromagnetic materials. An analytical oscillatory solution, which is qualitatively similar to that in Reference 39 has been derived.

As the standard method, based on modulation theory, does not yield results for DSW in focusing media, an alternative method, termed as *uniform soliton theory*, will be used. This was developed in Reference 41 for the KdV equation and widely applied to other nonlinear wave equations, such as the magma equation (Eq. 16.15). In Reference 42, an initial boundary value problem for magma flow in the Earth's mantle has been considered, which was formulated as

$$f_t + \left[f^3(1 - f_{tx})\right]_x = 0, \quad x > 0, \tag{16.15}$$

$$f(x, 0) = f_b(x), \quad x > 0, \quad f(0, t) = f_r(t), \quad t > 0,$$

where f is the magma fraction. Physically, a large reservoir of magma is located at $x = 0$ with liquid fraction f_r, and the magma then propagates into the semi-infinite domain, which has background liquid fraction f_b. They found modulation equations to describe the evolution of the magma DSW, by assuming slow variations in the traveling wave solution and averaging the two conservation equations for Equation 16.15. Figure 16.2 shows the development of a magma DSW for Equation 16.15. A partial DSW occurs with solitary waves at the leading edge, $x \approx 17$, whereas at $x = 0$, cnoidal waves are generated. As modulation theory gives averaged information about the DSW, the mean level and the wave peak and trough envelopes are plotted from the modulation theory solution. An excellent comparison between the modulation theory and numerical solutions was obtained.

Uniform soliton theory was also successfully applied to the magma evolution problem (Eq. 16.15) in Reference 42. The theory involves using conservation laws

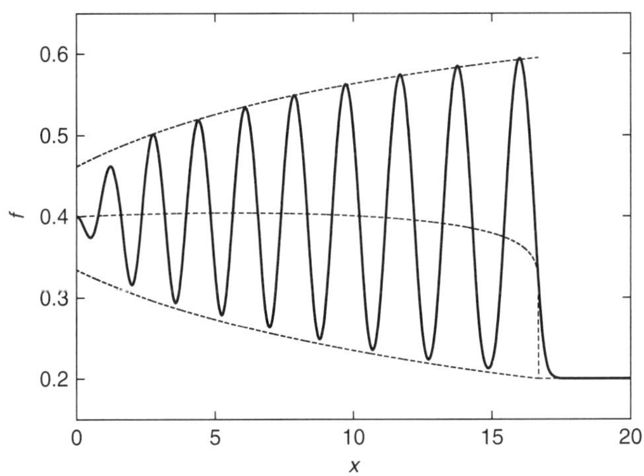

FIGURE 16.2 The liquid magma fraction f versus x at $t = 60$. Shown are the modulation theory (dashed lines) and numerical (solid lines) solutions. The modulation theory solution consists of the wave peak and trough envelopes and the mean height. *Source:* The figure is from Reference 42.

of the governing equation and also assuming that solitons of constant amplitude are generated at a constant rate. This then gave a transcendental equation

$$(f_r^2 + f_r f_b + f_b^2)H = 3f_b + 2a, \tag{16.16}$$

where

$$H = f_b^{-1} - \frac{2}{3}U^{-1}a + 2u^{-1}f_b\left(1 + f_b a^{-1}\right)^{1/2} \cosh^{-1}\left(1 + af_b^{-1}\right)^{1/2} - 2U^{-1}f_b.$$

Equation 16.16 implicitly determines the amplitude a of the solitary wave generated by the shock, in terms of f_r and f_b. Figure 16.3 shows the solitary wave amplitude a versus the background magma fraction level f_b. Compared are the solitary wave amplitude at the leading edge of the DSW, as given by the uniform soliton theory, modulation theory, and the numerical solution. It can be seen that the uniform soliton theory gives an excellent comparison with the numerical solutions and outperforms modulation theory as the background magma fraction f_b becomes very small. In this limit, as the solution is a train of solitary waves, the averaging process becomes invalid, as the wavelength is infinite, hence the modulation theory breaks down.

16.5.1 The 1 + 1 Dimensional Semianalytical Soliton

The uniform soliton theory discussed in Reference 41 is based on the exact soliton solution of the KdV equation. No such exact solitary wave solution of

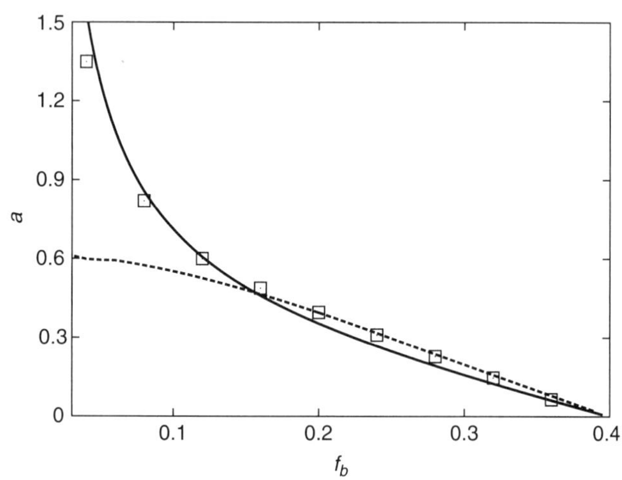

FIGURE 16.3 The solitary wave amplitude a versus liquid magma fraction f_b. Shown are the solutions from modulation theory (dashed lines), uniform soliton theory (solid lines), and numerical solutions (squares). *Source:* The figure is from Reference 42.

the Schrödinger–Poisson equations (Eq. 16.1) exists. Therefore, a variational approximation for this solitary wave will be found, which has been found to be accurate in previous work on nematicon evolution [43–45]. The variational approximation to the solitary wave in (1 + 1)D will be used in the uniform soliton theory. Equation 16.1 has the Lagrangian

$$L = \mathrm{i}\left(E^* E_z - E E_z^*\right) - |\nabla E|^2 + 4\theta|E|^2 - \nu|\nabla\theta|^2 - 2\theta^2, \tag{16.17}$$

for $q = 1$, where the * superscript denotes the complex conjugate. An approximate steady solitary wave solution in (1 + 1)D is sought using a trial function similar to that used by Minzoni et al. [45]

$$E = a \,\mathrm{sech}\, \frac{x - kz}{w}\, \mathrm{e}^{\mathrm{i}(kx+\sigma z)}, \quad \theta = \alpha\, \mathrm{sech}^2 \frac{x - kz}{\beta}. \tag{16.18}$$

The trial function for the electric field is based on the NLS soliton and has amplitude a and width w. The trial function for the director angle θ has amplitude α and width β and its form is chosen so that the relationship $\theta = |E|^2$ can be satisfied in the local limit, as $\nu \to 0$. The wave number k is chosen to be the same as the wave number in the cw initial profile. This results in the averaged Lagrangian

$$\mathcal{L} = -4a^2 w\sigma - 2k^2 a^2 w - \frac{2a^2}{3w} - \frac{16\nu\alpha^2}{15\beta} - \frac{8}{3}\alpha^2\beta + \alpha a^2 I, \tag{16.19}$$

$$\text{where } I(w, \beta) = \int_{-\infty}^{\infty} \mathrm{sech}^2\left(\frac{x}{\beta}\right) \mathrm{sech}^2\left(\frac{x}{w}\right) dx.$$

For $\beta \neq w$, the integral I cannot be evaluated in closed form. The variational equations, which give the solitary wave parameters, are then

$$\mathcal{L}_a = -8aw\sigma - 4k^2 aw - \frac{4a}{3w} + 2\alpha a I = 0, \tag{16.20}$$

$$\mathcal{L}_w = -4a^2\sigma - 2k^2 a^2 + \frac{2a^2}{3w^2} + \alpha a^2 I_w = 0, \tag{16.21}$$

$$\mathcal{L}_\alpha = -\frac{32\nu\alpha}{15\beta} - \frac{16}{3}\alpha\beta + a^2 I = 0, \tag{16.22}$$

$$\mathcal{L}_\beta = \frac{16\nu\alpha^2}{15\beta^2} - \frac{8}{3}\alpha^2 + \alpha a^2 I_\beta - 0. \tag{16.23}$$

Variational equations (Eqs 16.20–16.23) are four equations in the five parameters and represent a one-parameter family of solitary waves. In the local limit, when $\nu = 0$, variational equations (Eqs 16.20–16.23) give the exact NLS soliton solution

$$\alpha = a^2, \quad w = \beta = \frac{1}{2^{\frac{1}{2}} a}, \quad \sigma = a^2 - \frac{k^2}{2}. \tag{16.24}$$

16.5.2 Uniform Soliton Theory

The approximate solitary wave solution (Eq. 16.18) in (1 + 1)D will be used to find the amplitude of the solitary waves, generated in a DSW governed by the Schrödinger–Poisson equations (Eq. 16.1), using the method of [41]. This method is based on assuming that the shock is resolved by diffraction into a train of uniform solitary waves. The amplitude and spacing of these nematicons is calculated by using global conservation laws, which must hold no matter what form of solution is used [41].

The initial condition (Eq. 16.12) is considered with $b_m = 0$ so to the right of the bore the background light intensity is zero. The mass and energy conservation laws of Equation 16.1 in (1 + 1)D are

$$\mathrm{i}\frac{\partial}{\partial z}(|E|^2) + \frac{1}{2}\frac{\partial}{\partial x}(E^*E_x - EE_x^*) = 0, \tag{16.25}$$

$$\mathrm{i}\frac{\partial}{\partial z}(|E_x|^2 - 4\theta|E|^2 + \nu\theta_x^2 + 2\theta^2) + \frac{1}{2}\frac{\partial}{\partial x}$$
$$\times (E_x^*E_{xx} - E_xE_{xx}^* - 4\theta E^*E_x + 4\theta EE_x^* - 4i\nu\theta_x\theta_z) = 0, \tag{16.26}$$

respectively. These conservation laws are named from invariants of the Lagrangian (Eq. 16.17) [46] and do not correspond to these quantities in optical situations. For instance, the mass conservation law (Eq. 16.25) corresponds to conservation of power in the optical context.

Integrating the conservation equations (Eqs 16.25 and 16.26) over the bore then gives

$$\frac{\mathrm{d}}{\mathrm{d}z}\int_{-\infty}^{\infty} |E|^2\mathrm{d}x = ka_m^2, \tag{16.27}$$

$$\frac{\mathrm{d}}{\mathrm{d}z}\int_{-\infty}^{\infty} (|E_x|^2 - 4\theta|E|^2 + \nu\theta_x^2 + 2\theta^2)\mathrm{d}x = ka_m^2(k^2 - 4a_m^2). \tag{16.28}$$

Assuming that N solitary waves have been generated, the integrals on the left-hand side of Equation 16.28 are N times the integral for one solitary wave. From the trial functions (Eq. 16.18), the mass and energy for a single solitary wave are

$$\int_{-\infty}^{\infty} |E|^2\mathrm{d}x = 2a^2w, \quad \int_{-\infty}^{\infty} (|E_x|^2 - 4\theta|E|^2 + \nu\theta_x^2 + 2\theta^2)\,\mathrm{d}x$$

$$= \frac{2a^2}{3w} + \frac{16\nu\alpha^2}{15\beta} + 2k^2a^2w + \frac{8}{3}\alpha^2\beta - \alpha a^2 I. \tag{16.29}$$

Substituting Equation 16.29 into Equations 16.27 and 16.28 and taking the ratio of the two equations gives the transcendental equation

$$\frac{2a^2}{3w} + \frac{16\nu\alpha^2}{15\beta} + \frac{8}{3}q\alpha^2\beta - \alpha a^2 I = -8a_m^2 a^2 w, \tag{16.30}$$

for the amplitude a of the solitary wave. For a given height a_m of the initial shock and choice of ν, Equations 16.20–16.23 and Equation 16.30 are five equations for the five parameters describing the solitary waves generated at the shock.

In the local limit with $\nu = 0$, the uniform soliton theory has an exact analytical solution given by

$$a = \sqrt{6}a_m, \quad \alpha = 6a_m^2, \quad w = \beta = \frac{1}{2\sqrt{3}\,a_m}. \tag{16.31}$$

Hence, in this limit, the lead soliton in the DSW has an electric field amplitude that is 2.45 times the height of the initial shock and the director pulse has an amplitude of 6 times the initial shock (in θ). By way of contrast, for the KdV DSW, the lead soliton is twice the height of the initial shock.

16.5.3 Comparisons with Numerical Solutions

Here, numerical solutions of the focusing version of Equation 16.1 are compared with the results of the uniform soliton theory. Figure 16.4 shows the predictions of uniform soliton theory for Equation 16.1. Shown are the parameters of the solitary wave, a, α, w^{-1}, and β^{-1} versus ν. The initial shock has an amplitude $a_m = 0.5$. At $\nu = 0$, the solitary wave has an electric field amplitude of $a = 1.22$ and a director angle amplitude $\alpha = 1.5$. As ν increases, the electric field amplitude a, after a slight dip, increases, whereas the director angle, α, decreases. At $\nu = 0$, the width of both pulses is $\beta = w = 0.57$. As ν increases, the widths of both pulses increase, with β, the width of the director pulse, becoming much larger than the width of the electric field pulse w. This is consistent with the nonlocal limit, as for $\nu \to \infty$, the director pulse becomes much broader than the electric field.

Figure 16.5 shows the electric field and director angle amplitudes, a and α, versus ν. The other parameters are $a_m = 0.2$ and $k = 0$. Shown are the predictions of uniform soliton theory and the numerical solutions. Two different numerical estimates of the solitary wave amplitudes are shown, the amplitude of the first solitary wave generated by the shock and the maximum solitary wave amplitude in the bore, averaged over a range of z. Averaging is used as there is some oscillation in the solitary wave amplitude as the bore develops, because of the nonlocality. The general trend is that, as ν increases, the amplitude of the electric field increases and that of the director angle decreases. The comparison between the theoretical results and the numerical estimates of the maximum solitary wave amplitude is excellent. For the first solitary wave generated, the uniform soliton theory overestimates the numerical predictions by about 25% .

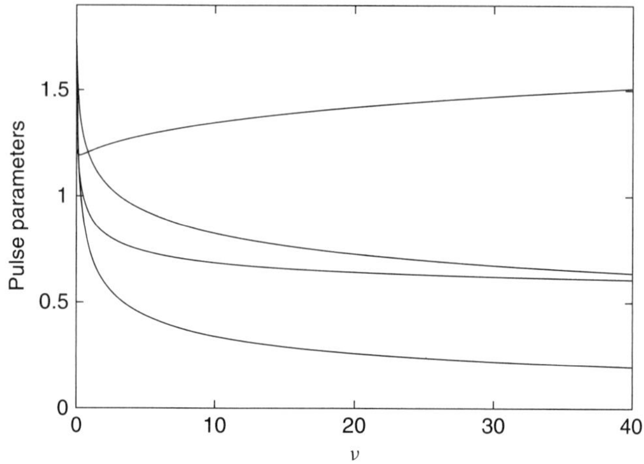

FIGURE 16.4 Solitary wave parameters versus ν. Shown is a (upper solid line), α (lower solid line), w^{-1} (upper dashed line), and β^{-1} (lower dashed line) from uniform soliton theory. The other parameter is $a_m = 0.5$. *Source:* The figure is from Reference 18.

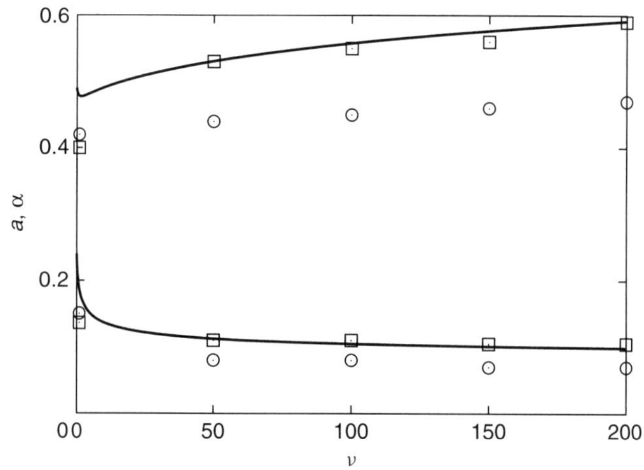

FIGURE 16.5 Solitary wave amplitudes versus ν. Shown is a, α (solid lines) from uniform soliton theory. Numerical estimates are for the first wave (circles) and the average maximum amplitude (squares). The other parameters are $a_m = 0.2$ and $k = 0$. *Source:* The figure is from Reference 18.

Figure 16.6 shows the numerical solutions for $|E|$ and θ versus x at $z = 300$. The other parameters are $\nu = 1$, $a_m = 0.2$, and $k = 0$. This figure shows a classical DSW, similar to the magma DSW of Figure 16.2. Solitary waves occur at the front of the DSW and linear waves of small amplitude at the rear. The lead wave has amplitude $a = 0.4$ and $\alpha = 0.14$. The uniform soliton theory gives $a = 0.48$ and $\alpha = 0.18$, which overestimate the numerical values by 20% and 28% , respectively.

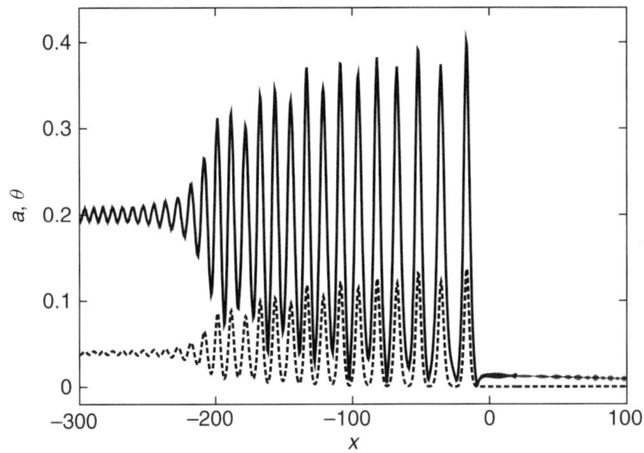

FIGURE 16.6 Numerical solutions of Equation 16.1 versus x at $z = 300$. Shown are $|E|$ (solid line) and θ (dashed line). The other parameters are $\nu = 1$, $a_m = 0.2$, and $k = 0$. *Source:* The figure is from Reference 18.

As this example is for the local limit, the director pulse has approximately the same width as the electric field. For this example, MI does not occur until $z \approx 600$.

Figure 16.7 shows the numerical solution of Equation 16.1 for $|E|$ and θ versus x at $z = 400$. The other parameters are $\nu = 50$, $a_m = 0.2$, and $k = 0$. A classical DSW, as seen in Figure 16.6, has a structure where the waves decrease in amplitude from the front to the rear. In this nonlocal case, this pattern disappears with the waves no longer ordered by amplitude. Also, it can be seen from the figure that there

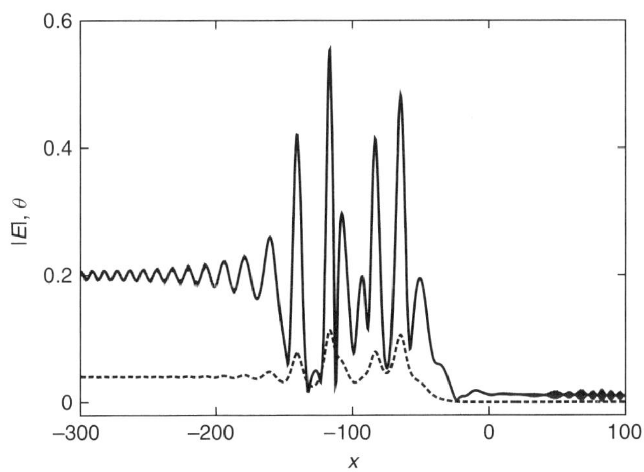

FIGURE 16.7 Numerical solutions of Equation 16.1 versus x at $z = 400$. Shown are $|E|$ (solid line) and θ (dashed line). The other parameters are $\nu = 50$, $a_m = 0.2$, and $k = 0$. *Source:* The figure is from Reference 18.

are less director peaks in the bore than electric field peaks. The larger nonlocality of this example has resulted in a broad director response where the director pulses are wider than the electric field pulses ($\beta = 6.9 > w = 2.9$). This generates a wide potential well enclosing all the solitary waves in the DSW, which wipes out many individual peaks. The solitary waves interact with each other in a nonlocal manner through this potential well, and the amplitude of the highest solitary wave varies in a complicated manner with z, because of this nonlocal interaction. However, the average maximum amplitude, $a = 0.55$ and $\alpha = 0.11$, are very close to the estimates from uniform soliton theory; $a = 0.53$, and $\alpha = 0.11$.

16.6 CONCLUSIONS

Experimental, numerical, and theoretical results have been reviewed for the development of DSW in nematic liquid crystal and a range of other optical media. A key focus has been the presentation of a method, uniform soliton theory, to obtain semianalytical results for DSW in a nonlocal focusing medium. Focusing media are subject to MI but a range of emerging experimental results show the establishment of a DSW, before MI destroys the bore structure.

The semianalytical solutions, which predict the amplitude of the largest wave in the DSW, were developed using uniform soliton theory that uses conservation laws, and the assumption that a train of uniform solitary waves is generated by the shock. The semianalytical predictions, for solitary wave amplitude, are found to be in excellent agreement with numerical solutions for $(1 + 1)$D bores.

This represents a novel method of finding analytical solutions for a focusing medium and will motivate additional experimental observations of DSW in a range of focusing and defocusing media. A key aim of future experimental studies should be measurement of the amplitude of solitary waves generated in a DSW, which will allow validation of the semianalytical theory. The theory can also be applied to other NLS-type equations of focusing type. These include NLS-type equations with saturable nonlinearity, a nonlocal version with a Gaussian or other nonexponential response function, and also the implicit nonlinearity that occurs for applications involving colloidal media.

Moreover, experiments are usually performed for DSW with both one- and two-dimensional geometries. The results presented here describe one-dimensional DSW but not two-dimensional ones [18] extended their one-dimensional results to obtain approximate results for two-dimensional DSW with circular symmetry, but the results were only valid for DSW with large radius. Future work could extend uniform soliton theory to two-dimensional DSW of arbitrary radius. This would involve using the variational approach to find an approximate annular solitary wave solution, which would have the appropriate symmetry for two-dimensional DSW.

There has been much experimental work on DSW in defocusing nonlocal media. Uniform soliton theory should also be a useful tool in predicting the amplitude of dark solitary waves, in the DSW generated in these media. The local defocusing NLS equation shows that the DSW is usually part of a more complicated structure involving rarefaction waves, vacuum points, and plateau regions. To accurately predict the amplitude of dark solitary waves in defocusing nonlocal media, all of the solution types need to be analytically modeled; a much more challenging task than in the case of nonlocal focusing equations.

REFERENCES

1. C. Conti, M. Peccianti, and G. Assanto. Observation of optical spatial solitons in a highly nonlocal medium. *Phys. Rev. Lett.*, 92:113902, 2004.
2. G. Assanto and M. Karpierz. Nematicons: Self-localized beams in nematic liquid crystals. *Liq. Cryst.*, 36:1161, 2009.
3. G. Assanto, M. Peccianti, and C. Conti. Nematicons: Optical spatial solitons in nematic liquid crystals. *Opt. Photon. News*, 14:44–48, 2003.
4. Y. S. Kivshar and G. P. Agrawal. *Optical Solitons. From Fibers to Photonic Crystals*. Academic Press, San Diego, CA, 2003.
5. A. Piccardi, A. Alberucci, U. Bortolozzo, S. Residori, and G. Assanto. Readdressable interconnects with spatial soliton waveguides in liquid crystal light valves. *IEEE Photon. Technol. Lett.*, 22:694–696, 2010.
6. C. Conti, M. Peccianti, and G. Assanto. Route to nonlocality and observation of accessible solitons. *Phys. Rev. Lett.*, 91:073901, 2003.
7. G. C. Duree, J. L. Shultz, G. J. Salamo, M. Segev, A. Yariv, B. Crosignani, P. Di Porto, E. J. Sharp, and R. R. Neurgaonkar. Observation of self-trapping of an optical beam due to the photorefractive effect. *Phys. Rev. Lett.*, 71:533–536, 1993.
8. R. Schiek, Y. Baek, and G. I. Stegeman. One-dimensional spatial solitary waves due to cascaded second-order nonlinearities in planar waveguides. *Phys. Rev. E*, 53:1138–1141, 1996.
9. A. Pasquazi, S. Stivala, G. Assanto, J. Gonzalo, and J. Solis. Transverse nonlinear optics in Heavy Metal Oxide Glasses. *Phys. Rev. A*, 77:043808, 2008.
10. C. Barsi, W. Wan, C. Sun, and J. W. Fleischer. Dispersive shock waves with nonlocal nonlinearity. *Opt. Lett.*, 32:2930–2932, 2007.
11. D. V. Grier. A revolution in optical manipulation. *Nature*, 424:810–816, 2003.
12. A. Hasegawa and F. D. Tappert. Transmission of stationary nonlinear optical pulses in dispersive dielectric fibres. I. Anomalous dispersion. *Appl. Phys. Lett.*, 23:142–144, 1973.
13. G. B. Whitham. *Linear and Nonlinear Waves*. Wiley, New York, 1974.
14. W. Wan, D. V. Dylov, C. Barsi, and J. W. Fleischer. Diffraction from an edge in a self-focusing medium. *Opt. Lett.*, 35:2819–2821, 2010.
15. M. Peccianti, C. Conti, and G. Assanto. Optical modulational instability in a non-local medium. *Phys. Rev. E*, 68:025602, 2003.
16. M. Peccianti, C. Conti, and G. Assanto. Optical multisoliton generation in nematic liquid crystals. *Opt. Lett.*, 28:2231–2233, 2003.
17. M. Peccianti, C. Conti, E. Alberici, and G. Assanto. Spatially incoherent modulational instability in a non local medium. *Laser Phys. Lett.*, 2:25–29, 2005.
18. G. Assanto, T. R. Marchant, and N. F. Smyth. Collisionless shock resolution in nematic liquid crystals. *Phys. Rev. A*, 78:063808, 2008.

19. S. Lopez-Aguayo, A. S. Desyatnikov, Y. S. Kivshar, S. Skupin, W. Krolikowski, and O. Bang. Stable rotating dipole solitons in nonlocal optical media. *Opt. Lett.*, 31:1100–1102, 2006.

20. P. D. Rasmussen, O. Bang, and W. Krolikowski. Theory of nonlocal soliton interaction in nematic liquid crystals. *Phys. Rev. E*, 72:066611, 2005.

21. A. I. Yakimenko, V. M. Lashkin, and O. O. Prikhodko. Dynamics of two-dimensional coherent structures in nonlocal nonlinear media. *Phys. Rev. E*, 73:066605, 2006.

22. T. R. Marchant and N. F. Smyth. Nonlocal validity of an asymptotic one dimensional nematicon solution. *J. Phys. A: Math. Theor.*, 41:365201, 2008.

23. W. Krolikowski, O. Bang, J. Rasmussen, and J. Wyller. Modulational instability in nonlocal nonlinear Kerr media. *Phys. Rev. E*, 64:016612, 2001.

24. Y. Y. Lin, R.-K. Lee, and Y. S. Kivshar. Suppression of soliton transverse instabilities in nonlocal nonlinear media. *J. Opt. Soc. Am. B*, 25:576–581, 2008.

25. G. Assanto and M. Peccianti. Spatial solitons in nematic liquid crystals. *IEEE J. Quantum Electron.*, 39:13–21, 2003.

26. M. Peccianti, C. Conti, G. Assanto, A. De Luca, and C. Umeton. Routing of anisotropic spatial solitons and modulational instability in liquid crystals. *Nature*, 432:733–737, 2004.

27. N. Ghofaniha, C. Conti, G. Ruocco, and S. Trillo. Shocks in nonlocal media. *Phys. Rev. Lett.*, 99:043903, 2007.

28. C. Conti, A. Fratalocchi, M. Peccianti, G. Ruocco, and S. Trillo. Observation of a gradient catastrophe generating solitons. *Phys. Rev. Lett.*, 102:083902, 2009.

29. A. Armaroli, S. Trillo, and A. Fratalocchi. Suppression of transverse instabilities of dark solitons and their dispersive shock waves. *Phys. Rev. A.*, 180:053903, 2009.

30. W. Wan, S. Jia, and J. W. Fleischer. Dispersive superfluid-like shock waves in nonlinear optics. *Nat. Phys.*, 3:46–51, 2007.

31. W. Wan, D. V. Dylov, C. Barsi, and J. W. Fleischer. Dispersive shock waves with negative pressure. *CLEO/QLES*, 1–5:2910–2911, 2009.

32. G. P. Agrawal and C. Headley. Kink solitons and optical shocks in dispersive media. *Phys. Rev. A*, 46:1573–1577, 1992.

33. Y. S. Kivshar and B. A. Malomed. Raman-induced optical shocks in nonlinear media. *Phys. Rev. Lett.*, 93:070485, 1993.

34. G. A. El, V. V. Geogjaev, A. V. Gurevich, and A. L. Krylov. Decay of an initial discontinuity in the defocusing NLS hydrodynamics. *Physica D*, 87:186–192, 1995.

35. G. A. El. Resolution of a shock in hyperbolic systems modified by weak dispersion. *Chaos*, 15:037103, 2005.

36. G. A. El, A. Gammal, E. G. Khamis, R. A. Kraenkel, and A. M. Kamchatnov. Theory of optical dispersive shock waves in photorefractive media. *Phys. Rev. A*, 76:053813, 2007.

37. E. G. Khamis, A. Gammal, G. A. El, Y. G. Gladush, and A. M. Kamchatnov. Nonlinear diffraction of light beams propagating in photorefractive media with embedded reflecting wire. *Phys. Rev. A*, 78:013829, 2008.

38. A. V. Gurevich and L. P. Pitaevskii. Nonstationary structure of a collisionless shock wave. *Sov. Phys. JETP*, 33:291–297, 1974.

39. G. A. El, A. V. Gurevich, V. V. Khodorovskii, and A. L. Krylov. Modulational instability and formation of a nonlinear oscillatory structure in a focusing medium. *Phys. Lett. A*, 177:357–361, 1993.
40. A. M. Kamchatnov, A. L Krylov, and G. A. El. On the asymptotic evolution of a localized perturbation of the one-dimensional Landau-Lifshits equation with unixial anisotropy. *Theor. Math. Phys.*, 109:1345–1351, 1996.
41. R. H. J. Grimshaw and N. F. Smyth. Resonant flow of a stratified fluid over topography. *J. Fluid Mech.*, 169:429–464, 1986.
42. T. R. Marchant and N. F. Smyth. Approximate solutions for magmon propagation from a reservoir. *IMA J. Appl. Math.*, 70:796–813, 2005.
43. C. García-Reimbert, A. A. Minzoni, and N. F. Smyth. Spatial soliton evolution in nematic liquid crystals in the nonlinear local regime. *J. Opt. Soc. Am. B*, 23:294–301, 2006.
44. C. García-Reimbert, A. A. Minzoni, N. F. Smyth, and A. L. Worthy. Large-amplitude nematicon propagation in a liquid crystal with local response. *J. Opt. Soc. Am. B*, 23:2551–2558, 2006.
45. A. A. Minzoni, N. F. Smyth, and A. L. Worthy. Modulation solutions for nematicon propagation in non-local liquid crystals. *J. Opt. Soc. Am. B*, 24:1549–1556, 2007.
46. C. J. Kaup and A. C. Newell. Solitons as particles, oscillators, and in slowly changing media: A singular perturbation theory. *Proc. R. Soc. London, Ser. A*, 361:413–446, 1978.

Index

Nematicons: Spatial Optical Solitons in Nematic Liquid Crystals, First Edition.
Edited by Gaetano Assanto.

WILEY SERIES IN PURE AND APPLIED OPTICS

Founded by Stanley S. Ballard, University of Florida

EDITOR: Glenn Boreman, University of Central Florida, CREOL & FPCE

ASSANTO · *Nematicons: Spatial Optical Solitons in Nematic Liquid Crystlas*
BARRETT AND MYERS · *Foundations of Image Science*
BEISER · *Holographic Scanning*
BERGER-SCHUNN · *Practical Color Measurement*
BOYD · *Radiometry and The Detection of Optical Radiation*
BUCK · *Fundamentals of Optical Fibers,* Second Edition
CATHEY · *Optical Information Processing and Holography*
CHUANG · *Physics of Optoelectronic Devices*
DELONE AND KRAINOV · *Fundamentals of Nonlinear Optics of Atomic Gases*
DERENIAK AND BOREMAN · *Infrared Detectors and Systems*
DERENIAK AND CROWE · *Optical Radiation Detectors*
DE VANY · *Master Optical Techniques*
ERSOY · *Diffraction, Fourier Optics and Imaging*
GASKILL · *Linear Systems, Fourier Transform, and Optics*
GOODMAN · *Statistical Optics*
HOBBS · *Building Electro-Optical Systems: Making It All Work,* Second Edition
HUDSON · *Infrared System Engineering*
IIZUKA · *Elements of Photonics, Volume I: In Free Space and Special Media*
IIZUKA · *Elements of Photonics, Volume II: For Fiber and Integrated Optics*
JUDD AND WYSZECKI · *Color in Business, Science, and Industry,* Third Edition
KAFRI AND GLATT · *The Physics of Moire Metrology*
KAROW · *Fabrication Methods for Precision Optics*
KLEIN AND FURTAK · *Optics,* Second Edition
MA AND ARCE · *Computational Lithography*
MALACARA · *Optical Shop Testing,* Third Edition
MILONNI AND EBERLY · *Lasers*
NASSAU · *The Physics and Chemistry of Color: The Fifteen Causes of Color,* Second Edition
NIETO-VESPERINAS · *Scattering and Diffraction in Physical Optics*
OSCHE · *Optical Detection Theory for Laser Applications*
O'SHEA · *Elements of Modern Optical Design*
OZAKTAS · *The Fractional Fourier Transform*
PRATHER, SHI, SHARKAWY, MURAKOWSKI, AND SCHNEIDER · *Photonic Crystals: Theory, Applications, and Fabrication*
SALEH AND TEICH · *Fundamentals of Photonics,* Second Edition
SCHUBERT AND WILHELMI · *Nonlinear Optics and Quantum Electronics*
SHEN · *The Principles of Nonlinear Optics*
STEGEMAN AND STEGEMAN · *Nonlinear Optics: Phenomena, Materials, and Devices*
UDD · *Fiber Optic Sensors: An Introduction for Engineers and Scientists*
UDD · *Fiber Optic Smart Structures*
VANDERLUGT · *Optical Signal Processing*
VEST · *Holographic Interferometry*
VINCENT · *Fundamentals of Infrared Detector Operation and Testing*
WEINER · *Ultrafast Optics*
WILLIAMS AND BECKLUND · *Introduction to the Optical Transfer Function*
WU AND REN · *Introduction to Adaptive Lenses*
WYSZECKI AND STILES · *Color Science: Concepts and Methods, Quantitative Data and Formulae,* Second Edition

XU AND STROUD · *Acousto-Optic Devices*
YAMAMOTO · *Coherence, Amplification, and Quantum Effects in Semiconductor Lasers*
YARIV AND YEH · *Optical Waves in Crystals*
YEH · *Optical Waves in Layered Media*
YEH · *Introduction to Photorefractive Nonlinear Optics*
YEH AND GU · *Optics of Liquid Crystal Displays,* Second Edition